RENEWALS 691-4574
DATE DUE

ILL: 2200383 SAE

WITHDRAWN
UTSA LIBRARIES

Cement Chemistry

Cement Chemistry

H. F. W. TAYLOR

Emeritus Professor of Chemistry, University of Aberdeen
Visiting Professor, Imperial College (University of London)

ACADEMIC PRESS
Harcourt Brace Jovanovich, Publishers
London San Diego New York
Boston Sydney Tokyo Toronto

This book is printed on acid-free paper.

ACADEMIC PRESS LIMITED
24–28 Oval Road
London NW1 7DX

United States Edition published by
ACADEMIC PRESS INC.
San Diego, CA 92101

Copyright © 1990, by
ACADEMIC PRESS LIMITED

All Rights Reserved
No part of this book may be reproduced in any form, by photostat,
microfilm or any other means, without written permission from the publishers

British Library Cataloguing in Publication Data is available

ISBN 0-12-683900-X

Filmset by Bath Typesetting Limited, Bath, Avon
Printed in Great Britain by St Edmundsbury Press Ltd., Bury St Edmunds, Suffolk

Contents

Preface		xv
1	**Portland cement and its major constituent phases**	1
1.1	Introduction	1
1.1.1	Portland cement: general	1
1.1.2	Types of Portland cement	2
1.1.3	Cement chemical nomenclature and other abbreviations	4
1.2	Alite	5
1.2.1	Polymorphism and crystal structure	5
1.2.2	Tricalcium silicate solid solutions	8
1.2.3	Compositions of alites in clinkers	8
1.2.4	Polymorphic modifications of the alites in clinkers	9
1.2.5	X-ray powder patterns and densities of tricalcium silicate and alites	13
1.2.6	Optical, thermal and other data	15
1.3	Belite	15
1.3.1	Polymorphism and crystal structure	15
1.3.2	Lamellar textures in clinker belites	19
1.3.3	Polymorphic types of belites in clinkers	20
1.3.4	Compositions of belites in clinkers	21
1.3.5	Cell parameters, X-ray powder patterns and other data	22
1.4	The aluminate phase	23
1.4.1	Crystal structure: cubic, orthorhombic and monoclinic modifications	23
1.4.2	Other modifications	25
1.4.3	Structural modifications of the aluminate phase in clinkers	26
1.4.4	Compositions of the aluminate phase in clinkers	27
1.4.5	X-ray powder data, densities and optical properties	28
1.5	The ferrite phase	28
1.5.1	Crystal structure and composition in the $Ca_2(Al_xFe_{1-x})_2O_5$ series	28
1.5.2	Compositions of the ferrite phase in clinkers	30
1.5.3	Crystal data and X-ray powder patterns for ferrite phase containing foreign ions	31
1.5.4	Optical, magnetic and other data	32
2	**High-temperature chemistry**	33
2.1	Introduction	33
2.2	Systems containing CaO with SiO_2 or Al_2O_3 or both	33
2.2.1	The $CaO-SiO_2$ system	33
2.2.2	The $CaO-Al_2O_3$ system	34

2.2.3	$C_{12}A_7$ and derived structures	36
2.2.4	C_5A_3, C_2A and C_4A_3	38
2.2.5	The CaO–Al_2O_3–SiO_2 system	38
2.2.6	Clinker formation in the CaO–Al_2O_3–SiO_2 system	39
2.3	Systems containing Fe_2O_3	41
2.3.1	The CaO–Al_2O_3–Fe_2O_3 system	41
2.3.2	The CaO–Al_2O_3–Fe_2O_3–SiO_2 system	43
2.3.3	Clinker formation in the CaO–Al_2O_3–Fe_2O_3–SiO_2 system	45
2.4	Systems containing MgO or FeO	48
2.4.1	General	48
2.4.2	Effect of MgO on equilibria in the CaO–Al_2O_3–Fe_2O_3–SiO_2 system	48
2.4.3	Phases structurally related to gehlenite	50
2.5	Systems containing alkalis or SO_3 or both	51
2.5.1	Phases	51
2.5.2	Equilibria	55
2.6	Systems with other components	56
2.6.1	Fluorides and fluorosilicates	56
2.6.2	Carbonates	58
2.7	Laboratory preparation of high-temperature phases	58
3	**The chemistry of Portland cement manufacture**	**60**
3.1	General considerations	60
3.1.1	Summary of the reactions in clinker formation	60
3.1.2	Lime saturation factor, silica ratio and alumina ratio	60
3.1.3	The Bogue calculation	62
3.1.4	Enthalpy changes in clinker formation	63
3.2	Raw materials and manufacturing processes	65
3.2.1	Raw materials and fuel	65
3.2.2	Dry and wet processes; energy requirements	66
3.2.3	The dry process; suspension preheaters and precalciners	67
3.2.4	The rotary kiln	69
3.2.5	Circulation of volatiles; dust; cooling of clinker	70
3.2.6	Other processes for clinker production; clinker grinding	70
3.3	Reactions below about 1300°C	71
3.3.1	Decomposition of carbonate minerals	71
3.3.2	Decomposition of clay minerals and formation of products	72
3.3.3	Sampling from cement kilns or preheater outlets	73
3.3.4	Reaction mechanisms	75
3.3.5	Condensation or reaction of volatiles	76
3.4	Reactions at 1300–1450°C	77
3.4.1	Quantity of liquid formed	77

3.4.2	Burnabilities of raw mixes	79
3.4.3	Nodulization	81
3.4.4	Formation and recrystallization of alite	81
3.4.5	Evaporation of volatiles; polymorphic transitions; reducing conditions	83
3.5	Reactions during cooling, grinding or storage	84
3.5.1	Solidification of the clinker liquid: indications from pure systems	84
3.5.2	Do Portland cement clinkers contain glass or $C_{12}A_7$?	85
3.5.3	Evidence from X-ray microanalysis	86
3.5.4	Effects of cooling rate on the aluminate and ferrite phases	87
3.5.5	Other effects of cooling rate	88
3.5.6	Crystallization of the sulphate phases	89
3.5.7	Changes during grinding or storage	91
3.6	Effects of minor components	92
3.6.1	General	92
3.6.2	Effects of s-block elements	93
3.6.3	Effects of p- and d-block elements	94
4	**Properties of Portland clinker and cement**	96
4.1	Macroscopic and surface properties	96
4.1.1	Unground clinker	96
4.1.2	Particle size distribution of ground clinker or cement	96
4.1.3	Specific surface area determination	98
4.1.4	Particle size distribution, phase composition and cement properties	99
4.1.5	Chemical analysis	100
4.2	Light microscopy	101
4.2.1	General	101
4.2.2	Effects of bulk composition, raw feed preparation and ash deposition	103
4.2.3	Effects of burning conditions and cooling rate	104
4.2.4	Applications of light microscopic investigations	104
4.3	Scanning electron microscopy, X-ray diffraction and other techniques	105
4.3.1	Scanning electron microscopy	105
4.3.2	X-ray diffraction	108
4.3.3	Chemical or physical methods for separation of phases	111
4.3.4	Other methods	112
4.4	Quantitative phase composition	113
4.4.1	General	113
4.4.2	Calculation of quantitative phase composition from bulk analysis	113

4.4.3	Estimation of sulphate phases	114
4.4.4	Estimation of major phases	115
4.4.5	Limitations and modifications of the calculation of phase composition	116
4.4.6	Comparison of results of different methods	118
4.5	Reactivities of clinker phases	119
4.5.1	Effect of major compositional variation	119
4.5.2	Effects of ionic substitutions, defects and variation in polymorph	121

5 Hydration of the calcium silicate phases — 123

5.1	Introduction	123
5.1.1	Definitions and general points	123
5.1.2	Experimental considerations	124
5.1.3	Calcium hydroxide	125
5.2	Compositions of C_3S and β–C_2S pastes	128
5.2.1	Calcium hydroxide content, thermal analysis and indirect determination of the Ca/Si ratio of the C–S–H	128
5.2.2	Water content of the C–S–H	130
5.2.3	X-ray microanalysis and analytical electron microscopy	131
5.3	Structural data for C_3S and β–C_2S pastes	133
5.3.1	Microstructure	133
5.3.2	Silicate anion structure	137
5.3.3	X-ray diffraction pattern, densities and other data	140
5.4	Structural models for C–S–H gel	142
5.4.1	General	142
5.4.2	1.4-nm tobermorite and jennite	142
5.4.3	C–S–H(I) and similar materials	145
5.4.4	Products formed in suspensions from C_3S or β–C_2S	147
5.4.5	Structural relationships	148
5.4.6	A mixed tobermorite–jennite-type model for C–S–H gel	150
5.5	Equilibria	153
5.5.1	Solubility relations	153
5.5.2	Species in solution	156
5.5.3	Thermochemistry and thermodynamics	157
5.5.4	Effects of alkalis	158
5.6	Kinetics and mechanisms	159
5.6.1	C_3S: experimental data	159
5.6.2	C_3S: the initial reaction	162
5.6.3	C_3S: the induction period	162
5.6.4	The main reaction (C_3S and β-C_2S)	164
5.6.5	Early hydration of β-C_2S	165

6	**Hydrated aluminate, ferrite and sulphate phases**	167
6.1	AFm phases	167
6.1.1	Compositional and structural principles	167
6.1.2	The C_4AH_x, $C_4A\bar{C}_{0.5}H_x$ and $C_4A\bar{C}H_x$ phases	169
6.1.3	The $C_4A\bar{S}H_x$ phases	169
6.1.4	Other AFm phases containing aluminium	173
6.1.5	AFm phases containing iron	175
6.1.6	XRD patterns, thermal behaviour, optical properties and IR spectra	175
6.2	AFt phases	177
6.2.1	Compositions and crystal structures	177
6.2.2	Properties	180
6.3	Other hydrated phases	181
6.3.1	Hydrogarnet phases	181
6.3.2	CAH_{10}	183
6.3.3	Brucite, hydrotalcite and related phases	184
6.3.4	Sulphate phases	186
6.4	Equilibria and preparative methods	188
6.4.1	The $CaSO_4$–H_2O, $CaSO_4$–$Ca(OH)_2$–H_2O and $CaSO_4$–K_2SO_4–H_2O systems	188
6.4.2	The CaO–Al_2O_3–H_2O, CaO–Al_2O_3–SiO_2–H_2O and CaO–Al_2O_3–SO_3–H_2O systems	189
6.4.3	Preparative methods	192
6.5	Hydration reactions of the aluminate and ferrite phases	193
6.5.1	Reaction of C_3A with water or with water and calcium hydroxide	193
6.5.2	Reaction of C_3A with water in the presence of calcium sulphate	194
6.5.3	Reaction of the ferrite phase	196
6.5.4	Enthalpy changes	197
7	**Hydration of Portland cement**	199
7.1	General description of the hydration process and products	199
7.1.1	Evidence from X-ray diffraction	199
7.1.2	Evidence from differential thermal analysis and infrared spectroscopy	201
7.1.3	Evidence from light and electron microscopy	203
7.2	Analytical data for cement pastes	204
7.2.1	Determination of unreacted clinker phases	204
7.2.2	Non-evaporable and bound water	206
7.2.3	Thermogravimetry and determination of calcium hydroxide content	207

x Contents

7.2.4	Determinations of hydrated aluminate and silicate phases	208
7.2.5	Analyses of individual phases	209
7.2.6	Silicate anion structure	212
7.3	Interpretation of analytical data	213
7.3.1	The nature of the cement gel	213
7.3.2	Fe_2O_3, SO_3 and minor oxide components in cement gel	215
7.3.3	The stoichiometry of cement hydration	216
7.4	Development of microstructure	221
7.4.1	The early period of hydration	221
7.4.2	The middle period of hydration	223
7.4.3	The late period of hydration	224
7.5	Calorimetry, pore solutions and energetics	226
7.5.1	The early and middle periods	226
7.5.2	Pore solutions after the first day	229
7.5.3	Energetics of cement hydration	230
7.6	Actions of calcium sulphate and of alkalis	231
7.6.1	Setting	231
7.6.2	Optimum gypsum	234
7.6.3	Effects of alkalis	236
7.7	Kinetics and modelling of the hydration process	237
7.7.1	Kinetics: experimental data	237
7.7.2	Interpretation of kinetic data	239
7.7.3	Mathematical modelling of the hydration process	241
8	**Structure and properties of fresh and hardened Portland cement pastes**	**243**
8.1	Fresh pastes	243
8.1.1	Workability	243
8.1.2	Rheology	243
8.1.3	Models of fresh paste structure	245
8.2	Hardened cement pastes: models of structure	246
8.2.1	The Powers–Brownyard model	246
8.2.2	Minimum water/cement ratio for complete hydration; chemical shrinkage	248
8.2.3	Calculation of volumetric quantities	250
8.2.4	Later models of hardened paste structure	252
8.3	Pore structure	254
8.3.1	Porosities obtained by calculation	254
8.3.2	Experimental methods: general points	256
8.3.3	Determination of porosities by pyknometry	257
8.3.4	Sorption isotherms; specific surface areas	258

8.3.5	Pore size distributions	260
8.3.6	Mercury intrusion porosimetry (MIP)	261
8.3.7	Other methods	263
8.4	Strength	265
8.4.1	Empirical relations between compressive strength and porosity	265
8.4.2	Relations between strength and microstructure or pore size distribution	266
8.4.3	Mechanisms of failure	268
8.5	Deformation	269
8.5.1	Modulus of elasticity	269
8.5.2	Drying shrinkage	270
8.5.3	Creep	272
8.6	Permeability and diffusion	273
8.6.1	Permeability to water	273
8.6.2	Diffusion of ions and gases	274
9	**Composite cements**	**276**
9.1	Introduction	276
9.2	Blastfurnace slag	277
9.2.1	Formation, treatment and use in composite cements	277
9.2.2	Factors affecting suitability for use in a composite cement	279
9.2.3	X-ray diffraction and microstructure of slags	280
9.2.4	Internal structures of slag glasses	280
9.2.5	Hydration chemistry of slag cements	282
9.2.6	X-ray microanalysis	284
9.2.7	Stoichiometry of slag cement hydration	286
9.2.8	Activation of slag glasses	287
9.2.9	Supersulphated cements	290
9.3	Pulverized fuel ash (pfa; fly ash) low in CaO	290
9.3.1	Properties	290
9.3.2	Factors governing suitability for use in composite cements	292
9.3.3	Rates of consumption of clinker phases and pfa, and contents of calcium hydroxide	293
9.3.4	Microstructure and compositions of the hydration products	296
9.3.5	The nature of the pozzolanic reaction	298
9.3.6	Stoichiometry of pfa cement hydration	298
9.4	Natural pozzolanas	299
9.4.1	Properties	299
9.4.2	Hydration reactions	302
9.5	Microsilica (condensed silica fume)	305

9.5.1	Properties	305
9.5.2	Hydration reactions	306
9.6	Other mineral additions	308
9.6.1	Class C fly ash	308
9.6.2	Other pozzolanic or hydraulic additions	311
9.6.3	Calcium carbonate and other fillers	312
9.7	Pore structures and their relation to physical properties	312
9.7.1	Porosities and pore size distributions	312
9.7.2	Relations between pore structure and physical properties	314
10	**Calcium aluminate, expansive and other cements**	**316**
10.1	Calcium aluminate cements	316
10.1.1	Introduction	316
10.1.2	Manufacture; chemical and mineralogical compositions	317
10.1.3	Reactivities of the phases and methods of studying hydration	319
10.1.4	Hydration reactions and products	320
10.1.5	Thermodynamic calculations	323
10.1.6	Setting times; mixing and placing	325
10.1.7	Microstructural development	326
10.1.8	Hardening; effects of conversion	326
10.1.9	Chemical admixtures	330
10.1.10	Mixtures with calcite, slag, gypsum or Portland cement	331
10.1.11	Reactions of calcium aluminate concrete with external agents	333
10.1.12	Refractory castables	334
10.2	Expansive cements	335
10.2.1	General	335
10.2.2	Types of expansive cement	335
10.2.3	Mechanism of expansion in Type K cements	337
10.3	Other cements	339
10.3.1	Very rapidly hardening cements	339
10.3.2	Low-energy cements	341
10.3.3	Alinite cements	343
11	**Admixtures and special uses of cements**	**345**
11.1	Introduction	345
11.2	Organic retarders and accelerators	345
11.2.1	Retarders	345
11.2.2	Mechanism of retardation	346
11.2.3	Practical retarders	349
11.2.4	Organic accelerators	350
11.3	Air-entraining agents and grinding aids	351

11.3.1	Air-entraining agents	351
11.3.2	Grinding aids	352
11.4	Water reducers and superplasticizers	352
11.4.1	Water reducers	352
11.4.2	Superplasticizers	353
11.4.3	Mode of action of water reducers and superplasticizers	354
11.4.4	Zeta potential, rheology and nature of the sorbent phases	355
11.4.5	Reasons for the enhanced dispersing power of superplasticizers	357
11.5	Inorganic accelerators and retarders	357
11.5.1	Accelerators of setting and hardening	357
11.5.2	Mode of action	358
11.5.3	Effects on the compositions and structures of the hydration products	360
11.5.4	Precipitation effects; inorganic retarders and setting accelerators	361
11.6	Effects of high or low temperatures at atmospheric pressure	362
11.6.1	Hydration at 25–100°C	362
11.6.2	Effects on kinetics and on the ultimate extent of hydration	363
11.6.3	Effects on hydration products	364
11.6.4	Low temperatures	365
11.7	High-pressure steam curing	365
11.7.1	General	365
11.7.2	Basic chemistry of autoclave processes	367
11.7.3	Mechanisms of reaction and equilibria	369
11.7.4	Characteristics of hydrothermally formed C–S–H and tobermorite	370
11.8	Oil-well cementing	371
11.8.1	General	371
11.8.2	Types of cement and of admixtures	372
11.8.3	Effects of temperature and pressure	373
11.9	Very high strength cement-based materials	374
11.9.1	General	374
11.9.2	Phase compositions, microstructures and causes of high strength	376
12	**Concrete chemistry**	377
12.1	Cement paste in concrete	377
12.1.1	Portland cement mortars and concretes	377
12.1.2	Backscattered electron imaging of the interfacial zone	378
12.1.3	The nature of the paste–aggregate bond	379
12.1.4	Composite cements and other topics	381
12.1.5	Effects at exposed surfaces	382

12.2	Durability—general aspects	383
12.3	Carbonation, chloride penetration and corrosion of reinforcement	383
12.3.1	General	383
12.3.2	Carbonation	384
12.3.3	Transport and reactions of chlorides	386
12.3.4	Corrosion	387
12.4	Alkali–silica reaction	388
12.4.1	General	388
12.4.2	Chemistry of alkali–silica reaction	392
12.4.3	The expansion process	393
12.4.4	ASR in mortars or concretes made with composite cements	394
12.5	Sulphate attack	396
12.5.1	General	396
12.5.2	Calcium sulphate	399
12.5.3	Magnesium sulphate	399
12.5.4	Effects of concrete properties and cement type	400
12.5.5	Composite cements and sulphate attack	401
12.5.6	Reactions involving sulphate and carbonate	401
12.6	Other forms of attack	402
12.6.1	Physical attack	402
12.6.2	Leaching	403
12.6.3	Miscellaneous forms of chemical attack	405
12.6.4	Sea water attack	406
12.6.5	Bacterial attack	406
12.6.6	Miscellaneous paste–aggregate reactions	407
12.6.7	Fire damage	408
References		409
Appendix	Calculated X-ray powder diffraction patterns for tricalcium silicate and clinker phases	447
Index		455

Preface

This book deals with the chemistry of the principal silicate and aluminate cements used in building and civil engineering. It is a completely re-written successor to the two volumes on 'The Chemistry of Cements' that I edited in 1964,* and, like them, is directed primarily to those whose background is in chemistry, materials science or related disciplines. Emphasis is placed throughout on the underlying science rather than on practical applications, which are well covered in other works.

The earlier volumes are outdated. Since 1964, several major areas previously unknown or neglected have been widely studied due to changes in the ways in which cement is made or used, or to increased awareness of the problems encountered in producing concrete of adequate durability. Alongside this, much information has accrued from the introduction of new and powerful experimental techniques, of which scanning electron microscopy and X-ray microanalysis have proved especially important. In 1964, some topics, though practically important, were so imperfectly understood from the standpoint of basic chemistry that I did not consider it useful to deal with them. Increased understanding now makes it possible, and indeed essential, to include them. On the other hand, descriptions of standard experimental techniques, included in the earlier book, have been omitted, except for some matters of specialist relevance to cement chemistry, as good textbooks on all of them are available.

All the cements considered in this book fall into the category of hydraulic cements; they set and harden as a result of chemical reactions with water, and if mixed with water in appropriate proportions continue to harden even if stored under water after they have set. Much the most important is Portland cement. Chapters 1 to 4 of the present work deal mainly with the chemistry of manufacture of Portland cement and with the nature of the resulting product. Chapters 5 to 8 deal mainly with the processes that occur when this product is mixed with water and with the nature of the hardened material. Chapters 9 to 11 deal with the chemistry of other types of cement, of admixtures for concrete and of special uses of cements. Chapter 12 deals with chemical and microstructural aspects of concrete, including ones relevant to processes that affect its durability or limit its service life.

The literature of cement chemistry is voluminous; the abstracting journal, *Cements Research Progress* has latterly listed about one thousand new contributions annually. Inevitably, coverage in the present work has been selective. I have tried to strike a balance between classical and recent contributions, in order to produce a book that will serve both as an

---

* *The Chemistry of Cements* (ed. H. F. W. Taylor), Vol. 1, 460 pp., Vol. 2, 442 pp., Academic Press, London and New York (1964).

introduction that assumes no previous specialist knowledge of cements, and as a guide to research. It is hoped that the more important contributions up to early 1989 have been covered.

Acknowledgments

I am most grateful to former colleagues at the University of Aberdeen and friends elsewhere with whom I have had productive discussions over a period of many years. More directly in relation to the present book, Mr C. P. Kerton, Dr G. R. Long, Dr G. K. Moir and Mr M. S. Sumner (Blue Circle Industries PLC) gave invaluable help, especially on the chemistry and technology of Portland cement production, composite cements, and clinker and cement properties. My thoughts on the hydration chemistry of calcium silicates owe much to discussions with Drs P. W. Brown, G. Frohnsdorff and H. M. Jennings, then all at the National Bureau of Standards (now National Institute of Standards and Technology), USA. Those on microstructural aspects of the hydration of calcium silicates and cement similarly owe much to discussions with Professor P. L. Pratt and Dr K. L. Scrivener, of Imperial College, London. Dr L. J. Parrott (British Cement Association) helped greatly with a discussion on the pore structures of cement pastes. Dr A. Capmas (Lafarge Fondu International, France) and Drs D. Ménétrier-Sorrentino and F. Sorrentino (Lafarge Coppée Recherche, France) gave quite exceptional help in providing a draft manuscript on calcium aluminate cements and with subsequent discussions, in which Dr C. M. George (Lafarge Calcium Aluminates, USA) also participated. Dr J. Bensted (The British Petroleum Company PLC) corrected some important errors on oil well cementing. For discussions and comments on various aspects of concrete durability, I am most grateful to Mr J. S. Lumley and Dr J. J. Kollek (Blue Circle Industries PLC). Many of those mentioned above gave help that included comments on parts of the manuscript, but any errors are mine.

Many people, noted in the reference list, generously made results available in advance of publication. Mr A. M. Harrisson and Mr N. B. Winter (WHD Microanalysis Consultants, Ltd), Dr D. C. Pomeroy (British Cement Association), Dr D. Ménétrier-Sorrentino and Dr K. L. Scrivener kindly provided some excellent electron or light micrographs, noted in the respective captions. Mr C. P. Kerton freely gave extensive bibliographical assistance, and I thank Mr S. Black and Ms J. Kerr (University of Aberdeen) for photography of the line drawings.

I thank publishers (as copyright holders; italics) and authors for permission to reproduce the figures noted below. Authors and sources, including copyright years, are given in the captions and reference list, and where it was requested, additional information is included below. *American Ceramic Society*, Fig. 5.4C and D (Ref. J10, Issue 10, Oct., on "Morphological development of hydrating tricalcium silicate as examined by electron microscopy techniques"); Fig. 5.6 (Ref. M38, Issue 12, Dec., on "Analytical electron microscopy of cement pastes: IV, β-dicalcium silicate pastes"); Figs. 5.8 and 5.9 (Ref. T24, Issue 6, June, on "Proposed structure for calcium silicate hydrate gel"); Fig. 5.12 (Ref. J15, Issue 8, Aug., on "Aqueous solubility relationships for two types of calcium silicate hydrate"); Fig. 9.3 (Ref. T44, Issue 12, Dec., on "Analytical study of pure and extended Portland cement pastes: II, fly ash- and slag-cement pastes"). *American Concrete Institute*, Fig. 11.6. *American Journal of Science*, Fig. 6.5. *British Cement Association*, Fig. 5.11. Cemento, Fig. 11.4. *Dunod* and *RILEM*, Figs. 5.10 and 5.13. *Editions Septima*, Fig. 10.7. *Elsevier Applied Science Publishers Ltd*, Figs. 10.1, 10.6 and 10.8. *International Cement Microscopy Association*, Figs. 4.3A, 4.3B and 7.2. *John Wiley and Sons Ltd*, Fig. 8.4. *Materials Research Society*, Figs. 8.7, 9.1, 9.2, 9.6 and 12.1. *Palladian Publications Ltd*, Fig. 7.1. *Pergamon Press PLC*, Figs. 1.2, 3.3, 5.3, 5.5, 5.14, 5.15, 8.8, 11.1, 11.5 and 12.5. *The Royal Society*, Fig. 7.7. *Sindicato Nacional da Indústria do Cimento* (Brazil), Fig. 6.8. *Society of Chemical Industry*, Fig. 5.1. *Stroyizdat*, Fig. 3.4. *Thomas Telford Ltd*, Figs. 12.4 and 12.6. *Transportation Research Board, National Research Council, Washington DC*, Fig. 7.8 (Ref. L32, on "Changes in composition of the aqueous phase during hydration of cement pastes and suspensions").

1

Portland cement and its major constituent phases

1.1 Introduction

1.1.1 Portland cement: general

Portland cement is made by heating a mixture of limestone and clay, or other materials of similar bulk composition and sufficient reactivity, ultimately to a temperature of about 1450°C. Partial fusion occurs, and nodules of clinker are produced. The clinker is mixed with a few per cent of gypsum and finely ground to make the cement. The gypsum controls the rate of set and may be partly replaced by other forms of calcium sulphate. Some specifications allow the addition of other materials at the grinding stage. The clinker typically has a composition in the region of 67% CaO, 22% SiO_2, 5% Al_2O_3, 3% Fe_2O_3 and 3% of other components, and normally contains four major phases, called alite, belite, aluminate phase and ferrite phase. Several other phases, such as alkali sulphates and calcium oxide, are normally present in minor amounts.

Alite is the most important constituent of all normal Portland cement clinkers, of which it constitutes 50–70%. It is tricalcium silicate (Ca_3SiO_5) modified in composition and crystal structure by incorporation of foreign ions, especially Mg^{2+}, Al^{3+} and Fe^{3+}. It reacts relatively quickly with water, and in normal Portland cements is the most important of the constituent phases for strength development; at ages up to 28 days, it is by far the most important.

Belite constitutes 15–30% of normal Portland cement clinkers. It is dicalcium silicate (Ca_2SiO_4) modified by incorporation of foreign ions and normally present wholly or largely as the β polymorph. It reacts slowly with water, thus contributing little to the strength during the first 28 days, but substantially to the further increase in strength that occurs at later ages. By one year, the strengths obtainable from pure alite and pure belite are about the same under comparable conditions.

The aluminate phase constitutes 5–10% of most normal Portland cement clinkers. It is tricalcium aluminate ($Ca_3Al_2O_6$), substantially modified in composition and sometimes also in structure by incorporation of foreign ions, especially Si^{4+}, Fe^{3+}, Na^+ and K^+. It reacts rapidly with water, and can cause undesirably rapid setting unless a set-controlling agent, usually gypsum, is added.

The ferrite phase makes up 5–15% of normal Portland cement clinkers. It is tetracalcium aluminoferrite (Ca_2AlFeO_5) substantially modified in composition by variation in Al/Fe ratio and incorporation of foreign ions. The rate at which it reacts with water appears to be somewhat variable, perhaps due to differences in composition or other characteristics, but in general is high initially and intermediate between those of alite and belite at later ages.

1.1.2 Types of Portland cement

The great majority of Portland cements made throughout the world are designed for general constructional use. The specifications with which such cements must comply are similar, but not identical, in all countries and various names are used to define the material, such as OPC (Ordinary Portland Cement) in the UK, or Type I Portland Cement in the USA. Throughout this book, we shall use the term 'ordinary' Portland cements to distinguish such general purpose cements from other types of Portland cement, which are made in smaller quantities for special purposes.

Specifications are, in general, based partly on chemical composition or physical properties such as specific surface area, and partly on performance tests, such as setting time or compressive strength developed under standard conditions. The content of MgO* is usually limited to either 4 or 5%, because quantities of this component in excess of about 2% are liable to occur as periclase (magnesium oxide), which through slow reaction with water can cause destructive expansion of hardened concrete. Free lime (calcium oxide) can behave similarly, and its potential formation sets a practical upper limit to the alite content of a clinker; in UK specifications, this is controlled by a compositional parameter called the lime saturation factor (LSF), which is discussed in Sections 2.3.3 and 3.1.2. Excessive contents of SO_3 can also lead to delayed expansion, and upper limits of 2.5–4% are usually imposed. Alkalis (K_2O and Na_2O) can undergo expansive reactions with certain aggregates, and some national specifications limit the content, e.g. to 0.6% equivalent Na_2O ($Na_2O + 0.66K_2O$). Other upper limits of composition widely used in specifications relate to matter insoluble

---

* Confusion can arise because the names or formulae of compounds can be used to denote either phases or components; this applies especially to CaO and MgO. Here and elsewhere, we shall generally use chemical or mineral names of oxides (e.g. calcium oxide, magnesium oxide, lime, periclase) for phases and formulae (e.g. CaO, MgO) for components. Mineral names or prefixed formulae (e.g. α-Al_2O_3) are never used for components.

in dilute acid, and loss on ignition. Many other minor components are limited in content by their effects on the manufacturing process, or the properties, or both, and in some cases the limits are defined in specifications.

Rapid-hardening Portland cements have been produced in various ways, such as varying the composition to increase the alite content, finer grinding of the clinker, and improvements in the manufacturing process, e.g. finer grinding or better mixing of the raw materials. The alite contents of Portland cements have increased steadily over the one and a half centuries during which the latter have been produced, and many present-day cements that would be considered normal today would have been described as rapid hardening only a few decades ago. In USA specifications, rapid-hardening Portland cements are called high early strength or Type III cements.

Destructive expansion from reaction with sulphates can occur not only if the latter are present in excessive proportion in the cement, but also from attack on concrete by sulphate solutions. The reaction involves the Al_2O_3-containing phases in the hardened cement, and in sulphate-resisting Portland cements, its effects are reduced by decreasing the proportion of the aluminate phase, sometimes to zero. This is achieved by decreasing the ratio of Al_2O_3 to Fe_2O_3 in the raw materials. In the USA, sulphate-resisting Portland cements are called Type V cements.

White Portland cements are made by increasing the ratio of Al_2O_3 to Fe_2O_3, and thus represent the opposite extreme in composition to sulphate-resisting Portland cements. The normal, dark colour of Portland cement is due to the ferrite phase, formation of which in a white cement must thus be avoided. It is impracticable to employ raw materials that are completely free from Fe_2O_3 and other components, such as Mn_2O_3, that contribute to the colour. The effects of these components are therefore usually minimized by producing the clinker under slightly reducing conditions and by rapid quenching. In addition to alite, belite and aluminate phase, some glass may be formed.

The reaction of Portland cement with water is exothermic, and while this can be an advantage under some conditions because it accelerates hardening, it is a disadvantage under others, such as in the construction of large dams or in the lining of oil wells, when a cement slurry has to be pumped over a large distance under pressure and sometimes at a high temperature. Slower heat evolution can be achieved by coarser grinding, and decreased total heat evolution by lowering the contents of alite and aluminate phase. Specifications in the USA include definitions of a Type II or 'moderate heat of hardening' cement, and a more extreme Type IV or 'low heat' cement. The Type II cement is also suitable for general use in constructions exposed to moderate sulphate attack. Heat evolution can also be decreased by partially replacing the cement by pfa (pulverized fuel ash; fly ash) or other materials

(Chapter 9), and this is probably today a more usual solution. The specialized requirements of oil well cements are discussed in Section 11.8.

1.1.3 Cement chemical nomenclature and other abbreviations

Chemical formulae in cement chemistry are often expressed as a sum of oxides; thus tricalcium silicate, Ca_3SiO_5, can be written as $3CaO \cdot SiO_2$. This does not imply that the constituent oxides have any separate existence within the structure. It is usual to abbreviate the formulae of the commoner oxides to single letters, such as C for CaO or S for SiO_2, Ca_3SiO_5 thus becoming C_3S. This system is often combined with orthodox chemical notation within a chemical equation, e.g.

$$3CaO + SiO_2 \rightarrow C_3S \tag{1.1}$$

or even within a single formula, as in $C_{11}A_7 \cdot CaF_2$ for $Ca_{12}Al_{14}O_{32}F_2$. The abbreviations most widely used are as follows:

C = CaO   S = SiO_2 A = Al_2O_3 F = Fe_2O_3
M = MgO   K = K_2O $\bar{S}$ = SO_3 N = Na_2O
T = TiO_2 P = P_2O_5 H = H_2O $\bar{C}$ = CO_2

The formulae of the simple oxide phases (e.g. CaO) are usually written in full. Other abbreviations and units used in this book are as follows:

Techniques

BEI = backscatterd electron imaging. BET method = Brunauer–Emmett–Teller. DTA = differential thermal analysis. EPMA = electron probe microanalysis. ESCA = electron spectroscopy for chemical analysis (X-ray photoelectron spectroscopy). GLC = gas-liquid chromatography. GPC = gel permeation chromatography. IR = infrared. MIP = mercury intrusion porosimetry. NMR = nuclear magnetic resonance. QXDA = quantitative X-ray diffraction analysis. SEM = scanning electron microscop(e,y). STEM = scanning transmission electron microscop(e,y). TEM = transmission electron microscope(e,y). TG = thermogravimetry. TMS = trimethylsilyl(ation). XRD = X-ray diffraction. XRF = X-ray fluorescence.

Materials

C–S–H = poorly crystalline or amorphous calcium silicate hydrate of unspecified composition. Ggbfs = ground granulated blast furnace slag. Hcp = hardened cement paste. Pfa = pulverised fuel ash (fly ash).

Properties or reactions

AR = alumina ratio (alumina modulus). ASR = alkali silica reaction. LSF = lime saturation factor. SR = silica ratio (silica modulus). C_x = analytical (total) concentration of x, irrespective of species. [x] = concentration of species x. {x} = activity of species x. RH = relative humidity. Na_2O_e = equivalent Na_2O (wt % Na_2O + 0.66K_2O). (+)2V, (−)2V = optic sign and optic axial angle.

Pressure units

1 MPa = 1 N mm^{-2} = 10 bar = 9.87 atm = 7500 torr = 145.0 lb in^{-2}.

1.2 Alite

1.2.1 Polymorphism and crystal structure

On being heated, pure C_3S undergoes a series of reversible phase transitions, which have been detected by a combination of DTA, high-temperature XRD and high-temperature light microscopy (B1,G1,Y1,R1,M1–M5):

$$T_1 \xrightleftharpoons{620°C} T_2 \xrightleftharpoons{920°C} T_3 \xrightleftharpoons{980°C} M_1 \xrightleftharpoons{990°C} M_2 \xrightleftharpoons{1060°C} M_3 \xrightleftharpoons{1070°C} R$$

(T = triclinic, M = monoclinic, R = rhombohedral).

The pure compound, when cooled to room temperature, is thus T_1. In production clinkers, due to the incorporation of foreign ions, the form present at room temperature is normally M_1 or M_3, or a mixture of these; rarely, T_2 is found (M1–M5,T1). There has been some uncertainty as to the number and nomenclature of these polymorphs; reported M_{1b} and M_{2b} forms appear to be identical with M_3, leaving reported M_{1a} and M_{2a} forms to be called simply M_1 and M_2, respectively (M4,M5).

Jeffery (J1) made the first determination of the crystal structure. He showed that the forms now known as R, T_1 and M_3 had closely similar structures, and determined the approximate or pseudostructure common to all three; it was built from Ca^{2+}, SiO_4^{4-} and O^{2-} ions, the latter being bonded only to six Ca^{2+} ions, as in CaO. Later, more exact determinations were reported for T_1 (G2), M_3 stabilized by Mg^{2+} (N1), R at 1200°C (N2) and R stabilized with Sr^{2+} (I1). Fig. 1.1 shows the structure of the R form. The known structures are all closely similar as regards the positions of the Ca^{2+} and O^{2-} ions and of the silicon atoms, but differ markedly in the orientations of the SiO_4^{4-} tetrahedra, which in some cases are disordered.

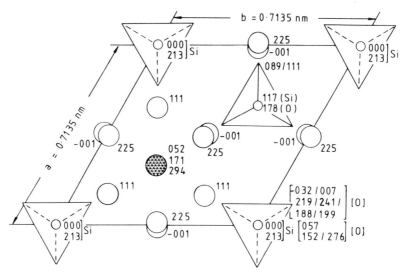

Fig. 1.1 Crystal structure of the R modification of C_3S, based on the results of Nishi and Takéuchi (N2) and showing calcium atoms (large, open circles), silicon atoms (small, open circles), oxide ions (large, hatched circle) and oxygen tetrahedra (triangles). Heights of atoms are in thousandths of the cell height ($c = 2.5586$ nm), slashes denoting statistical alternatives. Oxygen tetrahedra are shown in averaged orientations; in reality, each is tilted around its enclosed silicon atom statistically in any one of three directions so as to preserve the threefold symmetry. All tetrahedra point upwards excepting those surrounding the silicon atoms at height 213, of which 70% point downwards and 30% upwards. Only the bottom third of the cell is shown, the middle and top thirds being derivable from it by translations of 1/3, 2/3, 1/3 and 2/3, 1/3, 2/3 parallel to the a, b and c axes, respectively.

The structural differences between the polymorphs affect the coordination of the Ca^{2+} ions and the oxygen atoms of the SiO_4^{4-} tetrahedra. For each polymorph, there are several crystallographically distinct calcium sites, having different coordination, and for a given site, the coordination sometimes varies between individual atoms due to orientational disorder in the surrounding SiO_4^{4-} tetrahedra. Definitions of the calcium coordination numbers are somewhat arbitrary due to variations in the lengths of the bonds; for example, in the R form at 1200°C, the calcium atoms in one of the sites could be regarded as 7 coordinated if bonds as long as 0.296 nm are counted, and 5 coordinated if they are not (N2). If such abnormally long bonds are excluded, the mean coordination number of the calcium is 5.66 in the R polymorph, 6.15 in M_3 and 6.21 in T_1 (M5). In relation to reactivity towards water, the coordination of the oxygen atoms is possibly more important than that of Ca. This has not been discussed in detail in the literature, but mean oxygen coordination numbers may be expected to increase with those of calcium.

Table 1.1 Crystal data for the C_3S polymorphs

Polymorph	Space group	Cell parameters (a,b,c in nm)						Z	Ref.
		a	b	c	α	β	γ		
Pseudostructure	R3m	0.70	–	2.50	–	–	120°	9	J1
R (at 1200°C)	R3m	0.7135	–	2.5586	–	–	120°	9	N2
R (stabilized with Sr)	R3m	0.70567	–	2.4974	–	–	120°	9	I1
M_3 (stabilized with Mg)	Cm	3.3083	0.7027	1.8499	–	94.12°	–	36	N1
T_1	P$\bar{1}$	1.167	1.424	1.372	105.5°	94.3°	90.0°	18	G2

Table 1.1 gives those crystal data for the C_3S polymorphs that have been obtained using single crystal methods. The literature contains additional unit cell data, based only on powder diffraction evidence. Some of these may be equivalent to ones in Table 1.1, since the unit cell of a monoclinic or triclinic crystal can be defined in different ways, but some are certainly incorrect. Because only the stronger reflections are recorded, and for other reasons, it is not possible to determine the unit cells of these complex structures reliably by powder methods. The unit cells of the T_1, M_3 and R forms are superficially somewhat different, but all three are geometrically related; transformation matrices have been given (I2,H1).

1.2.2 Tricalcium silicate solid solutions

Hahn *et al.* (H2) reported a detailed study of the limits of substitution in C_3S of MgO, Al_2O_3 and Fe_2O_3, alone and in combination. For MgO, the limit was 2.0% at 1550°C, falling with temperature to 1.5% at 1420°C. The substitution was of Mg^{2+} for Ca^{2+}. For Al_2O_3, the limit was 1.0%, irrespective of temperature, and the substitution was partly of 2 Al^{3+} for $Ca^{2+} + Si^{4+}$ and partly of 4 Al^{3+} for 3 Si^{4+} and a vacant octahedral site. For Fe_2O_3, the limit was 1.1%, and substitution was of 2 Fe^{3+} for $Ca^{2+} + Si^{4+}$. The limit of substitution for MgO was not affected by the incorporation of Al^{3+} or Fe^{3+} or both, nor vice versa, but Al^{3+} and Fe^{3+} competed for the same sites and the limit for one was decreased in the presence of the other.

Incorporation of any of these oxides in sufficient quantity caused the higher temperature polymorphs to persist on cooling to room temperature. For each combination of substituents studied, increase in the total proportion of substituents led to the persistence of successively higher temperature polymorphs in the sequence $T_1 \rightarrow T_2 \rightarrow M_1$. In the system CaO–MgO–$Al_2O_3$–$SiO_2$, total contents of impurity oxide of about 1% caused T_1 to be replaced by T_2, and ones of about 2% caused T_2 to be replaced by M_1. Slightly higher total impurity contents were needed to effect each of these changes if Fe_2O_3 was present.

Woermann *et al.* (W1) found that up to 1.4% of either Na_2O or K_2O could be incorporated into C_3S at 1500°C and discussed mechanisms of substitution.

1.2.3 Compositions of alites in clinkers

Many workers have analysed alites from ordinary production clinkers by X-ray microanalysis (T2,T3,K1,G3,B2,G4,S1,H3,H4) or chemical analysis of separated material (Y1,U1); alites from white cement clinkers have also been analysed (B3). Considering the errors inherent in either method, the agreement between the various sets of data is generally good. The most important variations are in the contents of CaO, MgO and Fe_2O_3. Table 1.2 gives an average composition, with corrections to the contents of these components based on considerations given in this and the previous sections.

Pure C_3S contains 73.7% of CaO and 26.3% of SiO_2. Alites in clinkers typically contain 3-4% of impurity oxides, i.e. oxide components other than those present in the pure compound. Boikova (B4) found a positive correlation between the total percentage of impurity oxides in the alite (I_a) and the percentage of MgO in the clinker (M_c). Her results are approximately fitted by the equation $I_a = 0.7 M_c + 2.1$.

Correlations have also been found between the percentages of individual impurity oxides in the alite and in the clinker. In the following discussion, we shall use symbols of the type O_a and O_c to denote these quantities, where O is the cement chemical symbol for the oxide and the suffixes a and c indicate alite and clinker, respectively. For MgO, Kristmann's (K1) data indicate the approximate relations $M_a = 0.67 \times M_c$ for $M_c \leqslant 3.0$ and $M_a \simeq 2.0$ for $M_c > 3.0$. Yamaguchi and Takagi (Y1) and Terrier (T2) found similar correlations, but with slopes of 0.74 and 0.45, respectively. For Al_2O_3, Kristmann (K1) did not find any significant correlation, but for Fe_2O_3 his data indicate the approximate relation $F_a = 0.33 \times F_c$ for $F_c < 4\%$. Boikova (B4) reported ratios of 0.34 to 0.82 for M_a/M_c, 0.09 to 0.23 for A_a/A_c and 0.17 to 0.36 for F_a/F_c.

Several authors have discussed the types of substitution that occur in clinker alites (G1,G3,H3,B4). In general, the evidence as regards limits and patterns of substitution is consistent with the conclusions of Hahn et al. (H2) given in Section 1.2.2, but the contents of K_2O and Na_2O are much below the limits found by Woermann et al. (W1). Table 1.3 includes atomic ratios corresponding to the typical composition in Table 1.2, assuming 1.1% MgO. The cations are arranged in approximate sequence of ionic radius; in the absence of precise information from X-ray structure determinations, this, taken in conjunction with a knowledge of the known preferences of ions as regards coordination number, appears to provide the most rational basis for suggesting allocations of atoms to sites, and one that is consistent with the compositional data, both for alite and for the other clinker phases.

1.2.4 Polymorphic modifications of the alites in clinkers

Maki and co-workers (M2-M6) studied the crystallization of alite by high-temperature light microscopy of thin sections together with thermal and X-

Table 1.2 Typical compositions of phases in Portland cement clinkers (weight per cent)

	Note	Na_2O	MgO	Al_2O_3	SiO_2	P_2O_5	SO_3	K_2O	CaO	TiO_2	Mn_2O_3	Fe_2O_3
Alite	1	0.1	1.1	1.0	25.2	0.2	0.0	0.1	71.6	0.0	0.0	0.7
Belite	2	0.1	0.5	2.1	31.5	0.2	0.1	0.9	63.5	0.2	0.0	0.9
Aluminate (cubic)	3	1.0	1.4	31.3	3.7	0.0	0.0	0.7	56.6	0.2	0.0	5.1
(orthorhombic)	4	0.6	1.2	28.9	4.3	0.0	0.0	4.0	53.9	0.5	0.0	6.6
(low Fe)	5	0.4	1.0	33.8	4.6	0.0	0.0	0.5	58.1	0.6	0.0	1.0
Ferrite	6	0.1	3.0	21.9	3.6	0.0	0.0	0.2	47.5	1.6	0.7	21.4
Ferrite (low Al)	7	0.1	2.8	15.2	3.5	0.0	0.0	0.2	46.0	1.7	0.7	29.8

Notes
1. For a clinker with 1.65% MgO and 2.6% Fe_2O_3. More generally, % MgO = 0.667 × % MgO in clinker, but not >2.0%; % Fe_2O_3 = 0.25 × % Fe_2O_3 in clinker, but not >1.1%; % CaO = 71.6 − 1.4 × (% MgO in alite − 1.1). The composition should be normalized to a total of 100%.
2. SO_3 higher in clinkers of high SO_3/(Na_2O + K_2O) ratio.
3. Normal form in clinkers low in alkali and with Al_2O_3/Fe_2O_3 in the approximate range 1.0–3.0.
4. Orthorhombic or pseudotetragonal forms, found in clinkers high in alkalis. Na/K ratio varies with that of the clinker.
5. Tentative composition for aluminate phase in white cement clinkers.
6. Normal composition for clinkers with Al_2O_3/Fe_2O_3 in the approximate range 1.0–3.0.
7. Tentative composition for ferrite phase in sulphate-resisting clinkers.

Table 1.3 Atomic ratios for phases in Portland cement clinkers, calculated from the typical compositions in Table 1.2

	K	Na	Ca	Mg	Mn	Ti	Fe	Al	Si	P	O
Alite[a]	3	1	290	6			3	4	95	1	500
Belite	3	1	194 200	2	300		2	7	100 90	1	393
Aluminate (cubic)	4	9	273 296	9		1	17	166 200	100 17		600
Aluminate (orthorhombic)[a]	23	5	265 303	8		2	23	157 200	20		600
Aluminate (low Fe)	3	3	276 291	7		2	3	176 ≃200	20		600
Ferrite	1	1	198 200	17	2	5	62 200	100	14		500
Ferrite (low Al)	1		198 ≃200	17	2	5	90 200	72	14		500

[a] See notes to Table 1.2.

ray methods. It crystallizes from the clinker liquid at about 1450°C as the R polymorph, which inverts to lower temperature forms on cooling. Due to the lowering of symmetry, each crystal usually contains several optically distinct domains, recognizable in thin section using polarized light. At the higher temperatures of formation or recrystallization within the normal range, small crystals, relatively high in foreign ions, are formed, which on cooling invert to M_3. Crystals formed at lower temperatures are larger and poorer in foreign ions, and may undergo further transition, partial or complete, to M_1 or, rarely, to T_2.

The MgO and SO_3 contents of the clinker are especially important in determining whether the transformation into M_1 occurs. High MgO contents favour the formation of small crystals that invert to M_3, and high SO_3 contents that of large crystals that invert further to M_1 (Fig. 1.2), though the tendency to form M_1 is decreased if the ratio of alkalis to SO_3 is high, as the SO_3 is then largely combined as alkali sulphates. Crystals containing both M_3 and M_1 can occur, and are often zoned, with cores of M_1 and peripheral regions of M_3. This is because the clinker liquid becomes richer in MgO as crystallization proceeds, causing the material deposited later to persist as M_3. The M_3 to M_1 transformation is also affected by cooling rate, slow cooling favouring the transformation to M_1. In certain slowly cooled clinkers, further transformation may occur, giving T_2. This is likely to occur only if the alite is particularly low in substituents.

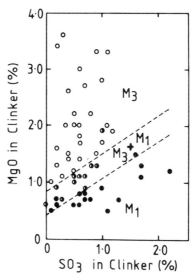

Fig. 1.2 Dependence of the polymorphic modification of alite in production clinkers on the MgO and SO_3 contents of the clinker. Maki and Goto (M6).

On prolonged heating at high temperatures, the large alite crystals grow at the expense of smaller ones, and in this process the content of foreign ions decreases, favouring inversion of M_3 to M_1 or T_2. Production clinkers are usually sufficiently high in MgO to prevent this inversion from taking place, even though considerable recrystallization may occur (M6). Recrystallization can thus markedly affect the relation between temperature of formation and crystal size.

Guinier and Regourd (G1) concluded from XRD evidence that the R polymorph occurred in certain clinkers. This view has been accepted by many other workers, but Maki and Kato (M4) concluded from thermal and other evidence that the supposed R form was really M_1, which had been misidentified because of the similarity of its XRD powder pattern to that of the R form. They gave two strong reasons for supposing the alites in question to be M_1 and not R. Firstly, they contained less MgO than M_3 alites; for R to persist at room temperature, it would have to contain more MgO than M_3, because it is a higher temperature form. Secondly, phase transitions were detected on heating; R, again because it is the high-temperature form, would not undergo these. Sinclair and Groves (S2) also described a case of the misidentification of an alite as rhombohedral; using single crystal electron diffraction, they found that some laboratory-prepared alites that had been regarded as rhombohedral were really either monoclinic or triclinic.

On present evidence, it is thus probable that the only alite polymorphs found in normal production clinkers are M_3, M_1 and, occasionally, T_2, and that those that have been described as R are really M_1. However, both XRD and thermal evidence show that some alites formed in the presence of fluoride are rhombohedral (M7). XRD and optical evidence indicates that the commonest polymorph in production clinkers is almost certainly M_3, the presence of which in such a clinker has also been demonstrated using single crystal electron diffraction (H1).

1.2.5 X-ray powder patterns and densities of tricalcium silicate and alites

A number of experimentally determined XRD powder patterns have been reported for C_3S and alite polymorphs (P1), but because of uncertainties in the indexing and sometimes in identification of the polymorph, these are usefully supplemented by patterns calculated from the crystal structures (Appendix). The patterns of the polymorphs are closely alike, and in clinkers, interpretation is complicated by the fact that many peaks overlap ones of other phases. The pattern depends not only on the polymorph or mixtures of polymorphs present, but also on the nature and amounts of the foreign ions present in each. It is therefore not always easy or even practicable to identify the polymorph or polymorphs in a clinker by the XRD powder pattern alone. The patterns differ principally in the fine structures of certain peaks (J1,G1,R1,M4,M5,B4). Fig. 1.3 shows characteristic portions of the patterns of T_1 C_3S and of some clinkers. M_1 alite, as present in clinkers, probably always has the maximum MgO content possible for this polymorph, and as such has a unit cell or pseudocell that is geometrically almost hexagonal. This causes the peak at about 51.7° 2θ (CuK$_\alpha$ radiation) to be almost a singlet; the corresponding peak of the M_3 alite is a well-defined doublet, and those of T_1 C_3S and T_2 alite are triplets. There are also differences in the peaks at 32° to 33° 2θ, but these are less useful because of overlaps with other phases. Because of their high relative intensities, care is needed with these latter peaks to avoid intensity errors from extinction.

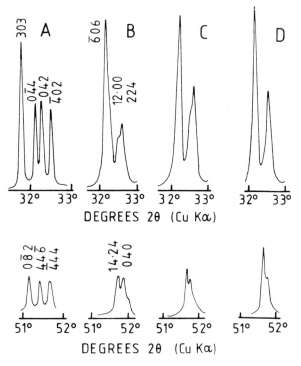

Fig. 1.3 Portions of XRD powder patterns of (A) the T_1 modification of C_3S at 605°C (after Regourd (R2)); (B), (C) and (D) clinkers containing, respectively, M_3, ($M_3 + M_1$) and M_1 alites (after Maki and Kato (M4)). Indexing of T_1 and M_3 patterns is based on the axes in Table 1.1 and calculated intensities.

The X-ray density of T_1 C_3S is 3120 kg m^{-3} (G2); that of M_3 alite, calculated assuming the atomic parameters of Nishi *et al.* (N1) and the composition $(Ca_{0.98}Mg_{0.01}Al_{0.0067}Fe_{0.0033})_3(Si_{0.97}Al_{0.03})O_5$, is 3150 kg m^{-3}.

1.2.6 Optical, thermal and other data

T_1 C_3S is biaxial with $(-)2V = 20\text{--}60°$, and $a > c_{hex} = 0\text{--}15°$ (O1); the refractive indices are $\alpha = 1.7139$, $\gamma = 1.7172$ for pure C_3S and $\alpha = 1.7158\text{--}1.7197$, $\gamma = 1.7220\text{--}1.7238$ for typical clinker alites, all determined using sodium light (B5). In thin sections in polarized light, M_1 and M_3 are most readily distinguished by the maximum birefringence, which is 0.005–0.006 for M_3 and 0.003 for M_1 (M2–M5). Other differences in optical properties include the optic orientation referred to the hexagonal axes of the R form, the shape of which persists through the transitions that occur on cooling, and the optic axial angle; Maki and Kato (M4) give details.

It is doubtful whether the polymorphs can be reliably distinguished by DTA, since both the M_1 and the M_3 forms are metastable at low temperatures and transform to T_2 at about $700°C$ (M4,M5). This effect is normally swamped by the endothermic β to α'_L transition of the belite. Due to the effect of substituents, neither T_3 nor M_2 appears to form on heating, and endothermic transitions occur at $790\text{--}850°C$, giving M_1, and at $950°C$, giving M_3. These could occur irrespective of the form initially present; however, clinkers high in MgO, in which the alite is present entirely as M_3, normally show no thermal effects attributable to the alite until more than $1000°C$, when transition to R occurs.

A substituted C_3S has been found in nature and given the mineral name hatrurite (G5). The rock containing it is interesting because it contains other phases present in clinker and appears to have been formed under conditions somewhat similar to those existing in a cement kiln.

1.3 Belite

1.3.1 Polymorphism and crystal structure

Thermal and X-ray evidence (R1,N3,B6) show that five polymorphs of C_2S exist at ordinary pressures, viz:

$$\alpha \xrightleftharpoons{1425°C} \alpha'_H \xrightleftharpoons{1160°C} \alpha'_L \xrightleftharpoons{630\text{--}680°C} \beta \xrightarrow{<500°C} \gamma$$
$$\underset{780\text{--}860°C}{\overset{690°C}{\uparrow\text{_____}|}}$$

(H = high; L = low)

The structures of all of them are built from Ca^{2+} and SiO_4^{4-} ions. The arrangements of these ions are closely similar in the α, $α'_H$, $α'_L$ and β polymorphs, but that in γ-C_2S is somewhat different. As with C_3S, the higher temperature polymorphs cannot normally be preserved on cooling to room temperature unless stabilized by foreign ions. γ-C_2S is much less dense than the other polymorphs, and this causes crystals or sintered masses of β-C_2S to crack and fall to a more voluminous powder on cooling, a phenomenon known as dusting. If the crystallites of β-C_2S are sufficiently small, the transformation does not occur, even if no stabilizer is present (Y2). In all normal Portland cement clinkers, the belite contains enough stabilizing ions to prevent the transformation to γ-C_2S from taking place.

Many of the earlier crystal data reported for the C_2S polymorphs were based on powder XRD patterns obtained either at high temperatures or from stabilized preparations at room temperature. As with the C_3S polymorphs, some such data have subsequently been shown to be incorrect, and the remainder must be considered unreliable. Table 1.4 gives data derived from single crystal evidence, which does not suffer from this limitation.

The structures of the α, $α'_H$, $α'_L$ and β polymorphs (Fig. 1.4A–C) all belong to a large family typified by that of glaserite, $K_3Na(SO_4)_2$ (M8). As is apparent from the data in Table 1.4, the unit cells of all four are closely related. High-temperature single crystal X-ray patterns of α-C_2S show symmetry consistent with the space group $P6_3/mmc$, but Udagawa et al. (U6,U7) found that this was due to the presence of a domain structure, individual domains being structurally similar to glaserite, for which the space group is $P\bar{3}1c$; this structure, without the domains, was originally proposed for α-C_2S by Bredig (B7). Structure determinations have also been reported for $α'_L$-C_2S (U8,I3,I2) and β-C_2S (M9,J2). $α'_H$-C_2S appears to have a structure similar to that of β-K_2SO_4 (B7). The $α'_L$-C_2S structure is a more complex variant on that of $α'_H$, in which the 0.69 nm axis is tripled.

The structures of the $α'_H$, $α'_L$ and β polymorphs are derived from that of α-C_2S by progressive decreases in symmetry, which arise from changes in the orientations of the SiO_4^{4-} tetrahedra and small movements of Ca^{2+} ions. The quantities of impurity ions needed to stablize the higher temperature polymorphs at room temperature decrease along the sequence from α- to β-C_2S. In all these respects, the situation parallels that existing with the C_3S polymorphs, described in Section 1.2.1. Also as in the latter case, and for the same reason, these C_2S polymorphs differ in Ca^{2+} ion coordination. In β-C_2S, some of the Ca^{2+} ions have 7, and others 8, oxygen atoms within 0.288 nm (J2).

The crystal structure of γ-C_2S is similar to that of olivine, $(Mg,Fe)_2SiO_4$ (Fig. 1.4D) (U4). The calcium is octahedrally coordinated. The unit cell and the arrangement of Ca^{2+} and SiO_4^{4-} ions show some similarities to those of

Table 1.4 Crystal data for the dicalcium silicate polymorphs

Polymorph	Note	Cell parameters (lengths in nm)			Angle[a]	Space group	Axial relation[b]	Z	Ref.
		a	b	c					
α	1	0.5579		0.7150	$\gamma = 120°$	$P\bar{3}1c$	$-ac$	2	U2, U3
α'_H	2	0.949	0.559	0.685		Pcmm	abc	4	S3
α'_L	3	2.0871 ($\sqrt{3} \times 0.548$)	0.9496 (3×0.6957)	0.5600		$Pna2_1$	bca	12	I2, I3
β	4	0.5502	0.6745	0.9297	$\beta = 94.59°$	$P2_1/n$	cab	4	J2
γ	–	0.5081	1.1224	0.6778		Pbnm	bac	4	U4

Notes
1. Stabilized with Ba. The space group is that of the individual domains.
2. Stabilized with P_2O_5 and examined at 1200°C.
3. Stabilized with Sr.
4. Crystal prepared from a $CaCl_2$ melt without additional stabilizer.

[a] Where not 90°.
[b] Sequence of axes respectively equivalent to a, b and c of the α'_H polymorph.

18 Cement Chemistry

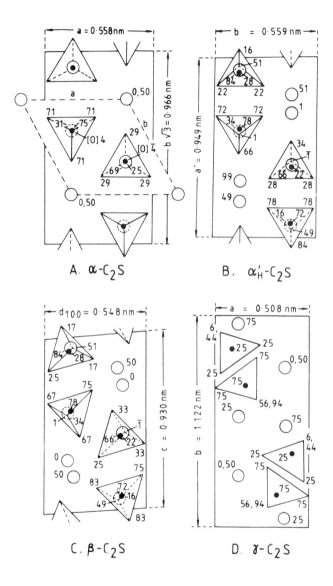

Fig. 1.4 Crystal structures of C_2S polymorphs, after Udagawa *et al.* (U2, U5). Large, open circles represent calcium atoms, small full circles, silicon atoms and triangles, tetrahedra of oxygen atoms. Heights of atoms are given as hundredths of the cell height (0.68–0.71 nm). For α-C_2S, the structure shown is that of an individual domain of space group P31c. The structures are shown in the relative orientations found when they are interconverted, an unconventional choice of origin being adopted for γ-C_2S to clarify the relation to the other polymorphs.

the other polymorphs, but also important differences. It has generally been supposed that transformations involving this form entail total reorganization of the structure, but single crystal high-temperature XRD evidence indicates that this is not correct; the products are formed topotactically (i.e. in a definite crystallographic orientation relative to the starting material; Fig. 1.4), but fragment because of the large volume change (U5,B8,G6). γ-C_2S scarcely reacts with water at ordinary temperatures.

1.3.2 Lamellar textures in clinker belites

The belite grains in Portland cement clinkers frequently show complex, striated structures. These have been studied over a long period by workers using light microscopy; Yamaguchi and Takagi (Y1) and Ono *et al.* (O1), who also used XRD and other methods, gave the first substantially complete interpretations. Groves (G7) and Fukuda and Maki (F1) extended these results, using single crystal electron diffraction and other methods.

A very common type of belite grain in production clinkers, called Type I belite, is rounded, typically 20–40 μm in mean dimension, and shows two or more sets of parallel striations. Within each set of striations, the lamellae extinguish simultaneously when examined in transmitted light between crossed polars and are therefore not twin lamellae; their origin is discussed in the next section. The lamellae in different sets do not extinguish simultaneously. Within each set, secondary striations, not parallel to the primary ones, have sometimes been observed using light microscopy, and are more clearly seen using electron microscopy. The secondary lamellae within a given set do not extinguish simultaneously, but alternately at different angles, in the manner characteristic of polysynthetic twinning.

Type I belite grains are those that have crystallized from the liquid at temperatures above about 1420°C and thus as α-C_2S. The primary striations arise on cooling through the α to α'_H transition, in which the symmetry decreases from hexagonal to orthorhombic, each set of striations thus representing a different orientation of the α'_H structure. The transition to α'_L-C_2S that occurs on further cooling does not further increase the number of orientations, but that from α'_L to β, in which the symmetry falls to monoclinic, causes each orientation to split into two. This is the cause of the secondary sets of striations described above.

Belite crystals have been observed that show only one set of striations, whose behaviour in transmitted light between crossed polars shows them to be twin lamellae. Such crystals are typically irregular in shape; they have been called Type II belites, and are rare in modern clinkers. The striations arise from the α'_L to β transformation. Such crystals can form either by crystallization from the liquid below the α to α'_H transition temperature, or by slow cooling of α-C_2S under conditions such that one of the twin components of the resulting α' modification has grown at the expense of the others. These two modes of formation can be distinguished morphologically, the second mode giving larger and less rounded grains (O1). Belite forms in clinkers by additional mechanisms, sometimes at lower temperatures than those discussed above, and may not then show any striations.

1.3.3 Polymorphic types of belites in clinkers

XRD powder evidence shows that in the majority of clinkers the belite is predominantly or entirely of β-C_2S structure (G1,Y1), though some peaks are broadened (G1) and the presence also of both α and α' (presumably α'_L) forms has been reported (G1,Y1,R1,O1). Characterization of the polymorphic form is rendered difficult by the similarities between their powder patterns (Fig. 1.5) and by overlaps between the peaks and ones of other phases, especially alite, but has been aided by examination of fractions in which the belite has been concentrated by chemical (R1) or heavy liquid (Y1) separation.

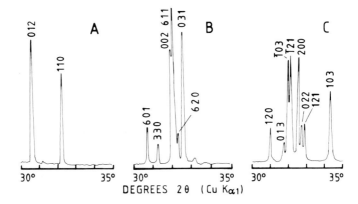

Fig. 1.5 Portions of XRD powder patterns (microdensitometer traces of Guinier photographs) of (A) α-C_2S at 1500°C, (B) α'_L-C_2S at 1000°C and (C) β-C_2S. Indices are based on axes used in Table 1.4. After Regourd and Guinier (R1).

The striations in the common, Type I grains of belite are zones into which impurity ions rejected during the transition from α to α' have been concentrated (Y1,O1,C1). In some cases, they appear to have retained the α-C_2S structure (Y1,O1,F1). In contrast, a study on synthetic belites showed that the material exsolved between the lamellae or at grain boundaries was often compositionally far removed from C_2S, and structurally either amorphous or composed of crystalline phases of entirely different structure, such as C_2AS, C_3A or C_3S_2 (C1). Chan *et al.* (C1) concluded from this study, in which SEM and TEM with diffraction and X-ray microanalysis were used, that β-C_2S cannot be stabilized by excess CaO alone, as some previous workers had supposed. They also concluded that the possibilities of compositional variation in β-C_2S are considerably narrower than has often been supposed, the grosser deviations from ideal composition that have been reported being attributable to exsolution of foreign ions in lamellae or at grain boundaries. They further concluded that the role of ionic substitution in stabilizing forms other than γ to room temperature needed to be reassessed, mechanical constraints associated with exsolution effects providing a possible, alternative explanation.

Studies on synthetic belites suggest the possible occurrence of modulated structures, i.e. ones in which compositional or structural variations occur with a statistical periodicity that is not a rational multiple of that of the basic structure (H5). Such effects might occur in clinker belites.

1.3.4 Compositions of belites in clinkers

Table 1.2 includes an average composition for belites in ordinary production clinkers, based on the results of many studies by X-ray microanalysis (T2,K1,G3,B2,G4,S1,H3,H4) or chemical analysis of separated material (Y1). As with alite (Section 1.2.3), most of the results from different laboratories are in relatively close agreement. Belites in white cement clinkers have also been analysed (B3). Pure C_2S, for comparison, contains 34.9% of SiO_2 and 65.1% of CaO.

Clinker belites typically contain 4–6% of impurity oxides, of which the chief ones are usually Al_2O_3 and Fe_2O_3. Correlations have been reported to exist between the MgO contents of belite and clinker (K1), the Fe_2O_3 contents of belite and clinker (K1), and the total impurity content of the belite and the MgO content of the clinker (B4). Early reports mention a compound $KC_{23}S_{12}$, but the K_2O content of 3.5% corresponding to this formula is well above those found in clinker belites, and recent electron optical work indicates that the limit of K_2O substitution is about 1.2% (C1).

In general, clinker belite compositions indicate ratios of other atoms to oxygen above the theoretical value of 0.75. As Regourd *et al.* (R3) noted, the C_2S structures are too densely packed for the presence of interstitial ions to appear likely, and it is more probable that vacancies occur in the oxygen sites, with possible concomitant shifts in the positions of other atoms. Table 1.3 includes atomic ratios, with a possible allocation of atoms to sites, corresponding to the typical composition given in Table 1.2. Any interpretation of belite analyses is, however, at present tentative because of the uncertainties arising from exsolution effects. From the practical standpoint, it may be necessary to regard the exsolved material as forming part of the belite.

1.3.5 Cell parameters, X-ray powder patterns and other data

Regourd et al. (R3) synthesized several belites of β-C_2S type with compositions similar to those found in clinkers and determined their cell parameters. With increased substitution, the crystallinity tended to decrease, causing broadening of XRD powder peaks. Two of their results are given in Table 1.5.

Table 1.5 Compositions of synthetic belites (R3)

Composition	a (nm)	b (nm)	c (nm)	β
$Ca_2Fe_{0.035}Al_{0.035}Si_{0.93}O_{3.965}$	0.5502	0.6750	0.9316	94.45°
$Ca_2Fe_{0.050}Al_{0.050}Si_{0.90}O_{3.950}$	0.5502	0.6753	0.9344	94.19°

Fig. 1.5 compares parts of the XRD powder patterns of the α, α'_L and β polymorphs. The appendix includes a calculated pattern for a typical clinker belite. Assuming $a = 0.550$ nm, $b = 0.675$ nm, $c = 0.933$ nm, $\beta = 94.4°$, with the atomic parameters of Jost et al. (J2) and site occupancies $(K_{0.01}Na_{0.005}Ca_{0.975}Mg_{0.01})_2(Fe_{0.02}Al_{0.06}Si_{0.90}P_{0.01}S_{0.01})O_{3.96}$, the X-ray density is 3300 kg m^{-3}. That for pure β-C_2S, based on Jost et al.'s data, is 3326 kg m^{-3}, and that for pure γ-C_2S, based on those of Udagawa et al. (U4), is 2960 kg m^{-3}.

Synthetic β-C_2S is biaxial positive with high 2V and $\alpha = 1.717$, $\gamma = 1.735$; γ-C_2S is biaxial negative with 2V = 60°, $\alpha = 1.642$, $\beta = 1.645$, $\gamma = 1.654$ (R4). In the Type I belites of clinkers, lamellae of β structure have mean refractive index 1.720 and birefringence 0.015–0.018, while for the intervening material the corresponding values are 1.700–1.710 and 0.000–0.003 (O1). Ono et al. (O1) and Yamaguchi and Takagi (Y1) give further data on the optical properties of clinker belites.

Guinier and Regourd (G1) summarized the thermal behaviour of C_2S polymorphs. On heating, β-C_2S shows endotherms beginning at 693°C and 1450°C, due respectively to the $\beta \to \alpha'_L$ and $\alpha'_H \to \alpha$ transitions; with γ-C_2S, the 693°C endotherm is replaced by a broad one beginning at about 748°C due to the transition to α'_L. The curve obtained on cooling α-C_2S from

Portland Cement and its Major Constituent Phases

1500°C shows exotherms beginning at about 1456°C, 682°C and 530°C, due respectively to the $\alpha \to \alpha'_H$, $\alpha'_L \to \beta$ and $\beta \to \gamma$ transitions. The $\alpha'_H \to \alpha'_L$ transition gives only a very weak effect at 1160°C.

β-C_2S, modified by solid solution, occurs in nature as larnite. Early identifications of α'-C_2S with bredigite were incorrect (B9,M10). α-C_2S has been assumed to resemble nagelschmidtite ($Ca_7Si_2P_2O_{16}$), but the unit cell of the latter seems to be more complex (G1).

1.4 The aluminate phase

1.4.1 Crystal structure: cubic, orthorhombic and monoclinic modifications

Pure C_3A does not exhibit polymorphism. It is cubic, with $a = 1.5263$ nm, space group Pa3 and $Z = 24$; the structure is built from Ca^{2+} ions and rings of six AlO_4 tetrahedra, of formula $Al_6O_{18}^{18-}$ (M11). These rings are highly puckered, such that the aluminium atoms lie near to six of the corners of a cube (Fig. 1.6). The unit cell is composed of 64 (4^3) subcells, of edge length 0.3816 nm. Of these subcells, 8 are occupied by Al_6O_{18} rings; the Ca^{2+} ions

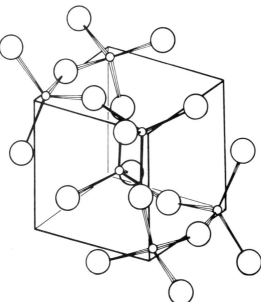

Fig. 1.6 Al_6O_{18} ring in the structure of C_3A, showing the situations of the aluminium atoms near to the corners of a cube; based on the results of Mondal and Jeffery (M11).

occur at the body centres of some of the remaining subcells and near to the corners of others. The coordination of those near the corners is octahedral, while that of those at the body centres is irregular, with either 5 or 6 oxygen atoms within 0.28 nm.

C_3A can incorporate Na^+ by substitution of Ca^{2+} with inclusion of a second Na^+ ion in an otherwise vacant site, thus giving solid solutions of general formula $Na_{2x}Ca_{3-x}Al_2O_6$ (F2,R5,R1,T4). The substitution occurs without change in structure up to a limit of about 1% Na_2O ($x \simeq 0.04$). Higher degrees of substitution lead to a series of variants of the structure (Table 1.6). In the absence of other substituents, the upper limit of Na_2O substitution is 5.7%.

Table 1.6 Modifications of the C_3A structure, of general formula $Na_{2x}Ca_{3-x}Al_2O_6$ (T4)

Approximate Na_2O (%)	Compositional range x	Designation	Crystal system	Space group
0–1.0	0–0.04	C_I	Cubic	Pa3
1.0–2.4	0.04–0.10	C_{II}	Cubic	$P2_13$
2.4–3.7	0.10–0.16	C_{II} + O	–	–
3.7–4.6	0.16–0.20	O	Orthorhombic	Pbca
4.6–5.7	0.20–0.25	M	Monoclinic	$P2_1/a$

The C_I structure (Table 1.6) is the C_3A structure, with up to about 1% of Na_2O present as described above. The C_{II} structure is a minor variant of it, of lower symmetry (T4); comparison of calculated XRD patterns (T5) shows that there is little possibility of distinguishing the two by powder XRD, unless by very precise determinations of the cell parameter, C_{II}, with 2.4% Na_2O having $a = 1.5248$ nm (T4). In C_{II}, and by analogy also in C_I, the additional Na^+ ions are located at the centres of the Al_6O_{18} rings. The O structure (originally called O_I) resembles C_I and C_{II} in having a unit cell that is composed of pseudocubic subcells with edge lengths of approximately 0.38 nm, which contain Ca^{2+} ions and Al_6O_{18} rings, but the arrangement of the rings within the true unit cell is entirely different (T4,N4). A preparation of composition $Na_{0.875}Ca_{8.375}Al_{5.175}Fe_{0.450}Si_{0.375}O_{18}$ (3.3% Na_2O) had $a = 1.0879$ nm, $b = 1.0845$ nm, $c = 1.5106$ nm, $Z = 4$ (T4). The M structure, which was originally considered to be orthorhombic and called O_{II} (R5), is a slightly distorted variant of the O structure; a preparation of composition $Na_{1.50}Ca_{8.25}Al_6O_{18}$ (5.7% Na_2O) had $a = 1.0877$ nm, $b = 1.0854$ nm, $c = 1.5135$ nm, $\beta = 90.1°$, $Z = 4$ (T4).

Substantial proportions of the aluminium in these structures can be replaced by other ions, of which Fe^{3+} and Si^{4+} are the most important. Lee et al. (L1) found the limits of substitution under equilibrium conditions to be around 2% for SiO_2 and 3–4% for Fe_2O_3, but higher degrees of substitution were obtainable under non-equilibrium conditions, such as crystallization from under cooled melts. If Si^{4+} is present, fewer large cations are needed to maintain charge balance; this extends the solid solution range of the O structure to lower Na_2O contents and that of the M structure to higher ones (T4). The literature contains many references to a supposed compound NC_8A_3. This formula corresponds to a Na_2O content of 7.6%, which cannot be attained in C_3A substituted only with Na^+, and it is reasonably certain that this compound does not exist (F2,R5,M12,T4); however, if Si^{4+} is also present, the upper limit of Na^+ substitution is near to this value (M12,T4).

Pollitt and Brown (P2) were unable to prepare an analogue of the orthorhombic phase with K^+ as the sole substituent, but obtained evidence that K^+ could stabilize it in clinker, probably because other substituents were also present. Maki (M12) also failed to prepare K forms of the orthorhombic or monoclinic phases under equilibrium conditions, but by moderately rapid cooling of melts he obtained orthorhombic crystals having cell parameters close to those of the corresponding sodium-containing phase. He considered that the presence of silicon in the clinker liquid would favour supercooling and thereby also non-equilibrium formation of the orthorhombic or monoclinic phase.

1.4.2 Other modifications

Two further modifications of the C_3A structure have been described. One was obtained as a high-temperature polymorph of the O and M forms and was considered on the basis of powder XRD evidence to be tetragonal (R1,R5). It appeared to be metastable at room temperature but could be preserved by quenching. Later studies, using thermal analysis, high-temperature light microscopy and high-temperature single crystal XRD (M13,M14,T4), showed, however, that the O modification was structurally unchanged up to its decomposition temperature and that the M modification was reversibly transformed on heating into the O form. A study of quenched material using powder XRD and single crystal electron diffraction (L2) confirmed that the supposed tetragonal form was really orthorhombic, though with a equal to b and thus geometrically tetragonal. The space group was considered to be probably Pcaa, which differs from that (Pbca) of the normal O form; however, three of the indices assigned to XRD powder peaks (023, 045 and 047) are incompatible with this space group, though not with Pbca. The high-temperature material may thus have the normal O structure and a composition such that the a and b axial lengths are equal to within experimental error. Its persistence to room temperature appears to be favoured by relatively high contents of Na_2O and Fe_2O_3 (L1).

The other C_3A modification is a disordered, poorly crystalline form, called 'proto-C_3A', which was obtained metastably from simulated clinker liquids, either by rapid cooling or by static crystallization at low temperatures (B10). It gave an XRD powder pattern with broadened peaks, corresponding to the systematically strong reflections of cubic C_3A indexable on the subcell with $a \simeq 0.39$ nm. Analytical electron microscopy of individual crystals showed it to be very high in substituents, one preparation, for example, having an Fe/Al ratio of 0.54 (H6).

1.4.3 Structural modifications of the aluminate phase in clinkers

Production clinkers have been found to contain cubic or orthorhombic forms of the aluminate phase, alone or in combination. The monoclinic modification has not been observed. The orthorhombic modification is also known as the prismatic, dark interstitial material, and is sometimes pseudotetragonal. It can arise only if sufficient alkali is available, but its formation appears to be favoured also by rapid cooling and by bulk compositions potentially able to yield a relatively high proportion of aluminate phase (M12).

Cubic aluminate phase in clinker is often finely grained and closely admixed with dendritic crystals of ferrite; when it forms larger crystals, these tend to be equidimensional. The XRD powder pattern is characterized by strong, singlet peaks at approximately 33.3°, 47.7° and 59.4° 2θ (CuK_α radiation; Fig. 1.7A), the indices of which are, respectively, 044, 008 and 448. Patterns obtained either from clinker or from material in which the aluminate and ferrite phases have been concentrated by chemical extraction of the silicates give $a = 1.5223-1.5255$ nm (R1). The slight decrease relative to the value of 1.5263 nm for pure C_3A agrees with the results for synthetic sodium-substituted preparations (R1,T4). These data are probably equally compatible with the presence of proto-C_3A, but the contents of substituents (Section 1.4.4) are considerably lower than those found in the latter material. There is probably not enough evidence to show whether the modification is C_I (Pa3) or C_{II} ($P2_13$).

The orthorhombic phase is recognizable in the light microscope or SEM by its occurrence as characteristic, lath-shaped crystals, which are often twinned (M12). An XRD powder pattern of material from which the silicate phases had been removed showed a splitting of the strong peak at 33.3° 2θ peak into a strong singlet at approximately 33.2° and a weaker, close doublet at 32.9°–33.0° (Fig. 1.7B). The unit cell parameters were $a = 1.0874$ nm, $b =$

1.0860 nm, $c = 1.5120$ nm (R1). Another clinker, in which the aluminate phase was pseudotetragonal, gave a strong singlet peak at 33.2° and a weaker singlet at 33.0° (Fig. 1.7C), and the cell parameters were $a = b = 1.0867$ nm, $c = 1.5123$ nm.

The aluminate phase in clinkers can also be characterized by its composition, determined by X-ray microanalysis; this is discussed in the next section.

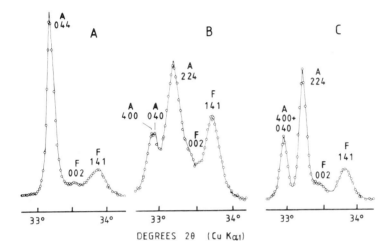

Fig. 1.7 Portions of XRD powder patterns of clinkers containing (A) cubic, (B) orthorhombic and (C) pseudotetragonal modifications of the aluminate phase. Peaks marked A and F are due to aluminate and ferrite phases, respectively, and are reindexed, where necessary, to correspond to axes in the text and Table 1.7, and to calculated intensities. After Regourd and Guinier (R1).

1.4.4 Compositions of the aluminate phase in clinkers

Because of the close admixture with other phases, which is often on a scale of 10 μm or less, X-ray microanalysis of the aluminate phase in clinkers is frequently difficult or unreliable. Data have been reported for cubic, orthorhombic, pseudotetragonal or unspecified forms of the aluminate phases in ordinary clinkers (R1,K1,B2,H3) and for aluminate phase (G3,G4,S1,B3) and glass (B3) in white cement clinkers. Tables 1.2 and 1.3 include, respectively, average compositions based on these somewhat scanty data, and suggested site occupancies based on them. The values in both tables take into account both the experimental data and the requirement of reasonable site occupancies.

Pure C_3A contains 62.3% CaO and 37.7% Al_2O_3. Substantial proportions of both calcium and aluminium are thus replaced, the total content of impurity oxides being typically around 13% for the cubic and up to about 20% for the orthorhombic modification. The content of equivalent Na_2O ($Na_2O + 0.66K_2O$) appears to be around 1% for the cubic form and 2–4% for the orthorhombic form. None of the analyses indicates alkali contents as high as that required by the formula NC_8A_3, even though considerable amounts of silicon are present.

1.4.5 X-ray powder data, densities and optical properties

The appendix includes calculated XRD powder patterns for cubic and orthorhombic clinker aluminates (T3). The X-ray density of the cubic modification is 3064 kg m^{-3}, assuming $a = 1.5240$ nm and composition $(K_{0.03}Na_{0.06}\text{-}Ca_{2.76}Mg_{0.08}Ti_{0.01})(Fe_{0.22}Al_{1.60}Si_{0.18})O_6$. That of the orthorhombic modification is 3052 kg m^{-3}, assuming $a = 1.0879$ nm, $b = 1.0845$ nm, $c = 1.5106$ nm and composition $(Na_{0.292}Ca_{2.792})(Fe_{0.150}Al_{1.725}Si_{0.125})O_6$. For pure C_3A with $a = 1.5263$ nm, the value is 3029 kg m^{-3}. Pure C_3A is optically isotropic, with refractive index 1.710 (R4); iron substitution can raise the value to 1.735 (H7). Maki (M13,M15) described the morphology and optical properties of the orthorhombic modification, including variations with composition.

1.5 The ferrite phase

1.5.1 Crystal structure and composition in the $Ca_2(Al_xFe_{1-x})_2O_5$ series

At ordinary pressures in the absence of oxide components other than CaO, Al_2O_3 and Fe_2O_3, the ferrite phase can be prepared with any composition in the solid solution series $Ca_2(Al_xFe_{1-x})_2O_5$, where $0 < x < 0.7$. The composition C_4AF is only a point in this series, with $x = 0.5$. The end member C_2A, with $x = 1$, has been prepared, but only at a pressure of 2500 MPa (A1). The series with $x < 0.7$ is not quite isostructural, as the space group changes near $x = 0.33$ (S4). Table 1.7 gives crystal data for various compositions. Care is needed in referring to the unit cell and space group, as some workers reverse the choice of a and c axes used here, with appropriate changes in space group symbol. There has also been uncertainty in the past as to the space group for the compositions with $x > 0.33$.

Table 1.7 Crystal data for ferrite phases in the series $Ca_2(Al_xFe_{1-x})_2O_5$

x	Unit cell parameters (nm)[a]			Space group	X-ray density (kg m^{-3})	References
	a	b	c			
0	0.55980	1.47687	0.54253	Pcmn	4026	B11
0.285	0.5588	1.461	0.5380	Ibm2	3862	C2
0.36	0.5583	1.458	0.5374	Ibm2	3812	C2
0.50	0.55672	1.4521	0.5349	Ibm2	3732	C3, M16
1	0.541	1.445	0.523	Ibm2 (?)	3480	A1

[a] The unit cell is orthorhombic, with $Z = 4$.

Plots of unit cell parameters or XRD powder spacings against x for the pure $Ca_2(Al,Fe)_2O_5$ series show changes in slope near $x = 0.3$, attributable to the structural change (S4). Such plots may be used to determine composition provided that oxide components other than CaO, Al_2O_3 and Fe_2O_3 are absent. They have often been applied to ferrites in clinkers, but this gives seriously inaccurate results because of the effects of other substituents (Sections 1.5.2 and 1.5.3).

Büssem (B12) determined the approximate crystal structure of C_4AF. Subsequent determinations or refinements were reported for preparations with $x = 0$ (C_2F) (B13,C4,B11), $x = 0.285$ and 0.36 (C2) and $x = 0.5$ (C3). Fig. 1.8 shows the structure for the compositional range with $0.33 < x < 0.7$. It is built from layers of corner-sharing octahedra similar to those in perovskite ($CaTiO_3$), alternating with layers composed of chains of tetrahedra, together with Ca^{2+} ions. The layers are perpendicular to the b axis and the chains are parallel to c. The composition of an individual octahedral layer in the ac cross-section of the unit cell is M_2O_8, and that of an individual chain is T_2O_6, where M and T denote octahedral and tetrahedral cations, respectively; two corners of each tetrahedron are shared with adjacent octahedra. Including also the Ca^{2+} ions, this gives the empirical formula Ca_2MTO_5. The structure for compositions with $x < 0.33$ differs from the above in that one half of the chains have the opposite polarity.

Each Ca^{2+} ion in C_4AF has 7 oxygen neighbours at 0.23–0.26 nm (C3). The aluminium and iron atoms are both distributed between octahedral and tetrahedral sites, the fraction of the aluminium entering tetrahedral sites under equilibrium conditions decreasing with temperature. For the three preparations with $x = 0.285$, 0.36 and 0.5 on which X-ray structure determinations were made, 75–76% of the total content of aluminium was found to be in tetrahedral sites. These preparations were shown to have been in equilibrium at about 750°C (C3); for a C_4AF preparation quenched from

1290°C, the Mössbauer spectrum indicated that only 68% of the aluminium was in tetrahedral sites (G8). It was suggested that x does not exceed the observed limit of about 0.7 because at this composition the tetrahedral sites are all occupied by aluminium (C2). There is evidence of clustering of aluminium and iron atoms to an extent depending on composition and conditions of formation (Z1).

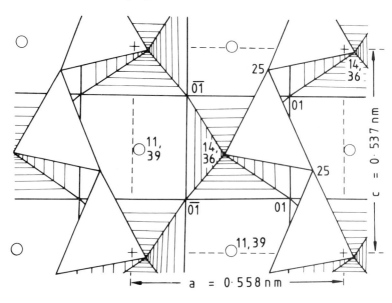

Fig. 1.8 Crystal structure of C_4AF, based on the results of Colville and Geller (C3) and showing calcium atoms (open circles), $(Al,Fe)O_4$ tetrahedra (triangles) and $(Fe,Al)O_6$ octahedra (hatched squares). Heights of atoms are given as hundredths of the cell height ($b = 1.452$ nm). Atoms at heights outside the range of -1 to $+39$ are not shown, their positions being derivable from those shown by translations of 1/2 parallel to each of the axes.

1.5.2 Compositions of the ferrite phase in clinkers

In many clinkers, the ferrite phase is closely mixed with aluminate; due to a similarity in cell parameters, oriented intergrowth can occur (M15). The close admixture often renders X-ray microanalysis difficult or unreliable. For ordinary Portland cement clinkers, the compositions found in different laboratories are nevertheless remarkably consistent. Table 1.2 includes an average value based on the results of investigations using X-ray microanalysis (H8,K1,B2,U1,H3,B4) or chemical analysis of separated material (Y1). Table 1.3 includes suggested site occupancies corresponding to these data.

As with the aluminate phase, the average compositions take into account the requirement that these site occupancies should be reasonable from the standpoint of crystal chemistry. There is no basis for allocating cations to octahedral and tetrahedral sites separately as the preferences of some of the cations, especially Mg^{2+}, in this structure are unknown, as is the temperature at which equilibrium is attained. This temperature probably varies between clinkers, and may be expected to affect the distribution.

The typical composition differs markedly from that of C_4AF (46.1% CaO, 21.0% Al_2O_3, 32.9% Fe_2O_3). It contains about 10% of impurity oxides and is much lower in Fe_2O_3. It approximates to $Ca_2Al\text{-}Fe_{0.6}Mg_{0.2}Si_{0.15}Ti_{0.05}O_5$, which is derived from C_4AF by replacing some of the Fe^{3+} by Mg^{2+} and an equal amount by Si^{4+} and Ti^{4+}. Harrisson et al. (H3) observed that, for data from several laboratories, the number of magnesium atoms per formula unit agreed with that of (Si + Ti); this supports the above interpretation.

Sulphate-resisting Portland cements have relatively high ratios of iron to aluminium, and the ferrite phase cannot have the composition given above if it contains most of the iron. Tables 1.2 and 1.3 include a tentative composition and atomic ratios corresponding to it, based on scanty data for the interstitial material as a whole (G3,G4) and the requirement of reasonable site occupancies.

1.5.3 Crystal data and X-ray powder patterns for ferrite phase containing foreign ions

Mn^{3+} can replace all the Fe^{3+} or up to 60% of the Al^{3+} in C_4AF (K2). Yamaguchi and Takagi (Y1) summarized data showing that incorporation of Mg^{2+} or Si^{4+} or both causes increases in cell parameters. Boikova (B4) determined the cell parameters of four preparations with compositions similar to those of typical clinker ferrites; the ranges were $a = 0.5535$–0.5554 nm, $b = 1.4492$–1.4642 nm, $c = 0.5288$–0.5333 nm. Marinho and Glasser (M17) found that Ti^{4+} substitution caused stacking changes in C_2F or C_4AF. It had negligible effects on the cell parameters and the charge was balanced by incorporation of additional oxygen atoms. The last conclusion could be irrelevant to clinker ferrites, since Ti^{4+} is a relatively minor substituent and the charge could be balanced in other ways.

Regourd and Guinier (R1) reported unit cell parameters for the ferrite phase in five clinkers. The ranges observed were $a = 0.5517$–0.5555 nm, $b = 1.455$–1.462 nm, $c = 0.5335$–0.5350 nm. Boikova (B4) reported XRD powder spacings for clinker ferrites which indicate similar values. The similarity of these cell parameters to those of the laboratory preparations

(B4) mentioned in the previous paragraph supports the results of X-ray microanalyses of clinker ferrites described in Section 1.5.2. Compared with pure C_4AF, typical clinker ferrites have smaller values of a and c, but larger values of b.

The XRD powder patterns of clinker ferrites are affected by the cooling rate (M18,O2). Ono and Nagasmima (O2) showed that the effect was associated with differing uptake of MgO and SiO_2. Extreme values of the cell parameters were $a = 0.557$ nm, $b = 1.462$ nm, $c = 0.532$ nm for a quenched sample, and $a = 0.543$ nm, $b = 1.465$ nm, $c = 0.533$ nm for one that was slowly cooled. The ferrite in the quenched samples was poorly crystalline, many peaks other than the three most intense (200, 141 and 202) disappearing. Broadening of peaks of clinker ferrites (M18,R1,B4) might be caused not only by poor crystallinity, but also by zoning. The cell parameters observed by Regourd and Guinier (R1) and Boikova (B4) are near to those of the quenched sample.

The appendix includes a calculated powder pattern for a ferrite phase having cell parameters and composition similar to those for the material in typical clinkers. The X-ray density, assuming $a = 0.5535$ nm, $b = 1.4642$ nm, $c = 0.5328$ nm and composition $Ca_2AlFe_{0.6}Mg_{0.2}Si_{0.15}Ti_{0.05}O_5$, is 3570 kg m^{-3}.

1.5.4 Optical, magnetic and other data

Ferrite phase of or near C_4AF composition and free from foreign ions is yellowish brown in transmitted light. The optical properties of C_4AF are as follows: biaxial, negative, with moderate 2V: α 1.96, β 2.01, γ 2.04 in lithium light; pleochroic, with γ brown, α yellowish brown; the refractive indices increase with Fe/Al ratio (H7). C_4AF–MgO solid solutions, in contrast, are grey or black, and also have higher electrical conductivity, both effects being attributable to the presence of electronic defects (M18). Normally cooled clinkers are grey or black, but quenching in water or cooling in nitrogen causes them to be yellow; this has been attributed to absence of oxygen during cooling (M18). The ferrite phase is ferromagnetic, and the more iron-rich members of the $Ca_2(Al_xFe_{1-x})_2O_5$ series have been studied for their magnetic properties, especially in regard to the coordinations of the iron atoms (C4,G9). Ferrite phase, with a composition near to C_4AF, has been found in nature. The name brownmillerite, already in use in cement chemistry, was adopted as the mineral name (G5).

2

High-temperature chemistry

2.1 Introduction

A knowledge of the relevant high-temperature phase equilibria is necessary for understanding the factors that govern acceptable bulk compositions for Portland cement clinker, the conditions under which the latter can be manufactured, and the phase composition and microstructure of the resulting material. This chapter deals with these equilibria and with the phases to which they relate, with the exception of the major clinker phases, which were described in Chapter 1. Some anhydrous phases primarily of interest in relation to other types of cement are also considered here. Principles underlying the preparation of anhydrous silicate, aluminate and other high-temperature phases are outlined.

2.2 Systems containing CaO with SiO_2 or Al_2O_3 or both

2.2.1 The CaO–SiO_2 system

Fig. 2.1 shows the phase diagram. For clarity, the polymorphism of C_3S and the distinction between α'_H- and α'_L-C_2S are omitted. Calcium oxide (CaO) has the sodium chloride structure, in which all ions are octahedrally coordinated; the unit cell is cubic, with $a = 0.48105$ nm, space group Fm3m, $Z = 4$, $D_x = 3345$ kg m^{-3} (S5). The refractive index is 1.837 (W3).

On equilibrium cooling below 1250°C, C_3S decomposes to give CaO and α'_H-C_2S; this process, which is always slow and below 700°C imperceptible, is considered in Section 3.5.5. β-CS, α-CS and C_3S_2 are commonly known by their mineral names of wollastonite, pseudowollastonite and rankinite, respectively; none reacts significantly with water at ordinary temperatures. The crystal structures of β-CS (O4) and α-CS (Y3) differ markedly from each other and the two polymorphs are easily distinguishable by powder XRD. Both exhibit polytypism (β-CS (H9,H10); α-CS (Y3)); the polytypes of a given polymorph are barely distinguishable by this method. The name 'parawollastonite' has been used for the 2M polytype of β-CS. The β–α transition is reversible but slow in the α to β direction, rendering α-CS easily preservable by quenching. The structure of rankinite is known (S6); a polymorph, kilchoanite, is known only as a natural mineral and as a product of hydrothermal reactions.

The polymorphs of silica relevant to cement chemistry are briefly considered in Section 3.3.2.

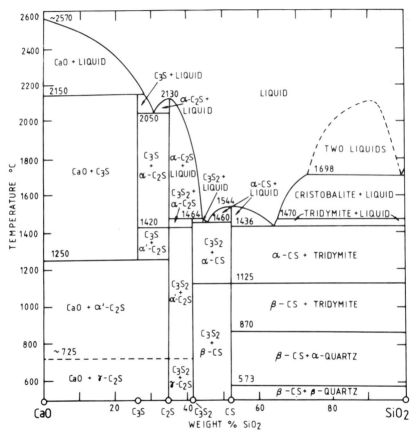

Fig. 2.1 The system $CaO-SiO_2$. After Day *et al.* (D1) with later modifications (R4, O3, W2, G10).

2.2.2 The $CaO-Al_2O_3$ system

There have been many phase and equilibrium studies on this system, and differences of opinion probably still exist on several points. In part, these differences arise because variations in the humidity and oxygen content of the furnace atmosphere markedly affect the phase relations for compositions in the region of $C_{12}A_7$. The phase diagram in Fig. 2.2 appears to provide the best description of the system modified by the presence of small amounts of water and oxygen, and thus represents the behaviour in air of ordinary humidity. With both CA and CA_2, opinions have differed as to whether melting is congruent or incongruent; Nurse *et al.* (N6), who found it to be incongruent in both cases, discussed earlier work. Of the five calcium aluminate phases appearing in Fig. 2.2, one, tricalcium aluminate (C_3A), was described in Section 1.4.

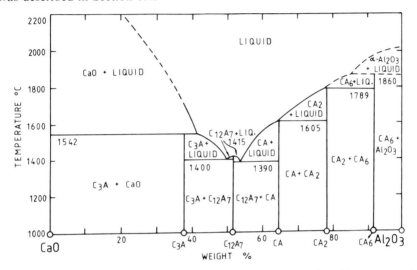

Fig. 2.2 The system $CaO-Al_2O_3$ modified by the presence of small amounts of H_2O and O_2, and thus representing the behaviour in air of ordinary humidity. After Rankin and Wright (R4), with later modifications (N5,N6,C5).

Monocalcium aluminate (CA), which is the main constituent of calcium aluminate cements (Section 10.1), reacts rapidly with water. It is monoclinic and pseudohexagonal, with $a = 0.8700$ nm, $b = 0.8092$ nm, $c = 1.5191$ nm, $\beta = 90.3°$, space group $P2_1/n$, $Z = 12$, $D_x = 2945$ kg m^{-3}, and has a stuffed tridymite structure, composed of Ca^{2+} ions and an infinite, three-dimensional framework of AlO_4 tetrahedra sharing corners (H11). It crystallizes as irregular grains, sometimes prismatic and often twinned, with $\alpha = 1.643$, $\beta = 1.655$, $\gamma = 1.663$, $(-)2V = 36°$ (R4). Formation from $CaCO_3$ and $Al(NO_3)_3$ in the presence of organic reducing agents is reported to proceed through an amorphous material at 500°C and an orthorhombic polymorph at 850°C (B14).

Calcium dialuminate (CA_2) occurs in some calcium aluminate cements. It reacts only slowly with water. The formula was at one time wrongly considered to be C_3A_5. CA_2 is monoclinic, with $a = 1.2840$ nm, $b = $

0.8862 nm, $c = 0.5431$ nm, $\beta = 106.8°$, space group C2/c, $Z = 4$, $D_x = 2920$ kg m^{-3}, and has a structure based on a framework of AlO$_4$ tetrahedra, in which some oxygen atoms are shared between two and others between three tetrahedra (B15,G11). It crystallizes as laths or rounded grains, with $\alpha = 1.6178$, $\beta = 1.6184$, $\gamma = 1.6516$, $(+)2V = 12°$ (B15). CA$_2$ has been found as a natural mineral (G5).

Calcium hexaluminate (CA$_6$) melts incongruently and does not react with water at ordinary temperatures. Its crystal structure is closely related to those of corundum and β-Al$_2$O$_3$.

2.2.3 C$_{12}$A$_7$ and derived structures

In air of ordinary humidities, a phase of approximate composition C$_{12}$A$_7$ is readily formed. It reacts rapidly with water and occurs in some calcium aluminate cements. A related phase, C$_{11}$A$_7$·CaF$_2$, is similarly reactive and occurs in certain special cements. In early work, C$_{12}$A$_7$ was wrongly assigned the formula C$_5$A$_3$ and called 'stable C$_5$A$_3$' to distinguish it from another phase of that composition, described later, which was called 'unstable C$_5$A$_3$'. C$_{12}$A$_7$ is cubic, with $a = 1.1983$ nm (J3), space group I$\bar{4}$3d and $Z = 2$ (B12). Studies on C$_{12}$A$_7$ or closely related phases (B16,W4,B17,W5,F3) show the crystal structure to be built from Ca^{2+} ions, an incomplete framework of corner-sharing AlO$_4$ tetrahedra and empirical composition Al$_7$O$_{16}^{11-}$, and one O^{2-} ion per formula unit distributed statistically between 12 sites.

At temperatures above about 930°C, a reversible equilibrium exists between C$_{12}$A$_7$ and water vapour in the furnace atmosphere (N5,R6,J3,N7). Material prepared in ordinary air and quenched from 1360–1390°C is almost anhydrous; on being gradually reheated, it takes up water until at 950°C a maximum content of about 1.3% is attained, corresponding to the formula C$_{12}$A$_7$H. On further increase in temperature, the water is lost again, until by the melting point in the region of 1400°C the material is almost anhydrous. The sorbed water affects the cell parameter and refractive index. There was early disagreement as to the direction of the variations, but it appears established that C$_{12}$A$_7$H has the smaller cell parameter and higher refractive index, Jeevaratnam et al. (J3) finding $a = 1.1983$ nm, $n = 1.611$ and $D_x = 2680$ kg m^{-3} for C$_{12}$A$_7$, and $a = 1.1976$ nm, $n = 1.620$ and $D_x = 2716$ kg m^{-3} for C$_{12}$A$_7$H.

Jeevaratnam et al. (J3) suggested that C$_{12}$A$_7$ can take up water because the 12 sites per formula unit that are occupied statistically by one O^{2-} ion can alternatively be occupied by two OH$^-$ ions, C$_{12}$A$_7$H thus having the constitution Ca$_6$Al$_7$O$_{16}$(OH). In support of this conclusion, they prepared the halide analogues C$_{11}$A$_7$·CaF$_2$ and C$_{11}$A$_7$·CaCl$_2$ of the latter compound.

$C_{11}A_7 \cdot CaF_2$ was found to have a cell parameter of 1.1964 nm, giving $D_x = 2732$ kg m^{-3}. It forms a continuous range of solid solutions with $C_{12}A_7$ (S7). Bromide, iodide, and sulphide analogues (T1,Z2) and a chloride analogue with partial replacement of aluminium by silicon, balanced by additional occupancy by chloride of the statistically available sites (F3), have also been described.

Imlach et al. (I4) found that they could prepare $C_{12}A_7$ reproducibly by slowly cooling melts in dry oxygen and that samples obtained in this way, or by cooling melts in moist air, contained excess oxygen, which could be detected chemically and was possibly present as peroxide ion. If similar melts were cooled in very dry nitrogen, mixtures consisting of CA, C_3A and, sometimes, C_5A_3, were obtained, with only poor yields of $C_{12}A_7$. These observations have since been confirmed and extended (B18,B19,Z2,Z3).

Zhmoidin and Chatterjee (Z2,Z3) concluded from density, viscosity and mass spectrometric evidence that melts of $C_{12}A_7$ composition contained regions of two kinds, differing in structure and density. The proportion of the less dense regions increased if the melt took up H_2O, O_2, CO, CO_2 or SO_3, or if CaF_2, $CaCl_2$, CaS, $CaSO_4$, or $CaCO_3$ was added; if this proportion was sufficiently high, $C_{12}A_7$ or a derivative was formed on freezing. The proportion of the denser regions in the melt was increased by maintaining a dry and slightly reducing atmosphere and by increase in temperature; if the less dense regions were thus eliminated, the melt yielded a homogeneous glass on quenching, or C_5A_3 on slow cooling. The molecules that stabilized the less dense regions acted by surrounding themselves with a particular type of open structure, from which $C_{12}A_7$ readily nucleated; analogies were noted with the roles of hydrated alkali cations in the formation of zeolites and of water or organic molecules in that of compounds of the noble gases. With the exception of water, the molecules thus occluded in crystalline $C_{12}A_7$ could not be removed without destroying the structure. $C_{12}A_7$ melts are considerably denser (2870–2910 kg m^{-3}) than solid $C_{12}A_7$, especially if they have been kept in reducing atmospheres.

The derivatives of $C_{12}A_7$ melt in normal ways, $C_{11}A_7 \cdot CaF_2$ congruently at 1577°C and $C_{11}A_7 \cdot CaS$ incongruently to give CaS and liquid at 1482°C (Z2). In contrast, $C_{12}A_7$ stabilized only by water shows anomalous behaviour, one study (N7) showing a sharp melting point of 1392°C over a range of compositions, and another (Z2,S7) showing melting over a range from 1380 to 1415°C with evolution of gas and incipient formation of CA, C_3A and C_5A_3. The diagram in Fig. 2.2 is thus approximate in this region. For practical purposes, the situation in air of ordinary humidities is more important than that in very dry and oxygen-free atmospheres, and the primary phase field of $C_{12}A_7$ will be included in the relevant ternary and quaternary phase diagrams presented later.

$C_{12}A_7$ or $C_{12}A_7H$ has been found as a natural mineral and named mayenite (H12,G5).

2.2.4 C_5A_3, C_2A and C_4A_3

The conditions of formation of C_5A_3 were discussed in the preceding section; it is probably an equilibrium phase in the strictly binary system $CaO–Al_2O_3$ but does not form in atmospheres of normal humidity and oxygen content. C_5A_3 is orthorhombic, with $a = 1.1253$ nm, $b = 1.0966$ nm, $c = 1.0290$ nm, space group $Cmc2_1$, $Z = 4$, $D_x = 3067$ kg m^{-3}, and a structure related to that of gehlenite (V1). The density is considerably higher than that of $C_{12}A_7$ and accords with formation of the compound from the denser regions present in the melt. C_5A_3 reacts rapidly with water (B20).

Two other anhydrous calcium aluminates are known. C_2A, a high-pressure phase, was described in Section 1.5.1. C_4A_3 is formed as the dehydration product of a hydrothermally produced phase, $C_4A_3H_3$. Its structure, which is similar to that of sodalite ($Na_4(Al_3Si_3O_{12})Cl$), is based on a framework of corner-sharing AlO_4 tetrahedra with Ca^{2+} and O^{2-} ions in interstices, giving the constitutional formula $Ca_4(Al_6O_{12})O$ (P3).

2.2.5 The $CaO–Al_2O_3–SiO_2$ system

The phase diagram was originally determined by Rankin and Wright (R4). Fig. 2.3 is based on that of Muan and Osborn (M19), with further amendments mainly following from studies on the bounding $CaO–SiO_2$ and $CaO–Al_2O_3$ systems discussed in the preceding sections. It relates to atmospheric pressure in an atmosphere of normal humidity, the primary phase field of $C_{12}A_7$ thus being shown. Some phases probably form solid solutions within the system, C_3S, for example, being able to accommodate some Al_2O_3; this is not shown.

Two ternary compounds exist stably in the system under these conditions. Gehlenite (C_2AS; $Ca_2Al_2SiO_7$) belongs to the melilite family. In its structure, layers of 8 coordinated Ca^{2+} ions alternate with ones of composition $Al_2SiO_7^{4-}$, in which both aluminium and silicon are tetrahedrally coordinated (L3). Gehlenite forms extensive solid solutions, e.g. with åkermanite (C_2MS_2; $Ca_2MgSi_2O_7$), but not within the $CaO–Al_2O_3–SiO_2$ system. It possibly forms as an intermediate compound in the manufacture of Portland cement clinker, but does not occur in the final product; it is present in some calcium aluminate cements. It is tetragonal, with $a = 0.7716$ nm, $c = 0.5089$ nm, space-group $P\bar{4}2_1m$, $Z = 2$, $D_x = 3006$ kg m^{-3} (L3) and refractive indices $\omega = 1.669$, $\varepsilon = 1.658$ (W3). It does not react with water at ordinary temperatures. Anorthite (a polymorph of CAS_2; $CaAl_2Si_2O_8$), which is less relevant to cement chemistry, is a triclinic felspar.

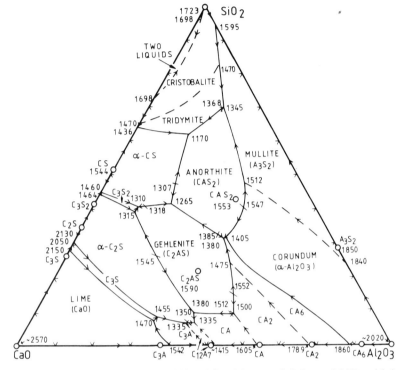

Fig. 2.3 The system $CaO-Al_2O_3-SiO_2$. After Muan and Osborn (M19), with later modifications.

Several other anhydrous calcium aluminosilicates are known, including grossular or garnet (C_3AS_3), which is a high-pressure phase, various dehydration products of zeolites, and various products formed metastably by crystallization from melts or glasses. Most are too acid in composition to be of clear relevance to cement chemistry, but some of the devitrification products, especially those with compositions near to CA and structures similar to those of nepheline ($Na_3KAl_4Si_4O_{16}$) or kalsilite ($KAlSiO_4$) (Y4), are of possible interest in relation to the formation of calcium aluminate cements.

2.2.6 Clinker formation in the $CaO-Al_2O_3-SiO_2$ system

The $CaO-Al_2O_3-SiO_2$ diagram provides a basis for a preliminary understanding of the chemistry underlying the formation of Portland cement clinker, in which all but the three most important oxide components are ignored. The approximations are least for white cements. In a cement kiln, the maximum temperature reached by the mix, called the clinkering temperature, is commonly 1400–1450°C; at this temperature, the mix is partly molten. For a mix in the pure $CaO-Al_2O_3-SiO_2$ system, a somewhat higher temperature is required to give a comparable situation. In this discussion, we shall assume that equilibrium is attained at a clinkering temperature of 1500°C and use the $CaO-Al_2O_3-SiO_2$ diagram to predict which solid phases will be present for various bulk chemical compositions. The processes by which the equilibrium is approached in cement making, and those taking place during the subsequent cooling, are discussed in Chapter 3.

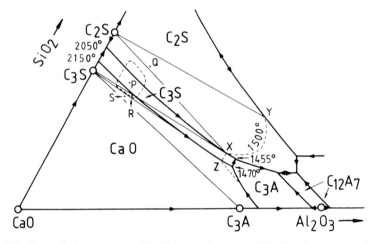

Fig. 2.4 Part of the system $CaO-Al_2O_3-SiO_2$, illustrating the formation of Portland cement clinker; for significance of lettered points and of thin full or broken lines, see text.

The compositions in the pure $CaO-Al_2O_3-SiO_2$ system that correspond most nearly to those of typical Portland cement clinkers lie within the ringed area surrounding point P in Fig. 2.4. This point lies within the triangle whose apices are the compositions of C_3S, C_2S, and point X, which lies at the intersection of the boundary of the C_3S and C_2S primary phase fields with the 1500°C isotherm. For any bulk composition in this triangle, the equilibrium phases at 1500°C will be C_3S, C_2S and liquid of composition X. This may be understood by considering what happens if a melt of composition P is cooled. C_3S first crystallizes, the liquid composition thus moving away from that of C_3S until the boundary between the primary phase fields of C_3S and C_2S is reached. C_3S and C_2S then both crystallize, and the liquid composition moves along this boundary until the 1500°C isotherm is reached. For the conditions that exist in cement making, the entire mix is never molten, so that the path to equilibrium differs, but the resulting phase assemblage must be the same. The crystals of C_3S and C_2S, being in contact with the melt, can grow to relatively large dimensions, which are typically around 30 μm. During the subsequent cooling, the liquid solidifies, producing the more finely grained interstitial material.

Some other bulk compositions may be considered in the same way. A composition such as Q, lying within the triangle C_2S–X–Y, will at equilibrium at 1500°C give C_2S and a liquid having a composition on the curve XY. One such as R, in the triangle C_3S–X–Z, will give C_3S and a liquid having a composition on XZ. Any composition lying below the line C_3S–Z, such as S, will give an assemblage that includes calcium oxide as a solid phase.

In making Portland cement, it is essential to avoid the presence of more than minimal amounts of free calcium oxide in the final product, and normally desirable to maximize that of C_3S. If equilibrium was continuously maintained during cooling, apart from decomposition of C_3S into C_2S and calcium oxide at sub-solidus temperatures, any bulk composition lying within the triangle C_3S–C_2S–C_3A would yield a final product consisting of these three phases. For bulk compositions below the line C_3S–Z, this would require that the crystals of calcium oxide present at the clinkering temperature be redissolved. This process is slow, and one cannot assume that it will be substantially completed in the conditions existing in the kiln. The line C_3S–Z therefore sets a practical limit to the CaO content of the mix. For the temperature range in which clinkering is practicable, it virtually coincides with a line joining the C_3S composition to the invariant point at 1470°C involving calcium oxide, C_3S, C_3A and liquid, and that line may therefore be used to define the upper limit of acceptable CaO contents. This approach, modified as discussed in Section 2.3.3 to allow for the presence of Fe_2O_3, leads to the definition of a quantity called the lime saturation factor, which can be used in practice as an important parameter of the bulk chemical composition.

2.3 Systems containing Fe_2O_3

2.3.1 The CaO–Al_2O_3–Fe_2O_3 system

This system includes the $Ca_2(Al_xFe_{1-x})_2O_5$ series of ferrite compositions. The bounding, binary system CaO–Fe_2O_3 includes three compounds, viz. C_2F, CF and CF_2. C_2F, as an end member of the above series, was discussed in Section 1.5. The other two compounds are of lesser importance to cement chemistry. Phillips and Muan (P4), in a study of the binary system, found that CF melted incongruently at 1216°C to give C_2F and liquid, and that CF_2 decomposed at 1155°C to give CF and hematite (α-Fe_2O_3). In this and other systems containing Fe_2O_3, iron-rich mixes tend to lose oxygen when heated in air above 1200–1300°C, with consequent replacement of hematite by magnetite (Fe_3O_4).

Fig. 2.5 shows part of the ternary system CaO–Al_2O_3–Fe_2O_3. C_3A, $C_{12}A_7$ and CA can all accommodate some Fe^{3+}; for C_3A under equilibrium conditions at 1325°C, the limit is about 4.5%, expressed as Fe_2O_3 (M20). The ferrite phase in equilibrium with iron-substituted C_3A can have compositions with x between 0.48 and 0.7 in the formula $Ca_2(Al_xFe_{1-x})_2O_5$; if CaO is also present, x is fixed at 0.48, i.e. the composition is close to C_4AF. Some reduction of Fe^{3+} to Fe^{2+} occurs when the ferrite phase is prepared from mixes with compositions in the $Ca_2(Al_xFe_{1-x})_2O_5$ series in air; it leads to the formation of minor amounts of other phases, which are not observed when similar experiments are carried out in oxygen (M20).

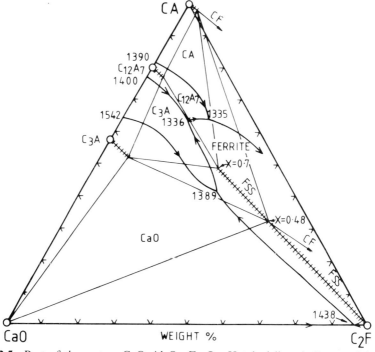

Fig. 2.5 Part of the system CaO–Al_2O_3–Fe_2O_3. Hatched lines indicate solid solutions. Ferrite phase compositions are indicated by the quantity x, which relates to the formula $Ca_2(Al_xFe_{1-x})_2O_5$. After Newkirk and Thwaite (N8) and Majumdar (M20).

Swayze (S8) noted that equilibrium was often difficult to achieve in this system. One effect was the tendency for crystals of the ferrite phase to be zoned. For bulk compositions in the $Ca_2(Al_xFe_{1-x})_2O_5$ series, the liquid is of higher Al/Fe ratio than the ferrite phase with which it is in equilibrium; the crystals that are initially deposited on cooling such a liquid therefore have a lower Al/Fe ratio than the bulk composition of the mix. On further cooling, the Al/Fe ratio of the material deposited progressively increases. Equilibrium within the ferrite crystals is difficult to attain, causing them to remain zoned, with cores richer in Fe^{3+} and outer regions richer in Al^{3+} than the mean composition.

A second effect was the tendency for protected phases to be formed. If a liquid having a composition somewhat on the CaO-rich side of the boundary between the CaO and C_3A primary phase fields (Fig. 2.5) is cooled calcium oxide is initially deposited and the liquid composition moves away from CaO and towards that boundary. When the latter is reached, and assuming that equilibrium were to be maintained, calcium oxide would redissolve, C_3A would be deposited, and the liquid composition would move along the boundary. In reality, the C_3A quickly surrounds the particles of calcium oxide, which thus form a protected phase, effectively removed from the system. This can markedly affect the composition of the ferrite phase which is subsequently formed on further cooling.

2.3.2 The $CaO-Al_2O_3-Fe_2O_3-SiO_2$ system

This system comprises the four major oxide components of Portland cement. Lea and Parker (L4,L5) made a classic study of the subsystem $CaO-C_2S-C_{12}A_7-C_4AF$, which is the part directly relevant to cement production. Following Lea (L6), the formula $C_{12}A_7$ is substituted in the present description of this and related investigations for C_5A_3, which was used by the original authors in accordance with contemporary opinion on the composition of the phase now regarded as $C_{12}A_7$. Any errors arising from this compositional difference will be small. Lea and Parker assumed the ferrite phase to have the fixed composition C_4AF.

Lea and Parker began by studying the bounding ternary subsystems $CaO-C_2S-C_4AF$ and $C_{12}A_7-C_2S-C_4AF$. The first of these, which is the more important, is shown in Fig. 2.6, modified in accordance with later work on the $CaO-SiO_2$ system. In general form, it resembles the lime-rich corner of the $CaO-Al_2O_3-SiO_2$ system (Fig. 2.4), with C_4AF in place of C_3A, the primary phase field of C_3S thus being an elongated area extending away from the $CaO-C_2S$ edge of the diagram.

The quaternary system may be represented on a tetrahedral model, each face of which represents one of the bounding, ternary systems. Within the

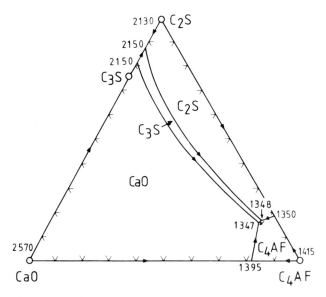

Fig. 2.6 The system CaO–C_2S–C_4AF. After Lea and Parker (L4), with later modifications.

tetrahedron, primary phase fields and subsolidus compatibility assemblages are represented by volumes, corresponding to the respective areas on the triangular diagram of a ternary system; correspondingly, isotherms are represented by surfaces. Fig. 2.7 shows some important features of the CaO–C_2S–$C_{12}A_7$–C_4AF subsystem, as found by Lea and Parker and modified to take account of later work (S8,N8,W2,G10,M20,S9).

No quaternary compounds exist in this subsystem. The tetrahedron can therefore be divided into three smaller ones representing subsolidus compatibility assemblages for CaO–C_3S–C_3A–C_4AF, C_3S–C_3A–C_2S–C_4AF and $C_{12}A_7$–C_3A–C_2S–C_4AF. For clarity, these are not shown on Fig. 2.7. The most important feature of the subsystem is the primary phase volume of C_3S, which is a thin sliver, roughly parallel to the C_2S–$C_{12}A_7$–C_4AF face of the tetrahedron and sandwiched between the larger phase volumes of CaO and C_2S. In accordance with the bounding subsystems (Figs 2.4 and 2.6), two of its edges lie on the CaO–C_2S–C_4AF and CaO–C_2S–$C_{12}A_7$ faces of the tetrahedron, but it cannot extend to the CaO–C_4AF–$C_{12}A_7$ face. Table 2.1 gives details of the eight invariant points involving C_3S in this subsystem.

Subsequent studies on the CaO–Al_2O_3–Fe_2O_3 (N8,M20) and CaO–Al_2O_3–Fe_2O_3–SiO_2 (S8,S9) systems have extended knowledge of the latter to a wider range of compositions, but have only slightly affected conclusions regarding the phase volume of C_3S. All indicate that the phase volume of

C_3A is larger, and that of the ferrite phase smaller, than was shown in Lea and Parker's diagrams, which in this region partly rested on early results for the CaO–$C_{12}A_7$–C_4AF system. To simplify the diagram, the volumes of phases other than C_3S are only indicated in general terms on Fig. 2.7.

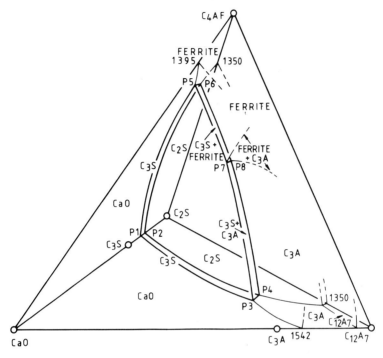

Fig. 2.7 The pseudosystem CaO–C_2S–$C_{12}A_7$–C_4AF, showing the primary phase volume of C_3S. For details of invariant points P1–P8, see Table 2.1. After Lea and Parker (L4), with later modifications.

2.3.3 Clinker formation in the CaO–Al_2O_3–Fe_2O_3–SiO_2 system

Lea and Parker (L5) derived a formula giving the maximum acceptable content of CaO in a Portland cement clinker composition. If equilibrium was maintained during cooling, apart from decomposition of C_3S into C_2S and CaO at subsolidus temperatures, any bulk composition within the tetrahedron C_3S–C_2S–C_3A–C_4AF would yield a clinker consisting of those four phases. However, as in the CaO–Al_2O_3–SiO_2 system, it cannot be assumed that any calcium oxide present at equilibrium at the clinkering temperature will subsequently be reabsorbed. Following reasoning similar to

Table 2.1 Invariant points involving C_3S in the $CaO-Al_2O_3-Fe_2O_3-SiO_2$ system and in the same modified by 5% MgO

Point	Solid phases in addition to C_3S	Liquid composition (weight %)					x in ferrite phase[a]	Type of point[b]	Temp. (°C)	Ref.
		CaO	Al_2O_3	Fe_2O_3	SiO_2	MgO				
P1	CaO	71.5	—	—	28.5	—	—	P	2150	G10
P2	C_2S	69.5	—	—	30.5	—	—	E	2050	G10
P3	$CaO + C_3A$	59.7	32.8	—	7.5	—	—	P	1470	R4
P4	$C_2S + C_3A$	58.5	32.9	—	8.6	—	—	P	1455	R4
P5	CaO + ferrite	52.8	16.2	25.4	5.6	—	0.50	E	1347	L4
P6	C_2S + ferrite	52.4	16.3	25.5	5.8	—	0.50	P	1348	L4
P7	$CaO + C_3A$ + ferrite	55.0	22.7	16.5	5.8	—	0.50	P	1341	L4
P7	$CaO + C_3A$ – ferrite	53.9	21.2	19.1	5.8	—	0.44	P	1342	S8
P7	$CaO + C_3A$ – ferrite + MgO	50.9	22.7	15.8	5.6	<5.0	0.47	P	1305	S10
P8	$C_2S + C_3A$ + ferrite	54.8	22.7	16.5	6.0	—	0.50	?	1338	L4
P8	$C_2S + C_3A$ + ferrite	53.5	22.3	18.2	6.0	—	0.57	P	1338	S8
P8	$C_2S + C_3A$ + ferrite + MgO	50.5	23.9	14.7	5.9	<5.0	0.67	P	1301	S10

[a] x in formula $Ca_2(Al_xFe_{1-x})_2O_5$.
[b] E = eutectic; P = peritectic.

that in Section 2.2.6, calcium oxide will not be present at the clinkering temperature of approximately 1450°C if the bulk composition lies on the lime-poor side of a plane defined by the compositions of C_3S, C_4AF and the invariant point P_3 (Fig. 2.7 and Table 2.1). The oxide ratios by weight for these three compositions are given in Table 2.2. One may then set up three simultaneous equations of the type

$$(SiO_2/CaO)x + (Al_2O_3/CaO)y + (Fe_2O_3/CaO)z = 1.0 \qquad (2.1)$$

where chemical formulae represent weight percentages of oxides in the clinker, and solve them to obtain an expression for the content of CaO at any point in the plane, viz:

$$CaO = xSiO_2 + yAl_2O_3 + zFe_2O_3 \qquad (2.2)$$

This gives $x = 2.80$, $y = 1.18$, $z = 0.65$.

Table 2.2 Oxide ratios by weight for points on the surface bounding the region in which calcium oxide is an equilibrium phase at 1450°C

Point on surface	SiO_2/CaO	Al_2O_3/CaO	Fe_2O_3/CaO
C_3S	0.357	0	0
C_4AF	0	0.456	0.714
P_3	0.126	0.549	0

The ratio $[CaO/(2.80SiO_2 + 1.18Al_2O_3 + 0.65Fe_2O_3)]$ is called the lime saturation factor (LSF). A mix having an LSF greater than unity will yield free calcium oxide at the clinkering temperature, and this phase is liable to persist in the final product, irrespective of the degree of mixing of the raw materials and the time during which the clinkering temperature is maintained. The calculation is approximate, because of the neglect of minor oxide components and of ionic substitutions in the solid phases and for other reasons. For values below unity, the LSF provides a measure of the extent to which the maximum attainable content of C_3S is approached. Values of 0.92–0.98 are typical of modern clinkers.

Lea and Parker considered that their calculations were valid only for Al_2O_3/Fe_2O_3 weight ratios in the clinker above that in C_4AF (0.64), but this does not appear to be correct (W6). The plane within the $CaO-C_2S-C_{12}A_7-C_4AF$ tetrahedron defined by the compositions of C_3S, C_4AF and the point P_3 passes close to the composition of C_2A, and therefore also to all other compositions in the $Ca_2(Al_xFe_{1-x})_2O_5$ series. Swayze's (S8) results, discussed in the next section, show that the extension of the $CaO-C_3S$

boundary surface to more iron-rich compositions remains close to this plane, which therefore serves also for Al_2O_3/Fe_2O_3 ratios below that in C_4AF.

2.4 Systems containing MgO or FeO

2.4.1 General

Portland cement raw materials contain small proportions of MgO; as noted in Section 1.1.2, these must be limited to avoid formation of more than a minor amount of periclase. The iron in Portland cement clinkers is normally present almost entirely as Fe^{3+}, but calcium aluminate cements may contain both Fe^{3+} and Fe^{2+}.

Table 2.3 lists some phases containing MgO that are in varying degrees relevant to cement chemistry. It is not a complete list of phases with essential MgO in the $CaO-MgO-Al_2O_3-SiO_2$ system. As seen in Chapter 1, some MgO is also taken up by all four of the major clinker phases, typical contents being 0.5–2.0% for alite, 0.5% for belite, 1.4% for the aluminate phase, and 3.0% for the ferrite phase. Magnesium oxide (periclase), like calcium oxide, has the sodium chloride structure; it is cubic, with $a = 0.4213$ nm, space group Fm3m, $Z = 4$, $D_x = 3581$ kg m^{-3} (S5) and refractive index 1.7366 (W3). FeO (wüstite) has the same structure, but always contains some Fe^{3+}, balanced by vacancies. It forms extensive solid solutions with MgO and can also accommodate some CaO.

Early studies of such systems as $MgO-C_2S-C_{12}A_7$ (H13), $CaO-MgO-C_2S-C_{12}A_7$ (M22) and $MgO-C_4AF$ (I5) indicated that periclase is the only phase with essential MgO liable to be formed in Portland cement clinkers. A later study on the $CaO-MgO-Al_2O_3-SiO_2$ system suggested that bredigite and 'Phase Q' could possibly occur in rapidly cooled clinkers (B9), but there is no convincing evidence that either of these phases is present in production clinkers.

2.4.2 Effect of MgO on equilibria in the $CaO-Al_2O_3-Fe_2O_3-SiO_2$ system

Swayze (S8,S10) extended the work of Lea and Parker (L4,L5) by considering the effects of variable ferrite composition and of the presence of MgO as an oxide component. In the work with MgO, a constant 5% of the latter was present in all the compositions examined. This was considered sufficient to saturate the liquid, as small amounts of periclase were detected in nearly all the products. Fig. 2.8 shows the results for the system in the presence of

Table 2.3 Some phases containing MgO relevant to cement chemistry

Name	Formula	Structure type	Note
Periclase	MgO	Sodium chloride	
Forsterite	M_2S	Olivine	
Monticellite	CMS		
Merwinite	C_3MS_2	Related to alkali sulphate structures	1
Bredigite	C_7MS_4		
Åkermanite	C_2MS_2	Melilite (C_2AS, etc.)	
Enstatite	MS	Pyroxene	2
Diopside	CMS_2		
Spinel	MA	Spinel	
–	C_3A_2M	Related to C_2AS and C_5A_3	3
–	C_7A_5M (?)		
'Phase Q'	$C_{20}A_{13}M_3S_3$		

Notes
1. 'Phase T', of approximate composition $C_{1.7}M_{0.3}S$ (S11, G12, S12, S13) is identical with bredigite (B9, M21).
2. Also the polymorphs clinoenstatite and protoenstatite, which too are of pyroxene type.
3. For references and descriptions, see Section 2.4.3. Phase Q is of variable composition; the formula given is a median.

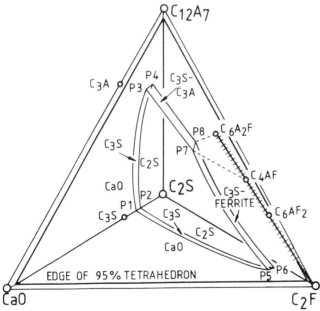

Fig. 2.8 The pseudosystem $CaO-C_2S-C_{12}A_7-C_2F$ modified by the presence of 5% of MgO, showing the phase volume of C_3S and tie lines for the ferrite phase, the compositional range of which is represented by the hatched line. For details of invariant points P1–P8, see Table 2.1. After Swayze (S10), with later modifications.

MgO, modified to take account of later work on the subsystems. The primary phase is thus always periclase, and the system is represented by using a tetrahedron whose edges correspond to a total weight percentage of 95%. Volumes within this tetrahedron relate to the second phase to crystallize.

The most important effects, whether of taking into account variable ferrite composition or the presence of MgO, concern the invariant points for C_3S, C_3A, ferrite phase, liquid and either CaO or C_2S, data for which are given in Table 2.1, where they may be compared with those of Lea and Parker. In the absence of MgO, the temperatures of the corresponding points agree well between the two studies, but the liquid compositions are slightly different. In the presence of MgO, the temperatures are each lowered by 37 deg C and the liquid compositions substantially affected, the Al_2O_3 concentrations being increased and those of CaO and Fe_2O_3 decreased. Swayze found the ferrite phase in equilibrium with C_3A, C_3S, C_2S and liquid to have a composition slightly more iron-rich than C_4AF if MgO was absent, and one close to C_6A_2F if MgO was present.

Swayze pointed out that these results had important consequences for the composition of the ferrite phase in Portland cement clinkers, the compositions of which normally include significant amounts of MgO. He concluded that, for Al_2O_3/Fe_2O_3 ratios above 1.6, the ferrite phase composition would be C_6A_2F. Probably a majority of Portland cement clinkers have Al_2O_3/Fe_2O_3 ratios in this range. For ratios below 0.84, assuming no free calcium oxide to be present, no C_3A would be formed, and the Al_2O_3/Fe_2O_3 ratio in the ferrite phase would be that of the bulk composition. For intermediate values, lack of information about the tie lines for the ferrite phase made it impossible to predict the composition of the ferrite phase. Analyses of the ferrite phase in typical clinkers (Section 1.5.2) show that these conclusions are essentially correct, though the situation is complicated by the fact that the ferrite phase contains significant proportions of minor oxide components, especially MgO, SiO_2 and TiO_2.

Swayze assumed that, for the bulk compositions that he studied, all the MgO was in the liquid phase at equilibrium, apart from the small proportions present as periclase. As noted in the preceding section, all four of the major clinker phases take up significant proportions of MgO. The contents of the latter component in the liquid phase were therefore probably substantially lower than 5%.

2.4.3 Phases structurally related to gehlenite

The phases C_3A_2M, C_7A_5M and 'Q' are structurally related to each other and to C_5A_3 and gehlenite (C_2AS); this is shown by unit cell determinations (M23) and, for Phase Q, by a structure determination (H14), which also yielded a probable structure for C_3A_2M. In all the known structures mentioned above, aluminium, magnesium and, where present, silicon are tetrahedrally coordinated. Phase Q was originally assigned the formula C_6A_4MS (P5), and it has also been suggested that two distinct quaternary phases exist, with compositions $C_{21}A_{14}M_7S$ and $C_{22}A_{13}M_3S_4$ (G13). The structure determination (H14) and an associated phase study (K3) indicated that there is only one such phase, and that its composition is $Ca_{20}Al_{32-2n}Mg_nSi_nO_{68}$ with n variable between 2.5 and 3.5, a median formula thus being $C_{20}A_{13}M_3S_3$. The unit cell, for $n = 3.0$, was found to be orthorhombic with $a = 2.7638$ nm, $b = 1.0799$ nm, $c = 0.5123$ nm, space group Pmmn, $Z = 1$ and $D_x = 2985$ kg m^{-3}.

A phase known as pleochroite, of fibrous or acicular morphology, occurs in some calcium aluminate cements. It appears to be Phase Q modified by partial replacement of Ca^{2+} by Fe^{2+} and of Al^{3+} by Fe^{3+} (P5,G13). X-ray microanalyses by Conjeaud (C6) appear compatible with the approximate composition $Ca_{20}Al_{22.6}Fe^{3+}_{2.4}Mg_{3.2}Fe^{2+}_{0.3}Si_{3.5}O_{68}$ which is a substituted form of that given by the structure determination and having $n = 3.5$. The results of another EPMA determination (M24) are of uncertain significance because they show improbably large deviations from charge balance.

Several phase equilibrium studies have been made on this part of the $CaO-MgO-Al_2O_3-SiO_2$ system (P5,B9,G13,K3). Phase Q has a primary phase volume in the system, but there are some uncertainties regarding the stable solid-phase assemblages that include it. C_3A_2M has a primary phase field in the $CaO-MgO-Al_2O_3$ and $CaO-MgO-Al_2O_3-SiO_2$ systems (W7,B9). C_7A_5M is a metastable phase, and the formula given may only be approximate.

2.5 Systems containing alkalis or SO_3 or both

2.5.1 Phases

Portland cement clinkers contain small amounts of alkalis and sulphates derived from the raw materials and fuel. Both alkalis and SO_3 can be present in the major clinker phases, but tend to combine preferentially with each other to form alkali or potassium calcium sulphates, and it is necessary to consider these components together. In addition, silicate and aluminate phases containing sulphate can form either as intermediates or in undesirable deposits in cement making, and a calcium aluminate sulphate is a major constituent of some expansive and other special cements.

Table 2.4 Crystal data for sulphate phases

Formula	Name	Crystal system[a]	Unit cell parameters (nm)			Space group	Z	D_x (kg m^{-3})	Ref.
			a	b	c				
K_2SO_4	Arcanite	Or	0.7476	1.0071	0.5763	Pnam	4	2668	M25
$KC_2\bar{S}_3$	Ca langbeinite	Or	1.0334	1.0501	1.0186	$P2_12_12_1(?)$	4	2683	S14
$K_3N\bar{S}_4$	Aphthitalite[b]	Tr	0.5680	–	0.7309	$P\bar{3}m1$	$\frac{1}{2}$	2703	O5
Na_2SO_4	Thenardite	Or	0.5861	0.9815	1.2307	Fddd	8	2665	M26
$CaSO_4$	Anhydrite	Or	0.7006	0.6998	0.6245	Amma	4	2952	K4
$C_4A_3\bar{S}$	'Aluminosulphate'	Cu	1.839	–	–	$I4_132$	16	2607	H15
$C_5S_2\bar{S}$	'Silicosulphate'	Or	1.0182	1.5398	0.6850	Pcmn	4	2973	B21

[a] Or = orthorhombic, Tr = trigonal ($\gamma = 120°$), Cu = cubic.
[b] Data are for composition given; cell parameters decrease with increasing Na/K ratio (P2).

Table 2.5 Optical properties of sulphate phases

Formula	Name	Refractive indices			Character	2V	Ref.
		α	β	γ			
K_2SO_4	Arcanite	1.4935	1.4947	1.4973	Biaxial +	67°	W3
$KC_2\bar{S}_3$	Ca langbeinite	1.522	1.526	1.527	Biaxial −	Low	S14
$K_3N\bar{S}_4$	Aphthitalite[a]	1.493	–	1.498	Uniaxial +	–	W3
Na_2SO_4	Thenardite	1.471	1.477	1.484	Biaxial +	84°	W3
$CaSO_4$	Anhydrite	1.5698	1.5754	1.6136	Biaxial +	43°	W3
$C_4A_3\bar{S}$	'Aluminosulphate'	–	1.569	–	Isotropic	–	H15
$C_5S_2\bar{S}$	'Silicosulphate'	1.632	1.638	1.640	Biaxial −	60°	P6

[a] Data are for composition given; refractive indices decrease with increasing Na/K ratio (W3).

The only phase containing essential alkali but not essential SO_3 known to occur in Portland cement clinkers in more than trace amounts is the orthorhombic aluminate phase described in Section 1.4. Trace amounts of alkali carbonates (P2) or potassium aluminate (F4) have been reported to occur in some clinkers, and some other alkali phases are formed as intermediates or deposits.

Tables 2.4 and 2.5 list the sulphate phases and give crystal data and optical properties, respectively. Arcanite is the polymorph (β) of $K\bar{S}$ stable at room temperature; it transforms reversibly to α-$K\bar{S}$ at 583°C and melts at 1069°C (R7). Thenardite is the polymorph (V) of $N\bar{S}$ stable in the presence of moisture at room temperature; on heating, it passes through a complex sequence of phase transitions and melts at 883°C (E1). Other polymorphs can persist to room temperature. Anhydrite is the polymorph (β) of $C\bar{S}$ stable at room temperature, formed on heating the hydrates or the metastable γ-$C\bar{S}$ ('soluble anhydrite'). A thermal effect at 1195°C, generally attributed to polymorphic change to an α-$C\bar{S}$, may be associated with onset of ionic rotation (F5). On heating in air, decomposition is detectable below 1000°C (K5), but slow below 1200°C; in a closed container, anhydrite melts at 1462°C (R7). Calcium langbeinite undergoes a minor phase transition at 940°C and melts incongruently to give $C\bar{S}$ and liquid at 1011°C (R7). Aphthitalite is essentially a solid solution phase of general composition $(K,Na)_2SO_4$. When it is formed stably at room temperature, the composition does not vary much from $K_3N\bar{S}_4$, but at higher temperatures compositions much higher in sodium are possible, up to a limit near $KN_3\bar{S}_4$, and the resulting solid solutions can be quenched to room temperature. Glaserite is a varietal name for aphthitalite of or near $K_3N\bar{S}_4$ composition.

$C_4A_3\bar{S}$ [$Ca_4(Al_6O_{12})(SO_4)$] is readily formed on heating mixtures of appropriate composition in air at 1350°C (R8,H15,H16). The crystal structure is a slightly distorted variant on that of sodalite [$Na_4(Al_3Si_3O_{12})Cl$]; it is composed of a three-dimensional framework of AlO_4 tetrahedra sharing corners, with Ca^{2+} and SO_4^{2-} ions in the cavities (H15). The distortion causes a doubling of the lattice parameter and change in space group (H16). $C_4A_3\bar{S}$ reacts readily with water, and melts at about 1600°C (H15).

$C_5S_2\bar{S}$ [$Ca_5(SiO_4)_2(SO_4)$] has sometimes been called sulphospurrite. The name is misleading, because the structure is unrelated to that of spurrite [$Ca_5(SiO_4)_2(CO_3)$] and elemental prefixes conventionally denote similarity not only in formula type, but also in structure. $C_5S_2\bar{S}$ is isostructural with silicocarnotite, $Ca_5(PO_4)_2(SiO_4)$ (B21). It is obtained on heating mixtures of appropriate composition in air of ordinary humidity (G14,P7), and in this environment is stable up to 1298°C (P8). Pliego-Cuervo and Glasser (P7) were unable to prepare it in the absence of water vapour and obtained evidence that it contained a small amount of essential OH^-. They considered that the atmosphere in a cement kiln was sufficiently moist to stabilize it in the appropriate temperature range, but that this might not hold true for the centres of clinker lumps.

2.5.2 Equilibria

Considering first systems of sulphates alone, phase equilibria have been reported for $K\bar{S}-N\bar{S}$ (E1), $K\bar{S}-C\bar{S}$ (R7,P9) and $K\bar{S}-N\bar{S}-C\bar{S}$ (B22). Melting of $K\bar{S}-N\bar{S}-C\bar{S}$ mixes begins below 800°C (R9). At high temperatures, $K\bar{S}$ is completely miscible with $N\bar{S}$ (E1) and accommodates up to 24 mole % of $C\bar{S}$ (R7) in solid solutions having the α-$K\bar{S}$ structure. With increase in temperature above ambient, the compositional range of aphthitalite solid solution rapidly broadens, extending much further in the direction of $N\bar{S}$ and contracting slightly in that of $K\bar{S}$; the maximum Na/K ratio of about 3 : 1 is attainable at 195°C (E1). At room temperature, all the phases other than metastable aphthitalites have essentially the fixed compositions shown in Tables 2.4 and 2.5.

A study of the CaO–$K\bar{S}$–$C\bar{S}$ system showed that the sulphate melts dissolved only a little CaO (P9). Alkali sulphate melts also show only limited mutual miscibility with oxide melts similar in composition to the main clinker liquid (T6,P7,P10,T1). For contents of alkali sulphates of 2–3% in the clinker liquid as a whole, the sulphate liquid is dispersed in the oxide liquid as microregions about 100 nm in diameter (T1), but with higher contents of alkali sulphates, the two liquids separate on a macroscopic scale (T1,G15). The partitioning of the components between sulphate melts and simulated clinker liquids at 1350°C has been studied (G15). SO_3 is virtually insoluble in the oxide liquid, and SiO_2, Al_2O_3 and Fe_2O_3 only dissolve to very minor extents in the sulphate liquid, but CaO, Na_2O and K_2O are appreciably soluble in both. For five mixes studied, the ratio of K_2O to Na_2O was between 1.9 and 4.6 times higher in the sulphate than in the oxide liquid.

Several early phase equilibrium studies on systems of calcium silicates or aluminates with alkalis were reported (N9), but their significance needs to be reassessed because they postulated the existence of the compounds NC_8A_3 or $KC_{23}S_{12}$, which in neither case is supported by more recent work.

Gutt and Smith (G16) reviewed earlier work on the effects of SO_3 on the formation of Portland cement clinker, and to augment the available information studied part of the CaO–Al_2O_3–Fe_2O_3–SiO_2–$CaSO_4$ system at 1400°C. They concluded that, if Al_2O_3 was present and MgO absent, even small proportions of SO_3 could restrict or prevent formation of C_3S. The effect was lessened, but not eliminated, if MgO was present, and it was noted that other minor components, especially alkalis, might further modify the situation.

Pliego-Cuervo and Glasser (P8) determined subsolidus phase assemblages in the CaO-rich part of the CaO–Al$_2$O$_3$–SiO$_2$–SO$_3$ system at 950–1150°C. Four-phase and three-phase assemblages defining the equilibria between CaO, C$_2$S, C$_3$A, C$_{12}$A$_7$, CA, C$_5$S$_2$$\bar{\text{S}}$ and C$_4$A$_3$$\bar{\text{S}}$ were established. This work was extended (P10) to cover parts of the CaO–Al$_2$O$_3$–SiO$_2$–K$\bar{\text{S}}$, CaO–Al$_2$O$_3$–K$\bar{\text{S}}$–C$\bar{\text{S}}$, CaO–Al$_2$O$_3$–SiO$_2$–K$\bar{\text{S}}$–C$\bar{\text{S}}$ and CaO–Al$_2$O$_3$–Fe$_2$O$_3$–K$\bar{\text{S}}$–C$\bar{\text{S}}$ systems, and the role of sulphur-containing species in the solid, liquid and vapour states in the formation of Portland cement clinker was discussed. Neither C$_5$S$_2$$\bar{\text{S}}$ nor C$_4$A$_3$$\bar{\text{S}}$ could occur in the final product, but both could play important parts in the kiln reactions.

Kaprálik et al. (K5,K6) studied equilibria in part of the CaO–Al$_2$O$_3$–SiO$_2$–Fe$_2$O$_3$–MgO–C$\bar{\text{S}}$–K$\bar{\text{S}}$ system at temperatures up to 1300°C with reference to the formation of clinkers designed to contain C$_4$A$_3$$\bar{\text{S}}$. Compatible phase assemblages were established. The authors noted that the subsystem C$_2$S–C$_4$A$_3$$\bar{\text{S}}$–C$\bar{\text{S}}$ included all the main compounds having hydraulic properties that were formed in clinkers of this type.

2.6 Systems with other components

2.6.1 Fluorides and fluorosilicates

Small amounts of F$^-$ may occur in Portland cement clinkers, arising from raw materials or deliberate additions to lower the clinkering temperature. Two ternary phases are known to exist in the CaO–C$_2$S–CaF$_2$ system, both of which were discovered by Bereczky (B23). One, now known to be of formula 2C$_2$S·CaF$_2$ or Ca$_5$(SiO$_4$)$_2$F$_2$, has a structure analogous to that of chondrodite [Mg$_5$(SiO$_4$)$_2$(OH,F)$_2$], with layers of γ-C$_2$S structure alternating with ones containing Ca^{2+} and F$^-$ (G17).

The second ternary phase has been described as 3C$_3$S·CaF$_2$ (G18), C$_{11}$S$_4$·CaF$_2$ (T7) and C$_{19}$S$_7$·2CaF$_2$ (G19), but its probable formula is Ca$_{6-x/2}$Si$_2$O$_{10-x}$F$_x$ ($x \simeq 1$), which has a structure related to that of C$_3$S with partial replacement of O^{2-} by F$^-$ balanced by omission of Ca^{2+} (P11,P12). The arrangement of Ca^{2+}, SiO$_4^{4-}$ and O^{2-} or F$^-$ ions differs from that in the C$_3$S polymorphs, and is the simplest possible in the broader family of structures to which the latter belong (P11). There is a strongly marked pseudocell, which (for $x = 1.0$) is hexagonal with $a = 0.7099$ nm, $c = 0.5687$ nm, space group P6$_3$/mmc and atomic contents 1/2 [C$_{10}$S$_4$·CaF$_2$], but the true cell is triclinic, with $a = 2.3839$ nm, $b = 0.7105$ nm, $c = 1.6755$ nm, $\alpha = 90.0°$, $\beta = 117.3°$, $\gamma = 98.56°$ and atomic contents

$5[C_{10}S_4 \cdot CaF_2]$ (P11,P12). The X-ray density, for $x = 1.0$, is 2936 kg m^{-3}. For brevity, we shall call this phase $C_{10}S_4 \cdot CaF_2$.

In the C_2S–CaF_2 system, the two end members form a eutectic at 1110°C. $2C_2S \cdot CaF_2$ can be obtained by solid-state reaction at 950°C and decomposes at 1040°C to give C_2S and CaF_2 (G18). Two phase equilibrium studies of the CaO–C_2S–CaF_2 system (T7,G20) (Fig. 2.9) are in good agreement. $C_{10}S_4 \cdot CaF_2$ melts incongruently at 1170°C to give C_3S and liquid. In the presence of CaF_2, C_3S can thus be obtained at this temperature through the successive formation and decomposition at lower temperatures of $2C_2S \cdot CaF_2$ and $C_{10}S_4 \cdot CaF_2$ (B23,G21,T7).

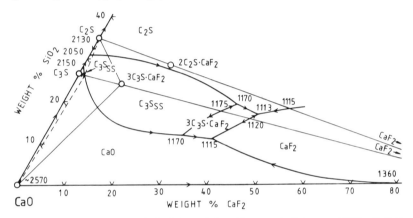

Fig. 2.9 The system CaO–C_2S–CaF_2. After Gutt and Osborne (G20; Building Research Establishment; Crown Copyright).

There is some uncertainty as to the extent to which F^- enters the C_3S structure (as opposed to that of $C_{10}S_4 \cdot CaF_2$) and also about the effectiveness of C_3S produced in presence of CaF_2 as a cement. Welch and Gutt (W8) found that addition of CaF_2 to the starting materials caused the polymorph to change from triclinic to monoclinic, and obtained results suggesting that 0.74% of F^- entered the C_3S; CaF_2 accelerated the formation of C_3S but the compressive strengths of mortars made with the latter were reduced. Tanaka et al. (T7) found that C_3S formed from $C_{10}S_4 \cdot CaF_2$ was trigonal (presumably rhombohedral), though it tended to invert to monoclinic on quenching. They found that $C_{10}S_4 \cdot CaF_2$ gave poor strengths and suggested that the poor hydraulic activity of Welch and Gutt's preparations might have been due to presence of this phase, formed from the C_3S during cooling. Maki et al. (M27) described an alite prepared from a melt containing F^-; it was monoclinic (M_3), and contained 0.9% F and 3.4% Al_2O_3.

In contrast to some of these results, Moir (M7) described a clinker containing 1.8% Al_2O_3, 2.3% SO_3 and 0.16% F, which gave high early strengths; the C_3S it contained was rhombohedral. Shame and Glasser (S15) found that little F^- could enter the C_3S structure in the absence of Al^{3+}, which permitted coupled substitution to occur, giving material of general formula $Ca_3(Al_xSi_{1-x})(O_{5-x}F_x)$. The upper limit of x was 0.15 (1.2% F). At $x = 0.15$, the material was rhombohedral, even if cooled rapidly, and gave high 28-day strengths. In the presence of Al^{3+} and F^-, C_3S could be made by solid-state reaction at temperatures as low as 1025°C, though reaction was slow below 1050°C. At higher temperatures, F^- not lost by volatilization was mainly present in the melt, and thus acted as a flux rather than as a mineralizer.

Another calcium silicate fluoride, cuspidine $[Ca_4(Si_2O_7)F_2]$, exists but is not formed under conditions relevant to clinker formation. Systems containing fluorosilicate ions $[SiF_6^{2-}]$ and the possible use of calcium fluorosilicate as a flux have been investigated (S16).

In the $CaO-Al_2O_3-CaF_2$ system, the compound $C_{11}A_7 \cdot CaF_2$ was described in Section 2.2.3. One other ternary phase, of compositon $3CA \cdot CaF_2$, is known; it is hexagonal, with $a = 1.729$ nm, $c = 0.701$ nm, and refractive indices $\omega = 1.628$, $\varepsilon = 1.618$ (L7). Several studies on parts of the system have been reported (G22,C5,C7,Z4).

2.6.2 Carbonates

Spurrite $[Ca_5(SiO_4)_2(CO_3)]$ is readily formed under suitably low partial pressures of CO_2; using oxalic acid as a source of CO_2, it has been obtained at 430°C (H17). At sufficiently low CO_2 pressures or sufficiently high temperatures, it decomposes to give C_2S and CaO. Some other syntheses, and conditions under which spurrite is formed or decomposed in the cement kiln, are mentioned in Sections 3.3.3 and 3.3.5. A pressure–temperature equilibrium curve for the reaction of wollastonite and calcite to give spurrite and CO_2 has been reported (H18). The crystal structure is known, and the unit cell is monoclinic, with $a = 1.049$ nm, $b = 0.6705$ nm, $c = 1.416$ nm, $\beta = 101.3°$, space group $P2_1/a$, $Z = 4$, $D_x = 3024$ kg m^{-3} (S17). The refractive indices are: $\alpha = 1.638$, $\beta = 1.671$, $\gamma = 1.676$; $(-)2V = 40°$ (H18). Another calcium silicate carbonate, tilleyite $[Ca_5(Si_2O_7)(CO_3)_2]$, occurs as a natural mineral and has been synthesized (H17), but is not known to form during the manufacture of clinker.

2.7 Laboratory preparation of high-temperature phases

In general, high-temperature phases are prepared in the laboratory by heating mixtures of appropriate composition at temperatures at which reference to the phase equilibria shows them to be stable. Platinum apparatus is normally required. Compounds having a lower temperature limit of stability (e.g. C_3S) may have to be cooled rapidly to a temperature at which they are kinetically stable. Some high-temperature polymorphs can be preserved by quenching, but with others it may be necessary to add a stabilizer. For β-C_2S, B_2O_3 (0.1–0.3%) has often been used, but it may be considered preferable to employ a composition closer to that of the material in clinker. As noted in Section 1.3.1, crystals of β-C_2S below a certain size do not invert to γ-C_2S on cooling. This makes it possible to prepare β-C_2S without any stabilizer by first preparing γ-C_2S and reheating at 1000°C (S18). For the same reason, β-C_2S is the normal product when hydrated calcium silicates of appropriate composition, including the major hydration product of C_3S or Portland cement, are decomposed at 800–1000°C.

As starting materials for high-temperature preparations, finely ground and well-mixed combinations of crystalline oxides (e.g. quartz) and carbonates (e.g. calcite) have been widely used. Mixtures containing carbonates must initially be heated at a relatively low temperature (900–1000°C for calcite) to avoid violent decarbonation. Repeated heating and regrinding, followed each time by examination of the product by light microscopy, XRD or other methods, is often necessary to produce an acceptable product. Intermediate products are often formed; for example, in the CaO–SiO_2 system, C_2S is formed initially, and reacts only slowly with any excess of calcium oxide or silica to give other phases. Some phases can be obtained by cooling of melts or devitrification of glasses. Particular care and choice of conditions may be needed if, as with the ferrite phase, protected phases or zoned crystals may be formed. More reactive starting materials can sometimes be used; for example, C_3S of 99% purity has been obtained by a single heating of a mixture of freshly precipitated calcium oxalate and hydrous silica (O6). It may also be possible first to prepare a hydrated compound and then to heat it; for example, C_3A can be obtained in this way via C_3AH_6.

If material can be lost by volatilization, as with phases containing alkalis, fluoride, or sulphate, it may be necessary to use a sealed platinum container and to test its effectiveness by chemically analysing the product. Control of the furnace atmosphere, e.g. by employing mixtures of CO and CO_2 to buffer the oxygen pressure, may be needed if variable oxidation states are possible, as with iron compounds. Special methods may be needed to make single crystals of sufficient size for X-ray structure determinations or other purposes. These are described in papers on X-ray structure determination. Single crystals of C_3S can be obtained from $CaCl_2$ melts (N10).

3

The chemistry of Portland cement manufacture

3.1 General considerations

3.1.1 Summary of the reactions in clinker formation

In the manufacture of Portland cement clinker, the raw materials, typically a limestone and a clay or shale, are intimately mixed and heated, ultimately to a temperature of about 1450°C. The principal reactions taking place are conveniently divided into three groups, *viz*:

(1) Reactions below about 1300°C, of which the most important are (a) the decomposition of calcite (calcining), (b) the decomposition of clay minerals, and (c) reaction of calcite or lime formed from it with quartz and clay mineral decomposition products to give belite, aluminate and ferrite. Liquid is formed only to a minor extent at this stage, but may have an important effect in promoting the reactions. At the end of this stage, the major phases present are belite, lime, aluminate and ferrite. The last two may not be identical with the corresponding phases in the final product.
(2) Reactions at 1300–1450°C (clinkering). A melt is formed, mainly from the aluminate and ferrite, and by 1450°C some 20–30% of the mix is liquid. Much of the belite and nearly all the lime react in the presence of the melt to give alite. The material nodulizes, to form the clinker.
(3) Reactions during cooling. The liquid crystallizes, giving mainly aluminate and ferrite. Polymorphic transitions of the alite and belite occur.

Fig. 3.1 shows these changes for a typical clinker. No attempt has been made to show a detailed sequence of phases below 1300°C, as sufficient data do not exist, and minor phases, including sulphates, are omitted. Quantitative phase compositions at various stages vary considerably with starting materials and other factors.

3.1.2 Lime saturation factor, silica ratio and alumina ratio

Chemical analyses of cements, clinkers and individual phases are commonly

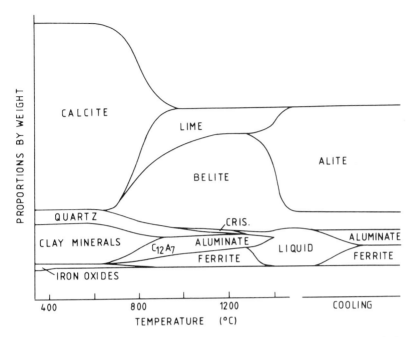

Fig. 3.1 Schematic diagram showing the variations in typical contents of phases during the formation of Portland cement clinker. Loosely based on a figure by Wolter (W9).

expressed in terms of oxide components, but it is often useful to employ quantities derived from these. In the equations that follow, chemical formulae also denote weight percentages. The following parameters are widely used in the UK and elsewhere:

Lime saturation factor (LSF) = $CaO/(2.8SiO_2 + 1.2Al_2O_3 + 0.65Fe_2O_3)$

(3.1)

Silica ratio (SR) = $SiO_2/(Al_2O_3 + Fe_2O_3)$ (3.2)

Alumina ratio (AR) = Al_2O_3/Fe_2O_3 (3.3)

The definition of LSF is theoretically based (Section 2.3.3), and in the form given above applies to clinkers; if corrected by subtracting $0.7SO_3$ from CaO, it may be applied to cements. It largely governs the ratio of alite to belite and also shows whether the clinker is likely to contain an unacceptable proportion of free lime, a value of 1.0 or above indicating that the latter will be present at equilibrium at the clinkering temperature and thus liable to persist in the product. In practice, values up to 1.02 may be acceptable; typical values for modern clinkers are 0.92–0.98. Other parameters similar in function to LSF are used in some countries. A suggested modification to the LSF definition, intended to allow for magnesium substitution in alite, replaces CaO by (CaO + 0.75MgO) for MgO ⩽ 2%, or by (CaO + 1.5) for MgO > 2% (S19). In the German literature, Kalkstandard II corresponds approximately to LSF and Kalkstandard III to this modification of it.

The SR and AR, also respectively called silica modulus and alumina modulus, are empirically based. For normal types of Portland cement clinker, SR is usually 2.0–3.0, and AR 1.0–4.0, but these ranges do not apply to special types, such as sulphate-resisting or white cement clinkers. The SR governs the proportion of silicate phases in the clinker. Increase in SR lowers the proportion of liquid at any given temperature in the kiln, and thus makes the clinker more difficult to burn. The AR governs the ratio of aluminate to ferrite phases in the clinker, which has important effects on cement properties, and also determines the quantity of liquid formed at relatively low temperatures; at 1338°C, for a given total content of Al_2O_3 and Fe_2O_3, the quantity of liquid theoretically passes through a maximum at an AR of 1.38.

The proportions of raw materials are commonly calculated on the basis of the above parameters, most obviously by setting up and solving simultaneous equations; to fix n parameters, $n + 1$ raw materials of appropriate composition are required. It is also necessary to consider the desired or allowable contents of minor components.

3.1.3 The Bogue calculation

In another approach, widely used in the USA and elsewhere, the quantitative phase composition is estimated using a procedure due to Bogue (B24). It is necessary also to know the content of free lime, which may be determined by a chemical extraction method (Section 4.3.3). The calculation is as follows:

(1) Assume that the compositions of the four major phases are C_3S, C_2S, C_3A and C_4AF.
(2) Assume that the Fe_2O_3 occurs as C_4AF.
(3) Assume that the remaining Al_2O_3 occurs as C_3A.
(4) Deduct from the CaO content the amounts attributable to C_4AF, C_3A and free lime, and solve two simultaneous equations to obtain the contents of C_3S and C_2S.

This leads to the following equations, in which CaO is assumed to have been corrected for free lime:

$$C_3S = 4.0710CaO - 7.6024SiO_2 - 6.7187Al_2O_3 - 1.4297Fe_2O_3 \quad (3.4)$$

$$\begin{aligned}C_2S &= -3.0710CaO + 8.6024SiO_2 + 5.0683Al_2O_3 + 1.0785Fe_2O_3 \\ &= 2.8675SiO_2 - 0.7544C_3S\end{aligned} \quad (3.5)$$

$$C_3A = 2.6504Al_2O_3 - 1.6920Fe_2O_3 \quad (3.6)$$

$$C_4AF = 3.0432Fe_2O_3 \quad (3.7)$$

As with LSF, the approach is applicable to cements if CaO is further corrected by deducting $0.7SO_3$. Because minor oxide components are ignored, the total for the four main phases plus free lime will not add up to 100%. It is implicit in the approach that all the MgO is assumed to occur as periclase. The phase composition calculated by Bogue's method is related to the LSF in that a ratio of C_2S to $C_3A \leqslant 0.546$, calculated without correcting the CaO content for free lime, corresponds to an LSF $\geqslant 1.0$ (D2).

The results of the Bogue calculation are often called potential phase compositions, because when the procedure was devised, it was generally considered that the principal source of error was failure to reach equilibrium during cooling. The results do indeed differ, probably often markedly, from the true phase compositions, notably in underestimating alite and overestimating belite (Section 4.4.6), and it is unlikely that equilibrium is maintained during cooling, but the direct source of error is that the compositions of the clinker phases differ considerably from those of the pure compounds. Bogue compositions are used in some specifications, and for proportioning by setting up and solving equations to calculate the relative amounts of raw materials needed to obtain given 'potential' contents of C_3S or other phases. They have often been misused for other purposes on the assumption that they are close to the actual phase compositions. Spohn *et al.* (S19) have indicated the dangers of such uncritical use.

3.1.4 Enthalpy changes in clinker formation

The enthalpy change on formation of Portland cement clinker cannot be calculated with high precision, mainly because of uncertainties associated with the clay minerals in the raw material. Table 3.1 gives data for the main thermochemical components of the reaction, almost all of which have been calculated from a self-consistent set of standard enthalpies of formation, and which are therefore likely to be more reliable than other values in the literature. The conversion of the clay minerals into oxides is an imaginary reaction, but valid as a component in a Hess's law calculation. Few reliable thermochemical data exist for clay minerals; those for pyrophyllite and kaolinite can probably be used with sufficient accuracy, on a weight basis, for other 2:1 and 1:1 clay minerals, respectively. The data for the pure compounds may similarly be used for the clinker phases. Table 3.2 illustrates the use of these data. The quantitative phase composition of the clinker was estimated from the bulk composition by the method described in Section 4.4. The overall enthalpy change is often calculated approximately from the bulk composition using a formula (Z5), and an experimental method for its determination has been described (C8).

Table 3.1 Standard enthalpies of reaction

Reaction	ΔH (kJ)	For 1 kg of
$CaCO_3$ (calcite) $\rightarrow$ CaO + CO_2(g)	+1782	$CaCO_3$
AS_4H (pyrophyllite) $\rightarrow$ α-Al_2O_3 + $4SiO_2$ (quartz) + H_2O(g)	+224	AS_4H
AS_2H_2 (kaolinite) $\rightarrow$ α-Al_2O_3 + $2SiO_2$ (quartz) + $2H_2O$(g)	+538	AS_2H_2
$2FeO\cdot OH$ (goethite) $\rightarrow$ α-Fe_2O_3 + H_2O(g)	+254	$FeO\cdot OH$
2CaO + SiO_2 (quartz) $\rightarrow$ β-C_2S	−734	C_2S
3CaO + SiO_2 (quartz) $\rightarrow$ C_3S	−495	C_3S
3CaO + α-Al_2O_3 $\rightarrow$ C_3A	−27	C_3A
6CaO + 2α-Al_2O_3 + α-Fe_2O_3 $\rightarrow$ C_6A_2F	−157	C_6A_2F
4CaO + α-Al_2O_3 + α-Fe_2O_3 $\rightarrow$ C_4AF	−105	C_4AF

Values for starting materials and products at 25°C and 0.101 MPa, calculated from the data of Wagman et al. (W10) excepting those for the formation of C_6A_2F (N11) and C_4AF (T8). The value for C_4AF is for 20°C.

Table 3.2 Enthalpy of formation of 1 kg of a Portland cement clinker

Starting materials treated as		Products treated as	
Calcite	1.20 kg	C_3S	0.673 kg
Quartz	0.10 kg	β-C_2S	0.133 kg
Pyrophyllite	0.15 kg	C_3A	0.118 kg
Kaolinite	0.04 kg	C_6A_2F	0.064 kg
Goethite	0.03 kg	CaO	0.010 kg

Component of reaction	ΔH (kJ)
$CaCO_3$ $\rightarrow$ CaO + CO_2	+2138
AS_4H (pyrophyllite) $\rightarrow$ α-Al_2O_3 + $4SiO_2$ (quartz) + H_2O(g)	+34
AS_2H_2 (kaolinite) $\rightarrow$ α-Al_2O_3 + $2SiO_2$ (quartz) + $2H_2O$(g)	+21
$2FeO\cdot OH$ (goethite) $\rightarrow$ α-Fe_2O_3 + H_2O(g)	+9
3CaO + SiO_2 (quartz) $\rightarrow$ C_3S	−333
2CaO + SiO_2 (quartz) $\rightarrow$ β-C_2S	−98
3CaO + α-Al_2O_3 $\rightarrow$ C_3A	−3
6CaO + 2α-Al_2O_3 + α-Fe_2O_3 $\rightarrow$ C_6A_2F	−7
Total	+1761

The overall enthalpy change in forming clinker is dominated by the strongly endothermic decomposition of calcite. The component reactions for the replacement of clay minerals by oxides are endothermic, because the heat required for dehydroxylation exceeds that liberated on forming the products.

The formation of β-C_2S from lime and quartz is moderately exothermic, but that of C_3S from lime and β-C_2S is endothermic, with $\Delta H = +59$ kJ kg^{-1}. All these calculations refer to reactants and products at 25°C and 0.1 MPa. The enthalpy changes at the temperatures at which the reactions occur are somewhat different, because the specific heats of reactants and products are not the same. The reaction of lime with C_2S giving C_3S changes from endothermic to exothermic at 1430°C (J4). For the decomposition of calcite at 890°C, ΔH is $+1644$ kJ kg^{-1} (L6).

3.2 Raw materials and manufacturing processes

3.2.1 Raw materials and fuel

These, and manufacturing processes, will be considered only to the extent needed for a basic understanding of their chemistry: fuller accounts are given elsewhere (D3,G23,K7,P13).

The raw mix for making Portland cement clinker is generally obtained by blending a calcareous material, typically limestone, with a smaller amount of an argillaceous one, typically clay or shale. It may be necessary to include minor proportions of one or more corrective constituents, such as iron ore, bauxite or sand, to increase the proportions of Fe_2O_3, Al_2O_3 or SiO_2, respectively. On the other hand, some argillaceous limestones and marls have compositions near to that required, making it possible to use a blend of closely similar strata from the same quarry.

Limestones vary in physical characteristics from compact rocks of low porosity to friable and highly porous ones, such as chalk, which may contain up to 25% of water. All consist essentially of calcium carbonate, normally in the polymorphic form of calcite. Other naturally occurring forms of $CaCO_3$, such as shell deposits, are sometimes used. Many limestones contain significant amounts of minor components, either as substituents in the calcite or in accessory phases, some of which are deleterious if present in amounts exceeding a few per cent (e.g. MgO, SrO), a few tenths of a per cent (e.g. P_2O_5, CaF_2, alkalis) or even less (some heavy metals).

Suitable shales and clays typically have bulk compositions in the region of 55–60% SiO_2, 15–25% Al_2O_3 and 5–10% Fe_2O_3, with smaller amounts of MgO, alkalis, H_2O and other components. Mineralogically, their main constituents are clay minerals, finely divided quartz and, sometimes, iron oxides. The dominant clay minerals are usually of the illite and kaolinite families, but small amounts of smectites (montmorillonite-type minerals) may also be present. Kaolinite has a 1:1 layer silicate structure and ionic constitution $Al_2(Si_2O_5)(OH)_4$. Illites and smectites have 2:1 layer silicate structures, derived from those of pyrophyllite $[Al_2(Si_2O_5)_2(OH)_2]$ or talc $[Mg_3(Si_2O_5)_2(OH)_2]$ by various ionic substitutions and incorporation of interlayer cations and, in smectites, water molecules. In place of clays or shales, other types of siliceous rocks, such as schists or volcanic rocks of suitable compositions, are sometimes used.

Pulverized coal is the most usual fuel, though oil, natural gas and lignite are also used. The contributions of the fuel to the clinker composition must be taken into account, especially with coal or lignite, which produce significant quantities of ash broadly similar in composition to the argillaceous component. The fuels also contribute sulphur.

Due to the increasing cost of energy, the need to preserve the environment, and the non-existence or exhaustion of suitable natural raw materials in some areas, industrial and other waste materials are of interest as possible raw materials or supplementary fuels or both. Energy can be saved if even a part of the CaO can be provided by a material, such as blastfurnace slag, that does not require decarbonation. Supplementary fuels include such materials as used or reject tyres and pulverized household refuse, which can be introduced into the system in various ways. Some materials, such as pulverized fuel ash (pfa; fly ash) can serve as raw materials that also possess some fuel content. Other wastes that have been used include calcium silicate residues from aluminium extraction, mining residues, and precipitated calcium carbonate from various industries.

3.2.2 Dry and wet processes; energy requirements

Comminution and mixing of the raw materials may be carried out either dry or wet. The theoretical amount of heat needed to produce 1 kg of clinker from typical raw materials is about 1750 kJ (Section 3.1.4). Additional heat is required because heat is retained in the clinker, kiln dust and exit gases after they leave the system, lost from the plant by radiation or convection, and, in the wet process, used to evaporate water. Table 3.3 compares the heat requirements of the wet process and a modern version of the dry process, in each case assuming typical raw materials and proper plant design and operating conditions. The greater thermal efficiency of the dry process occurs largely because there is no added water to be evaporated. The total energy requirement of a cement works includes also electrical energy used in operating the plant, to which grinding of raw material and of clinker make important contributions. This is higher for the dry process than for the wet process, typical values being 120 kWh and 77 kW per tonne of cement (432 kJ kg^{-1} and 277 kJ kg^{-1} respectively, including that used for grinding the clinker in each case (D4)) but the difference is small compared with that in the amounts of fuel required.

The wet process has certain advantages if the raw materials are already soft and moist, but its low fuel efficiency has rendered it obsolete, and the following description is of a modern version of the dry process.

Table 3.3 Heat requirement in dry and wet process kilns (kJ/kg of clinker)

	Dry	Wet
Theoretical heat requirement for chemical reactions	1807	1741
Evaporation of water	13	2364
Heat lost in exit gases and dust	623	753
Heat lost in clinker	88	59
Heat lost in air from cooler	427	100
Heat lost by radiation and convection	348	682
Total	3306	5699

Dry kiln with suspension preheater. Data adapted from Ziegler (Z6).

3.2.3 The dry process; suspension preheaters and precalciners

The raw materials first pass through a series of crushing, stockpiling, milling and blending stages, which yield an intimately mixed and dry raw meal, of which typically 85% passes through a 90-μm sieve. With automated, computer-controlled procedures, the LSF, SR and AR can be maintained constant to standard deviations of 1%, 0.1 and 0.1, respectively.

The raw meal passes through a preheater and frequently also a precalciner before entering a rotary kiln. A preheater is a heat exchanger, usually of a type called a suspension preheater in which the moving powder is dispersed in a stream of hot gas coming from the kiln. Fig. 3.2 shows a common arrangement, which employs a series of cyclones. Heat transfer takes place mainly in co-current; the raw material passes through the preheater in less than a minute, and leaves it at a temperature of about 800°C. These conditions are such that about 40% of the calcite is decarbonated. It is possible to introduce a proportion of the fuel into a preheater, with a corresponding reduction in the proportion fed to the kiln; a precalciner is a furnace chamber introduced into the preheater into which 50–65% of the total amount of fuel is introduced, often with hot air ducted from the cooler. The fuel in a precalciner is burned at a relatively low temperature; heat transfer to the raw meal, which is almost entirely convective, is very efficient. The material has a residence time in the hottest zone of a few seconds and its exit temperature is about 900°C; 90–95% of the calcite is decomposed. Ash from the fuel burned in the precalciner is effectively incorporated into the mix.

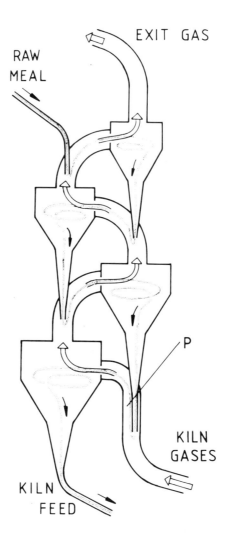

Fig. 3.2 Schematic diagram of a four-stage cyclone-type suspension preheater. P indicates the position at which a precalciner burner may be incorporated.

Because less heat is supplied to the kiln, precalcination allows the rate at which material can be passed through a kiln of given size to be greatly increased, thus saving on capital cost. Alternatively, the rate of providing heat can be reduced, which lengthens the life of the refractory lining. Less NO_x is formed, since much of the fuel is burnt at a low temperature, and with some designs, NO_x formed in the kiln may be reduced to N_2. Low-grade fuels can often be used in the precalciner. Precalcination may also make it economically feasible to deal with the problem, discussed in the next section, of formation of deposits in the duct carrying the hot gases from the kiln to the preheater.

3.2.4 The rotary kiln

The rotary kiln is a tube, sloping at 3–4% from the horizontal and rotating at 1–4 rev min^{-1}, into which the material enters at the upper end and then slides, rolls or flows counter to the hot gases produced by a flame at the lower or 'front' end. In a system employing a precalciner, the kiln is typically 50–100 m long and its ratio of length to diameter is 10–15. The maximum material temperature, of about 1450°C, is reached near to the front end of the kiln in the 'burning zone', also called the clinkering or sintering zone, in which the material spends 10–15 min. The kiln is lined with refractory bricks, of types that vary along its length in accordance with the varying gas and material temperatures. The bricks become coated with a layer of clinker, which plays an essential part in the insulation and in extending their life.

Nodules of clinker, typically 3–20 mm in diameter, are formed in a semisolid state in the burning zone, and solidify completely on cooling, which begins in a short cooling zone within the kiln, and continues in a cooler. In modern plants, when the nodules leave the kiln, their internal temperatures are around 1350°C, but their surface temperatures are considerably lower.

Liquid or pulverized solid fuels are blown into the kiln through a nozzle with 'primary' air. Additional 'secondary' air is drawn into the kiln through the clinker cooler. The flame in the rotary kiln must meet several requirements. The clinker must be correctly burned, so as to minimize its content of free lime, with the least expenditure of fuel. The ash from a solid fuel must be uniformly absorbed by the clinker. For normal Portland cements, the conditions must be sufficiently oxidizing that the iron is present as Fe^{3+}; however, for white cements, mildly reducing conditions may be preferable. Proper flame control also extends the life of the refractory lining of the kiln. Computer-aided or fully automated control of kiln operating conditions is increasingly used.

3.2.5 Circulation of volatiles; dust; cooling of clinker

In the hotter parts of the kiln, the potassium, sodium, sulphur, chlorine and some other elements provided by the raw material or fuel are partly or wholly volatilized. The fractions that are not pass into the clinker. The volatilized material is swept back with the hot gases to the cooler parts of the system, where, assuming that a suspension preheater or precalciner is used, it is largely reabsorbed, so that cycles are set up. Material that is not reabsorbed may leave the system through the preheater as gas or dust, but some can form solid or semisolid deposits. Deposits formed in this and other ways can seriously restrict the movement of material and gases; in the kiln, they form rings. They can form in the duct carrying the hot gases from the kiln to the preheater. With some raw materials or fuels, it is necessary to divert part of the gas through a bypass and to filter out the dust that it carries, and to accept the consequent losses of heat and material, but changes in the raw materials or fuel may sometimes be a preferable solution (D5).

Dust from the kiln is largely captured in a suspension preheater; that which is not is removed by an electrostatic precipitator before the gas passes up the chimney. The flow of gas is aided by an exhaust fan. The dust is as far as possible returned to the system by mixing with the raw meal or with the fuel (insufflation), though the extent to which either is possible may be restricted by its content of alkali sulphates or chlorides, which if high could lead to an unduly high recirculating load, conducive to the formation of deposits or undesirably high concentrations in the clinker.

The clinker cooler is essentially a heat exchanger that extracts heat from the clinker for return to the system; also, a cooled clinker is more readily transported, ground or stored. Rapid cooling from the clinkering temperature down to $1100°C$ produces a better quality clinker (Section 3.5.5), and the clinker should be effectively air quenched as soon as it leaves the burning zone. When the clinker enters the grinding mill, its temperature should preferably be below $110°C$. Current practice favours grate coolers, in which the clinker passes over moving grates through which air is blown. More air is needed for cooling than for combustion in the kiln, especially with a fuel-efficient process. Some of the excess ('tertiary air') may be ducted to a precalciner or used to dry the coal and raw materials, but a quantity normally remains for which it has proved difficult to find economically viable uses, and which is therefore cleaned and exhausted; steam raising for electricity generation has been suggested as a possible use.

3.2.6 Other processes for clinker production; clinker grinding

In the wet process, the raw material is fed as a slurry directly to a rotary kiln, which typically has a length to diameter ratio of about 30 and may be up to some 200 m long. In surviving wet process kilns, the slurry typically has a water content of 30–35%. A system of chains near the back end assists heat transfer. Cycling of volatiles and of dust is much less pronounced than with preheater systems. In the semi-wet process, some improvement in fuel efficiency is obtained by pressing out part of the water to produce a cake containing somewhat below 20% of water which can be fed to the kiln directly or through certain types of preheater. In the semi-dry or Lepol process, the raw materials are made into nodules with just over 10% water content which are fed to a moving grate preheater. The fuel efficiency is lower than that obtainable with a suspension preheater. Combined processes have been brought into operation in various countries for making cement and other materials, such as aluminium, iron or sulphuric acid (G24). While there are problems associated with such processes, the latter may become increasingly important as classical raw materials become scarcer and environmentally acceptable ways of disposing of industrial wastes more difficult.

Portland cement clinker has a long storage life, and while it may be ground immediately, there are often good economic reasons for grinding it intermittently, and not necessarily in the same plant. To produce Portland cement, it is ground together with gypsum. Natural gypsum, which is generally used, commonly contains significant proportions of such impurities as anhydrite, quartz, calcite and clay minerals. In some countries, byproduct gypsums from various industrial processes are used. A proportion of anhydrite may also be added. Some national specifications permit the addition of materials other than calcium sulphate. To make interground composite cements, widely varying proportions of granulated blastfurnace slag, pfa or other materials are added (Chapter 9). Grinding typically requires a power consumption of 35–50 kWh tonne^{-1} (125–180 kJ kg^{-1}), almost all of which is converted into heat, and cooling is required.

3.3 Reactions below about 1300°C

3.3.1 Decomposition of carbonate minerals

The dissociation pressure of calcite reaches 0.101 kPa (1 atm) at 894°C (S20) and the decarbonation reaction is highly endothermic (Section 3.1.4). The rate of decarbonation becomes significant at 500–600°C if a sufficiently low partial pressure of CO_2 is maintained or if the calcite is intimately mixed with materials, such as quartz or clay mineral decomposition products, that react with the calcium oxide. Even in a precalciner, such mixing occurs, aided by agglomeration caused by the presence of low-temperature sulphate melts.

In the absence of other substances, dolomite [$CaMg(CO_3)_2$] begins to decompose rapidly in air at about 750°C, giving initially periclase and a carbonate of higher Ca/Mg ratio. The decomposition temperature is much affected by the presence of other substances.

The mechanism and kinetics of calcite decomposition have been much studied. The reaction proceeds by the movement inwards from the surface of an interface, behind which the material is converted into lime, thus producing a highly porous pseudomorph. The interface moves at a constant rate, implying that the rate at any instant is proportional to the area of the interface. In principle, the rate is controlled by the slowest of the following five steps:

(1) Transfer of heat to the exterior surface.
(2) Transfer of heat from the exterior surface to the interface.
(3) Chemical reaction at the interface.
(4) Transfer of CO_2 to the exterior surface.
(5) Transfer of CO_2 away from the exterior surface.

Hills (H19) showed that, contrary to some earlier opinions, it was not possible to determine which is the rate-controlling step either from the observed relationship between the extent of reaction and the time, or from the apparent energy of activation, obtained by applying the Arrhenius equation to the rate constant determined at different temperatures. For large particles and low CO_2 pressures, the temperature within the sphere could be as much as 45 deg C below that of the surrounding gas, indicating that heat transfer controlled the rate. For the conditions existing in a rotary kiln, in which decomposition occurs in a deep, moving mass of material, heat transfer determines the rate (B25). A rotary kiln is well suited to the later stages of clinker production, but much less so for calcination, which if carried out in it requires some tens of minutes for completion.

Very different conditions exist in a precalciner, where the raw meal is dispersed in hot gas. The reaction still proceeds by the movement of an interface inwards, but the rate is controlled by the chemical reaction, and the temperature within the particle is virtually that of the surrounding gas (V2,B26). The rate is much higher than in a rotary kiln, and, as seen in Section 3.2.3, decomposition is normally 90–95% complete within a few seconds.

3.3.2 Decomposition of clay minerals and formation of products

Laboratory experiments in which single substances or mixtures are heated in stationary containers in air provide a starting point for understanding the reactions that occur in manufacturing clinker, though caution is needed in applying the results because of the important differences in conditions.

The behaviour of clay minerals on heating depends on their structure, composition, crystal size and degree of crystallinity. In general, any interlayer or adsorbed water is lost at 100–250°C; dehydroxylation begins at 300–400°C and is rapid by 500–600°C. Clay minerals in which Al^{3+} is the predominant octahedral cation do not form phases of radically different structure as the immediate products of dehydroxylation, but undergo varying degrees of structural modification and disorganization due to the replacement of $2OH^-$ by O^{2-}. Kaolinite gives a poorly crystalline product, called metakaolin; with illites, the structural disorganization is less marked, perhaps because less water has to be lost. Above about 900°C, all begin to give new, crystalline phases, which are, typically, Al–Si spinel, cristobalite and, above 1100°C, mullite.

Quartz undergoes a minor, rapid and reversible phase transition to α-quartz at 573°C. It is unstable relative to tridymite at 867–1470°C, and to cristobalite above 1470°C. These changes are slow in the absence of other materials, but many admixed materials accelerate them, and lead to formation not of tridymite, but of cristobalite at temperatures over about 1000°C.

As noted in Section 3.3.1, the decarbonation of calcite is greatly favoured by intimate mixing with quartz or clay minerals. Under these conditions, much CO_2 is lost before any free lime can be detected. The formation of C_2S as an early product is well established, but the situation with the calcium aluminate phases is more complex, no one phase being dominant as an initial product (W11). The aluminate or aluminosilicate phases most often reported as early or intermediate products in laboratory experiments with pure chemicals or raw meals have been CA, $C_{12}A_7$ and, less frequently, gehlenite (W12,D6–D8,W13,L8,L9,I6,C9,C10,R10). C_3A can form by 850°C (L9,C10) but seems usually to be a later product. Ferrite phase forms readily, and is initially of low Al/Fe ratio (D7,R10).

The reactions are notably accelerated by water vapour (C10). In the cement kiln or precalciner, agglomeration may influence their course. Glasser (G25,G26) reviewed kinetic studies. The complexity of the mechanisms has not always been recognized, and it is doubtful whether kinetic data suffice to determine them, but electron microscopy and X-ray microanalysis are beginning to provide reliable information (R11).

3.3.3 Sampling from cement kilns or preheater outlets

Several investigators have examined samples from rotary kilns during normal operation. In one such study, Weber (W13) obtained samples, and measured temperatures, mainly using sampling ports in the kiln walls. In all cases, the raw materials were limestone and marl, and the fuel was pulverized coal. CA and $C_{12}A_7$, but not gehlenite, were found as early products. As in the laboratory studies, much calcite decomposed before any free lime was detected. The free lime content passed through a maximum value of up to 17%, referred to the weight of clinker, about 15 m from the front end of the kiln. At this point, the measured temperature was about 1000°C. Laboratory studies (I6) and analyses of kiln input meals from precalciners, described later, show that the maximum content of free lime reached varies greatly with the nature of the raw material.

One may also examine samples from kilns that have been shut down, though uncertainties inevitably exist due to possible phase changes and movement of material during cooling, which is carried out with the kiln still rotating. Moore (M28) reported an XRD study of charge and coating samples taken from numerous points along the lengths of several kilns. Sulphate phases were prominent in both types of sample taken from the cooler parts of the kiln, and comprised anhydrite, $C_4A_3\bar{S}$, $C_5S_2\bar{S}$, calcium langbeinite and hydroxyl-ellestadite ($C_{10}S_3\bar{S}_3H$). KCl was sometimes found. Spurrite ($C_5S_2\bar{C}$), β-C_2S, $C_{12}A_7$ and ferrite were usually also found before all the calcite had decomposed. Gehlenite was usually found in coating samples at intermediate temperatures, but rarely in the charge; CA was never found. C_3A was a late product. Rarely, ferrite was found only as a late product, being replaced earlier by iron-containing spinel, presumably indicating the existence of reducing conditions at the cooler end of the kiln. Moore concluded that sulphates and chlorides were important mineralizers and that both spurrite and $C_4A_3\bar{S}$ played parts in the normal reaction sequence. Later results probably do not exclude this latter conclusion, but both phases can undoubtedly be undesirable artifacts.

Wolter (W14) determined the phase compositions of kiln inlet meals from about 20 plants using cyclone preheaters, usually with precalciners. XRD showed that the decomposition products of the clay minerals were especially reactive, though some reaction of quartz also occurred. Phases detected, and some notable absences, were as follows:

(1) Unreacted phases from the raw meal: calcite, quartz, minor mica and felspars, traces of clay minerals.
(2) Product phases: free lime, periclase, poorly crystalline belite (β-C_2S), ferrite, $C_{12}A_7$ assuming that enough Al_2O_3 was present, and often spurrite. Gehlenite was possibly present, but if so, was minor compared with belite. Less basic silicates and aluminates (e.g. CA, CA_2) were absent.
(3) Condensed volatiles: arcanite ($K\bar{S}$), sylvine (KCl), aphthitalite (rare), NaCl (exceptionally), anhydrite (for high SO_3/alkali ratios, and then

less than expected). Absences included thenardite, calcium langbeinite, alkali carbonates, $C_5S_2\bar{S}$ and $C_4A_3\bar{S}$.
(4) Clinker dust and coal ash: alite, pfa (rare).
(5) Secondary products formed on storage or treatment: CH, hemihydrate.

The quantitative composition varied widely, depending on the degree of calcination, heterogeneity of the raw meal, and whether kiln dust or volatiles were recirculated. $C_{12}A_7$ and ferrite formed easily. At high degrees of calcination, virtually all the Al_2O_3 and Fe_2O_3 had reacted, 10–60% of the total SiO_2 content was present as belite or spurrite and the content of free lime was 5–40%. The proportion of the CaO released from the calcite that was combined in other phases was high at low degrees of calcination, and decreased as the latter proceeded; at 40–70% calcination, it was 30–75%, or in rare cases over 90%.

3.3.4 Reaction mechanisms

The mechanisms of the reactions in cement raw meals (C9,I6,C11) and lime–silica composites (W15) have been studied using light and electron microscopy, XRD, high-temperature XRD and high-temperature light microscopy, EPMA and thermal analysis, and free lime determinations. Below about 1200°C reactions occur largely in the solid state, the first visible effect in cement raw meals being the appearance of reaction rims around the larger quartz grains, which later become replaced by clusters of belite crystals. Chromý (C9) found that the rims comprised zones of cristobalite, isotropic material of refractive index 1.515–1.525 and (furthest out) belite. By 1200°C, the belite layer was up to 2–3 µm thick, and the isotropic material, which formed a thicker layer, was tending to break through, to form necks linking adjacent quartz grains. The isotropic material contained calcium and silicon, and had Ca/Si < 1.0; it gave a diffuse XRD peak at 0.37 nm and was probably a glass. XRD and optical evidence also showed the presence of wollastonite and a little tridymite; in samples cooled from 1280°C, both were detectable optically, as needles dispersed in the amorphous material and as crystals on the border of the latter with the quartz, respectively. The wollastonite had presumably formed from α-CS on cooling.

At 1200–1300°C, high-temperature light microscopy showed movement of the silica cores and CS crystals; the amorphous material was softening. Where the belite shell had burst, clusters of belite crystals with a central pore usually replaced the original quartz grain, but where it had not, these clusters only formed at a higher temperature and were more compact, with very little interstitial material. Results at higher temperatures are described in Section 3.4.4.

Rapid increase in temperature is desirable at temperatures below those at which substantial liquid formation occurs (C9,B27,S21,C11,W9,G26). Most of the belite, and almost all of the other product phases, subsequently either melt or react in the presence of the melt, and there is no merit in promoting crystal growth or removal of imperfections, which would impede these processes. Slow heating may also allow the decomposition products of the clay minerals to transform into less reactive phases. It can also lead to the formation of microstructures unfavourable to the later reactions; Chromý (C9) found that it allowed the belite shells around the silica particles to thicken, producing composites slow to react with lime. In contrast, rapid heating increases movement of the liquid phase, when this forms, and thus improves the mixing of the calcareous and siliceous constituents (C11).

3.3.5 Condensation or reaction of volatiles

Bucchi (B28) reviewed the cycling of volatiles in kilns and preheaters. Condensation or reaction of species from the gas stream takes place below about 1300°C. The sulphate and chloride phases produced were noted in Section 3.3.3. Some of the material thus deposited is liquid, even at quite low temperatures: melting in the $N\bar{S}$–$K\bar{S}$–$C\bar{S}$–KCl system begins below 700°C (R9). These low-temperature melts have both good and bad effects. Deposition on the feed can promote reactions, as noted earlier. Deposition in the preheater or kiln can cause serious obstructions, and condensation of volatiles is responsible for the formation of kiln rings and deposits in preheaters at temperatures below 1300°C (A2,S22,S23,B28,C12). Dust particles become coated with films of liquid condensate which cause them to adhere to obstacles or cooler surfaces. The temperature within a coating or deposit gradually falls as more material is deposited; this allows the formation of compounds within the body of the coating that differ from those present on the hot face.

Spurrite, if formed, reinforces deposits by producing a mass of interlocking crystals; $C_4A_3\bar{S}$ and $C_5S_2\bar{S}$ behave similarly. The kiln atmosphere is high in CO_2, and spurrite is readily formed on heating either clinker (A2) or raw meal (S22) in CO_2 at 750–900°C. It decomposes by 950°C (A2). In SO_2, either raw meal or clinker gives anhydrite at 550–1150°C, and silicosulphate at 1050–1150°C; in mixed gases, the reactions with SO_2 are favoured over those with CO_2, and if K_2O is present in the solid, calcium langbeinite is formed at 600–900°C (A2). Formation of spurrite appears to be favoured by the presence of K_2O or chlorides and inhibited by phosphates (S22), which are, however, undesirable constituents of clinker. Fluoride also appears to favour formation of spurrite and silicosulphate.

Fluorides and compounds of zinc, cadmium, thallium and lead also undergo cycling and can be deposited in preheaters (K8). Emission of the toxic heavy metals must be avoided. Fluoride phases that have been reported in deposits include $KCa_{12}(SO_4)_2(SiO_4)_4O_2F$ (F6) and an apatite phase of composition $K_3Ca_2(SO_4)_3F$ (T9,P14).

3.4 Reactions at 1300–1450°C

3.4.1 Quantity of liquid formed

The processes described in this section occur in the presence of substantial proportions of liquid. They comprise:

(1) Melting of the ferrite and aluminate phases, and some of the belite.
(2) Nodulization.
(3) Reaction of the free lime, unreacted silica and some of the belite, to give alite.
(4) Polymorphic change of belite to the α form.
(5) Recrystallization and crystal growth of alite and belite.
(6) Evaporation of volatiles.

At the clinkering temperature, the principal phases present at equilibrium are alite, belite and liquid. For the pure $CaO-C_2S-C_3A-C_4AF$ system, this phase assemblage will, except at high Al_2O_3/Fe_2O_3 ratios, be reached by 1400°C (Section 2.3.3). Following Lea and Parker (L5), the quantity of liquid formed in a mix of given composition at this temperature may be calculated as follows.

(1) All the Fe_2O_3 is present in the liquid; hence the percentage of liquid is $100 \times Fe_2O_3/(Fe_2O_3)_1$, where Fe_2O_3 and $(Fe_2O_3)_1$ are the percentages of that component in the mix and in the liquid, respectively.
(2) The composition of the liquid is governed only by the Al_2O_3/Fe_2O_3 ratio of the mix, being represented by the point on the C_3S-C_2S surface where the 1400°C isotherm is intersected by a plane corresponding to the value of that ratio. The intersection of the 1400°C isotherm with the C_3S-C_2S surface runs near and approximately parallel to the boundaries between that surface and the C_3A and C_4AF surfaces. Table 3.4 gives liquid compositions for four values of the Al_2O_3/Fe_2O_3 ratio.
(3) The Al_2O_3/Fe_2O_3 ratios and Fe_2O_3 contents in these compositions are related by the empirical equation

$$Al_2O_3/Fe_2O_3 = -0.746 + 33.9/(Fe_2O_3)_1 \qquad (3.8)$$

which may be rearranged to give

$$Fe_2O_3/(Fe_2O_3)_l = (Al_2O_3 + 0.746Fe_2O_3)/33.9 \quad (3.9)$$

whence the percentage of liquid in the mix is $2.95Al_2O_3 + 2.20Fe_2O_3$.

(4) Lea and Parker (L5) corrected for the contributions to the liquid of MgO and alkalis. Assuming that the solubility of MgO in the liquid is 5–6%, and that the percentage of liquid is about 30%, the contribution from MgO was assumed to be the MgO content of the mix or 2%, whichever was smaller. The liquid was assumed to contain all the K_2O and Na_2O.

Table 3.4 Compositions of liquids in equilibrium with C_3S and C_2S in the CaO–Al_2O_3–Fe_2O_3–SiO_2 system at 1400°C

Weight ratio Al_2O_3/Fe_2O_3	Weight percentages			
	CaO	Al_2O_3	Fe_2O_3	SiO_2
6.06	56.6	30.3	5.0	8.0
2.62	56.4	26.2	10.0	7.4
0.94	55.1	18.8	20.0	6.1
0.64	53.9	15.3	24.0	6.8

Lea and Parker (L5), with corrections supplied by F. M. Lea and noted by Bogue (B29).

Similar calculations were made at other temperatures. At lower temperatures, the relationship differs according to whether the Al_2O_3/Fe_2O_3 ratio is above or below 1.38, the value for the C_3S–C_2S–C_3A–C_4AF invariant point. Formulae for the percentage of liquid for various combinations of temperature and Al_2O_3/Fe_2O_3 ratio are given below:

1450°C	$3.00Al_2O_3 + 2.25Fe_2O_3 + MgO^* + K_2O + Na_2O$
1400°C	$2.95Al_2O_3 + 2.20Fe_2O_3 + MgO^* + K_2O + Na_2O$
1338°C ($Al_2O_3/Fe_2O_3 \geq 1.38$)	$6.10Fe_2O_3 + MgO^* + K_2O + Na_2O$
1338°C ($Al_2O_3/Fe_2O_3 \leq 1.38$)	$8.20Al_2O_3 - 5.22Fe_2O_3 + MgO^* + K_2O + Na_2O$

where MgO* denotes an upper limit of 2.0% to the term for MgO. Typical percentages of liquid (for a mix with 5.5% Al_2O_3, 3.5% Fe_2O_3, 1.5% MgO and 1.0% K_2O) are thus 24% at 1340°C, 26% at 1400°C and 27% at 1450°C.

Such results, omitting the contributions of MgO and alkalis, are conveniently represented as contours on plots of Al_2O_3 against Fe_2O_3 (Fig. 3.3). At 1400°C, the effects of the two oxides on the quantity of liquid are additive and fairly similar weight for weight. At 1338°C, the effects are not additive, the maximum amount of liquid for a given total content of Al_2O_3 and Fe_2O_3 being obtained at an Al_2O_3/Fe_2O_3 ratio of 1.38.

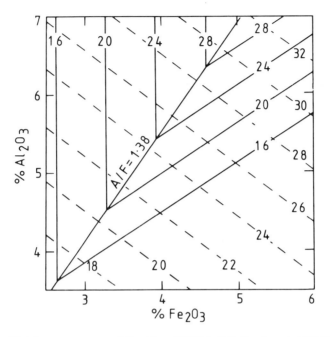

Fig. 3.3 Weight percentages of liquid at 1338°C (full lines) and 1400°C (broken lines) in the $CaO–Al_2O_3–Fe_2O_3–SiO_2$ system. Banda and Glasser (B30).

The effects of the minor components on the quantity of liquid have, however, not been thoroughly studied. Swayze's results (S10) indicate that the MgO depresses liquidus temperatures by about 50°C and increases the Al_2O_3/Fe_2O_3 ratio in the liquid at the $C_3S–C_2S–C_3A–C_4AF$ eutectic to 1.63. Alkalis and sulphate may depress the liquidus temperatures further, but the effects of liquid immiscibility must be considered.

3.4.2 Burnabilities of raw mixes

The 'burnability' of a mix denotes the ease with which free lime can be reduced to an acceptable value in the kiln. For laboratory use, it has usually been defined by either the quantity of free lime present after a specified regime of heat treatment, or by the time of such treatment needed to reduce the free lime content to some specified value, though some investigators have used dynamic heating or other regimes in attempts to simulate kiln conditions more closely. As Bucchi (B27) and Sprung (S21) have noted, none of the measures so far proposed has more than comparative significance, as they do not take into account such factors as the atmosphere, movement of material or rate of heat transfer in the kiln. The rate at which the free lime reacts depends not only on characteristics of the raw mix but also on the effects on the latter of the heat treatment received before entering the clinkering zone and the conditions existing in that zone itself. The measures can strictly be justified only for a given kiln and set of operating conditions.

Many studies have shown that burnability decreases with increasing LSF or increasing SR. Increase in LSF implies more CaO that has to react, and increase in SR implies less liquid at a given temperature. These relationships have been discussed quantitatively (C13). The AR is relevant at low clinkering temperatures, because it then greatly affects the quantity of liquid (Section 3.4.1) and the temperature at which substantial formation of liquid begins, but at 1400°C and above its effect is smaller; however, in the extreme case of white cements, the AR is very high and burnability is low. The temperature at which melting begins is also affected by the contents of minor components, such as MgO, and by the addition of fluxes. Burnability is affected not only by the quantity of liquid, but also by its physical properties, some of which are considered in Section 3.6.

Burnability is affected by the particle size of the raw mix, and especially by the contents of coarse particles (H20,L9,C14). During calcination, large particles of calcite and of siliceous materials are converted into aggregates of lime crystals and clusters of belite crystals, respectively. The size distributions of these large particles determine the time required during the last stages of clinkering, when the free lime content is falling to its final value. The fine end of any distribution likely to occur in practice is less important. For assessing burnability at 1400°C, the proportions of calcite particles larger than 125 µm, and of quartz particles larger than 44 µm, are particularly important (C14).

Burnability also depends on the nature, microstructures and intimacy of mixing of the raw materials and on the contents of minor components. Silica present as quartz is less reactive than that present in clay minerals, and a limestone high in silica or silicate minerals is likely to be more reactive than one that is nearly pure calcite. Inadequate mixing of the particles of the different constituent rocks present in the raw mix has effects broadly similar to those of too high a proportion of coarse particles. Petersen and Johansen (P15) discussed the mixing of particles from a statistical viewpoint.

Several workers have obtained expressions relating burnability to compositional and other parameters (L9,C13–C15,F7). Fundal (F7), defining burnability (C_{1400}) by the percentage of free lime remaining after 30 min at 1400°C, found that

$$C_{1400} = 0.33[(\text{LSF} - (107 - 5.1\text{SR})] + 0.93 S_{44} + 0.56 C_{125} + 0.2 A_q$$
(3.10)

where SR denotes silica ratio, S_{44} and C_{125} are the percentages of siliceous particles above 44 µm and of calcareous particles above 125 µm, respectively, and A_q is the percentage of acid-insoluble material, assumed to be quartz. This relation was considered to hold for LSF values of 88–100% and silica ratios of 2–6.

3.4.3 Nodulization

Nodule formation occurs through the sticking together of solid particles by liquid. Petersen (P16,P17) showed that the nodules grow partly by coalescence and partly by accretion, and developed a theory to predict the size distribution. Timashev (T1) regarded the process as one in which an initially relatively open assemblage of solid particles was compacted through the action of the liquid, initially by rearrangement of particles, and increasingly later by dissolution and crystallization processes. Nodulization requires an adequate proportion of liquid, and is favoured by low viscosity and high surface tension of the liquid and by small particle size of the solid (T1,P17). If there is not enough liquid, the outer parts of the nodules become enriched in silicates, some of which form dust (P15). Overburning can have a similar effect (L10). If the friable skin of large alite crystals remains on leaving the kiln, the clinker has a sparkling appearance. Much is abraded in the kiln, to form a dust that can cause rings or other deposits to form. It can also reaggregate to form porous, alite-rich lumps, which can incorporate small clinker nodules.

Nodulization, chemical reactions and evaporation of volatiles are interdependent. Alite formation and loss of volatiles affect nodulization, and the compaction that the latter entails must affect the kinetics of the former processes. Alite formation thus occurs in a changing environment. Initially, the clinker liquid is not a continuous and uniform medium enveloping the grains of lime, belite and other solid phases; rather, it fills separate pores and capillaries, forming thin films on the particles, and there are large local variations in composition (T1). As compaction proceeds, it becomes more continuous and more uniform.

3.4.4 Formation and recrystallization of alite

In the study described in Section 3.3.4, Chromý (C9) showed that a melt of low Ca/Si ratio began to form at 1200–1300°C around the original quartz grains, which were being converted into clusters of belite crystals. From 1300°C, a liquid also began to form around the grains of lime. It differed in reflectance from that surrounding the silica or belite crystals and was considered to be of higher Ca/Si ratio. Where it bordered on the belite clusters, alite was precipitated, initially as a compact layer. This later recrystallized, allowing the liquids to mix. Alite now formed more rapidly; the belite clusters contracted, and the crystals that remained recrystallized.

In further studies, Chromý and co-workers (C11,C15,C16) distinguished three stages in the reaction at clinkering temperatures. In the first, melt and belite are formed. In the second, belite continues to form; this stage yields a material consisting of isolated clusters of lime crystals having thin layers of alite on their surfaces and dispersed in a matrix of belite crystals and liquid. In the third, the clusters of lime react with the surrounding belite to give alite, by a mechanism based on diffusion of CaO through the liquid. The temperature increases rapidly during the first stage, but the subsequent stages are isothermal.

Several workers have discussed the kinetics of the clinker-forming reactions (J5,B31,T1,C11,C15,C16,G26). Because the material is not uniform, different factors may dominate at different times and in different regions (T1) but the principal one appears to be diffusion of Ca^{2+} through the liquid present between the alite crystals in the layers coating the lime clusters (C16).

From the standpoint of 28-day strength, the optimum size of the alite crystals appears to be about 15 μm, larger crystals containing fewer defects and thus being less reactive (L10). Ono (O7) considered that cannibalistic growth of alite crystals was unimportant, and that the size depended essentially on the rate of heating to the clinkering temperature; if this was low, the alite formed at relatively low temperatures as large crystals, and if it was high, it formed at higher temperatures as smaller crystals. This view has been only partially accepted by other workers, Hofmänner (H21), Long (L10) and Maki and co-workers (M29,M30,M5) all having found that alite crystals grow larger through recrystallization with increase in either time or temperature in the clinkering zone, and even (H21) during cooling. The rate of heating to the clinkering temperature is thus only one factor affecting alite size. In practice, few production clinkers contain alite with an average size as low as 15 μm.

Long (L10) noted that alite size is indirectly affected by both bulk composition and fineness of the raw feed. Increases in LSF, SR or AR all make a clinker harder to burn, which tends to increase the size, though with increased SR or AR, the effect is counteracted by decreases in the quantity and mobility of the liquid, both of which tend to decrease the size. The presence of coarse particles, especially of silica, also demands harder burning and tends to increase the size. Coarse silica particles lead to increased alite size both for this reason and because they yield large, dense clusters of belite crystals, which on subsequent reaction with lime give large crystals of alite.

3.4.5 Evaporation of volatiles; polymorphic transitions; reducing conditions

Miller (M31) summarized factors affecting the evaporation of volatile species in the burning zone. Much of the alkali and SO_3 is often present as sulphate melts. Alkali sulphates volatilize slowly because of their low vapour pressures, because diffusion within the clinker nodules is slow, and because the nodules do not spend all their time on the surface of the clinker bed. Evaporation is favoured by increased residence time or temperature, decreased size or increased porosity of the clinker nodules. Low partial oxygen pressure also favours evaporation through decomposition to give SO_2, O_2 and alkali oxides. Chloride favours it through formation of the much more volatile alkali halides. A high partial pressure of H_2O favours it through formation and volatilization of alkali hydroxides. SO_3 present as anhydrite is more easily lost, because decomposition to SO_2 and O_2 occurs, and this is also true of sulphur present in the fuel as FeS_2 or organic compounds. Alkali present in the silicate or aluminate phases also appears to be more easily lost than that present as sulphates.

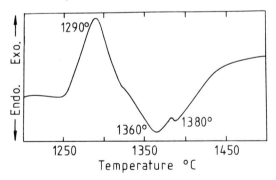

Fig. 3.4 DTA curve for a white cement clinker raw mix, obtained in air at $20°C\,min^{-1}$. From Chromý (C9).

Alite may be expected to form in clinkers as the R polymorph, and belite as α'_L, α'_H or α, according to the temperature. Chromý (C9) and Hofmänner (H21) discussed DTA curves of raw meals (Fig. 3.4). The main exotherm is due to belite formation, and the main endotherm to melting. A small superimposed endotherm near 1380°C was attributed to the α'_H–α transition of the belite, at a temperature lowered by solid solution (C9). Curves for different raw meals show considerable variations.

One would expect some loss of oxygen to occur at the burning temperature, but under normal conditions, this appears to be superficial and reversed during cooling. Long (L11) summarized the effects of a reducing atmosphere in the burning zone. Decomposition of the alite is promoted, and more aluminate phase can be formed at the expense of ferrite, with consequent acceleration of setting and change in colour of the ground cement from the normal greenish grey to brown. Some of the Fe^{2+} that is formed replaces Ca^{2+} in the solid phases, making burning more difficult and decreasing reactivity towards water. Severe reduction can lead to formation of iron(0). The sulphate phases may also be reduced; moderate reduction leads only to formation of SO_2 without O_2, but more severe reduction can give CaS and $KFeS_2$. The fate of the alkalis is not clear, but such clinkers become deliquescent if quickly cooled.

For normal cements, the effects of reducing conditions are all undesirable, but for white cements, which contain very little iron, reducing conditions are preferred because they yield a whiter product. Locher (L12) concluded that the bad effects of reducing conditions were avoided so long as the clinker left the kiln at a temperature of at least 1250°C and thereafter cooled rapidly in air; however, reoxidation of Fe^{2+} in alite below 1300°C has been observed to cause the formation of exsolution lamellae consisting of C_2F and belite (L11).

3.5 Reactions during cooling, grinding or storage

3.5.1 Solidification of the clinker liquid: indications from pure systems

In principle, the clinker liquid can behave in any of three ways during cooling, with intermediate possibilities. In the first, equilibrium between the liquid and the pre-existing solid phases (alite and belite) is continuously maintained. This implies the possibility of material transfer between these solids and the liquid in either direction for as long as any liquid remains. We shall call this situation 'equilibrium crystallization'. In the second, the liquid does not crystallize but forms a glass. In the third, the liquid crystallizes independently, i.e. without interacting with the solid phases already present. Intermediate modes include, for example, equilibrium crystallization down to a certain temperature and independent crystallization below it.

Lea and Parker (L5) found that, in the pure $CaO–Al_2O_3–Fe_2O_3–SiO_2$

system, the compositions of the liquids in equilibrium with C_3S and C_2S at 1400°C lie within the C_3S–C_2S–C_3A–C_4AF subsystem if the AR is between 0.7 and 1.7; for these compositions, equilibrium crystallization and independent crystallization lead to identical quantitative phase compositions. If the AR exceeds 1.7, the liquid composition is deficient in CaO relative to the above subsystem; this results in formation of a little $C_{12}A_7$ at the expense of C_3A if crystallization is independent, and in reaction with some of the alite to form belite if it is not. Conversely, for an AR below 0.7, the liquid composition has a surplus of CaO relative to the C_3S–C_2S–C_3A–C_4AF subsystem, resulting in formation of a little free lime if crystallization is independent and in reaction with some of the belite to form alite if it is not. Equations were derived for calculating the quantitative phase composition for various cooling conditions (L5,D2,B29).

3.5.2 Do Portland cement clinkers contain glass or $C_{12}A_7$?

The applicability of these conclusions to production clinkers requires examination. The compositions of the aluminate and ferrite phases differ markedly from C_3A and C_4AF, respectively (Chapter 1) and, because of the presence of minor components, especially MgO, TiO_2 and alkalis, the compositions of the liquid phase differ from those in the pure system. We shall consider in turn: first, whether glass is formed, and if so, under what conditions; second, whether either $C_{12}A_7$ or free lime crystallizes from the liquid phase under appropriate conditions; and third, whether there is evidence of either equilibrium or independent cooling.

The early literature contains many references to the presence in production clinkers of glass, often in substantial proportions. This view was based partly on observations by light microscopy; however, this method cannot distinguish glass from crystalline solids of the cubic system unless crystals with distinct faces have been formed, nor from crystalline materials of any kind if the crystals are below a certain size. It was also found that if clinkers believed to contain glass were annealed, their heats of solution in an acid medium increased, and this method was used to obtain approximate estimates of the glass content (L13). This evidence, too, is inconclusive, because the same effect would arise from the presence of small or structurally imperfect crystals.

Neither quantitative XRD, SEM nor X-ray microanalysis supports the view that modern clinkers contain significant quantities of glass, except in rare cases. Regourd and Guinier (R1) found that a few clinkers gave a strong, diffuse XRD peak at 30–35° 2θ (CuK_α), indicating the presence of glass or other amorphous material, but concluded that the amount of glassy or microcrystalline phase did not exceed a few per cent. An SEM and X-ray microanalytical study showed that white cement clinkers contained glass if quenched in water, but not if cooled in air (B3). A laboratory study on the cooling of simulated clinker liquids showed that glass, amorphous to X-rays, is formed on quenching small samples in water (B10). Somewhat slower cooling, from 1300°C to 700°C in a few seconds, gave 'proto-C_3A', a poorly crystalline, iron-rich form of that phase; still slower cooling (1300–1100°C in 6 min) gave normally crystalline aluminate and ferrite phases. Material probably identical with proto-C_3A was earlier obtained as a non-equilibrium quench growth (L13), and appears to form easily. The viscosity of the clinker liquid is near the lower limit for glass formation (B10). These observations suggest that glass may occur in clinkers that have been quenched in water, but is unlikely to be found with slower methods of cooling.

Neither XRD nor electron or light microscopy indicates that $C_{12}A_7$ is present in any modern clinkers, and when free lime is found, microscopic examination does not suggest that it has formed by precipitation from the liquid.

If clinkers of relatively high AR are cooled slowly, the effect of equilibrium cooling is detectable by light microscopy (L11). The alite crystals are eroded, and a layer of belite forms on their surfaces; 'pinhead' crystals of belite separate from the liquid. At ARs > 2.5, slow cooling can cause a reduction of up to 10% in the concrete strength, attributable to the lowered content of alite. These effects are not observed at the moderately fast cooling rates more usual in practice. This evidence suggests that under normal conditions crystallization approximates to independent, and that slower cooling is needed to achieve equilibrium crystallization. Production clinkers thus behave differently from mixes in the pure, quaternary system, which at AR > 1.7 give $C_{12}A_7$ on independent crystallization.

3.5.3 Evidence from X-ray microanalysis

X-ray microanalyses by many investigators (Chapter 1) also provide no indication that $C_{12}A_7$ occurs in normal clinkers. Ghose's (G4) results are of particular interest because he did not attempt to distinguish between the individual phases within the interstitial material, of which he thus obtained bulk analyses. Taken in conjunction with the results of other investigations in which individual phases were analysed, they are consistent with the hypothesis that for ARs between about 1 and 4 the interstitial material consists of aluminate and ferrite having the normal compositions given in Table 1.2, together with 4–10% of belite, the amount of which increases with

AR. For lower ARs, the interstitial material consists essentially of ferrite of higher iron contents, and in white cements it consists of aluminate very low in iron, possibly together with glass. In each case, small amounts of silicate phases are probably also present.

The reason why mixes with AR > 1.7 do not yield any $C_{12}A_7$ on independent crystallization is that the solid phases are not pure C_3A, C_4AF and C_2S. For AR = 2.71, the quaternary liquid in equilibrium with C_3S, C_2S and C_3A at 1400°C contains 55.7% CaO, 27.1% Al_2O_3, 10.0% Fe_2O_3 and 7.2% SiO_2 (S8). This composition can be closely matched by a mixture of aluminate (63%), ferrite (30%) and belite (7%) with the normal compositions given in Table 1.2, the bulk composition of this mixture being 54.4% CaO, 26.4% Al_2O_3, 9.7% Fe_2O_3, 5.6% SiO_2 and 1.8% MgO, with <1% each of TiO_2, Mn_2O_3, Na_2O and K_2O. Independent crystallization can thus yield a mixture of the three phases. The liquid composition cannot be matched by a mixture of pure C_3A, C_4AF and C_2S, which is relatively too high in CaO, so that if no ionic substitutions occurred, some $C_{12}A_7$ would also be formed. A strict comparison would be with the actual composition of the clinker liquid, which is modified by minor components, but lack of adequate data precludes this.

3.5.4 Effects of cooling rate on the aluminate and ferrite phases

It follows from the above discussion that the compositions of the aluminate and ferrite phases in typical clinkers are controlled by that of the clinker liquid and the conditions under which the latter crystallizes. For clinkers with AR 1–4, their compositions are relatively constant because both phases are present, and because cooling rates are normally moderately fast. The compositions must nevertheless depend to some extent on that of the raw mix, especially in regard to minor components. The observation that resorption of alite occurs on slow cooling implies that their compositions, or their relative amounts, or both, depend on the cooling rate. They could also be affected by the clinkering temperature, and the time during which it is maintained. XRD evidence shows that the composition of the ferrite, at least, is affected by the cooling rate (O2,B32), but the lack of data on the relation between composition and cell parameters except for compositions in the pure $C_2(A,F)$ series precludes detailed interpretation. IR evidence indicates the occurrence of structural and compositional variations in both phases dependent on cooling rate (V3). X-ray microanalyses of the aluminate and ferrite phases in similar clinkers cooled in different ways would provide data of interest.

The conditions of crystallization affect the ferrite and aluminate in ways other than in their mean compositions. Slow cooling produces relatively large crystals of each phase, while fast cooling produces close intergrowths; textures are discussed further in Section 4.2.1. Zoning occurs readily in the ferrite (Section 2.3.1), and may also occur in the aluminate. The distribution of atoms between octahedral and tetrahedral sites in the ferrite depends on the temperature at which internal equilibrium within the crystal has been achieved (Section 1.5.1). The degree of crystallinity of both phases appears to vary with cooling rate (V3). All these effects, and perhaps others, may affect the behaviour of the interstitial material on hydration.

3.5.5 Other effects of cooling rate

The behaviour of MgO in clinker formation depends markedly on the cooling rate (L11). If a clinker is burned at a high temperature ($>1500°C$), relatively high contents of MgO can enter the liquid, and on rapid cooling, much of it remains in the aluminate and ferrite phases and only a small quantity of periclase separates as small crystals. On slow cooling from high temperatures, only around 1.5% of MgO (referred to the clinker) is taken into solid solution, the excess forming large periclase crystals. At temperatures below 1450°C, the MgO is less readily taken into solution, and clusters of periclase crystals can remain; the crystals grow with time and even with temperature in this range.

The polymorphic transitions of the alite and belite that occur during cooling were discussed in Chapter 1. The belite of normal clinkers contains sufficient proportions of substituent ions to prevent dusting to give the γ-form; the mechanical constraint provided by the surrounding material may also contribute. The cooling rate affects the way in which the belite transitions occur, with effects recognizable microscopically (O2). Cooling rate is also reported to affect alite crystal size, rapid cooling giving smaller crystals, which are more easily ground (S24).

If a clinker is cooled too slowly in the region from 1250°C to about 1100°C, the alite begins to decompose. An intimate mixture of belite and lime forms as pseudomorphs after the alite, and the decomposition is especially likely to occur if the clinker has been fired under reducing conditions (L11). Mohan and Glasser (M32) reviewed and extended knowledge of the kinetics of this process. The kinetics follow a sigmoidal curve, in which an initial induction period merges into an acceleratory phase, followed by a deceleratory one as the process nears completion. These results can be explained by assuming that the rate is controlled by nucleation of the products on the surfaces of the C_3S grains; in accordance with this, the reaction is accelerated, and the induction period largely eliminated, if the

C_3S is sufficiently intimately mixed with either C_2S or CaO. Some substituent ions affect the rate: Al^{3+} has little effect, Fe^{3+} markedly accelerates, Na^+ slightly retards, and Mg^{2+} markedly retards decomposition. Where both Mg^{2+} and Fe^{3+} are present, the accelerating effect of the Fe^{3+} is dominant. The decomposition is also accelerated by water vapour and, very strongly, by contact with sulphate melts; with a $CaSO_4-Na_2SO_4$ melt, decomposition was complete in 2 h at 1025°C.

All the effects described above indicate that rapid cooling is desirable: the aluminate phase reacts more slowly with water when finely grained and intimately mixed with ferrite, making it easier to control the setting rate (S24), decrease in alite content either from reactions involving the interstitial material or from decomposition is avoided, a higher MgO content can be tolerated, and the clinker is easier to grind.

3.5.6 Crystallization of the sulphate phases

Much of the SO_3 is present at the clinkering temperature in a separate liquid phase, immiscible with the main clinker liquid. The alkali cations are distributed between the two liquids and the alite and belite. During cooling, some redistribution of alkali cations and sulphate ions between the liquids may be expected to occur, the sulphate liquid finally solidifying below 900°C to yield alkali or potassium calcium sulphates.

In most Portland cement clinkers the K_2O and Na_2O are in excess of the equivalent amount of SO_3, and K_2O is usually more abundant than Na_2O. Early work, reviewed by Newkirk (N9), indicated that the SO_3 combined as alkali sulphates, the excess of alkalis entering the silicate and aluminate phases. Sulphate phases shown or believed to occur in clinkers were potassium sulphate and aphthitalite. Subsequently, calcium langbeinite ($C_2K\bar{S}_3$) and anhydrite were also found, and Pollitt and Brown (P2) made a systematic study of the distribution of alkalis and SO_3 in a number of works- and laboratory-prepared clinkers. Fig. 3.5 and the following discussion are based on their conclusions and the tabulated data in their paper. In this work, the water-soluble alkalis and SO_3 were those present in alkali sulphates or $C_2K\bar{S}_3$, the remaining fractions of these components being present in anhydrite or in the silicate or aluminate phases.

The distributions of all three components are largely controlled by the molar ratio of total SO_3 to (total K_2O + total Na_2O) in the clinker, especially for the important range of this ratio below 1.0. K_2O has a greater tendency than Na_2O to combine with SO_3, the fraction of the total K_2O that is water soluble being consistently about double the corresponding quantity for Na_2O.

90   *Cement Chemistry*

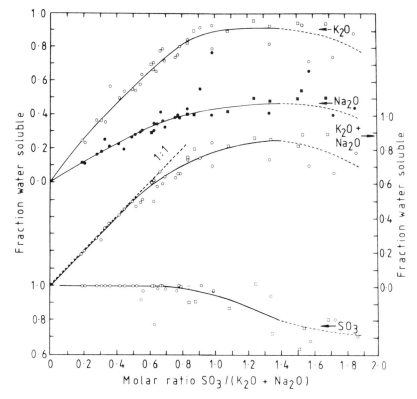

Fig. 3.5 Fractions of the K_2O, Na_2O, $K_2O + Na_2O$ (molar) and SO_3 that are water soluble, plotted against the molar ratio of total SO_3 to (total K_2O + total Na_2O) for production clinkers (circles) and laboratory clinkers (squares). Recalculated from the data of Pollitt and Brown (P2).

For $SO_3/(K_2O + Na_2O)$ molar ratios below about 0.5, the contents of total SO_3, water-soluble SO_3 and water-soluble $K_2O + Na_2O$ are approximately equal, showing that virtually all the SO_3 occurs as alkali sulphates, which also contain up to 55% of the K_2O and 27% of the Na_2O. For ratios between about 0.5 and 1.0, almost all the SO_3 is usually still water soluble, but the content of water-soluble $K_2O + Na_2O$ falls below that of water-soluble SO_3, showing that $C_2K\bar{S}_3$ is also formed. For ratios above 1.0, appreciable fractions of the SO_3 are not water soluble, and thus occur in the silicate or aluminate phases or as anhydrite. The fractions of the K_2O and Na_2O that are water soluble approach 1.0 and 0.5, respectively, at ratios around 1.5; at higher ratios, they are somewhat erratic and show some tendency to decrease again. This decrease is unexplained, but in other respects these results are broadly consistent with those of the phase equilibrium studies on pure systems described in the previous section.

From the smoothed curves in Fig. 3.5, and given the total contents of K_2O, Na_2O and SO_3 in the clinker, one may readily estimate the probable water-soluble contents of these components, and thence, making certain assumptions, the contents and compositions of the individual sulphate phases. The curves may also be represented by empirical equations (Section 4.4.3). Based on the data in the preceding section, we shall assume that water-soluble SO_3 not balanced by water-soluble K_2O and Na_2O occurs as $C_2K\bar{S}_3$, and the remaining water-soluble K_2O and Na_2O as aphthitalite if the K/Na ratio is within the appropriate range, or as aphthitalite of the appropriate end member composition mixed with $K\bar{S}$ or $N\bar{S}$ if it is not. As an example, for a clinker with 0.80% K_2O, 0.30% Na_2O and 0.60% SO_3, this gives: water-soluble K_2O, 0.50%; water-soluble Na_2O, 0.09%; water-soluble SO_3, 0.60%; $C_2K\bar{S}_3$, 0.1%; $K_{1-x}Na_x\bar{S}$, 1.0% and $x = 0.25$; $K\bar{S}$, 0.1%.

For most normal clinker compositions, the major sulphate phase will be aphthitalite with its maximum K/Na ratio of 3.0, with minor amounts of $K\bar{S}$ or $C_2K\bar{S}_3$ or both; only with unusually low values of K/Na in the clinker would one expect $N\bar{S}$ to be formed. It is only possible to estimate a maximum value for the content of anhydrite, based on the amount of SO_3 that is not water soluble; this will often be zero, as in the example given above. Anhydrite may, however, be determined by QXDA (Section 4.3.2). Rarely, the amount of water-soluble SO_3 is insufficient to balance that of water-soluble alkalis; this has been attributed to the presence of traces of alkali carbonates (P2), or of potassium oxide and potassium aluminate (KA) (F4).

3.5.7 Changes during grinding or storage

Méric (M33) reviewed the grinding and storage of clinker. Factors influencing the ease with which a clinker can be ground include the size distribution of the nodules, the texture and phase composition, the cooling rate and the conditions under which it has been stored. Grindability increases with alite content and decreases with that of the interstitial phases. Hornain and Regourd (H22) defined a brittleness index, which they showed to have values of 4.7 for alite, 2.9 for aluminate, about 2 for ferrite and 1.8 for belite. They also found that microcracking produced by rapid cooling from 1250°C increased grindability. Some specifications allow the use of grinding aids (Section 11.3.2). The particle size distributions of cements are discussed in Section 4.1.

The heat produced on grinding can cause partial conversion of the gypsum into hemihydrate ($2CaSO_4 \cdot H_2O$) or $\gamma\text{-}CaSO_4$. The extent to which this occurs depends on the temperature reached and the relative humidity within the mill. The conversions affect the rate at which the calcium sulphate dissolves on hydration. Partial conversion to hemihydrate may be desirable, as the water present in gypsum can cause the particles of cement to adhere during storage, with formation of lumps (L12). To obtain an optimum rate of dissolution on hydration, part of the gypsum may be replaced by anhydrite. Lumpiness on storage can also result from the reaction of gypsum and potassium sulphate to form syngenite ($CK\bar{S}_2H$) (L11). Maultzsch et al. (M34) reported on the effects of prolonged storage; minor amounts of gypsum (from hemihydrate), syngenite, calcite and normal hydration products were present.

3.6 Effects of minor components

3.6.1 General

Minor or trace components derived from raw materials, fuel, refractories or other plant materials, or added deliberately, can affect the reactions of clinker formation, or the properties of the product, or both. Their effects can be beneficial or harmful. Beneficial effects include acceleration of the clinkering reactions or lowering of the temperature at which they occur, or increase in the reactivity of the product leading to more rapid strength development. Harmful effects include decrease in alite content, volatilization in the kiln with consequent formation of kiln rings or other deposits, decrease in the durability of concrete made with the cement, or the introduction of poisonous elements. Some elements have beneficial effects at low concentrations and harmful ones at higher concentrations. Bucchi (B33,B28) reviewed some of the effects on the manufacturing process.

The main clinker liquid can be affected through alteration in the temperature at which it begins to form or in its quantity or properties at a given temperature. A flux is an agent that promotes a reaction by increasing the quantity of liquid at a given temperature; Al_2O_3 and Fe_2O_3 are fluxes for the formation of alite. Butt, Timashev and co-workers studied the properties of clinker liquids (B31,T1). They found that the viscosity and surface tension of the liquid are markedly affected by relatively small additions of other ions. Species containing strongly electropositive elements increase the viscosity and ones containing strongly electronegative elements decrease it. This was attributed to changes in the distribution of Al^{3+} and Fe^{3+} between octahedral and tetrahedral coordination. The effects of adding more than one minor component are not additive, and can be complex, e.g. alkali sulphates decrease the viscosity, but even a small stoichiometric excess of alkali oxides over SO_3 increases it. Many species lower the surface tension, because ions that do not easily fit into the structure of the liquid tend to become concentrated near to an interface, but the situation may be complicated by liquid immiscibility. Viscosity has a particularly important effect, a low value favouring alite formation by accelerating dissolution of lime and belite and diffusion through the liquid. Decrease in surface tension may be expected to favour wetting of the solids and penetration of the liquid into porous aggregates of lime crystals.

Effects of minor components on the solid phases are of various kinds. A mineralizer is an agent that promotes the formation of a particular solid phase by affecting the equilibria through incorporation in one or more of the solid phases. CaF_2 acts both as a mineralizer and as a flux in promoting the formation of alite. The mechanism of an observed effect is often not clear, and perhaps for this reason, the term 'mineralizer' is often used loosely. Other examples of action of minor components on solid phases include the stabilization of M_3 C_3S and β-C_2S at room temperature, the modification of the aluminate structure resulting from incorporation of alkali, and the non-formation of $C_{12}A_7$ in situations where it would be formed in the pure CaO–Al_2O_3–Fe_2O_3–SiO_2 system. Minor components can affect the properties of a clinker phase in other ways, such as by causing changes in crystal size, morphology or structural perfection, and can also modify the microstructure of the clinker as a whole.

The first 11 minor components of raw mixes, in sequence of decreasing median concentration, are MgO, K_2O, SO_3, Na_2O, TiO_2, Mn_2O_3, P_2O_5, SrO, fluorides, chlorides, and Cr_2O_3 (B27).

3.6.2 Effects of s-block elements

The effects of K_2O and Na_2O in producing low-temperature melts and volatile species have already been discussed. K_2O and Na_2O lower the temperature of formation of the melt but, in the absence of SO_3, increase its viscosity. If present in sufficient concentration, they lead to formation of orthorhombic or pseudotetragonal aluminate; this can have undesirable effects on the rheology of fresh concrete. High K_2O contents can lead to syngenite formation and consequent lumpiness on storage. High contents of either K_2O or Na_2O can lead to destructive reactions with certain aggregates (Section 12.4). Alkalis increase early strength but often decrease late strength (Section 7.6.3).

MgO also lowers the temperature at which the main melt begins to form and increases the quantity of liquid; it slightly decreases the latter's viscosity (B31,T1). Its presence is probably the basic cause of the non-formation of $C_{12}A_7$ in normally cooled clinkers of high AR. It stabilizes M_3 alite (Section 1.2.4). It retards the decomposition of alite during cooling. These effects are beneficial, but the formation of periclase, discussed in Section 3.5.5, limits the permissible content, which most national specifications set at 4% or 5%.

Some limestones contain substantial proportions of SrO. Small contents favour alite formation, but phase equilibrium studies show that it increases the range of compositions at which free lime, with SrO in solid solution, is present at the clinkering temperature (B34,G27,G28). This limits the permissible content of SrO in mixes of normal LSFs to about 3%.

3.6.3 Effects of p- and d-block elements

The behaviour of CaF_2 as a flux and mineralizer was described in Section 2.6.1. In addition to promoting formation of alite at markedly lower temperatures, CaF_2 accelerates the decomposition of calcite. Volatilization does not cause serious problems in the kiln, and it has often been added to raw meals, typically in proportions below 0.5%. It retards setting, possibly through formation of $C_{10}S_4 \cdot CaF_2$ during cooling, and this limits its use. As with admixtures to the raw meal in general, it is also necessary to consider whether the resulting savings outweigh the extra costs. This is most likely to be the case with raw materials that would otherwise require particularly high burning temperature.

Chloride contents must be kept low to avoid corrosion of steel in reinforced concrete (Section 12.3) and formation of kiln rings and preheater deposits. Contents below 0.02% are preferred, though higher ones can be acceptable if a sufficient proportion of the kiln gases is bypassed or in less energy-efficient (e.g. wet process) plants.

The effects of SO_3 include formation of alkali sulphates and other low melting point phases and volatile species, discussed earlier in this chapter and in Section 2.5.2. Although much of the SO_3 is present in a separate liquid phase, the main melt begins to form at a lower temperature and its viscosity is reduced, rendering $CaSO_4$ an effective flux. Against this, it leads to the stabilization of belite and lime at clinkering temperatures, and consequent decrease in content or even disappearance of alite, though this effect is lessened if MgO is present (G16). Sulphate can be removed by increasing the burning temperature, but the inclusions of lime in alite that have formed at lower temperatures are not easily lost (L14).

Portland cement clinkers typically contain around 0.2% of P_2O_5; higher contents lead to decreased formation of alite, and if above about 2.5% to presence of free lime. An early conclusion that α-C_2S forms a continuous series of solid solutions with α ('super α') C_3P above 1450°C (N12) was modified by later work, which showed that a miscibility gap exists at 1500°C (G29). Solid solutions also exist between other combinations of C_2S and C_3P polymorphs, lowering the temperatures of the C_2S transitions and, for some compositions, stabilizing α or α' forms at room temperature. In the CaO–C_2S–C_3P system (G29), C_3S has a primary phase field, but at a clinkering temperature of 1500°C, mixes with more than a few per cent of P_2O_5 do not yield C_3S. The tolerance for P_2O_5 is somewhat increased if F^- is also present (G21).

The 3d elements from titanium to zinc all lower the viscosity of the clinker liquid in varying degrees, but their effects on clinker formation and properties present a confusing picture. It has been claimed that the content of alite in clinkers decreases with increase in that of TiO_2, but the effect is much less marked if the latter is regarded as replacing SiO_2 (B27). Chromium can enter the clinker from raw materials or refractories; cements high in it have been reported to give high early strengths, but toxicity renders their use impracticable. The enhanced early strength is possibly due to high defect concentrations in the alite (B27), in which the chromium occurs as Cr^{4+} (J6). Both manganese and titanium occur in their highest concentrations in the ferrite phase, and probably for this reason yield darker coloured clinkers.

The cycling of zinc, cadmium, thallium, lead, sulphur, fluorine and chloride was noted in Sections 3.2.5, 3.3.5 and 3.4.5.

4

Properties of Portland clinker and cement

4.1 Macroscopic and surface properties

4.1.1 Unground clinker

This chapter deals with chemical and physical properties other than ones for which the nature of the hydration products must be considered, which are treated in Chapters 5 to 8. In general, properties of the whole clinker or cement are alone considered, those of the constituent phases having been dealt with in Chapter 1, but factors affecting the reactivities of these phases are included as a link with the following chapters on hydration.

Portland cement clinker emerges from a dry process kiln as rounded pellets, or from a wet process kiln as irregularly shaped lumps, in either case typically of 3–20 mm dimensions. Typical clinkers are greenish black, the colour being due to ferrite phase that contains Mg^{2+} (M18); in the absence of Mg^{2+}, they are buff. Reducing conditions in the kiln typically produce clinkers that are yellowish brown, especially in the centres of the lumps, where less reoxidation has occurred. Light colours can also arise from underburning.

The bulk density of a clinker, determined under standard conditions by rejecting material passing a 5-mm sieve, pouring the remainder into a conical container, and weighing, is called the litre weight. This quantity, which is typically 1.25–1.35 kg l^{-1}, gives some indication of the operating conditions. Its optimum value depends on the composition.

The measured true density of clinker is typically 3150–3200 kg m^{-3}; the value calculated from the X-ray densities and typical proportions of the individual phases is about 3200 kg m^{-3}. The difference is probably due mainly to the presence of pores inaccessible to the fluids used in the experimental determination. Using mercury porosimetry, Butt *et al.* (B35) determined pore size distributions and discussed their relations with burning and cooling conditions and grindability.

4.1.2 Particle size distribution of ground clinker or cement

Allen (A3) discussed methods suitable for determining this important property. Sieving is suitable only for determining the proportions of particles in ranges above about 45 μm. In airjet sieving, jets speed the passage of material through the sieve and prevent the mesh from becoming blocked. For the entire range of size, the most widely used methods are ones based on either sedimentation in liquids or diffraction of light. In the X-ray sedigraph, the progress of sedimentation is monitored by the absorption of an X-ray beam and the results calculated using Stoke's law. Because highly concentrated suspensions are used, there is some doubt as to whether this law is strictly applicable. An older method employs the Andreason pipette, by which samples are withdrawn from a suspension at a fixed depth after various times, so that their solid contents may be determined. In laser granulometry, light from a laser passes through a suspension and the particle size distribution (PSD) is calculated from the resulting diffraction pattern. Light microscopy can provide information on the distribution of both particle size and particle shape, and SEM with image analysis (Section 4.3.1) can, in principle, provide much additional information.

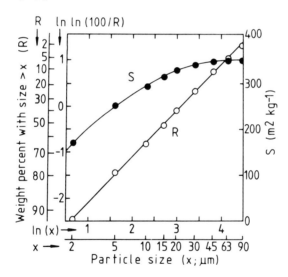

Fig. 4.1 Rosin–Rammler plot of the particle size distribution of a typical Portland cement. S = specific surface area attributable to particles of size smaller than x. Open circuit grinding; based on the data of Sumner et al. (S25).

With all the methods, complications or uncertainties of interpretation arise due to effects associated with particle shape and with flocculation, so that the results given by different methods do not necessarily agree and usually have relative rather than absolute significance. Fig. 4.1, curve R, shows the cumulative PSD, obtained using an X-ray sedigraph, for a typical Portland cement ground in an open circuit mill (S25). In modern ordinary Portland cements, some 7–9% of the material is typically finer than 2 μm and 0–4% coarser than 90 μm.

Méric (M33) discussed analytical expressions used to represent the PSD of cements. Probably the most successful is the Rosin–Rammler function, which may be written in logarithmic form as

$$\ln(100/R) = n(\ln x - \ln x_o) \qquad (4.1)$$

where R is the weight percentage of material of particle size greater than x. The constants x_o and n describe the distribution; x_o is a measure of average size, 36.79% of the material being of size greater than x_o, and n is a measure of breadth of distribution, increasing as the latter becomes narrower. For the example shown in Fig. 4.1, x_o is 22 μm and n is 1.03. For open circuit milling, the values of n obtained using an X-ray sedigraph are 0.99–1.07; for closed circuit milling, in which the fine material is separated and only the coarser material further ground, n can be as high as 1.23–1.28 if the separation is efficient (S25). A narrow PSD can also be obtained using a vertical spindle mill. Laser granulometry typically gives values of n lower by 0.15–0.20 (S25).

4.1.3 Specific surface area determination

The specific surface area of cement is commonly determined directly by air permeability methods. In the Lea and Nurse method (L15), a bed of cement of porosity 0.475 is contained in a cell through which a stream of air is passed, and steady flow established. The specific surface area is calculated from the density of the cement, the porosity and dimensions of the bed of powder, the pressure difference across the bed, and the rate of flow and kinematic viscosity of the air. In the Blaine method (B36), a fixed volume of air passes through the bed at a steadily decreasing rate, which is controlled and measured by the movement of oil in a manometer, the time required being measured. The apparatus is calibrated empirically, most obviously using a cement that has also been examined by the Lea and Nurse method. The two methods gave closely similar results. The Blaine method, though not absolute, is simpler to operate and automated variants of it have been devised.

Other methods that have been used to determine specific surface areas of cements include the Wagner turbidimeter (W16) and BET (Brunauer–Emmett–Teller) gas adsorption. The former, as conventionally used, gives very low results because of a false assumption that the mean diameter of the particles smaller than 7.5 μm is 3.8 μm, which is much too high. The BET method gives results two to three times higher than the air permeability methods, because it includes internal surfaces present in microcracks or in pores open at only one end.

The specific surface area, like the PSD, is thus a quality whose value depends on how it is defined, and is liable to be affected by any pretreatment or conditions affecting the degree of flocculation. In practice, air permeability methods are widely used. Typical values are 300–350 m² kg^{-1} for modern ordinary Portland cements and 400–450 m² kg^{-1} for rapid-hardening Portland cements.

From a PSD curve, one may calculate the specific surface area, S:

$$S = 6 \times 10^6 \times F \times \Sigma f/(d \times \rho) \tag{4.2}$$

where f is the weight fraction of material consisting of grains assumed to have a diameter or edge length d (in μm), ρ is the density in kg m^{-3}, and the result is in m² kg^{-1}; F is an empirical constant which takes into account the differing assumptions regarding surface shape that are made implicity or explicity in the determination of PSD and the definition of specific surface area. This formula applies to both cubic and spherical grains. For a calculation starting from a PSD obtained using an X-ray sedigraph to give a specific surface area comparable to that obtained by an air permeability method, F is typically about 1.13. Fig. 4.1, curve S, shows the contributions to the calculated specific surface area of particles of differing sizes, assuming that particles smaller than 2 μm have a mean size of 1 μm. About 40% of the total is provided by particles smaller than 2 μm.

In principle, the specific surface area may also be calculated from the PSD using analytical expressions derived from such functions as the Rosin–Rammler expression. In practice, this is not always satisfactory because the specific surface area is so highly dependent on the bottom end of the distribution.

4.1.4 Particle size distribution, phase composition and cement properties

Because the constituent phases of a cement are not equally easy to grind, different particle size fractions differ in composition. Gypsum, and its dehydration products, are concentrated in the finer fractions. Osbaeck and Jøns (O8) concluded that each 1% of gypsum contributed about 10 m² kg^{-1} to the specific surface area; in a typical case, some 15% of the total specific surface area is thus due to gypsum. The content of alite decreases, and that of belite increases, with increasing particle size (R12,G30), the contents of aluminate and ferrite phases being little affected.

The depth of reaction of the cement grains after a given time has been determined by light microscopy; it may be expected to vary with the composition and microstructure of the clinker. Anderegg and Hubbell (A4,A5) found values of 0.43–0.47 µm at 1 day, 1.7–2.6 µm at 7 days, 3.5–5.4 µm at 28 days and 6.1–8.9 µm at 150 days. Zur Strassen (Z7) gave somewhat higher values (e.g. 6–7 µm at 28 days). From the depth of reaction and the PSD curve, and assuming the grains to be spherical, one may calculate the fraction, α, of cement that has reacted after a given time:

$$\alpha = \Sigma f \times [1 - (1 - 2h/d)^3] \qquad (4.3)$$

where h is the depth of reaction in µm and Σ, f and d have the same meanings as in equation 4.2 and the quantity in square brackets cannot be less than zero. For the cement to which the data in Fig. 4.1 refer, and using the means of Anderegg and Hubbell's values, this gives 0.28 at 1 day, 0.63 at 7 days, 0.80 at 28 days and 0.90 at 150 days. These values are in reasonable agreement with those typically found using XRD (Section 7.7.1). Ritzmann (R12) determined the compressive strengths of pastes of a number of cements of known PSD, made by mixing separated fractions. For a given cement and water/cement (w/c) ratio, the strength was linearly related to the fraction of cement reacted, calculated in this way.

At 1 day, the reaction is primarily of the smaller particles, and the fine end of the PSD is therefore the important one for the strength. As the age increases, the coarser part of the distribution becomes progressively more important; by 28 days the particles smaller than about 10 µm have reacted completely, and the content of particles up to about 45 µm in size is of major importance. For a given specific surface area, narrowing the PSD increases the degree of reaction at 28 days, and thus also the strength (L16,S25,K9). The relation may be quantified using equation 4.3. Narrowing the PSD also saves energy in grinding (S25), but increases the amount of water needed to produce a mix of given workability (S25,S26); this latter effect can be compensated by adjusting the temperature during milling so as to increase the extent to which the gypsum is dehydrated to hemihydrate or soluble anhydrite (S25).

4.1.5 Chemical analysis

Ground clinker or cement may be analysed by standard methods, such as X-ray fluorescence (XRF). Determination of free lime is discussed in Section 4.3.3. With XRF or other methods in which no fraction of the material is excluded from the analysis, insoluble residue (determined separately) is included in the oxide components and therefore should be omitted from the total. It can often be assumed that two-thirds of it is SiO_2 and the rest

Al_2O_3. Classical, wet analysis gives inaccurate results for Al_2O_3 unless the effects of P_2O_5 and TiO_2 are allowed for.

Chemical analysis, microscopy, XRD and other methods of examination should be carried out on the same, representative sample of material. A procedure for obtaining such a sample is described in the next section.

4.2 Light microscopy

4.2.1 General

Light microscopy provides much information on clinker microstructure and thereby on the conditions existing at various stages of the manufacturing process. Much the most widely used technique has been the examination of polished and etched sections in reflected light, and the following sections refer to this unless otherwise stated, but additional information has been obtained from the examination of thin sections or powder mounts in transmitted light. Fuller accounts are given elsewhere (G31,H21,L10,C17).

It is vital to obtain a representative sample; the following procedure (L10) is typical. If a kiln is known to give a reasonably consistent product, samples taken at intervals over a suitable period are aggregated, but if not, they are examined separately. Each sample examined comprises at least 15 kg of clinker, and is first examined visually to distinguish kiln deposits or refractories. It is then crushed to below 6 mm and a representative sample obtained, e.g. by quartering. The 2–4 mm fraction of this sample is used. Ten to twelve samples of uncrushed nodules, covering the range of nodule sizes, are also examined, together with a sample of the material smaller than 2 mm, if much of this is present in the original clinker. The relative amounts of nodules of different sizes, and of the material below 2 mm, are noted.

For examination in reflected light, the sample is mounted in a plastic resin and is cut and polished using only non-aqueous lubricants. The most generally useful etchant is probably HF vapour, which has the merit of not removing alkali sulphates; the slide is inverted over a vessel containing 40% aqueous HF for 10–20 seconds. Treatment with water followed by 0.25% HNO_3 in ethanol and finally ethanol is also widely used, and many other etchants have been employed for specific purposes. The etchants produce thin films of decomposition products, which yield interference colours when viewed in reflected light.

Fig. 4.2 shows a polished and etched section of a production clinker. Crystals of alite and belite are embedded in a matrix of aluminate and ferrite phases. If HF vapour is properly used, the alite is yellowish brown, the belite is blue or red, and the aluminate is grey; the ferrite, which is unattacked, is brightly reflective. The alite crystals are angular, often pseudohexagonal; the belite crystals are rounded and normally striated, as described in Section 1.3.2. Ideally, the alite crystals are 15–20 μm in average size, though in practice they are usually much larger, and there should be no clustering of either alite or belite. The average size of the belite crystals is typically 25–40 μm.

Fig. 4.2 Reflected light micrograph of a polished and etched section of a Portland cement clinker, showing crystals of alite (dark, angular) and belite (less dark, rounded) embedded in a matrix of interstitial material, itself composed mainly of dendritic ferrite (light) and aluminate (dark). Courtesy Materials Science Department, British Cement Association.

A good clinker should have been rapidly cooled, and this will cause the ferrite and aluminate phases to be mixed on a scale of a few micrometres, the ferrite often forming dendrites or needles in a matrix of aluminate. In transmitted light, the ferrite phase is pleochroic. The normal, cubic aluminate tends to fill the spaces between the other phases and thus to exhibit no definite form; it is isotropic in transmitted light. In clinkers with an excess of alkali over SO_3, cubic aluminate may be partly or wholly replaced by the orthorhombic variety. This forms relatively large, prismatic crystals, which occur in a matrix of the other interstitial phases and are birefringent in transmitted light.

There should be little free lime. What there is should occur as rounded grains, typically 10–20 µm in size, and associated with alite and interstitial material. Lime appears cream in sections etched with HF vapour. Its presence may be confirmed by a microchemical test using White's reagent (5 g of phenol in 5 ml of nitrobenzene + 2 drops of water); long, birefringent needles of calcium phenate are formed. The test also responds to CH. Alkali sulphates occur in the clinker pore structure; they are etched black with HF vapour, and inhibit the etching of silicate phases with which they are in contact.

The phases may be determined quantitatively by point counting (H21,C18). The areas of phases on a polished section of a cement clinker are proportional to the volumes in the material. Random and systematic errors arise from imperfect sampling, counting statistics and incorrect identifications of phases by the operator. Error from counting statistics depends on the number of points, the size of the grid and the grain sizes of the phases; 4000 points normally gives reasonable precision (H21). Systematic errors from incorrect identification of phases generally exceed errors from counting statistics, and give rise to differences in results between operators (C18). With many clinkers, it is impracticable to determine the aluminate and ferrite phases separately by this method.

4.2.2 Effects of bulk composition, raw feed preparation and ash deposition

The bulk composition affects the relative amounts of phases, the LSF, SR and AR affecting primarily the ratios of alite to belite, silicate phases to interstitial material and aluminate to ferrite phase, respectively. An LSF that is too high, either absolutely or in relation to the burning conditions, also gives rise to an excessive content of free lime. Belite may occur in such clinkers only as inclusions in alite.

Inadequate grinding of the raw meal is shown by the presence of large clusters of free lime or belite, up to several hundred micrometres in size and with defined outlines reflecting those of the original particles of calcareous or siliceous material, respectively. Coarse fragments of dolomitic limestones similarly yield localized concentrations of periclase, usually intimately mixed with lime. Periclase crystals are characteristically angular, up to about 20 µm in size, and are unattacked by normal etchants. Due to their hardness, they stand out in high relief in unetched sections. Less sharply defined segregation indicates inadequate mixing, and is typically shown by the presence of regions rich in belite and in alite plus lime, in each case together with interstitial material. In extreme cases, entire nodules can depart seriously from the mean composition.

Incomplete assimilation of coal ash is shown by the existence of large regions high in belite, and containing also interstitial phases. The locations of these regions indicate the stage of the burning process during which the ash was deposited; if it was late, the belite-rich material forms coatings on clinker nodules, in which small nodules can be encapsulated, whereas if it is deposited earlier, it can form part of the kiln coating, which may later break away and form the centres of nodules. If the ash contains unburnt matter, localized reducing conditions can occur. Insufflation of materials into the kiln can produce similar effects (L10).

4.2.3 Effects of burning conditions and cooling rate

Underburning is shown by the presence of abundant free lime and a low content of alite, the crystals of which are very small ($< 10\,\mu m$). The average size and other characteristics of the alite and belite crystals provide much further information about the burning conditions; factors governing them were discussed in Chapters 1 and 3. Ono (O7) considered that alite size was an indication of the rate of heating to the clinkering temperature, which if high led to the formation of smaller crystals, but other studies indicate that many other factors are involved, and that overburning, in particular, can cause increase in size through recrystallization (Section 3.4.4). Another indication of excessively hard burning is the presence of material high in large alite crystals and low in interstitial material, formed by the withdrawal of the liquid into the centres of the clinker nodules (L10).

The effects of slow cooling (Section 3.5) are readily detectable by light microscopy. The most general are coarsening of the texture of the interstitial material and a change in the belite, when viewed in transmitted light, from colourless to yellow. The belite crystals may also develop ragged or serrated edges. Slow cooling can also cause resorption of alite, with deposition of small crystals of belite as fringes on the alite and in the body of the interstitial material, increase in alite crystal size and, if it occurs below 1200°C, decomposition of alite to an intimate mixture of lime and belite. The effects of reducing conditions (Section 3.4.5) are similarly detectable by light microscopy (L11,L14).

4.2.4 Applications of light microscopic investigations

Light microscopy can be used to find the causes of unsatisfactory clinker quality or to determine what modifications in composition or plant operation are needed to change the clinker properties in a desired direction. It has also been used to predict strength development. Ono (O7) described results obtained from examinations of powder mounts. The values of four parameters, indicated in Table 4.1, were each estimated on a scale of 1–4 and the strength, R, in MPa, of a mortar at 28 days then predicted using the regression equation:

$$R = 23.6 + 0.59\text{AS} + 2.04\text{AB} + 0.37\text{BS} + 2.00\text{BC} \qquad (4.4)$$

The four parameters AS, AB, BS and BC were considered to be measures of heating rate, maximum temperature, time at that temperature and cooling rate, respectively. As noted in the preceding section, the first of these assignments is questionable, but this does not necessarily invalidate the results obtained by the method, which was tested by routine application in many plants of a major cement company over a long period. Some of the other workers using the method have, in contrast, had little success with the method (e.g. Ref. S27), and the generality of the procedure, or of the coefficients, remains to be established. The constant term in the equation probably subsumes terms relating to composition and cement fineness, which would have to be taken into account if a wider range of samples was considered.

Table 4.1 Parameters of Ono's method (O7) for predicting cement strength

AS (alite size, μm)	15–20	20–30	30–40	40–60
AB (alite birefringence)	0.008–0.010	0.006–0.007	0.005–0.006	0.002–0.005
BS (belite size, μm)	25–40	20–25	15–20	5–10
BC (belite colour)	Clear	Faint yellow	Yellow	Amber
Hydraulic activity	Excellent	Good	Average	Poor
Value of parameter	4	3	2	1

4.3 Scanning electron microscopy, X-ray diffraction and other techniques

4.3.1 Scanning electron microscopy

Scanning electron microscopy (SEM) of polished sections of clinkers provides information essentially similar to that given by light microscopy, with two important additions. First, phases can be analysed chemically. Second, the image can be stored electronically and subsequently processed or analysed in various ways. The specimen is polished and coated with an electrically conducting layer, normally 30 nm of carbon, but not etched. The most generally useful technique is backscattered electron imaging, using a detector designed to maximize contrast from compositional differences and to minimize that from topography, combined with provision for X-ray microanalysis. This should include facilities both for obtaining X-ray images giving semiquantitative information on the spacial distributions of individual elements, and for obtaining complete and quantitative chemical analyses in regions of micrometre dimensions. Techniques are described in standard texts (e.g. Ref. G32), and the results of analyses of individual clinker phases were considered in Chapter 1.

The intensity of the electrons backscattered from a particular region of a specimen depends approximately on the mean atomic number of the material of which that region is composed; more precisely, it is represented by the backscattering coefficient, η, which may be calculated from the formulae:

$$\eta = \Sigma\, C_i \eta_i \tag{4.5}$$

$$\eta_i = -0.0254 + 0.016 Z_i - 1.86 \times 10^{-4} Z_i^2 + 8.3 \times 10^{-7} Z_i^3 \tag{4.6}$$

where the summation is over elements and C_i is the weight fraction of an element having atomic number Z_i (G32). This equation assumes the accelerating voltage to be 20 kV. Assuming the typical compositions given in Table 1.2 for the individual phases, this gives values of 0.171 for alite, 0.167 for belite, 0.168 for cubic aluminate and 0.178 for ferrite. The differences between these values are such that the grey levels of the alite, belite and ferrite phases on micrographs are distinguishable to the eye, but those of the belite and aluminate are not; however, the phase boundaries are normally visible, presumably due to slight topographical contrast or compositional irregularity. Fig. 4.3 shows backscattered electron images of a typical clinker and of a polymineralic cement grain present in a fresh paste (S28).

The resolution available is limited by the size of the region from which the backscattered electrons are produced, which is somewhat below 1 μm in each direction. Much higher resolution is obtainable using the secondary electron image, but this depends mainly on topographical contrast and thus gives, at best, poor definition of the phases. The resolution given by the X-ray image is poorer than that given by the backscattered electron image.

Fig. 4.3 Backscattered electron images of polished sections of (A) a Portland cement clinker and (B) grains of a Portland cement in a fresh paste. In both sections, alite is the predominant clinker phase. In (A), the relatively large, darker areas are of belite, and the interstitial material consists of dendritic ferrite (light) in a matrix of aluminate (dark); cracks and pores (black) are also visible. In (B), the belite forms well-defined regions, which are rounded, striated and darker than the alite; the interstitial material, present, for example, in a vertical band left of centre within the larger grain, consists mainly of ferrite (light) and aluminate (dark). Scrivener and Pratt (S28).

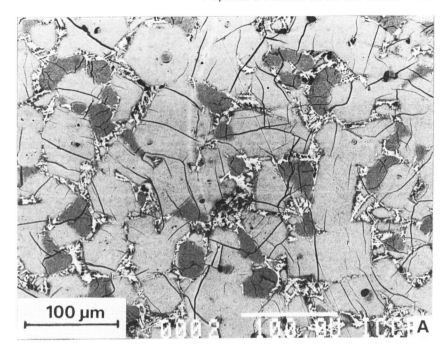

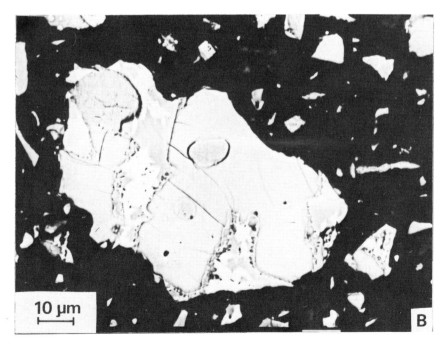

SEM with image analysis has been used in the study of clinker and ground cement (L17,S29). Scrivener (S29) stored the backscattered electron image and X-ray images for several elements from the same area of a specimen. She did not find it possible to distinguish the clinker phases from the backscattered electron images alone, but by computer combination of data from these and the X-ray images, the silicate phases could be distinguished from the aluminate phases. The method was used to study the phase distributions within individual grains of cement and to compare the volume and surface distributions. For the cements studied, the large grains consisted wholly, or almost wholly, of belite. Almost all the remaining grains appeared to be polymineralic, even including those that were only a few micrometres in size (Fig. 4.3B), but grains smaller than about 0.5 µm were not easily studied and may have been monomineralic. Overall, fracture during grinding appeared to be predominantly through alite, and to avoid phase boundaries. The proportions of interstitial phases exposed on the surfaces of the grains were significantly greater than those by volume in the material. The volume fractions of silicate and interstitial phases were compared with those determined by QXDA or obtained by the Bogue calculation.

The application of image analysis to SEM is a rapidly developing technique, and further development may be expected. In addition to giving areas of phases, it can provide information on such matters as the distributions of shape, size and surroundings of particles and of phases and pores within them. It is, of course, essential to examine a sufficient number of samples and fields to ensure that the results are representative, but the labour and subjectivity associated with light microscopic studies of a comparable nature are largely eliminated.

4.3.2 X-ray diffraction

The potential uses of XRD powder diffraction in the study of clinker or anhydrous cement include the qualitative and quantitative (QXDA) determination of phase composition, and the determination of polymorphic modification, state of crystallinity and other features of individual phases. In principle, information on compositions of phases is obtainable through cell parameters, but, due to the lack of adequate reference data, XRD is generally less satisfactory for the clinker phases than X-ray microanalysis. Table 4.2 gives the pattern of a typical Portland cement, with indications of the assignments of peaks to phases.

Alite is normally so much the predominant phase that its pattern tends to swamp those of the other phases. All the stronger peaks of belite are overlapped by ones of alite. Other difficulties in quantitative analysis are caused by variability of patterns of phases due to compositional or polymorphic variation, and for the ferrite phase, peak broadening due to compositional zoning or imperfect crystallinity. Other problems in quantitative work are those of QXDA in general; e.g. the material must be ground sufficiently finely that primary extinction of the more intense peaks is minimized, and that adequate mixing with an internal standard is possible, if one is used, but not so strongly that crystallinity is reduced. Milling to below 5 μm in an agate ball mill with cyclohexane as grinding aid, followed by backloading into the sample holder of a diffractometer, has been found satisfactory for specimen preparation (G30).

Table 4.2 XRD powder pattern of a typical Portland cement (T5)

$2\theta°$	d (nm)	I_{pk}	Main phases	$2\theta°$	d (nm)	I_{pk}	Main phases
11.7	0.756	5	Gyp	34.4	0.2607	83	Ali, Bel
12.1	0.731	6	Fer	36.7	0.2449	6	Ali
14.9	0.595	6	Ali	37.4	0.2404	2	Bel
20.7	0.429	7	Gyp	38.8	0.2321	12	Ali
21.9	0.406	2	Alu	39.5	0.2281	5	Bel
23.0	0.3867	7	Ali	41.3	0.2186	41	Ali, Bel
23.4	0.3802	3	Bel	41.6	0.2171	16	Ali, Bel
24.4	0.3648	3	Fer	44.1	0.2053	6	Fer
25.3	0.3520	4	Ali	44.5	0.2036	3 (B)	Bel
26.4	0.3376	2	Bel	44.7	0.2027	2	Bel
27.6	0.3232	2	Bel, Ali	45.8	0.1981	10	Ali, Bel
28.1	0.3175	4	Ali, Bel	47.0	0.1933	11	Ali
29.1	0.3069	5	Gyp	47.4	0.1918	8 (B)	Fer
29.4	0.3038	60	Ali	47.8	0.1903	7	Alu
30.1	0.2969	19	Ali	49.9	0.1828	5	Ali, Fer
31.1	0.2876	4	Bel, Gyp	51.7	0.1768	33	Ali
32.2	0.2780	100	Ali, Bel, Fer	51.8	0.1765	35	Ali
32.6	0.2747	85	Ali, Bel	56.0	0.1642	2	Ali
33.2	0.2698	40	Alu, Ali	56.6	0.1626	18	Ali
33.9	0.2644	23	Fer	59.4	0.1555	3	Alu
				59.9	0.1544	6	Ali

CuK$_\alpha$ radiation. I_{pk} = relative peak height (B = broad). Phases are given in decreasing order of their contributions to each peak, ones making only minor contributions being omitted. Ali = alite, Bel = belite, Alu = aluminate, Fer = ferrite, Gyp = gypsum.

Struble (S30) summarized QXDA studies on clinkers and cements. Most investigators have used some form of the internal standard method (K10). Rutile (TiO$_2$) has been widely used as the internal standard, but other substances, such as silicon, KBr, CaF$_2$ and corundum, have also been employed; corundum has the advantage of giving no overlaps with important cement peaks up to 60° 2θ. The method in its conventional form, in which only a single peak each of the standard and the phase to be determined are used, is moderately effective for alite, using the peak at approximately 51° 2θ, assuming that a means of accommodating the splitting of this peak is found. It is less satisfactory for the aluminate and ferrite, and useless for belite, because of the weakness of their peaks and interference from those of alite.

With any form of the internal standard method, it is necessary to employ reference standards that are as near as possible to the phases in the material as regards composition, polymorphism and degree of crystallinity. This implies both knowledge of the characteristics of the particular clinker under examination and the ability to prepare the necessary specimens, or alternatively to isolate them physically from the clinker itself. One solution (G30) has been to prepare reference patterns for a wide range of specimens of each phase and to use those that appear to correspond most closely to those present in the clinker under examination. Clinkers for which results had been obtained by point counting have also been used (K11).

Gutteridge (G30) described a method, based on earlier work by Berger *et al.* (B37), in which a least squares fit is obtained between a diffractometer trace of the sample with internal standard and a simulated trace summing the contributions from the component phases. The trace is obtained over the range of 24–39° 2θ by step counting at 0.05 degree intervals using CuK$_{\alpha 1}$ radiation, and stored in a computer. Auxiliary procedures were devised for dealing with background levels, adjusting the peak widths and, to some extent, peak positions in the reference patterns to match those in the clinker, and matching the observed and computed patterns. Minor phases, such as lime, periclase, calcium sulphate phases and mineral additions (Chapter 9), could be included in the analysis. This method is probably the most satisfactory yet described, but still depends on the choice of standards. Its power could probably be increased, at some expense of apparatus and computer time, by extending the range of 2θ values used. Some workers have used similar methods, but matching observed and calculated intensities integrated over selected, wider ranges of 2θ (K11). This could avoid problems arising from variations in the patterns of the component phases, though at some loss of sensitivity.

The Appendix gives calculated patterns based on the results of X-ray structure determinations for the clinker phases. For each pattern, a 'Reference Intensity Ratio' (H23) is included. This is the integrated intensity of the strongest individual reflection (which may be a component of an overlap) relative to that of the strongest peak of corundum in a 1:1 mixture by weight. In principle, it is possible by using this ratio to dispense with a reference standard. The internal standard used need not be corundum, provided that its own reference intensity ratio is known. This method has not been tested with clinkers, but has obvious possibilities. Given the necessary crystal structure data, it would permit the calculation and use in QXDA of patterns corresponding to any desired composition and polymorph for each phase.

4.3.3 Chemical or physical methods for separation of phases

Partial separations of the clinker phases can be effected by chemical extraction methods, which may be combined with XRD to identify and determine minor constituents. Of several reagents that have been used to dissolve the silicate phases, the most successful has been a solution of salicylic acid in methanol (SAM), introduced by Takashima (T10). This reagent dissolves free lime, alite and belite, at rates decreasing in that order (H24). By adjusting its composition, it is possible either to remove much of the alite, so facilitating investigation of the belite, or all of the alite and belite, facilitating that of the remaining phases. The procedure is as follows (H24).

The clinker is ground in propan-2-ol to an average particle size of 5 μm. After evaporating off the propan-2-ol, the sample (5 g) is treated for 2 h at room temperature with the SAM reagent (300 ml), with constant stirring. The solution, which is now red, is filtered on a No.4 sintered glass filter, and the residue washed with methanol and dried at 90°C. If the residue is to be examined by QXDA, the internal standard (e.g. rutile) may be mixed with the sample before the extraction. For removal of alite and belite, the salicylic acid content of the reagent is 20 g per 300 ml. For complete removal of the alite, leaving the maximum quantity of belite undissolved, it is 10–15 g, depending on the alite content; five times the weight of C_3S in the sample, as given by the Bogue calculation, was found to be satisfactory.

A solution of maleic acid in methanol has also been used to dissolve the silicate phases (T11). This is quicker than the SAM method, but if the methanol contains water, ettringite may be formed and water-soluble phases, such as K_2SO_4, lost, while if it is anhydrous, gelling occurs; the SAM method is thus preferable (S31). Extraction of the silicate phases by the trimethylsilylation method of Tamás et al. (T12) (Section 5.3.2) also leaves a residue containing the aluminate and ferrite phases.

Gutteridge (G33) found that a solution of sucrose in aqueous KOH (KOSH reagent) dissolves the aluminate and ferrite, leaving the silicates and minor phases. The clinker or cement (10 g) is reduced to below 5 μm by grinding in cyclohexane (20 ml) for 40 min in an agate ball mill. After evaporating off the cyclohexane in a stream of N_2, the sample (9 g) is added to the KOSH solution (30 g of KOH and 30 g of sucrose in 300 ml of water) at 95°C and stirred for 1 min. The solution is then filtered and the residue washed in water (50 ml) followed by methanol (100 ml) and dried at 60°C. By using a combination of the SAM and KOSH procedures, it is possible to produce a residue consisting largely of belite.

Various solvents may be used to dissolve free lime, which can be determined by titrating the resulting solution with a suitable acid. In one procedure (J7), the clinker or cement (1.0 g), ground to about $300 \, m^2 \, kg^{-1}$ (Blaine), is treated with anhydrous ethylene glycol (50 ml) at 80–100°C for 5 min, using a magnetic stirrer. The solution is then filtered under suction and the residue washed at least twice with a further 10–15 ml of hot ethylene glycol. The filtrate and washings are diluted with 25 ml of water and titrated to phenolphthalein with 0.05 M HCl. Other procedures employ mixtures of ethanol and glycerol (L18) or of acetoacetic ester and butan-2-ol (Franke solvent) (F8). If properly used, all give similar results. They also dissolve CH, and have been used to determine this phase in hydrated cements, but their use for this purpose is questionable as they also appear to attack other hydration products.

Physical methods are of limited effectiveness for separating the phases in cement clinker because of the intimate scale on which the latter are mixed; however, Yamaguchi and Takagi (Y1) had some success with the use of dense liquids. Some concentration of the ferrite phase can be effected by magnetic separation (M35,Y1).

4.3.4 Other methods

Bensted (B38) described the use of TG in the study of clinkers or unhydrated cements. At $4°C \, min^{-1}$ in N_2 with cements, losses normally occur at 100–200°C from gypsum or hemihydrate, 400–500°C from CH and 500–800°C from $CaCO_3$. If a normal, open sample cup is used, the losses from conversion of gypsum to hemihydrate and from dehydration of hemihydrate are poorly resolved, making interpretation difficult. Better separation can be achieved by using a sample cup with a top that is closed except for a narrow exit (S32).

Syngenite, if present, decomposes at 250–300°C. Small amounts of CH can occur due to hydration, and of $CaCO_3$ due to carbonation during milling or storage, impurity in the gypsum, or, where specification allows it, deliberate addition. The increasing tendency to allow additions of substances other than calcium sulphate may lead to the presence of other phases detectable by TG. A weight gain at 300–400°C in air that does not occur in N_2 normally indicates that the clinker was made under reducing conditions, but could also conceivably be due to the presence of an excessive quantity of metallic iron incorporated during milling. In this latter case, it would, of course, not be shown by the clinker.

Applications of IR and Raman spectroscopy to the study of clinkers and unhydrated cements have been reviewed (B39,B40). The laser Raman microprobe, with which regions of micrometre dimensions on a polished surface may be examined, has been used to investigate structure and crystallinity, especially of the alite and belite (C19). Spectroscopic methods for studying the surface structures and compositions of cements are considered in Section 5.6.2.

4.4 Quantitative phase compositions

4.4.1 General

The methods that have proved effective for quantitative determination of phases in clinkers are based on light microscopy (Section 4.2.1), X-ray diffraction (Section 4.3.2) and calculation from the bulk chemical analysis. The last two of these are applicable also to cements. SEM with image analysis (Section 4.3.1) shows promise, but other methods that have been investigated, such as IR spectroscopy, appear to have little potential. Sections 4.4.2 to 4.4.5 describe a method of calculation from the bulk analysis and Section 4.4.6 compares the results of the various methods.

4.4.2 Calculation of quantitative phase composition from bulk analysis

As will be seen in Section 4.4.6, there is wide agreement that the Bogue calculation gives seriously incorrect results, especially for alite. It has often been stated that the errors arise because equilibrium is not maintained during cooling, but this is only partly correct. Even if equilibrium was maintained, all the major phases would contain significant proportions of substituent ions. The calculation gives incorrect results because the phases do not have the compositions assumed for them, and the absence of equilibrium conditions during cooling merely alters the compositional differences from the pure compounds.

For a calculation of the quantitative phase composition of a clinker from the bulk chemical analysis to give correct results, the following conditions are necessary and sufficient:

(1) The analysis, including any auxiliary data such as the content of free lime, must be free from errors.
(2) The analysis must be corrected for any of the four major components not present in the four major phases. This is routinely done for free lime, but CaO may also be present in sulphate phases, and, assuming modern analytical methods (e.g. XRF) to have been used, some of the SiO_2 and Al_2O_3 will normally be present in an acid-insoluble residue.
(3) The contents of the four major components in each of the four major phases must be known. If necessary, weighted mean values must be used to allow for variable composition, arising, for example, from zoning in alite or ferrite, or from the presence of both cubic and orthorhombic aluminate. Belite presents a special problem because of the probable occurrence of material exsolved between the lamellae. In the present state of knowledge, use of an average composition including this material, such as is given by X-ray microanalysis, is the most practical solution.

The Bogue calculation is a solution of four linear simultaneous equations for four unknowns. Following Yamaguchi and Takagi (Y1) and Harrisson et al. (H3), one may modify it by using the best available estimates of the compositions of the four phases instead of those of the pure compounds. If these compositions were free from error, and the other conditions noted above were satisfied, this procedure would necessarily give correct results. For modern, rapidly cooled clinkers of ordinary Portland cements, the compositions of the phases do not seem to vary excessively, apart from some variation in the contents of MgO, CaO and Fe_2O_3 in alite (Chapter 1). By using typical compositions, such as those given in Table 1.2, one may expect to obtain results that are substantially more accurate than those given by the Bogue calculation. The following procedure (T13) is based on this approach, and in the form described in Sections 4.4.3 and 4.4.4 applies to most modern clinkers. Limitations and modifications for cements or other types of clinkers are noted in Section 4.4.5. It is illustrated by an example, in which the following analysis is assumed:

Na_2O	MgO	Al_2O_3	SiO_2	P_2O_5	SO_3	K_2O	CaO	TiO_2	Mn_2O_3	Fe_2O_3	Loss	Total
0.22	1.2	5.7	21.7	0.20	0.37	0.45	65.9	0.29	0.06	3.2	0.6	99.9

Free lime, 0.6%; insoluble residue, 0.3%; main analysis by XRF.

4.4.3 Estimation of sulphate phases

The procedure given below provides an estimate of the amounts of K_2O, Na_2O, CaO and SO_3 present in sulphate phases. It is based on the results of Pollitt and Brown (P2) (Section 3.5.6), and more specifically on the curves in

Fig. 3.5, which embody them. These curves are represented by empirical equations. We shall use the following symbols, which refer to quantities expressed as moles per 100 g of clinker: K = total K_2O; N = total Na_2O; $\bar{S}$ = total SO_3; K_s, N_s, C_s and $\bar{S}_s$ = K_2O, Na_2O, CaO and SO_3 present as sulphates.

(1) Calculate the ratios $R = \bar{S}/(K+N)$ and $k = K/N$.
(2) Calculate $\bar{S}_s$:
   (a) If $R \leqslant 0.8$, $\bar{S}_s = \bar{S}$
   (b) If $0.8 < R \leqslant 2.0$, $\bar{S}_s = [1.0 - 0.25(R - 0.8)]\bar{S}$
   (c) If $R > 2.0$, $\bar{S}_s = 0.7\bar{S}$
(3) Calculate K_s and N_s:
   (d) If $R \leqslant 0.8$ and $k < 3.67$, $K_s = 1.12RK$ and $N_s = 0.56RN$
   (e) If $R > 0.8$ and $k < 3.67$, $K_s = 0.9K$ and $N_s = 0.45N$
   (f) If $k \geqslant 3.67$, proceed as in (d) or (e) but multiply each result by $(K+N)/(1.12K + 0.56N)$. This ensures that $(N_s + K_s) \leqslant \bar{S}_s$
(4) Calculate $C_s = \bar{S}_s - K_s - N_s$.

If the contents of water-soluble K_2O and Na_2O are known, it is necessary only to calculate K_s and N_s from them, and R, and then to carry out stages 2 and 4. From K_s, N_s and C_s, the quantities of individual sulphate phases may be estimated as described in Section 3.5.6. In the example given, the calculation indicates that the sulphate phases account for all the SO_3, 0.07% of Na_2O, 0.28% of K_2O and 0.04% of CaO. These probably occur as aphthitalite (0.6%) and calcium langbeinite (0.1%).

4.4.4 Estimation of major phases

From the CaO content, deduct free lime and any CaO present as sulphate or $CaCO_3$. In the absence of more detailed data, assume that 67% of the insoluble residue is SiO_2 and that 33% is Al_2O_3, and deduct accordingly. This gives Al_2O_3, 5.6%; SiO_2, 21.5%; CaO, 65.3%.

Assume that the four major phases have the compositions given in Table 1.2, that of alite being adjusted in accordance with the MgO and Fe_2O_3 contents of the clinker as indicated in Note 1 to that table.

Set up four linear, simultaneous equations represented in matrix form as $\mathbf{A.X = B}$, where a_{ij} is the weight fraction of oxide component i in phase j, x_j is the weight percentage of phase j in the clinker, and b_i is the corrected weight percentage of oxide component i in the clinker:

$$0.7185x_1 + 0.635x_2 + 0.566x_3 + 0.475x_4 = b_1 = 65.3 \quad (4.7)$$
$$0.2515x_1 + 0.315x_2 + 0.037x_3 + 0.036x_4 = b_2 = 21.5 \quad (4.8)$$

116   Cement Chemistry

$$0.0100x_1 + 0.021x_2 + 0.313x_3 + 0.219x_4 = b_3 = 5.6 \quad (4.9)$$
$$0.0080x_1 + 0.009x_2 + 0.051x_3 + 0.214x_4 = b_4 = 3.2 \quad (4.10)$$

The solution to these equations, obtained by inverting the matrix **A**, is

$$x_1 = 4.5131b_1 - 8.5947b_2 - 6.8987b_3 - 1.5116b_4 = 66.5 \quad (4.11)$$
$$x_2 = -3.6133b_1 + 10.0842b_2 + 5.1743b_3 + 1.0286b_4 = 13.1 \quad (4.12)$$
$$x_3 = 0.1320b_1 - 0.3961b_2 + 3.6483b_3 - 3.9598b_4 = 7.9 \quad (4.13)$$
$$x_4 = -0.0482b_1 - 0.0084b_2 - 0.8292b_3 + 5.6298b_4 = 10.0 \quad (4.14)$$

The calculated percentages of alite, belite, aluminate and ferrite are given by x_1–x_4, respectively. The Bogue calculation gives 60.3%, 16.1%, 9.4% and 9.7%, assuming that the SiO_2 and Al_2O_3 contents have been corrected for insoluble residue.

Any standard method of matrix inversion, such as the Gauss–Jordan method (N13), may be used to solve the equations. The coefficients in equations 4.11–4.14 may be used without serious error for most ordinary Portland cement clinkers in which the alite composition is not too different from that assumed here. As a byproduct of the calculation described in this section, and using the full compositions of the phases given in Table 1.2, one may calculate a mass balance table (Table 4.3) showing the distributions of all the oxide components among the phases.

4.4.5 Limitations and modifications of the calculation of phase composition

The procedure described above should give satisfactory results for most modern ordinary Portland cement clinkers, but the following points should be noted:

(1) If the aluminate phase is orthorhombic, the appropriate composition (Table 1.2) should be substituted in the matrix **A**. If both cubic and orthorhombic forms are present, intermediate values may be used.
(2) If the result for aluminate is negative, this phase is absent and its composition should be replaced in the matrix **A** by that of the low-aluminium ferrite phase given in Table 1.2. From the compositions and proportions found for the two ferrite components, the content and composition of the ferrite may be estimated. Similarly, if the result for ferrite is negative, the ferrite composition is replaced by that of low-iron aluminate. These conditions will occur at ARs below about 1.0 and above about 4.0, respectively. The data for the low-aluminium

Table 4.3 Distribution of oxide components among phases in a typical Portland cement clinker, calculated from the bulk chemical analysis

	Na$_2$O	MgO	Al$_2$O$_3$	SiO$_2$	P$_2$O$_5$	SO$_3$	K$_2$O	CaO	TiO$_2$	Mn$_2$O$_3$	Fe$_2$O$_3$	Total
Alite	0.1	0.5	0.7	16.8	0.1	0.0	0.1	48.1	0.0	0.0	0.5	66.9
Belite	0.0	0.1	0.3	4.2	0.0	0.0	0.1	8.4	0.0	0.0	0.1	13.2
Aluminate	0.1	0.1	2.5	0.3	0.0	0.0	0.1	4.5	0.0	0.0	0.4	7.9
Ferrite	0.0	0.3	2.2	0.4	0.0	0.0	0.0	4.8	0.2	0.1	2.2	10.1
Free lime	0.0	0.0	0.0	0.0	0.0	0.0	0.0	0.6	0.0	0.0	0.0	0.6
Periclase	0.0	0.2	0.0	0.0	0.0	0.0	0.0	0.0	0.0	0.0	0.0	0.2
K$_3$N$\bar{\text{S}}_4$	0.1	0.0	0.0	0.0	0.0	0.3	0.3	0.0	0.0	0.0	0.0	0.6
C$_2$K$\bar{\text{S}}_3$	0.0	0.0	0.0	0.0	0.0	0.1	0.0	0.0	0.0	0.0	0.0	0.1
Insol. res.	0.0	0.0	0.1	0.2	0.1	0.0	0.0	0.0	0.1	0.0	0.0	0.3
Other	−0.1	0.0	−0.1	0.0	0.0	0.0	−0.1	0.0	0.0	0.0	0.0	0.0
Total	0.2	1.2	5.7	21.9	0.2	0.4	0.5	66.4	0.3	0.1	3.2	100.0

Results given as percentages on the ignited weight. Periclase calculated by difference. 'Other' comprises errors arising from rounding, incorrect assumptions regarding compositions of phases, and contributions from any phases not included in the calculation.

118 *Cement Chemistry*

and low-iron aluminate in Table 1.2 are tentative, and no high accuracy may be expected.

(3) The results will be less accurate for slowly cooled clinkers, as the compositions of the ferrite and possibly also the aluminate phases may differ significantly from those assumed here. At present, there are not enough data to deal with this problem. The method is not applicable without major modification to clinkers made under reducing conditions. It is doubtful whether the procedure is applicable to white cements, both for this reason and because they may contain glass.

The calculation of phase composition from bulk analysis for cements presents no special difficulties if the content and composition of the gypsum and any other admixture are known, as the analysis can then readily be corrected to give the composition of the clinker. If they are not known, an assumption must be made as to the amount of SO_3 in the clinker, so that the contribution of the gypsum to the CaO content of the cement may be estimated. Further assumptions and corrections may have to be made to allow for impurities in the gypsum or other additives (e.g. calcite).

4.4.6 Comparison of results of different methods

From the results of a study in which six clinkers were examined in a number of different laboratories, Aldridge (A6) concluded that light microscopy gave accurate results for the quantitative determination of phases in clinkers, assuming an experienced operator and adequate sampling. Results obtained using QXDA varied widely between laboratories, and it was concluded that, as carried out in most of them, the method was of low accuracy. One laboratory, however, obtained results that agreed well with those given by microscopy, and which were considered to be accurate. In general, the contents of alite found by microscopy were higher than those given by the Bogue calculation.

The results of another cooperative study, in which microscopy and extraction methods were used, showed that the Bogue calculation always gives low results for alite and generally gives high ones for belite when compared with microscopy (C18). For the aluminate and ferrite phases, agreement was reasonably good. The total contents of silicates, determined by chemical extraction, agreed reasonably well with those found by microscopy.

Several early studies using QXDA gave results in fair agreement with those of the Bogue calculation, but in the light of the above and other recent evidence are of doubtful validity. Odler *et al.* (O9) found that alite contents found either by microscopy or by QXDA were always greater than those given by the Bogue calculation. Gutteridge (G30) reached the same conclusion for QXDA, and also noted that the total amounts of the crystalline phases did not vary significantly from 100%. Kristmann (K12) concluded from a study of 39 clinkers that alite and belite were determined more satisfactorily by microscopy than by QXDA, but that aluminate and ferrite, due to their close admixture, were better determined by QXDA or by determining the total content of interstitial material by microscopy and the relative amounts of its constituent phases by QXDA of a residue from SAM extraction (Section 4.3.3). His results indicated that the values given by the Bogue calculation are low for alite and aluminate, high for belite, and reasonably satisfactory for ferrite. The mean discrepancy for alite was 8%.

The method for calculating phase composition from bulk analysis described in Sections 4.4.2–4.4.4 was tested using Kristmann's data (T13). Agreement was substantially better than with the Bogue calculation for alite, belite and aluminate, and about as good for ferrite. For alite and belite, there was no significant bias in either direction. For aluminate, there was a possible tendency to give high results; for ferrite, it was not possible to tell whether there was bias, due to inadequacies in the experimental data. Good agreement was also obtained in tests on a wide range of cements, using experimental data obtained by Gutteridge (G34) using QXDA.

Taken as a whole, the evidence indicates that all four major phases can be satisfactorily determined by QXDA if an adequate experimental procedure is employed (e.g. Ref. G30) or by calculation. Alite, belite and the total content of interstitial material can also be determined by microscopy. With any of these methods, assuming the best available procedures to be used, the absolute accuracy is probably 2–5% for alite and belite, and 1–2% for aluminate and ferrite.

4.5 Reactivities of clinker phases

4.5.1 Effect of major compositional variation

The ability of a substance to act as a hydraulic cement depends on two groups of factors. First, it must react with water to a sufficient extent and at a sufficient rate; γ-C_2S, which is virtually inert at ordinary temperatures, has no cementing ability. Second, assuming that an appropriate ratio of water to cement is used, the reaction must yield solid products of very low solubility and with a microstructure that gives rise to the requisite mechanical strength, volume stability, and other necessary properties. C_3A reacts rapidly and completely with water, but the products that are formed when no other substances are present do not meet these criteria, and its ability to act as a hydraulic cement is very poor. C_3S, in contrast, satisfies both sets of conditions, and is a good hydraulic cement.

In this section, we consider only the factors that control reactivity. Any understanding of them must be based on a knowledge of the mechanisms of the reactions with water. It is highly probably that these are in all cases ones in which an essential step is transfer of protons to the solid form from the water, which is thus acting as a Brønsted acid (D9,B41). As an initial approximation, one may consider the problem purely in terms of the idealized crystal structures. The reactivities of the oxygen atoms towards attack by protons depend on their basicities, i.e. the magnitudes of the negative charges localized on them. Any structural feature that draws electrons away from the oxygen atoms renders them less reactive, the basicity thus depending on the electronegativities of the atoms with which they are associated. In this respect, the nearest neighbours are the most important, but the effects of atoms further away are far from negligible. This hypothesis leads to the following predictions, all of which agree with the facts:

(1) Oxygen atoms attached only to atoms of a single element will be less reactive as the electronegativity of that element increases. Thus, C_3S and lime both contain oxygen atoms linked only to calcium, and are more reactive than β-C_2S, in which all the oxygen atoms are also linked to silicon, while α-Al_2O_3, in which all are linked to aluminium, is inert. However, the reactivity of lime decreases if it is strongly burned; this is due to effects discussed later.

(2) Oxygen atoms forming parts of silicate, aluminate or other anionic groups will be less reactive as the degree of condensation with other such groups increases. β-C_2S, in which the tetrahedra do not share oxygen atoms, is more reactive than any of the C_3S_2 or CS polymorphs, in which they do.

(3) Oxygen atoms forming parts of anionic groups similar as regards degree of condensation will be less reactive as the central atoms in the groups become more electronegative. In contrast to the C_3S_2 and CS polymorphs, CA, $C_{12}A_7$ and C_3A, in which the AlO_4 tetrahedra share corners, are all highly reactive. On the other hand, C_3P, which contains isolated tetrahedra, is unreactive.

Other hypotheses based on considerations of crystal structure have been proposed, and in varying degrees may account for second-order effects. Thus, Jeffery (J1) suggested that irregular coordination of calcium was responsible for the high reactivity of C_3S as compared with γ-C_2S; however, lime reacts with water, but the coordination of its calcium atoms is highly symmetrical. Jost and Ziemer (J8), who noted the weaknesses in this and other earlier hypotheses, considered that high reactivity was associated with the presence of face-sharing CaO polyhedra, but lime contains no such groupings, yet can be highly reactive.

4.5.2 Effects of ionic substitutions, defects and variation in polymorph

The considerations discussed in the preceding section suffice only to explain the effects on reactivity of major variations in composition. They do not, in general, explain the effects of variations in polymorph or introduction of foreign ions. These effects are often complex, and while in some cases the results appear clearcut, there are many others in which they present a confusing and sometimes apparently contradictory picture. The probable reason is that many variables are involved which are difficult or impossible to control independently. These include, besides chemical composition and polymorphic change: concentrations and types of defects; particle size distribution; textural features such as crystallite size and morphology, mechanical stress, presence of microcracks and the mono- or polycrystalline nature of the grains; and influences of reaction products, whether solid or in solution. In a cement as opposed to a pure phase, a foreign ion may not only substitute in a particular phase, but also influence the microstructure of the material in ways unconnected with that substitution, e.g. by altering the physical properties of the high-temperature liquid present during its formation. For both microstructural and chemical reasons, the reactivity of a given phase in a cement may not be the same as that observed when it is alone.

Defects play a particularly important role. Sakurai et al. (S33) noted that C_3S, alite, C_2F and C_4AF could all take up Cr_2O_3 and that this greatly accelerated reaction at early ages. Electron microscopic studies showed that reaction began at grain boundaries and at points of emergence of screw dislocations, the concentration of which was greatly increased by Cr_2O_3 substitution. It was also shown that the substituted materials were semiconductors, and a mechanism of attack based on electron transfer processes was suggested. Fierens and co-workers (F9,F10) also found that the influence of foreign ions on the reactivity of C_3S was due to the presence of defects, which could be studied using thermoluminescence, and which could also be introduced by suitable heat treatment.

The complex relations between content of foreign ions, polymorphism, defects and reactivity are well illustrated by the results of Boikova and co-workers (B42,B4) on ZnO-substituted C_3S. With increasing ZnO content, triclinic, monoclinic and rhombohedral polymorphs were successively stabilized. Curves in which the extents of reaction after given times were plotted against ZnO content showed maxima corresponding approximately to the T_2–M_1 and M_2–R transitions, with intermediate minima. As in the work mentioned above, it was shown that these compositions were also ones in which the numbers of defects of a particular type were at a maximum. Reactivity thus depends not so much on the amount of substituent or nature of the polymorph, as on the types and concentrations of defects.

Studies on the comparative reactivities of β-, α'- and α-C_2S, reviewed by Skalny and Young (S34), have given contradictory results. As these authors conclude, reactivity probably depends on specimen-specific factors other than the nature of the polymorph, and in the lower temperature polymorphs may be affected by the exsolution of impurities. This latter hypothesis receives strong support from subsequent observations that attack begins at exsolution lamellae and grain boundaries, and that these may contain phases, such as C_3A, that are much more reactive than the C_2S (C1). Without these lamellae, β-C_2S might be much less reactive than it is.

There is wide agreement that substitution of alkali metal ions retards the early reaction of the aluminate phase, which is thus less for the orthorhombic than for the cubic polymorphs (S35,B43,R13). The effect has been attributed to structural differences, but the early reaction of pure C_3A is also retarded by adding NaOH to the solution, and the OH^- ion concentration in the solution may be the determining factor (S35). The reaction of C_3A is also retarded by iron substitution and by close admixture with ferrite phase; formation of a surface layer of reaction products may be a determining factor, at least in later stages of reaction, and the retarding effect of such a layer may be greater if it contains Fe^{3+} (B44).

The reactivities of ferrite phases in the pure $Ca_2(Al_xFe_{1-x})_2O_5$ series have been widely found to increase with Al/Fe ratio, and material of composition similar to that in clinkers is more reactive than either C_4AF or C_6A_2F (e.g. Refs. D10,B43 and B44). Anomalies in the relation between rate and Al/Fe ratio found by some workers can possibly be attributed to zoning or other variations in crystallinity. As with the aluminate phase, the decrease in reactivity with iron content is possibly explainable by differences in the properties of a retarding layer of product. The relatively high reactivity of the clinker material may be related to its semiconducting properties (M18) and to disorder arising from the complex ionic substitutions (B44).

5

Hydration of the calcium silicate phases

5.1 Introduction

5.1.1 Definitions and general points

In cement chemistry, the term 'hydration' denotes the totality of the changes that occur when an anhydrous cement, or one of its constituent phases, is mixed with water. The chemical reactions taking place are generally more complex than simple conversions of anhydrous compounds into the corresponding hydrates. A mixture of cement and water in such proportions that setting and hardening occur is called a paste, the meaning of this term being extended to include the hardened material. The water/cement (w/c) or water/solid (w/s) ratio refers to proportions by weight; for a paste, it is typically 0.3–0.6. Setting is stiffening without significant development of compressive strength, and typically occurs within a few hours. Hardening is significant development of compressive strength, and is normally a slower process. Curing means storage under conditions such that hydration occurs; conditions commonly employed in laboratory studies include storage in moist air initially and in water after the first 24 h, storage in air of 100% relative humidity and, less favourable for reaction, storage in a sealed container.

Because Portland cement is a relatively complex mixture, many studies having the aim of elucidating its hydration chemistry have been made on its constituent phases. For a given particle size distribution and w/s ratio, tricalcium silicate or alite sets and hardens in a manner similar to that of a typical Portland cement. Using XRD and other methods, it may be shown that about 70% of the C_3S typically reacts in 28 days and virtually all in 1 year, and that the products are calcium hydroxide (CH) and a nearly amorphous calcium silicate hydrate having the properties of a rigid gel. β-C_2S behaves similarly, but much less CH is formed and reaction is slower, about 30% typically reacting in 28 days and 90% in 1 year. In both cases, reaction rates depend on particle size distribution and other factors. Development of compressive strength runs roughly parallel to the course of the chemical reactions, and the strengths at 1 year are comparable to those of Portland cements of the same w/s ratio and cured under the same conditions.

The calcium silicate hydrate formed on paste hydration of C_3S or β-C_2S is a particular variety of C–S–H, which is a generic name for any amorphous or poorly crystalline calcium silicate hydrate. The dashes indicate that no particular composition is implied, and are necessary because CSH in cement chemical nomenclature denotes material specifically of composition $CaO \cdot SiO_2 \cdot H_2O$. The term 'C–S–H gel' is sometimes used to distinguish the material formed in cement, C_3S or $\beta\text{-}C_2S$ pastes from other varieties of C–S–H.

This chapter deals primarily with the reactions and products of hydration of C_3S and $\beta\text{-}C_2S$ in pastes at ordinary temperatures. Except where stated, a temperature of 15–25°C is assumed. Background information has been obtained from studies of C_3S or $\beta\text{-}C_2S$ hydration at higher w/s ratios, i.e. in aqueous suspensions, or on C–S–H prepared in other ways, and the results of such studies are also considered.

5.1.2 Experimental considerations

In all laboratory studies on the hydration of cements or of their constituent phases, or of C–S–H or other hydrated phases formed by other methods, rigorous exclusion of atmospheric CO_2 is essential. Contamination by atmospheric CO_2 ('carbonation') is apt to be much more serious with small laboratory samples than with practical concrete mixes of smaller surface to volume ratio. All operations, such as mixing, curing or drying of pastes, or preparation, filtration or drying of solid phases in suspensions, must be carried out in a CO_2-free atmosphere, either in a dry box or in apparatus otherwise designed to isolate the materials from the atmosphere. Wet or moist materials are especially susceptible to attack, but dry ones are not immune. Water must be deionized and freshly boiled. Prolonged contact with glass must also be avoided; wet curing may be carried out in the presence of a small excess of water in a sealed plastic container. Despite all precautions, hydrated materials will inevitably contain some CO_2, analysis for which will often be required. Jones's (J9) method, based on decomposition with dilute acid and absorption of the CO_2 in $Ba(OH)_2$ solution, which is then back-titrated with HCl, is simple and accurate. TG has often been used to determine CO_2, but is not satisfactory because the loss at 700–900°C is not wholly due to CO_2. Absence of calcite peaks from an XRD pattern is not a proof that CO_2 is absent, as CO_2 can occur in other forms, e.g. as vaterite ($\mu\text{-}CaCO_3$), in calcium aluminate hydrate phases and possibly within the C–S–H structure.

Especially with relatively young pastes, the problem often arises of removing excess water after a specified time of curing to stop the hydration reactions and to render the material less susceptible to carbonation. For this purpose, soaking or grinding in polar organic liquids, such as methanol, propan-2-ol or acetone, has often been employed, but should be used only in situations where it will not invalidate the results of subsequent tests. Organic substances can be completely removed from C_3S pastes only by drastic procedures that seriously alter the material. Their presence is especially undesirable if any form of thermal analysis (e.g. TG or DTA) is to be carried out, since on heating they react with C-S-H to give CO_2 (D11,T14). Acetone has especially complex effects, since in the alkaline medium it rapidly undergoes aldol condensations at room temperature, giving mesityl oxide, phorone and other relatively involatile products (T14). Some workers consider that primary aliphatic alcohols alter the pore structures of C_3S or cement pastes or of synthetic C-S-H preparations (Section 8.3.2) and it has also been claimed that methanol reacts with CH at ordinary temperatures, giving calcium methoxide (B45).

In choosing a procedure to remove excess water, it is necessary to consider the likely effects on the tests subsequently to be performed. Retention of organic substances is not the only possible complication; unduly drastic drying procedures, e.g. heating at 105°C or equilibration to constant weight in atmospheres of low relative humidity, such as 'D-drying' (Section 5.2.2), partially dehydrate the C-S-H and, with cement pastes, can also decompose crystalline hydrated calcium aluminate or aluminate sulphate phases. For some procedures, such as XRD, the presence of organic material may not matter, but even here, the possibility of decomposing or altering hydrated aluminate phases must be considered. For electron microscopy, including the preparation of polished sections for SEM, freeze drying has been found satisfactory (J10). For many purposes, including thermal analysis above about 150°C, pumping with a rotary pump for 1 h is suitable. The drying of samples for studies on pore structure is considered in Section 8.3.2.

All analytical data on hydrated cements or cement constituents should normally be referred to the ignited weight. The original weight includes an arbitrary amount of water, and results referred to it, or to the weight after some arbitrarily chosen drying procedure (e.g. to constant weight at 105°C), are often impossible to interpret in detail.

5.1.3 Calcium hydroxide

CH has a layer structure (Fig. 5.1) (B46,P18). The calcium atoms are octahedrally, and the oxygen atoms tetrahedrally, coordinated. The interlayer forces are weak, with negligible hydrogen bonding, thus giving good (0001) cleavage. The unit cell is hexagonal, with $a = 0.3593$ nm, $c = 0.4909$ nm, space group $P\bar{3}m1$, $Z = 1$, $D_x = 2242$ kg m^{-3}. The optical properties are: uniaxial $-$, $\omega = 1.573$, $\varepsilon = 1.545$ (S5). Calcium hydroxide occurs as a natural mineral, known as portlandite.

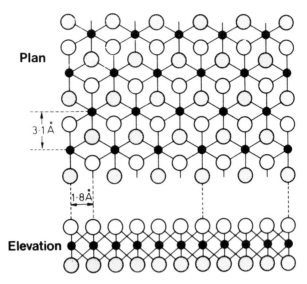

Fig. 5.1 Structure of a single layer of $Ca(OH)_2$. Small, full circles represent calcium atoms, large open or shaded circles oxygen atoms; a hydrogen atom, not shown, completes the tetrahedral coordination of each oxygen atom. Taylor (T15).

Under ideal conditions of crystallization, CH forms hexagonal plates. Admixtures greatly affect the morphology, and especially the $c:a$ aspect ratio (B47). Euhedral crystals are observed in young pastes and in pores of older pastes, but as hydration proceeds, the main deposits of CH become massive and of indeterminate shape, though the good cleavage persists. Several studies on the CH of cement pastes by light microscopy or X-ray microanalysis have indicated the presence of small amounts of SiO_2 and other components, but it is uncertain whether these occur as ionic substituents in the CH structure or in inclusions (B48). There does not appear to be any convincing experimental evidence for claims that some of the CH in pastes is amorphous. A TEM study of ion-thinned sections of a C_3S paste of normal w/s ratio showed the CH to occur as large, imperfect crystals; no finely dispersed, microcrystalline material was observed, though some crystals were only tens of nanometres thick (G35,G36); cryptocrystalline CH was found only in pastes of abnormally low w/s ratio (G37). A reported occurrence of amorphous CH in a Renaissance lime plaster (M36) was disproved (N14), but a natural occurrence of amorphous CH has been described (N15).

It has been reported that the 0001 peak of CH is anomalously strong relative to the $10\bar{1}1$ peak in XRD powder patterns of calcium silicate or cement pastes obtained using a diffractometer, to an extent that increases with the w/s ratio and decreases with age (G38,Z8). Grudemo (G38) attributed the effect to overlap between 0001 and a peak from C–S–H. Although care was taken to minimize preferred orientation in these studies, judgement on the results must be reserved. Because of the good cleavage, preferred orientation occurs very readily, and if a diffractometer is used, greatly increases the relative intensity of 0001. Such orientation could be produced in various ways, e.g. on sawing a hardened paste to prepare a flat surface. Stacking faults can also affect the pattern (S36).

The dissolution of CH in water is exothermic ($\Delta H = -13.8$ kJ mol^{-1} at 25°C) (G39), and the solubility therefore decreases with temperature (Table 5.1). The data given are for large crystals; higher apparent solubilities, up to at least 1.5 g CaO l^{-1} at 25°C, readily occur with finely crystalline material.

Table 5.1 Solubility of calcium hydroxide (S; g CaO l^{-1}) (Bassett, B49)

T°C	S	T°C	S	T°C	S	T°C	S	T°C	S	T°C	S
0	1.30	10	1.25	20	1.17^a	30	1.09	50	0.92	80	0.66^a
5	1.28	15	1.21^a	25	1.13	40	1.00	60	0.83^a	99	0.52

^{a} Interpolated values.

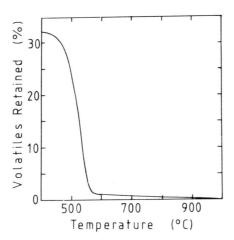

Fig. 5.2 Thermogravimetric curve of Ca(OH)$_2$ (0.2% CO$_2$); 50 mg sample heated at 10 deg C min^{-1} in dry, CO$_2$-free N$_2$ flowing at 15 ml min^{-1}. Taylor (T5).

Halstead and Moore (H25) reported thermal decomposition equilibria. The decomposition pressure reaches 101 kPa (1 atm) at 512°C and the mean enthalpy of decomposition at 300–510°C is 102.9 kJ mol^{-1}. When heated in a vacuum or in a stream of dry N_2, CH begins to decompose at an easily detectable rate at about 370°C. Under typical conditions for TG, decomposition is 98% complete at the end of the step at 580°C (Fig. 5.2). An earlier conclusion that only about 94% has decomposed at this point (T16) is incorrect. The finely divided CH available as a laboratory or analytical reagent normally contains several per cent of $CaCO_3$, and the curve in Fig. 5.2 is for coarsely crushed large crystals containing 0.2% CO_2.

5.2 Compositions of C_3S and β-C_2S pastes

5.2.1 Calcium hydroxide content, thermal analysis and indirect determination of the Ca/Si ratio of the C–S–H

The Ca/Si ratio of the C–S–H in a fully reacted C_3S or β-C_2S paste may be calculated if the contents of CH and CO_2 are known, though an assumption must be made as to whether the CO_2 occurs in the C–S–H or as $CaCO_3$. If the content of unreacted C_3S or β-C_2S is also known, the method can be extended to incompletely reacted pastes, though small errors in the contents of CH or unreacted starting material greatly affect the result at low degrees of reaction. Quantitative XRD is probably the only method presently available for determining unreacted C_3S or β-C_2S; the precision is probably not better than $\pm 3\%$ (O10). Methods used to determine CH have included TG, DTG, semi-isothermal DTG, thermal evolved gas analysis, DTA, differential scanning calorimetry (DSC), QXDA, IR spectroscopy, image analysis of backscattered electron images and extraction methods using either aqueous media or organic solvents similar to those used to determine free lime in anhydrous cements.

Some workers have found significant differences between the results of thermal methods and QXDA, the latter usually giving lower results for the CH content (M37,O10,B50,R14,D12), but others have found no differences outside experimental error (L19,T17). There is broad agreement that extraction methods give higher results than either thermal methods or QXDA (O10,M37,B50,R14). It has been suggested that QXDA gives low results due to the presence of amorphous CH (R14). Thermal methods have also been considered to give low results for the same reason (O10), or because of CH adsorbed on the C–S–H (B50) or present as interlayer material (S37). Sources of error in CH determination by TG have been discussed (T16). As noted earlier, major errors can arise from the use of thermal methods on material that has been in contact with organic liquids (T14). Preliminary results from image analysis did not correlate well with those from TG (S38).

In the writer's view, adsorbed or interlayer material is not CH, but part of the C–S–H. The absence of evidence for amorphous CH, or, except at very low w/s ratios, of cryptocrystalline CH, was noted earlier (Section 5.1.3). The higher results given by extraction methods are probably attributable to partial decomposition of the C–S–H. Of all the methods, TG and DTG are probably the least open to criticism, though precision and accuracy are limited by the difficulty of distinguishing sharply between the effects due to CH and the other phases present. DTA is somewhat suspect, because the peak area for CH is reported to depend markedly on particle size (W17).

Fig. 5.3 shows the TG and DTA curves of a fully reacted C_3S paste, obtained under the same conditions as the TG curve of CH in Fig. 5.2. The paste had been prepared at w/s = 0.45 and stored wet, in a sealed container, for 25 years at 25°C; it contained 0.8% of CO_2, referred to the ignited weight. The height of the CH step (mean from seven determinations and estimated as shown in the inset) was 8.9%, referred to the ignited weight. Assuming that this step represents the decomposition of 98% of the CH and that the CO_2 occurs as $CaCO_3$, there is 1.15 mol CH and 0.04 mol $CaCO_3$ per mole C_3S, and the Ca/Si ratio of the C–S–H is 1.81.

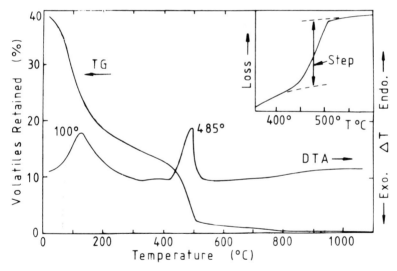

Fig. 5.3 Thermogravimetric and differential thermal analysis curves for a fully reacted C_3S paste; experimental conditions as for Fig. 5.2. After Taylor and Turner (T14).

This result is similar to those of several other recent investigations, which have indicated Ca/Si ratios of 1.99 using QXDA or 1.60 using an extraction method (O10), 1.7 ± 0.1 (L19) or 1.8 (implicitly) (P19) for mature pastes of C_3S or alite and of around 1.8 (K13,F11) or 2.0 (S18) for ones of β-C_2S.

There is evidence of a lower Ca/Si ratio during the first few hours of reaction (O10), but with that reservation the more recent work (O10,L19,P19) does not support the conclusion of some earlier workers that the ratio changes markedly during the course of the reaction.

5.2.2 Water content of the C–S–H

Evidence considered in Sections 5.3.1 and 8.3 indicates that the C–S–H gel of calcium silicate or cement pastes has a layer structure, and that, together with a pore solution, it forms a rigid gel in which the pores range in size from macroscopic to enlarged interlayer spaces of nanometre dimensions. One can therefore define a water content only in relation to a specified drying condition. Three such conditions will be considered.

The most highly hydrated state is that existing in a saturated paste, i.e. one in which the pores are totally filled with water. According to the theory of Powers and Brownyard (P20) (Section 8.2.1), the C–S–H in a cement paste can form only if sufficient space is available to permit it to be accompanied by a defined proportion of pore space. There is thus a certain w/c ratio below which complete hydration is impossible, and for which a mature paste consists entirely of hydration product, including this essential pore space. The total water content of such a paste in the saturated condition is typically 42–44%, referred to the ignited weight. For nearly saturated C_3S pastes, Young and Hansen (Y5) concluded that the composition of the C–S–H was approximately $1.7CaO \cdot SiO_2 \cdot 4H_2O$. Taking into account the water present in the CH, this corresponds to a w/c ratio of 0.42, which is similar to the values found for cement.

Chemically bound water is most reasonably defined as including that present in interlayer spaces, or more firmly bound, but not that present in pores larger than interlayer spaces. As will be seen in Chapter 8, the distinction between interlayer space and micropores is not sharp; water adsorbed on surfaces of pores further blurs the definition. From the experimental standpoint, the determination is complicated by the fact that the amount of water retained at a given RH depends on the previous drying history of the sample and on the rate at which water is removed. An approximate estimate is obtained by equilibrating a sample, not previously dried below saturation, with an atmosphere of 11% RH (F12,F13,F14). Saturated aqueous $LiCl \cdot H_2O$ gives the required RH (partial pressure of water vapour = 2.7 torr at 25°C). To achieve apparent equilibrium in a reasonable time (several days), the sample must be crushed and the system evacuated; the salt solution should be stirred, at least intermittently. Young and Hansen (Y5) found the composition of the C–S–H in C_3S paste thus equilibrated to be $1.7CaO \cdot SiO_2 \cdot 2.1H_2O$. For C_3S hydrated in suspension, which probably gives a similar product, Feldman and Ramachandran (F14) found the C–S–H composition to be $1.64CaO \cdot SiO_2 \cdot 1.96H_2O$.

As noted earlier, treatment of C_3S pastes with methanol leaves some of the latter strongly sorbed. The methanol is only removable on heating at temperatures at which it reacts with the C–S–H, causing carbonation. However, the total weight of volatiles retained in a C_3S paste that has been soaked in methanol and then pumped for about 1 h with a rotary pump is near to that obtained on equilibration with saturated $LiCl \cdot H_2O$ (T14). Methanol treatment can thus be used as a rapid, though approximate, method of determining chemically bound water. Other organic liquids could possibly be used in a similar way.

In a procedure known as 'D-drying', the sample is equilibrated with ice at $-79°C$ by continuous evacuation with a rotary pump through a trap cooled in a mixture of solid CO_2 and ethanol (C20). The partial pressure of water vapour is 5×10^{-4} torr. Heating to constant weight at 105°C in an atmosphere of uncontrolled humidity, but free from CO_2, reduces the water content to approximately the same value, although it is not a strictly defined drying condition. The corresponding temperature on a TG curve obtained in dry and CO_2-free N_2 at $10 \deg C \min^{-1}$ is about 145°C, but depends somewhat on the rate of gas flow and design of the apparatus. Water retained in pastes subjected to D-drying or equivalent procedures is known as 'non-evaporable' water. Contrary to some statements in the literature, it is not a measure of chemically bound water, as much water is lost from interlayer spaces under these conditions. It can nevertheless be used as an empirical measure of the degree of reaction. Typical values in the literature for the water contents of fully reacted C_3S pastes dried by these procedures are 20.4–22.0% (O10), corresponding to C–S–H compositions of $1.7CaO \cdot SiO_2 \cdot 1.3$–$1.5H_2O$.

Powers and Brownyard (P20) called the water lost from the C–S–H on passing from the saturated to the D-dry condition 'gel water'. It comprises part of the pore water plus an arbitrarily defined fraction of the interlayer water. Whatever procedure is used to dry or equilibrate a paste, CO_2-free conditions are essential, and the amount of water retained can be obtained accurately only if the ignition loss and CO_2 content are also determined. Calculations based on the initial w/s ratio are often unreliable due to changes in the water content during curing.

5.2.3 X-ray microanalysis and analytical electron microscopy

SEM of polished sections of mature C_3S or β-C_2S pastes shows well-defined regions of C–S–H and CH (Section 5.3.1). Taylor and Newbury (T18) reviewed X-ray microanalysis (SEM or EPMA) studies of such sections and reported new results. The recorded Ca/Si ratios for the C–S–H vary irregularly from point to point and increase with the voltage used to accelerate the electrons. As in all analyses of heterogeneous materials by this method, the choice of voltage rests on a compromise: it must be high enough to excite all the elements adequately, but low enough to produce an interaction volume sufficiently small that single phases can be analysed. The mean or median Ca/Si ratios reported for C_3S pastes range from 1.4 at 6 kV to around 2.0 at 25 kV. It is questionable whether 6 kV is sufficient to ensure satisfactory analysis for calcium, and at 25 kV the volume analysed, taking into account the effect of porosity, is several micrometres in each direction, which may be too large to ensure absence of CH or unreacted C_3S. Working at 10 kV, Taylor and Newbury obtained mean values of 1.72 and 1.78 for mature pastes of C_3S and β-C_2S, respectively; these values are near to those obtained using the indirect methods described in the previous section. The range of values was approximately 1.5–2.0 in both cases.

The totals obtained in SEM or EPMA microanalyses of C–S–H in C_3S or β-C_2S pastes, after including oxygen equivalent to the calcium and silicon, should in theory fall below 100% by an amount equal to the water content under the conditions existing during the analysis, i.e. high vacuum and electron bombardment. The evidence from thermal dehydration and intensive drying at room temperatures (Section 5.2.2) indicates that this content is unlikely to exceed about 15% of the weight of material analysed, implying an analysis total of at least 85%. The totals are sometimes as high as this, but are more often 70–75%, and sometimes even lower (T18). A similar effect is observed with the C–S–H of cement pastes (Section 7.2.5), and appears to increase with the porosity of the region analysed. It has not been fully explained, and more than one cause may operate. Organic material remaining from liquids used in preparing the polished section, or carbonaceous residues formed from it under the action of the electron beam, may remain in the pores. Superficial carbonation is a possibility (R15). Electrons may lose energy through some ill-understood effect during passage through microporous material (H26).

Analytical electron microscopy (X-ray microanalysis in transmission) would be expected to allow smaller volumes, in the order of tens of nanometres in each direction, to be analysed. The volume analysed is not significantly affected by the exciting voltage within the range (80–200 kV) typically used. Existing techniques give atom ratios but not absolute contents of individual elements. Early studies on ground and redispersed samples of C_3S and β-C_2S pastes (e.g. G40,M38) indicated variability of Ca/Si ratio on a micrometre scale, with ranges of 1.2–2.0 for C_3S pastes and 1.1–

1.6 for β-C_2S pastes and means of 1.5 and 1.4, respectively. In a study using ion-thinned sections, Groves et al. (G35) found that a high-resolution probe caused the apparent Ca/Si ratio to decrease with time. The problem was avoided by scanning a 5-μm square raster, or by using a high-voltage TEM, which gave a spot size of comparable dimensions. The mean values of Ca/Si thus obtained, for three groups of analyses, were 1.66–1.93, and the earlier values of 1.4–1.5 are thus probably incorrect. The marked local variability of Ca/Si ratio was confirmed. No significant compositional differences were observed between 'inner' and 'outer' products (Section 5.3.1). Crystals approximately 100 nm in size, identified by electron diffraction as calcite, were present.

5.3 Structural data for C_3S and β-C_2S pastes

5.3.1 Microstructure

The first effective studies of microstructural development in calcium silicate or cement pastes were made using light microscopy of thin sections (B51). They showed that the CH grows from relatively few centres in the water-filled space, where it forms isolated masses typically some tens of micrometres in size. Much more CH is observed with C_3S than with β-C_2S. At the same time, gel forms around the anhydrous grains and spreads into the remainder of the water-filled space. Early TEM studies, made on ground and redispersed material, showed the presence of acicular and rounded, platey particles, but could not show where these came from in the microstructure (G41,C21). SEM studies of fracture surfaces (C22,W18,D13,G42) confirmed these conclusions and showed also that the gel formed in situ from the larger anhydrous grains, called 'inner product', differed in texture from the 'outer product' formed in the water-filled space. It was massive and seemingly almost structureless, whereas the outer product appeared to form columns or fibres radiating from the anhydrous grains.

Diamond (D13) distinguished four morphological types of C–S–H gel visible on fracture surfaces of cement pastes; similar forms have been observed in calcium silicate pastes (Fig. 5.4). Type I, prominent at early ages, was the fibrous material, the fibres being up to about 2 μm long. Type II, described as forming honeycombs or reticular networks, was considered to be very rare in pure calcium silicate pastes, but later work shows that material resembling it is a normal early product. Type III, prominent in older pastes, was a more massive form, apparently consisting of tightly packed equant grains up to 300 nm across. Type IV, still more featureless and massive, was the inner product, and was also observed in older pastes.

Fig. 5.4 (A,B) Types I and II C–S–H, respectively (SEM of fracture surfaces; courtesy K. L. Scrivener). (C,D) SEM/STEM pair of ion beam thinned section, showing Type III C–S–H (top, right) and Type IV C–S–H (top, left and bottom, right; Jennings *et al.* (J10)). (A) is of an ordinary Portland cement paste, w/c = 0.5, aged 10 h. (B) is of a paste of an oil well cement, w/c = 0.44, with 2.4% of $CaCl_2$ on the weight of cement, aged 1 day. (C) and (D) are of a C_3S paste, w/c = 0.47, aged 330 days.

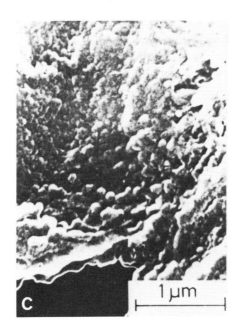

SEM of fracture surfaces has thus provided valuable information, but its utility is limited by the inherently unrepresentative nature of a fracture surface and by the effects of dehydration in the high vacuum of the instrument. Further information has come from the use of other techniques, especially high-voltage transmission electron microscopy (HVTEM) with environmental or 'wet' cells to reduce water loss (D14,J11,G43), examination of ion-thinned sections by TEM or scanning transmission electron microscopy (STEM) (D15,G35), and BEI of polished sections with detectors that minimize contrast due to topography and maximize that due to compositional differences (S28,S38,S39). Scrivener and Pratt (S40) briefly described these techniques. Other methods have included X-ray imaging of polished surfaces (T18,T19), SEM of polished surfaces impregnated and etched to reveal porosity (J12,J13) and the use of a cryo stage in SEM to avoid dehydration (B52). Each of these methods has its own limitations, but together they are beginning to yield a reasonably clear picture.

Unless a wet cell or cryo stage is used, the fine microstructure is much altered by dehydration in the instrument (J10,S41). However, localized drying occurs in any paste even before it is placed in a high vacuum, as soon as the RH falls below saturation. The water is lost initially from the wider pores, which are probably represented disproportionately on fracture surfaces. The state of the cement paste in a practical concrete may thus vary on both a macro and a micro scale between dry and saturated.

Jennings *et al.* (J10) distinguished three principal stages of product formation in C_3S pastes: an early product formed during the first 4 h, a middle product formed at 4–24 h, and a late product formed subsequently. Several workers studying the early hydration of C_3S in pastes or compacted powders placed in water have described the early product as consisting of foils, flakes or honeycombs, which appear to exfoliate from the C_3S surface, probably together with some CH (F15,M39,M40,J10,S42,G36). The honeycomb material is similar to 'Type II' C–S–H, and the foils have also been called 'Type E' C–S–H (J10). On drying, they collapse, crumple or roll up, forming fibres (J10). Broadly similar results have been obtained for C_3S hydrated in CH solution (M40) or in the presence of gypsum (M41), sodium silicate or sodium aluminate (J14). The early hydration of β-C_2S appears to follow a similar course to that of C_3S, but reaction is slower and begins preferentially at grain boundaries and exsolution lamellae (M42,G43).

Ménétrier *et al.* (M39,M40) reported that the C_3S surfaces were attacked unevenly, and that the first detectable product consisted of minute particles, some of which were probably CH. They found no continuous layer of product. Such layers have been reported, but probably only at w/s ratios above those of normal pastes (G44,L19,G36). Ings *et al.* (I7) placed single crystals of C_3S in an alite slurry with w/s = 2, and later removed them for

SEM examination. A layer of apparently structureless material built up to a thickness of 5 µm in 30 min, together with cryptocrystalline CH; this later shrank, and acicular outgrowths were formed. The product layer contained aluminium, and had thus formed in part from the surrounding alite.

The middle stage of hydration is characterized by rapid formation of C–S–H and CH. In fracture surfaces of calcium silicate pastes, Type I C–S–H is prominent, but HVTEM using wet cells (S39) showed the undried product at 1 day to consist of foils, with no evidence of fibrillar outgrowths, which were secondary forms resulting from ageing or drying. The latter will certainly occur unless a wet cell or cryo stage is used, and may do so even if it is, since ageing may cause local exhaustion of water. Fibrillar material may also form during fracture (J10).

The characteristic products of the late stage of hydration are Types III and IV C–S–H, and more CH. STEM examination of Type III material in ion-thinned section (J10) shows that it, too, consists of interlocked and interleaved thin foils (Fig. 5.4). Type IV material is almost featureless even at the 100-nm level, though a fine pore structure has been observed (D13,G35,G36).

Types I, II, III and E C–S–H thus all appear to have an underlying foil morphology, which may be modified or disguised by compaction or drying, and this could well apply also to Type IV. The pastes may therefore contain only one type of C–S–H at the nanometre level. Against this, a less structured form of C–S–H has been observed as the very first product at high w/s ratios. A similar material, described as a viscous, gelatinous product and called 'Type 0' C–S–H, has also been reported to precede the formation of the other varieties in the middle stage of hydration (J10).

BEI of pastes of C_3S mixed with C_3A shows that no spaces develop between the hydrating C_3S particles and the surrounding hydrate (S41). In this respect, C_3S differs from Portland cement (Section 7.4.2).

5.3.2 Silicate anion structure

Because C–S–H gel is nearly amorphous, X-ray diffraction has given only very general indications of its structure. The nature of the silicate anions has been determined from the kinetics of the reaction with molybdate (S43), and, in greater detail, by trimethylsilylation (TMS) and ^{29}Si NMR. In TMS methods, the sample is treated with a reagent that converts the silicate anions into the corresponding silicic acids, which then react further with replacement of SiOH by $SiOSi(CH_3)_3$. The resulting TMS derivatives can be identified and semiquantitatively determined by various procedures, of which the most widely used have been differential evaporation to isolate the higher molecular weight species, gas liquid chromatography (GLC) and gel permeation chromatography (GPC). It has to be assumed that conversion of the silicic acids into their TMS derivatives is sufficiently rapid to preclude side reactions in which they are altered by hydrolysis or condensation. In practice, these side reactions have never been totally eliminated, though the technique of Tamás *et al.* (T12) was a major advance on that originally used by Lentz (L20). Because of side reactions, and because it is rarely possible to account for more than 80–90% of the total silicon, the absolute accuracy of percentages of individual anionic species is at best a few per cent.

In chemical formulae of silicate ions in the following discussion, negative charges and any attached hydrogen atoms are omitted. Most of the investigations on C_3S pastes hydrated for more than a few hours at 15–25°C have shown that any monomer (SiO_4) present is attributable within experimental error to unreacted C_3S, and that the products formed during the first few days contain dimer (Si_2O_7), which is first supplemented, and later replaced, by larger species, collectively called polymer (L20,T12,B53,D16,M43, H27,W19,M44,M45). Fig. 5.5 gives typical results. The percentage of the silicon present as dimer passes through a maximum at about 6 months, but even after 20–30 years is still around 40%.

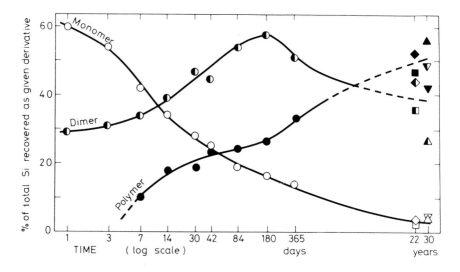

Fig. 5.5 Percentages of the total silicon in C_3S or alite pastes recovered as monomer (open symbols), dimer (half-open symbols) and polymer (closed symbols) by a trimethylsilylation method. Differently shaped symbols denote different C_3S or alite specimens. Mohan and Taylor (M43).

GLC, GPC, mass spectrometry and chemical analyses of TMS derivatives established the nature and size distribution of anion sizes within the polymer (D16,U9,M43,H27). From chemical analysis, one may calculate the mean connectivity of the SiO_4 tetrahedra, i.e. the number of other SiO_4 tetrahedra with which each one shares oxygen atoms; typical values are 1.6–2.0 (M43,M44). The anions are linear, with pentamer (Si_5O_{16}) the most important single component at all ages, the next most important being probably linear octamer (Si_8O_{25}). As the paste ages, the mean size of the anions increases, this process continuing long after all the C_3S has reacted. At ages up to 6 months, the polymer fraction consists largely of pentamer and octamer, and even at 20–30 years these species account for a substantial fraction of the total silicon, though larger anions, containing some tens of SiO_4 tetrahedra, are also present (M43).

Studies at 2–5°C have given divergent results. Two investigations showed more monomer at early ages than could be attributed to unreacted C_3S, suggesting that the initial product formed at higher temperatures also contained monomeric silicate ions, which rapidly dimerized (B53,S43), but later investigations did not confirm this (P21,H27). With increase in temperature above ambient, condensation is increasingly rapid, and by 65°C larger anions, containing ≥ 100 SiO_4 tetrahedra, are formed (B53,S43,H27). Early conclusions that admixtures, such as $CaCl_2$, affect the degree of polymerization for a given degree of reaction of the C_3S were not confirmed (H27). The behaviour of β-C_2S is similar to that of C_3S, apart from the rate of reaction (M44). A conclusion that three-dimensional cluster ions are formed (C23) is incompatible with the observed connectivities and with the NMR evidence discussed in the next paragraph, and can probably be attributed to occurrence of side reactions.

The use of ^{29}Si 'magic angle' NMR complements the trimethylsilylation method in two ways. First, it is a purely physical technique, in which there is no possibility of altering the structure through side reactions. Except with very young samples, no preliminary drying is needed. Second, it gives slightly different data, relating not directly to the fractions of the silicon present in different anionic species, but to the fractions present in different environments. The terms $Q^0 \ldots Q^4$ refer to the connectivities, Q^0 thus denoting isolated tetrahedra, Q^1 end group tetrahedra (and thus including silicon present as Si_2O_7), Q^2 middle groups, and so on. For C_3S pastes, the content of Q^0 begins to decrease after a few hours, with formation of Q^1, later accompanied by Q^2; no Q^3 or Q^4 is detected (L21,C24,R16). This result is compatible with the formation of dimer, later accompanied or replaced by either single chains or rings, but not with that of double chains, clusters or other more complex species. Clayden *et al.* (C24) found high correlations between the relative intensities of the Q^0 and Q^1 peaks in the

NMR spectrum, the amount of CH given by TG and the cumulative heat evolution.

The TMS method is insufficiently precise to exclude the possibility that the hydration products include a small fraction of monomer. Cross-polarization NMR can distinguish between silicon atoms in tetrahedra that carry one or more hydrogen atoms and ones that do not, and thus between monomer present in the anhydrous compounds and that present in a hydration product, which is almost certain to carry hydrogen. Using this technique, Young (Y6) found only small amounts of hydrated monomer in C_3S pastes cured at 2–65°C, which he attributed to water adsorbed on the C_3S surface before mixing. In contrast to this observation, Rodger *et al.* (R16) found the product formed during the first few hours in a C_3S paste to contain only monomeric anions, dimer only beginning to form later. The monomer did not disappear when the larger species were formed; it accounted for 2% of the silicon at 24 h and was still detectable at 6 months. The initial formation of monomer agrees qualitatively with some of the results at low temperatures mentioned earlier (B53,S43).

Early conclusions from TMS studies that the average size of the silicate anions in pastes increases on drying (M46,B54) were not supported by later work (P21). An NMR study on the effect of drying conditions on the anion structures of C–S–H prepared in suspension (M47) (Section 5.4.4) has important implications for pastes.

5.3.3 X-ray diffraction pattern, densities and other data

Fig. 5.6 shows the XRD powder pattern of a 23-year-old paste of β-C_2S. Patterns of fully reacted C_3S pastes are similar, except that the CH peaks are relatively more intense. The only effects definitely attributable to C–S–H are the diffuse peak at 0.27–0.31 nm and the somewhat sharper one at 0.182 nm. Attempts to obtain selected area electron diffraction patterns from the C–S–H of calcium silicate or cement pastes have usually failed, but, occasionally, particles present in ground and redispersed samples have yielded poorly defined patterns (G41,C25) (Section 5.4.6). A later study by this method (M48) has been severely, and in the writer's opinion justifiably, criticized (G45).

The density of C–S–H gel depends on the water content. For strongly dried samples, it also depends on how the solid volume is defined, causing observed values to vary with the fluid used in their determination. Widely differing values can thus be equally meaningful, but for a given application the appropriate one must be used. Since the C–S–H in C_3S or β-C_2S pastes is mixed with CH, it is necessary either to correct for this phase ($D =$

2242 kg m^{-3}) or to remove it. A procedure for the latter has been described (R14).

At about 11% RH on first drying, the interlayer spaces are full of water and larger spaces are empty; H_2O/SiO_2 is approximately 2.0 (Section 5.2.2). Using helium pyknometry, Feldman (F16) obtained values of 2350–2360 kg m^{-3} for fully reacted C_3S pastes; correction for CH gave 2430–2450 kg m^{-3} for the C–S–H. Determinations on C–S–H samples made by hydrating C_3S in suspension showed that substitution of methanol or aqueous CH for helium as the fluid had no significant effect on the result. For a fully hydrated C_3S paste with w/s = 0.4 from which the CH had been removed, Hansen (H28) found 2180 kg m^{-3}. For all these values, the empty pores are excluded from the solid volume.

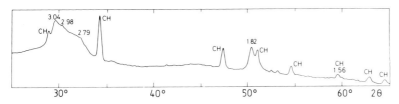

Fig. 5.6 XRD powder diffraction pattern (microdensitometer trace of Guinier film; CuK$_{\alpha 1}$ radiation: spacings in Å) of a fully reacted β-C_2S paste. Mohan and Taylor (M38).

The C–S–H present in a saturated paste is assumed to include that part of the porosity, and its content of water, without which it cannot be formed. The H_2O/SiO_2 ratio is approximately 4.0. Assuming the water in excess of that present at 11% RH to have a density of 1000 kg m^{-3}, Hansen (H28) calculated the density of the C–S–H to be 1850–1900 kg m^{-3}. This value is near those of 1900–2100 kg m^{-3} for cement pastes under saturated conditions (P20), if allowance is made for the CH and other hydrated phases which these contain.

Helium pyknometry of D-dried samples has given values of 2320–2350 kg m^{-3} for C_3S pastes (F16) and of 2440 kg m^{-3} for cement pastes (P20); correction for CH and other phases suggests values of 2400–2500 kg m^{-3} for the C–S–H. For a C_3S paste with w/s 0.4 from which the CH had been removed, Hansen (H28) found 2390 kg m^{-3}. With D-dried samples, methanol gives similar results to helium, but water or CH solution give much higher values, Brunauer and Greenberg (B55) obtaining values around 2860 kg m^{-3}, after correction for CH, for the C–S–H of C_3S and β-C_2S pastes. Determinations using water on D-dried cement pastes, based on the difference between the weights of the D-dried and saturated materials, typically give values, uncorrected for CH, of 2500–2600 kg m^{-3}. Relis and

Soroka (R17) found these to increase with the degree of hydration, an upper limit of 2700 kg m^{-3} being found for a specific cement.

Feldman (F16) explained the difference between the values for D-dried samples obtained with aqueous and other fluids. When the material is dried, the layers partially collapse, and some empty spaces remain between them. Water penetrates into this space, and the density obtained, after correction for CH and any other phases present, is that of the layers themselves. Organic liquids do not penetrate, and helium penetrates only slowly. The empty interlayer space is counted as part of the solid volume, with a consequent decrease in the measured density.

The IR absorption spectra of C_3S, alite or β-C_2S, pastes (H29,L22,S44) are relatively diffuse, as would be expected from the poor crystallinity. Si–O stretching is represented by a single, broad band peaking at about 970 cm^{-1}; this is compatible with the existence of condensed silicate anions. Small amounts of carbonate are readily detectable by the absorption near 1400 cm^{-1}.

5.4 Structural models for C–S–H gel

5.4.1 General

Many crystalline calcium silicate hydrates are known. Most are formed only under hydrothermal conditions, i.e. in aqueous systems under pressure above 100°C (Section 11.7), but afwillite ($C_3S_2H_3$) is probably the stable phase in contact with CH and solution at ordinary temperatures (L23). It can be prepared from C_3S by ball milling with water at ordinary temperature (K14) or seeding (D17), but shows no similarities to C–S–H gel and nothing resembling it is formed in normal cement pastes. Two other crystalline phases that can be prepared in aqueous suspensions below 100°C show much closer relationships. These are 1.4-nm tobermorite ($C_5S_6H_9$ approx.) and jennite ($C_9S_6H_{11}$). Various types of semicrystalline calcium silicate hydrate are intermediate in structure between these compounds and C–S–H gel. Two relatively well defined ones are C–S–H(I) and C–S–H(II), which are closely related to 1.4-nm tobermorite and jennite, respectively. A gelatinous calcium silicate hydrate, called plombierite, occurs in nature; it is of variable composition, and some specimens resemble C–S–H(I) (M49).

5.4.2 1.4-nm tobermorite and jennite

1.4-nm tobermorite has a layer structure and the prefix relates to the layer thickness. On being heated at 55°C, it loses interlayer water and undergoes unidimensional lattice shrinkage, giving 1.1-nm tobermorite ($C_5S_6H_5$ approx.), often called, simply, tobermorite. 1.4-nm tobermorite occurs as a natural mineral (F17) and may be synthesized from CH and silicic acid in aqueous suspensions at 60°C (K15,H30). Table 5.2 gives crystal data. The crystal structure has not been directly determined, but may be inferred from that of 1.1-nm tobermorite (M50,H31), aided by evidence from dehydration and IR studies (F17), broad line proton NMR and silicate anion determination by the molybdate method (W20) and ^{29}Si NMR (W21,K16). Each layer consists of a central part of empirical formula CaO_2, of which all the oxygen atoms are shared with Si–O chains, which form ribs covering each surface; between these layers are water molecules and additional calcium atoms (Fig. 5.7A and B). The idealized constitutional formula is possibly $Ca_5(Si_6O_{18}H_2)\cdot 8H_2O$, but the atomic contents of the unit cell of the natural mineral show that only 5.5 silicon atoms are present where 6 would be expected (M51), and synthetic studies also indicate that the Ca/Si ratio is higher than 0.8 (H30). Many silicate tetrahedra thus appear to be missing from the chains, and the actual formula may be nearer to $Ca_5Si_{5.5}O_{17}H_2\cdot 8H_2O$, corresponding to a mean chain length of 11 tetrahedra. The conclusion from ^{29}Si NMR (K16) that little SiOH is present appears to be of doubtful validity.

Table 5.2 Crystal data for 1.4-nm tobermorite, jennite and related phases

Phase	1.4-nm tobermorite	C–S–H(I)	Jennite	C–S–H(II)
Molar ratios				
CaO	5	5	9	9
SiO_2	5.5	5	6	5
H_2O	9	6	11	11
Pseudocell parameters				
a (nm)	0.5264	0.560	0.996	0.993
b (nm)	0.3670	0.364	0.364	0.364
c (nm)	2.797	2.5	2.136	2.036
α	90.0°	90.0°	91.8°	90.0°
β	90.0°	90.0°	101.8°	106.1°
γ	90.0°	90.0°	89.6°	90.0°
Lattice type	I	I	A	A
Z	1	1	1	1
D_x (kg m^{-3})	2224	2250	2332	2350
Reference	F17, T5	T20, T5	G46, T5	G47, T5

All data are for the pseudocells and compositions stated. The true cells have doubled values of b and can be in varying degrees disordered; a and c must often be differently defined. The structure of C–S–H(I) has little more than two-dimensional order, and that of C–S–H(II) has less than full three-dimensional order.

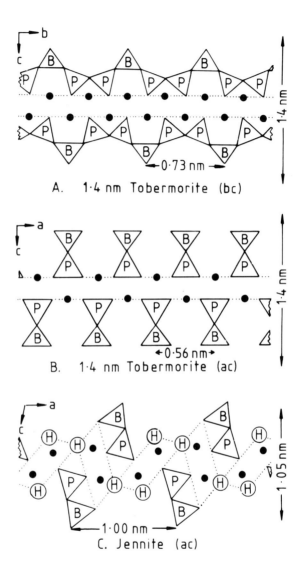

Fig. 5.7 (A) and (B) Structure of a single layer of 1.4-nm tobermorite in *bc*- and *ac*-projections, respectively. In (B), the chains are seen end on. (C) Suggested structure for a single layer of jennite, in *ac* projection; the chains are seen end on and the Ca–O sheets edge on, parallel to their corrugations, and circled 'H's denote hydroxyl groups. In (A), (B) and (C), full circles denote calcium atoms, P and B denote paired and bridging tetrahedra, respectively. Jennite axes relate to monoclinic pseudocell with $a = 1.00$ nm, $b = 0.36$ nm, $c = 2.14$ nm, $\beta = 101.9°$.

The Ca–O parts of the layers have a pseudohexagonal, centred lattice with $a = 0.56$ nm, $b = 0.36$ nm, and may be regarded as layers of CH extremely distorted due to replacement of their OH groups by SiO_4 tetrahedra, the O–O distances of which have to be satisfied. CH, referred to orthogonal axes, has $a = 0.62$ ($0.36 \times \sqrt{3}$), $b = 0.36$ nm. The Si–O chains are of a type called dreierketten, i.e. they are kinked so as to repeat at intervals of three tetrahedra. This conformation is imposed by the coordination requirements of the calcium ion, and occurs in many calcium silicates, including wollastonite (Section 2.2.1). In 1.4-nm tobermorite, they are single, i.e. not condensed with other similar chains to form double chains or more complex Si–O structures, and have the empirical formula Si_3O_9H. 1.1-nm tobermorite, when formed by dehydration of the 1.4-nm form, differs from 1.4-nm tobermorite only in having lost some of its interlayer water molecules, and thus also contains single dreierketten. Another variety of it, formed under hydrothermal conditions, contains double chains (Section 11.7.4).

Jennite ($C_9S_6H_{11}$) occurs as a natural mineral (C26,G46), and, like 1.4-nm tobermorite, may be synthesized from CH and hydrous silica in aqueous suspensions below 100°C (H32,H33). It forms most readily at 80°C from mixes of CH with reactive silica at Ca/Si ratios of 1.1–1.2, though on prolonged reaction these yield 1.4-nm tobermorite (H33). In nature, it occurs in contact with 1.4-nm tobermorite (C26), with which it is probably in metastable equilibrium. Like 1.4-nm tobermorite, it has a layer structure, and at 70–90°C loses water with unidimensional lattice shrinkage, giving metajennite ($C_9S_6H_7$) (C26,H32). The layer thickness is 1.05 nm in jennite and 0.87 nm in metajennite (G46). Table 5.2 includes crystal data for both phases. The crystal structure has not been determined, but its major features (Fig. 5.7C) may be inferred from the crystal data, dehydration behaviour, IR spectral, ^{29}Si NMR and other evidence. As with 1.4-nm tobermorite, NMR evidence against the presence of SiOH groups is weak, and the constitutional formula is probably $Ca_9(Si_6O_{18}H_2)(OH)_8 \cdot 6H_2O$, the structure somewhat resembling that of 1.4-nm tobermorite but with alternate chains replaced by rows of OH groups. The unit cell parameters show that while each layer could be based on a central part of empirical formula CaO_2, the distortion of the CH layer needed to produce it would differ greatly from that in 1.4-nm tobermorite. It is probably folded into corrugations parallel to the b axis (T21).

5.4.3 C–S–H(I) and similar materials

Attempts to synthesize 1.4-nm tobermorite or jennite from aqueous suspensions at ordinary temperatures normally yield semicrystalline products.

These vary in crystallinity from ones whose XRD powder patterns approach those of 1.4-nm tobermorite or, much less frequently, jennite, to materials no more ordered than those formed in pastes. Reaction between CH and hydrous silica gives a product called C–S–H(I), first adequately described by zur Strassen and Strätling (Z9) and more fully by Taylor (T20). Similar but often less crystalline products are formed on mixing solutions of sodium silicate and calcium salts, or under certain conditions from C_3S (T20,B56,J15,G48) and also as initial or intermediate products of hydrothermal reactions at temperatures up to at least 180°C. Table 5.3 gives the XRD powder pattern of a relatively highly crystalline specimen of C–S–H(I). It consists largely of the hk0 reflections, or hk band heads, of 1.4-nm tobermorite, together with a broad basal reflection. The latter, which is absent in less ordered forms, corresponds to the mean layer thickness. C–S–H(I) may be prepared with Ca/Si ratios varying from 0.8 to about 1.5. With increase in Ca/Si ratio, the degree of crystallinity tends to decrease, and the basal spacing, if present, also decreases, from about 1.3 nm at Ca/Si = 0.8 to about 1.1 nm at Ca/Si = 1.3 (T22). The basal spacing also decreases on heating, values of 0.91–1.13 nm being obtained at 108°C (T23). TEM shows C–S–H(I) to consist of crumpled foils a few nanometres thick, with a tendency to elongation or fibrous character at the higher Ca/Si ratios (G41).

Table 5.3 X-ray powder diffraction data for C–S–H(I)

d(nm)	I_{rel}	hkl	d(nm)	I_{rel}	hkl	d(nm)	I_{rel}	hkl	d(nm)	I_{rel}	hkl
1.25	vs	002	0.24	w,d	?	0.152	vw	22.	0.111	w	42.
0.53	vvw	10.	0.21	w,d	?	0.140	w	40.	0.107	vw	51.
0.304	vs	11.	0.182	s	02.	0.123	vw	?			
0.280	s	20.	0.167	mw	31.	0.117	vw	13.			

Pattern is for a relatively highly crystalline variety; all peaks are broad. s = strong, w = weak, m = moderately, v = very, d = diffuse. Indices relate to pseudocell in Table 5.2; hkl denotes a band head given by an essentially two-dimensional lattice (T22).

Stade and co-workers (S45–S47) studied silicate anion structures in preparations of 'C–S–H(di,poly)', obtained by reactions in suspension at either 80°C or 150°C and which gave XRD patterns generally similar to those of C–S–H(I). They used the molybdate and TMS methods and ^{29}Si NMR. The results showed that for Ca/Si ratios of 1.1–1.5, the anion size distribution varied little with either Ca/Si ratio or temperature of preparation, 40–50% of the silicon being in dimeric, and 50–60% in polymeric, single chain anions. At Ca/Si = 1.0, the average anion size was greater, and at Ca/Si = 2.0, a higher proportion of end groups, found using NMR, indicated an increased proportion of dimer. Hydroxyl groups were attached to both end group (Q^1) and middle group (Q^2) tetrahedra. Changes in silicate anion structure on heating were also studied.

The silicate anion structures of C–S–H(I) preparations appear to be affected by how long the material remains in contact with its mother liquor and by how strongly it is subsequently dried. Experiments using the molybdate method showed that the anions in precipitates made from $CaCl_2$ and sodium silicate solutions at 0°C were mainly those present in the silicate solution, and thus monomeric if the latter is sufficiently dilute (S45). By allowing such products to stand in contact with their mother liquors at 0°C, and drying at −10°C, preparations with Ca/Si ratios of 1.2–1.5 were obtained that contained only dimeric silicate anions.

Macphee et al. (M47) made an NMR study of C–S–H samples prepared in aqueous suspension. In preparations with Ca/Si = 1.8 that had been in contact with mother liquor for 88 weeks the anions were almost wholly dimeric, irrespective of drying condition between 10% and 50% RH, but in one with Ca/Si = 1.1 that had been in contact with the solution for only 4 weeks the ratio of Q^2 to Q^1 increased on drying. The C–S–H with Ca/Si = 1.8 thus differed markedly in silicate anion structure from that of similar composition formed in C_3S pastes. The authors suggested that, in freshly formed C–S–H, drying caused increased condensation. If the material was formed in suspension, the free supply of water ensured that little condensation beyond dimer occurred in the presence of the mother liquor, and the aged material was less susceptible to the effect of drying. In a C_3S or cement paste, the demand for water caused localized drying in the inner product, with a consequent increase in condensation. The inner or late product could thus be less hydrated and more highly condensed than the outer product.

DTA curves of C–S–H(I) preparations show endotherms at 100–200°C and exotherms at 835–900°C, the latter temperatures increasing with Ca/Si ratio (K17). C–S–H(di,poly) preparations gave somewhat similar results (S46), but the temperature of the exotherm varied erratically with Ca/Si ratio and, except at Ca/Si = 0.8, an additional endotherm occurred at 332–372°C. Solvents that dissolve lime, such as ethyl acetoacetate, have little action on preparations with Ca/Si ratios of 1.25–1.33, but remove CaO from ones of higher Ca/Si ratio until that composition is reached (K17,S46).

5.4.4 Products formed in suspensions from C_3S or β-C_2S

Many studies have been reported on the hydration of C_3S or β-C_2S in aqueous suspensions, often termed 'bottle hydration'. If the reaction is prolonged, so that no unreacted starting material remains, and the w/s ratio is sufficiently high that the Ca/Si ratio of the product is below about 1.5, the product is C–S–H(I), essentially similar to that obtained by other methods (T20). At lower w/s ratios, such that the product contains CH, the C–S–H is similar in composition and crystallinity to that formed in pastes (F13). The products of bottle hydration for short times, such that unreacted starting material remains, are discussed in Section 5.5.1.

Under some conditions, prolonged reaction at room temperature gives a product differing from any of the above, which has been called C–S–H(II). This was obtained from C_3S by shaking with water, and repeatedly removing proportions of the liquid until the bulk Ca/Si ratio had fallen to about 2.0 (T20), and by bottle hydration of β-C_2S (B55). It has a Ca/Si ratio somewhat below 2.0, is seen by TEM to consist of bundles of fibres (G41) and gives a characteristic XRD powder pattern (Table 5.4), which differs from that of C–S–H(I) and shows similarities to that of jennite. The unit cell parameters, determined by electron diffraction, are also unlike those of 1.4-nm tobermorite or C–S–H(I) and near those of jennite (Table 5.2). The silicate anion structure comprises dimeric and single chain anions in comparable proportions, and the structure as a whole appears to be derived from that of jennite by omission of many tetrahedra from the silicate chains, giving a constitutional formula approximating to $Ca_9(Si_3O_9H)(Si_2O_7H)(OH)_8 \cdot 6H_2O$ (G47). Attempts to repeat these preparations have failed, and the conditions giving C–S–H(II) are obscure.

Table 5.4 X-ray powder diffraction data for C–S–H(II)

d(nm)	I_{rel}	hkl	d(nm)	I_{rel}	hkl	d(nm)	I_{rel}	hkl
0.96	85	002,100	0.312	100	$\bar{1}$13	0.209	12	3$\bar{1}$3(?)
0.80	20	$\bar{1}$02	0.307	95	$\bar{3}$04	0.203	25	4$\bar{1}$3,$\bar{5}$02, +
0.48	12	004,200	0.294	90	$\bar{2}$11	0.183	75	020,$\bar{4}$17
0.358	10	011	0.283	95	$\bar{2}$13	0.1635	5	$\bar{6}$02
0.340	12	$\bar{1}$11	0.243	20	$\bar{4}$04,$\bar{3}$11, +	0.1565	20	$\bar{3}$24, +
0.335	25	$\bar{3}$02(?)	0.224	5	$\bar{4}$06	0.1212	5	031
0.329	40	111,006	0.216	10	$\bar{4}$02(?)	0.1174	5	$\bar{2}$31,$\bar{8}$15, +

All peaks are broad; indexing, which relates to the pseudocell in Table 5.2, is approximate as order is not fully three-dimensional. Data are from Ref. P1, Card 29–374, with peaks of $I_{rel} < 5$ omitted.

5.4.5 Structural relationships

The XRD powder pattern of C–S–H gel (Fig. 5.6) suggests structural similarities to 1.4-nm tobermorite and jennite. The band at 0.27–0.31 nm and the peak at 0.182 nm could be derived from the sharper peaks in corresponding parts of the patterns of either or both of those phases. The patterns of C–S–H(I) and C–S–H(II), which are phases of intermediate crystallinity, provide links and thus strengthen the hypothesis. The C–S–H(I) peaks at 0.304 nm, 0.280 nm and 0.182 nm (Table 5.3) represent the three shortest repeat distances in the plane of a distorted CH layer (compare Fig. 5.1), and those in the same regions of the C–S–H(II) pattern have a similar significance for a different type of distortion. Slegers *et al.* (S44) attempted to interpret the radial distribution function; this approach might repay further investigation.

Early studies suggested that the increase in Ca/Si ratio in C–S–H(I) above that of tobermorite could arise from a combination of omission of tetrahedra from the chains and incorporation of additional calcium in the interlayer (T23). In a dreierkette (Fig. 5.8), the tetrahedra are of two types, which we shall call 'paired' and 'bridging'. In calcium silicate structures, the paired tetrahedra share oxygen atoms with columns of Ca–O polyhedra, which in 1.4-nm tobermorite and, by analogy, also in jennite, are themselves condensed into CaO_2 layers. If all the bridging tetrahedra are missing, a series of dimeric anions results; more generally, if some or all are missing, chains containing 2, 5, 8, $(3n - 1)$ silicon atoms result (G47,T15). This accounts for the observed sequence of chain lengths in C_3S pastes. The hypothesis that increase in Ca/Si ratio is due to removal of parts of the structure is also consistent with the observation that it is associated with decrease in crystallinity and layer thickness.

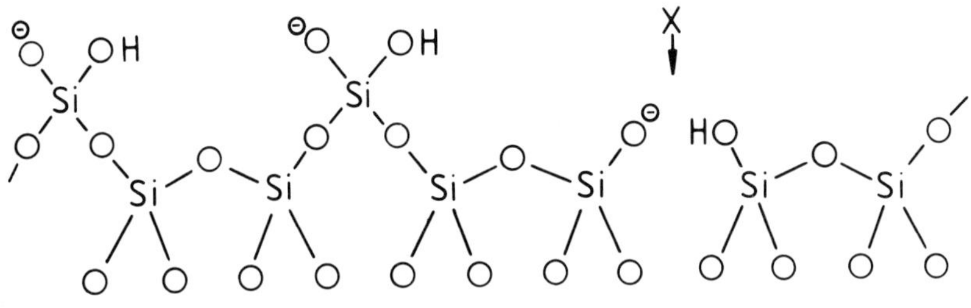

Fig. 5.8 Silicate chain of the type present in 1.4-nm tobermorite and jennite (dreierkette). In the tobermorite structure, the oxygen atoms at the bottom of the figure are also part of the central CaO_2 layer. The tetrahedra in the lower row are described as paired and those in the upper row as bridging. A bridging tetrahedron is missing (at X). Suggested positions of hydrogen atoms and negative charges balanced by interlayer cations are included. Taylor (T24).

At one time, the relation between C–S–H gel and tobermorite was widely believed to be very close, and the name 'tobermorite gel' was in common use. The Ca/Si ratio and poor crystallinity of C–S–H gel do not justify this name. Assuming the idealized formula $C_5S_6H_9$ for 1.4-nm tobermorite, removal of all the bridging tetrahedra would only raise the Ca/Si ratio to 1.25, and in pastes more than a few days old, not all of these tetrahedra are absent. Omission of tetrahedra is therefore not the only cause of increased Ca/Si ratio. The other cause can only be incorporation of extra Ca^{2+} ions, which could be balanced by omission of H^+ or incorporation of OH^- or both.

From their studies on 'C–S–H(di,poly)', Stade and Wieker (S45) concluded that Ca^{2+} and OH^- ions could both be present in the interlayer region of a tobermorite-like structure. Stade (S46) suggested that in the products containing both dimeric and polymeric ions, one surface of each tobermorite-type layer was composed of dimeric, and the other of polymeric, ions, thus accounting for the observed near-constancy of dimer/polymer ratio. In the purely dimeric material isolated at $-10°C$, both surfaces were composed of dimeric ions.

From the crystal chemical standpoint, it is difficult to see how anions, as well as cations, could be present in the interlayer of a tobermorite-type structure. Another possibility is that regions of jennite-type structure are present (T21,T24). If the structure suggested in Section 5.4.2 for jennite is correct, this is equivalent from the compositional standpoint to adding the elements of CH to a tobermorite-type structure. The difference between this hypothesis and that of Stade and Wieker may be largely semantic, since the probable corrugation of the layers in a jennite-type structure could bring the additional Ca^{2+} and OH^- ions into what could alternatively be described as part of the interlayer region.

5.4.6 A mixed tobermorite–jennite-type model for C–S–H gel

Taylor (T24) suggested that C–S–H gel, assumed to be equilibrated at 11% RH, contains elements of structure derived from both 1.4-nm tobermorite and jennite. He suggested that these were present in separate layers, but transition between the structures might also occur within a layer if the boundary is parallel to the chain direction. In Fig. 5.9, the Ca/Si ratio is plotted against a function of chain length for each structure, assuming that omission of bridging tetrahedra is compositionally equivalent to removal of SiO_2. The observed Ca/Si ratio of about 1.8 could arise from a mixture of the two structures, both with dimeric anions, or from a jennite-type structure with, on the average, pentameric anions. Transition from the first to the second would explain the observed changes in anion distribution with time. Because it is a change from a more to a less imperfect structure, it might also explain the direction taken by the process. However, as noted in Section 5.4.3, the increase in mean anion size may be at least in part a consequence of local dehydration.

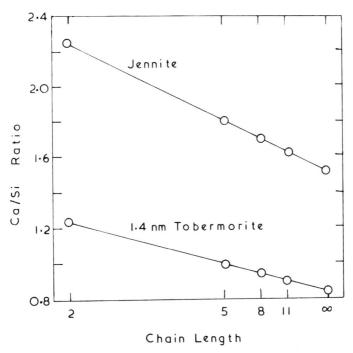

Fig. 5.9 Calculated Ca/Si ratio plotted against a function of chain length for jennite and 1.4-nm tobermorite modified by omission of bridging tetrahedra. Taylor (T24).

Several other lines of evidence were cited in support of this hypothesis. The TG curve of C–S–H gel (Fig. 5.3), expressed in terms of H_2O/Ca ratio, was shown to be intermediate between those of 1.4-nm tobermorite and jennite. The densities and H_2O/Ca ratios of C–S–H gel are similar to those of 1.4-nm tobermorite, jennite and structurally related minerals of comparable H_2O/Ca ratios (Table 5.5). The XRD evidence has already been noted; of the few selected area electron diffraction patterns that have been obtained from particles of C–S–H gel, some were shown to resemble ones of tobermorite minerals, and others that of C–S–H(II). Finally, the occurrence of two types of structure, with differing compositions, could explain the local variability in composition observed in electron optical analyses.

The same hypothesis can be applied to preparations made in suspension. In C–S–H(I), increase in Ca/Si ratio above the theoretical minimum value of 5:6 probably results entirely from omission of bridging tetrahedra until a value of about 1.0 is reached. Relatively highly ordered materials can be obtained. Further increase in Ca/Si ratio could result either from omission of more bridging tetrahedra, or from the presence of regions of jennite-type structure, a given ratio thus being obtainable by different combinations of the two effects, leading to possible variability in properties. The purely dimeric materials would, in general, contain regions of both structural types. C–S–H(I) is thus material predominantly, but not always entirely, of tobermorite-type structure. The mixing of structural types markedly impairs the crystallinity, and at Ca/Si ratios above about 1.5, it is impossible to obtain material giving the tobermorite-type, C–S–H(I) XRD pattern. At higher ratios, preparations sufficiently ordered to give a distinctive pattern give one of C–S–H(II), or, at higher temperatures, jennite.

Table 5.5 Water/calcium ratios and observed densities (kg m^{-3})

Material	H$_2$O/Ca	Density	Ref.[a]
C–S–H gel (saturated)	2.3	1850–1900	H28
Plombierite (0.8CaO·SiO$_2$·xH$_2$O; saturated)	2.5	2000–2200	M49
C–S–H gel (11% RH)	{1.2 ?}	2180 / 2430–2450	H28 / F16
Plombierite (11% RH)	1.64	?	M49
Jennite	1.22	2320	C26, G46
1.4-nm tobermorite	1.80	2200	F17
Tacharanite (Ca$_{12}$Al$_2$Si$_{18}$O$_{69}$H$_{36}$)	1.50	2360	C27
C–S–H gel (110°C)	0.85	2600–2700	[b]
Plombierite (110°C)	1.07	?	M49
Metajennite	0.78	2620	G46
1.1-nm tobermorite	1.00	2440	M49
Oyelite (Ca$_{20}$Si$_{16}$B$_4$O$_{83}$H$_{50}$)	1.25	2620	K18

[a] Values are in some cases calculated from data in references given.
[b] Typical values obtained using water, corrected for CH.

Some other models for the structure of C–S–H gel that have been proposed are incompatible with the evidence. A proposal identifying it with natural tobermorite, based on IR and extraction results (S37), appears to ignore both composition and degree of crystallinity. Another, assuming it to be closely related to the CH structure, with incorporation of monomeric silicate ions (G38), is inconsistent with the observed silicate anion structure. As noted earlier, one assuming three-dimensional anionic clusters (C23) is inconsistent with the ^{29}Si NMR evidence, and with the overwhelming proportion of the other evidence on silicate anion structure.

The monomeric material formed as the initial product, either in pastes or in suspensions, is unlikely to have a structure specifically related to that of 1.4-nm tobermorite or jennite, as both require the presence of dreierketten or Si$_2$O$_7$ groups, though its structure might still be based on Ca–O sheets.

5.5 Equilibria

5.5.1 Solubility relations

Equilibria in the $CaO-SiO_2-H_2O$ system at ordinary temperatures have been widely studied for the light they may cast on cement hydration and on leaching of concrete by ground waters. Steinour (S48,S49) reviewed early work and Jennings (J15) discussed later studies. If C-S-H(I) is placed in water or CH solutions, its Ca/Si ratio changes until equilibrium is reached. Except at low CaO concentrations, the SiO_2 concentrations in solution are very low, so that transfer between solid and solution is almost entirely of CaO. The data may be plotted schematically on a triangular diagram (Fig. 5.10), but because of the low concentrations it is more useful to employ separate plots of the Ca/Si ratio of the solid or C_{SiO_2} against C_{CaO} (Figs 5.11 and 5.12).

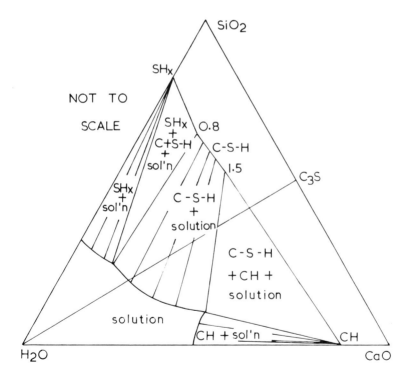

Fig. 5.10 The system $CaO-SiO_2-H_2O$ at ordinary temperatures (schematic). Taylor *et al.* (T25).

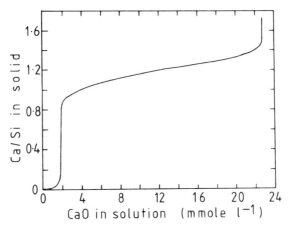

Fig. 5.11 Metastable equilibrium curve relating the Ca/Si ratios of C–S–H preparations to the analytical concentration of CaO in the solution, selected as typical from the results of many investigations. After Steinour (S49).

In most of the investigations, the solid starting materials were either CH and hydrous silica, or C–S–H prepared either from those materials or by mixing solutions of sodium silicate and a soluble calcium salt, followed by washing to remove the soluble product. In some, anhydrous calcium silicates were used. The various sets of data represented by Figs 5.11 and 5.12 show considerable divergencies; a few studies suggested the existence of a step in the curve of Fig. 5.11 at Ca/Si $\simeq$ 1.1 (S48). With the important exceptions of some of the data obtained using C_3S or C_2S, none of these differences is clearly related to ones in starting material. They may, nevertheless, be associated with differences in structure, discussed in Section 5.4.6. XRD powder evidence shows that the central part of the curve in Fig. 5.11 ($C_{CaO} \simeq$ 2–22 mol l^{-1}) and curve A of Fig. 5.12 represent the solubilities of C–S–H(I) preparations and that there are invariant points for hydrous silica, C–S–H(I) with Ca/Si $\simeq$ 0.8 and solution with $C_{CaO} \simeq$ 1 mmol l^{-1} and $C_{SiO_2} \simeq$ 2 mmol l^{-1}, and for CH, C–S–H(I) with Ca/Si $\simeq$ 1.4 and solution with $C_{CaO} \simeq$ 22 mmol l^{-1} and $C_{SiO_2} \simeq$ 1 μmol l^{-1} (S49,T20,J15). Fig. 5.12 includes metastable solubility curves for hydrous silica and CH.

Jennings (J15) noted that observed points on plots of C_{SiO_2} versus C_{CaO} tended to fall on one or other of two curves, with relatively few in between (Fig. 5.12). Of those that fell between, some were obtained using β-C_2S and others were suspect from carbonation or for other reasons. Those falling on the upper curve, B, had all been obtained by reaction of C_3S with water for periods not exceeding a few days; probably in all cases, unreacted C_3S was present. Some of the experiments using C_3S, and all of those using starting materials other than C_3S or C_2S, gave points lying on the lower curve, A. Of the experiments using C_3S that gave points on curve A, some were of durations of 4 h or less (B56), but most appear to have been of long duration, and in at least one case, points lying on curve A were obtained only if no unreacted C_3S remained (T20).

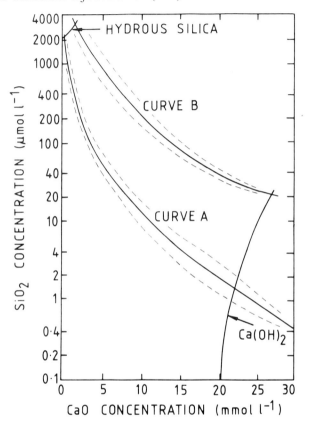

Fig. 5.12 Concentration data for C–S–H based on the results of many investigations; for explanations of curves, see text. Broken lines give an indication of the scatter of the data. After Jennings (J15).

Curve A is clearly the solubility curve of C–S–H(I); scatter at the CaO-rich end may arise because a given Ca/Si ratio can be associated with more than one structural arrangement. However, any form of C–S–H that is structurally derived from 1.4-nm tobermorite or jennite, or perhaps more widely that is based on Ca–O sheets, probably has a solubility lying on or close to this curve. There are no solubility data for C–S–H(II), 1.4-nm tobermorite or jennite, but one might expect the curves for the crystalline phases to lie below the relevant parts of curve A, and that of C–S–H(II) to lie near or slightly below the high-CaO portion of that same curve.

Curve B is most unlikely to represent the metastable solubility of anhydrous C_3S; thermodynamic calculations (S50) indicate that such a curve would be very much higher, as would be expected from the high reactivity. Jennings (J15,J16) and Gartner and Jennings (G48) considered it to be the metastable solubility curve of a product formed as a layer on the C_3S surface. Thermodynamic calculations based on this assumption indicated Ca/Si ratios of 1.1 at low and 1.65 at high CaO concentrations (G48); these values are broadly compatible with ESCA evidence (Section 5.6.2). Barret and Bertrandie (B57,B58) considered that curve B was not a solubility curve, but represented a quasisteady state in which the concentrations were such as to equalize the rate of dissolution of a hydroxylated surface layer on the C_3S and the rate of precipitation of C–S–H. Both views associate the curve, though in different ways, with the presence of an altered surface layer on the C_3S that is metastable relative to the C–S–H whose solubility is represented by curve A. Experiments using β-C_2S gave compositions lying on a curve between curves A and B (B57,G49) and which may similarly be associated with an altered surface layer on the β-C_2S.

There are few data on the concentrations of CaO and SiO_2 in the pore solutions of C_3S pastes (w/s < 1.0), as opposed to suspensions, and those that exist are conflicting as regards C_{SiO_2} (B56,O11). The values of C_{CaO} after long times (e.g. $27 \text{ mmol } l^{-1}$) (O11) appear to be significantly above the solubility of CH; this has been attributed to incorporation of SiO_2 in the CH (W22), but might also be due to the small crystal size of some of this phase.

5.5.2 Species in solution

Barby *et al.* (B59) summarized data for silicate species in alkali silicate solutions. Monosilicic acid, H_4SiO_4, has $pK_{a1} = 9.46$, $pK_{a2} = 12.56$, and pK_{a3} probably about 15. Monomeric species are overwhelmingly predominant at C_{SiO_2} below about $2 \text{ mmol } l^{-1}$ at pH values up to 9, and up to higher concentrations at higher pH. The solubility product of CH, in terms of activities, is $9.1 \times 10^{-6} \text{ mol}^3 l^{-3}$ at $25°C$ (G39), and the stability constant of the complex $CaOH^+$ is $20 \text{ mol } l^{-1}$ (S51). Undissociated CH is not a significant species in saturated CH solution (G48). In saturated CH solution at $25°C$, the activities of Ca^{2+} and $CaOH^+$ are roughly equal, and the pH is 12.45. The stability constants of $CaH_2SiO_4^0$ and $CaH_3SiO_4^+$ are reported to be $1240 \text{ mol } l^{-1}$ and $2.5 \text{ mol } l^{-1}$, respectively (S51), but some doubt has been cast on these values by the results of thermodynamic calculations using them (G48).

Using the equilibrium constants given above, together with standard expressions to calculate ionic strength and activity coefficients, one may calculate the species concentrations and species activities in a system of given bulk composition. For $C_{CaO} = 28.0 \text{ mmol l}^{-1}$, $C_{SiO_2} = 20.0 \text{ }\mu\text{mol l}^{-1}$, and assuming that calcium silicate complexes are not present, the species concentrations (in mol l^{-1}) are: Ca^{2+}, 1.90×10^{-2}; $CaOH^+$, 8.98×10^{-3}; OH^-, 4.70×10^{-2}; H_4SiO_4, 4.19×10^{-9}; $H_3SiO_4^-$, 6.82×10^{-6}; $H_2SiO_4^{2-}$, 1.30×10^{-5}; $HSiO_4^{3-}$, 1.46×10^{-7} (G48). The activity of OH^- is 3.78×10^{-2} mol l^{-1}; this corresponds to a pH of 12.58. If the complexes $CaH_2SiO_4^0$ and $CaH_3SiO_4^+$ are assumed to occur, with stability constants of 2000 and 20 mol l^{-1}, respectively, the concentrations of the SiO_2-containing species become (mol l^{-1}): $CaH_3SiO_4^+$, 2.01×10^{-7}; $CaH_2SiO_4^0$, 1.61×10^{-5}; H_4SiO_4, 7.75×10^{-10}; $H_3SiO_4^-$, 1.26×10^{-6}; $H_2SiO_4^{2-}$, 2.41×10^{-6}; $HSiO_4^{3-}$, 2.70×10^{-8}, the neutral complex $CaH_2SiO_4^0$ thus being the principal SiO_2-containing species. The concentrations of Ca^{2+}, $CaOH^+$ and OH^- are not significantly affected.

5.5.3 Thermochemistry and thermodynamics

Using a heat of solution method, Lerch and Bogue (L24) found the enthalpies of hydration of C_3S and β-C_2S in pastes to be -115 kJ mol^{-1} and -45 kJ mol^{-1}, respectively. Similar values have been reported by other investigators, e.g. -118 kJ mol^{-1} and -45 kJ mol^{-1} by Brisi (B60) and -121 kJ mol^{-1} and -46 kJ mol^{-1} by Fujii and Kondo (F18), who recalculated data obtained by Brunauer et al. (B61) for dried products. Brunauer et al. noted that, if the hydration products of C_3S and β-C_2S were identical apart from the quantities of CH formed, the enthalpy of hydration of C_3S should equal the sum of the enthalpies of hydration of β-C_2S, hydration of CaO ($\Delta H_{298} = -65.2$ kJ mol^{-1}) (W10), and reaction of β-C_2S and CaO to give C_3S ($\Delta H_{298} = -13.4$ kJ mol^{-1}) (W10). For the moist material, this sum is -124 kJ mol^{-1}, which is near the observed value.

Fujii and Kondo (F18) calculated the thermodynamic properties of C–S–H, using solubility data and properties of 1.4-nm tobermorite and CH given by Babushkin et al. (B62). The C–S–H was treated as equivalent to a solid solution of these two compounds. For the composition $1.7CaO \cdot SiO_2 \cdot 2.617H_2O$, they found $\Delta H_f^\circ = -2890$ kJ mol^{-1}, $\Delta G_f^\circ = -2630$ kJ mol^{-1} and $S^\circ = 200$ J mol^{-1} K^{-1}. They showed that these values were consistent with data for the enthalpies of hydration of C_3S and β-C_2S. The free energy of formation of C–S–H from 1.4-nm tobermorite and CH was shown to become increasingly negative with increasing Ca/Si ratio, but to tend towards a limit at ratios approaching 2.0; this is consistent with the observed upper value of this quantity.

Glasser et al. (G50) calculated the activities of the principal ionic species in solution in equilibrium with C–S–H of various Ca/Si ratios, and solubility products defined in terms of those species. Using values of $\Delta G_f°$ for the ionic species (B62), values of $\Delta G_f°$ for the solids were then calculated. The results depended somewhat on the set of solubility data used, but were broadly similar to those obtained by Fujii and Kondo (F18). Using one set of data, values ranged from -1720 kJ mol^{-1} at Ca/Si = 0.93 to -2350 kJ mol^{-1} at Ca/Si = 1.63, referred to formulae containing one silicon atom in each case.

The Gibbs–Duhem equation has been used to calculate Ca/Si ratios of solids from the CaO and SiO$_2$ concentrations in solutions with which those solids are in equilibrium (F19,G48). Gartner and Jennings (G48) showed that, at the low concentrations involved, this equation reduces to the approximate form $(Ca/Si)_{solid} = -d\mu_S/d\mu_C$, where μ_S and μ_C are the chemical potentials of SiO$_2$ and CaO, respectively. The calculations were carried out for curve B and three variants of curve A on Fig. 5.12. For points along each curve, activities of the ionic species in solution were calculated, and from these, values of μ_S and μ_C; these were plotted against each other to give curves on a chemical potential phase diagram. From the slopes at points along these curves, the Ca/Si ratios of the solids were obtained. For one variant of curve A, the calculated Ca/Si ratios at $C_{CaO} = 3$ mmol l^{-1} and 14.4 mmol l^{-1} were 1.0 and 1.45 respectively. These values agree well with typical experimental results shown in Fig. 5.11. For curve B, the calculated Ca/Si ratios were 1.1 at 5 mmol l^{-1}, 1.65 at 13 mmol l^{-1} and 1.65 at 22 mmol l^{-1}.

5.5.4 Effects of alkalis

Several studies on the quaternary systems of CaO–SiO$_2$–H$_2$O with Na$_2$O or K$_2$O have been reported (K19,S52,M52). Alkali greatly lowers the concentrations of CaO in the solution and raises those of SiO$_2$. The solid phase compositions are difficult to study. Determinations based on changes in concentration on adding CH to alkali silicate solutions are subject to considerable experimental errors, while direct analyses of the solid are difficult to interpret because the alkali cations are easily removed by washing. Suzuki et al. (S52) considered that they were adsorbed. Macphee et al. (M52) reported TEM analyses of the C–S–H in washed preparations obtained by reaction of C$_3$S (10 g) in water or NaOH solutions (250 ml). The C–S–H obtained with water had a mean Ca/Si ratio of 1.77; that obtained with 0.8 M NaOH had a mean Ca/Si ratio of 1.5 and a mean Na$_2$O/SiO$_2$ ratio of 0.5. These results do not appear to be directly relevant to cement pastes. The pore solutions of the latter may be 0.8 M or even higher in alkali hydroxides (Section 7.5.2), but the ratios of $(Na_2O + K_2O)$ to SiO_2 in the C–S–H are 0.05 or below (Section 7.2.5).

5.6 Kinetics and mechanisms

5.6.1 C_3S: experimental data

Fig. 5.13 shows the general form of the curve relating the fraction of C_3S consumed (α) to time in a paste of w/s ≃ 0.5 at about 25°C and with moist curing. Such curves have been determined using QXDA for unreacted C_3S (e.g. Refs K20,O10), though the precision is low for values of α below about 0.1. At low values of α, other methods are available, such as conduction calorimetry (e.g. Ref. P22), aqueous phase analyses (e.g. Ref. B63) or determinations of CH content or of non-evaporable water. At very early ages, it may be necessary to allow for the fact that the property determined depends on the nature of the hydration products; e.g. precipitation of C–S–H begins before that of CH.

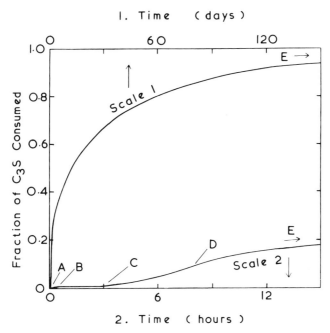

Fig. 5.13 General form of curves relating the fraction of C_3S consumed to time in a paste. AB: initial reaction. BC: induction period. CD: acceleratory period. DE: deceleratory period and continuing, slow reaction. Taylor *et al.* (T25).

The rate, as followed from that of heat evolution (Fig. 5.14), passes through an initial maximum, decreases to a minimum during a so-called induction period, passes through a second maximum and then gradually declines. Kondo and Ueda (K20) described five periods of reaction, viz: (1) the initial reaction, (2) the induction period, (3) the acceleratory period, in which the main reaction first begins to occur rapidly, (4) the deceleratory period and (5) a period of slow, continued reaction. Periods 1 and 2 correspond to the early stage of reaction as defined from microstructural studies (Section 5.3.1), periods 3 and 4 to the middle stage, and period 5 to the late stage. Setting takes place during the acceleratory period.

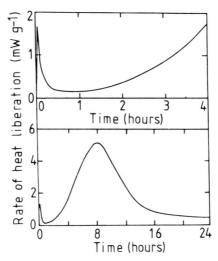

Fig. 5.14 Typical calorimetric curve relating the rate of heat liberation to time in a C_3S paste to time. After Brown et al. (B64).

Many investigators have followed the changes in C_{CaO} and C_{SiO_2} or [OH$^-$] with time in the course of the reaction of water with C_3S or β-C_2S in suspensions (e.g. B63,B56,G51; Fig. 5.15). The curves for C_{CaO} or [OH$^-$] typically show a steep rise during the initial reaction and a more gradual one of uniform slope during the induction period. Significant precipitation of CH begins just before the maxima are reached, and is rapid when they decrease, during the acceleratory period (W22). Some of the curves reported in the literature show a plateau in the early part of the induction period, followed by a rise (e.g. Ref. B57). In the most marked cases of this behaviour, the maximum is unusually high ($\simeq$ 40 mmol l^{-1}) and unusually delayed (24–36 h; e.g. Ref. F20). As noted in Section 5.5.1, the final values reported for pastes tend to be well above the solubility of CH. The SiO_2

concentration passes through a maximum in the order of 1 mmol l^{-1} during the initial reaction. Brown *et al.* (B56) found the concentrations to lie exclusively on curve A (Fig. 5.12), but other workers have found the paths followed by the concentrations to cross curve A, passing through maxima near to (G51) or beyond (B57,B58) curve B and then falling back to curve A.

At early ages, $d\alpha/dt$ increases markedly with w/s ratio above 0.7 (B56). Moderate variations in specific surface area have little effect on the length of the induction period, but with finer grinding, $d\alpha/dt$ during the acceleratory period increases (K20,O12,B56). The rate of reaction increases with temperature up to the end of the acceleratory period, but is much less affected thereafter (K21), suggesting a change from chemical to diffusion control. Introduction of defects into the C_3S shortens the induction period (M53,F20,O12).

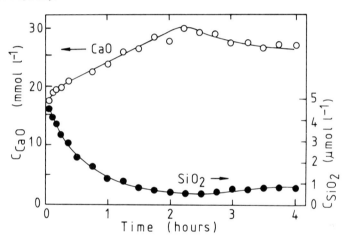

Fig. 5.15 Plots of the analytical concentrations of CaO and SiO_2 against time for a C_3S paste (C_3S 352.5 m^2 kg^{-1}, w/s = 0.7, 24°C). After Brown *et al.* (B56).

The results of experiments at high w/s ratios have been interpreted assuming superficial hydroxylation of the C_3S followed by congruent dissolution and subsequent precipitation (B63,B57). Reaction in pastes must also occur at least partly by dissolution and precipitation, since much of the product is deposited at a distance from the starting material, and because both C_{CaO} and C_{SiO_2} pass through maxima. Topochemical mechanisms, defined as ones in which the material does not pass through a true solution phase, may also occur, especially in the late stage of reaction (T26).

The kinetics up to the middle of the acceleratory period are discussed in the following sections. Those of the later stages have been more thoroughly studied with cement (Section 7.7) and only some aspects are considered here.

5.6.2 C_3S: the initial reaction

Two ESCA studies of the hydration of C_3S pastes or compacts showed an initial surface Ca/Si ratio of 3.0, which within the first minute fell sharply and then increased again; it then fell more slowly, reaching a value of 1.5–2.0 in 15–30 min (T27,M39,R18). Allowing for the fact that the region analysed will include unreacted C_3S until the layer of hydration product is sufficiently thick and continuous, these results may be interpreted in terms of the formation of an initial hydrate of low Ca/Si ratio, which rapidly increases (R18). Another study (B65), in contrast, showed an unreacted C_3S sample to have a low surface Ca/Si ratio. ESCA results also indicate a change in the environment of the silicon atoms at the end of the induction period, possibly associated with the formation of dimeric silicate ions (R18). Secondary ion mass spectrometry (SIMS) shows promise as an alternative method for studying surface compositions, but has not yet given information on that of the hydrated material (G52). Electron diffraction shows that an amorphous surface layer can be produced by grinding (U10,K20); this may explain some of the reported differences in behaviour between apparently similar specimens.

Calculations of α from concentrations in the bulk solution indicate values of around 0.3% at the start of the induction period (B63,B56). These are likely to be minimum estimates, because they assume that no CH is formed; in reality, small amounts may be, on account of concentration gradients, which will themselves introduce an error in the same direction. Reported values of α at the end of the induction period, summarized by Odler and Dörr (O10), range from 0.1% to an improbable 8%. Their own estimate of 1–2% is supported by subsequent evidence from cross-polarization ^{29}Si NMR (R16). Assuming a specific surface area of 300 m^2 kg^{-1}, a value of 1% corresponds to a mean depth of attack of about 10 nm.

5.6.3 C_3S: the induction period

The causes of the induction period and of its termination have been the subject of much debate. Hypotheses have been reviewed in a collaborative paper (T25). The main ones are as follows:

(1) The product of the initial reaction forms a protective layer on the C_3S particles; the induction period ends when this is destroyed or rendered more permeable by ageing or phase transformation (S53,B66,J15).
(2) The product of the initial reaction forms a semipermeable membrane which encloses an inner solution and is destroyed by osmotic bursting (D18).

(3) The rate of reaction in the induction and acceleratory periods is controlled by nucleation and growth of the C-S-H formed in the main reaction, the induction period ending when growth begins (S54,O13,F20,B63).

(4) The induction period occurs because the CH nuclei are poisoned by SiO_2 and cannot grow, and ends when the level of supersaturation is sufficient to overcome this effect (G53,T28,W22).

The induction period is shortened by adding prehydrated C_3S (O13), but additions of lime or CH, including that formed from C_3S, are variously reported to be ineffective (O13,B67) or to lengthen it, though shortening it with cement (U11). In cement mixes, additions of pfa or some other finely divided materials accelerate hydration after the first day, apparently by acting as nucleation sites for C-S-H (Section 9.3.3). Additions of reactive silica markedly accelerate hydration (S53). Most of this evidence supports hypothesis 3 and tells against hypothesis 4. Hypothesis 3 does not exclude hypothesis 1, as the breakdown of a protective layer could be associated with formation of a new product.

The recent microstructural evidence (Section 5.3.1) gives no indication that a membrane or other product distinct from that formed later is formed during the initial reaction in C_3S pastes, though a gelatinous coating is formed in cement pastes, which show an induction period similar to that observed with C_3S (Section 7.5.1). For C_3S pastes, this evidence excludes hypothesis 2, and gives no positive support to hypothesis 1. It does not exclude the formation of an altered layer on the C_3S surface, no more than a few nanometres thick. Tadros *et al.* (T28) postulated the formation of a SiO_2-rich layer with chemisorbed Ca^{2+}, and Barret *et al.* (B63) that of a superficially hydroxylated C_3S, formed by protonation of the O^{2-} and SiO_4^{4-} ions, balanced by loss of Ca^{2+}.

The calculated solubility of C_3S is about 1 molal (S50). In the absence of an altered or protective layer, the observed low rate of reaction during the induction period could be explained only if the concentrations close to the C_3S surface were of this order of magnitude. It is doubtful whether the very high concentration gradients that this implies could exist (J15,G49).

The balance of the evidence favours a combination of hypotheses 1 and 3. The following model is essentially that of Gartner and Gaidis (G49), but has features in common with those of Barret and Bertrandie (B57) and of Grutzeck and Ramachandran (G51).

(1) In the initial reaction, C_3S dissolves and a material is deposited which will be termed Product B. The dissolving C_3S probably has a hydroxylated surface, as proposed by Barret *et al.* (B63). Under typical paste conditions, Product B forms an overlying surface layer about 1 nm in

average thickness, though it is likely to be non-uniform and thicker around active sites. It may be capable of existence only on a C_3S surface and similar to a passive layer on a metal. Within 30 s, the C_3S is almost isolated from the solution, and an unstable equilibrium, represented by curve B of Fig. 5.12, is established between the solution and Product B.

(2) Product B is unstable with respect to the material represented by curve A of Fig. 5.12, for which we shall reserve the term C–S–H. During the early part of the induction period, C–S–H nucleates and begins to grow. This implies dissolution of Product B, and increased access of solution to the C_3S, which thus dissolves more rapidly. The rate of reaction in this stage is controlled by the growth of C–S–H, the rate of which increases with the amount already formed.

The amount of Product B at any given stage in the reaction must depend on the relative rates of its formation and dissolution. It may or may not disappear completely when the latter is sufficiently high. The NMR evidence of Rodger *et al.* (R16) (Section 5.3.2) indicates that the hydrated material formed up to the end of the induction period contains only monomeric silicate ions and that the latter persist alongside larger ions when the latter are formed. This suggests that the silicate ions are monomeric both in Product B and in C–S–H in the early stages of formation of the latter product. The Ca/Si ratio of Product B is possibly about 1 initially, increasing to about 1.7 as equilibrium with CH is approached.

5.6.4 The main reaction (C_3S and β-C_2S)

From the kinetic standpoint, there is probably no distinction between the induction period and the early part of the acceleratory period; indeed, as Gartner and Gaidis (G49) noted, there is strictly no induction period in the absence of retarding admixtures. Throughout both stages, growth of C–S–H proceeds at an increasing rate, which is reflected in increasingly rapid dissolution of the C_3S. The Ca^{2+} and OH^- concentrations in the bulk solution increase steadily, and when a sufficient degree of supersaturation has been reached, CH begins to precipitate in quantity, causing these concentrations to fall. Its formation is not normally rate determining, but could become so in the presence of admixtures that retard its growth. The kinetic data for this stage can be fitted by Avrami-type equations, but the assumptions underlying the latter (formation of a product within an initially homogeneous material) do not correspond to this situation existing in the hydration of C_3S or cement.

The smaller grains of C_3S probably react completely, by dissolution and precipitation, during the acceleratory period. In the final, slow periods of hydration, the remaining grains of C_3S, which are relatively large, are gradually replaced by C–S–H through the inward movement of an interface. It is difficult to believe that such a process occurs by dissolution into a true liquid phase, followed by precipitation. Following earlier studies (M49,F21,K22,K20), Taylor (T26) suggested that a topochemical reaction may occur with both C_3S and β-C_2S. Comparison of the numbers of atoms of each element in the initial and product phases shows that to convert either C_3S or β-C_2S into C–S–H, relatively little oxygen must be gained or lost; much of the calcium, and smaller proportions of the silicon, must be lost, and equivalent amounts of H^+ gained (Table 5.6). At the interface, a narrow zone could exist, in which the necessary atomic rearrangements take place. It is not necessary to postulate any migration of water molecules through the inner product; instead, Ca^{2+} and Si^{4+} move outwards, and H^+ inwards. At the surface, the calcium and silicon enter the solution, and ultimately precipitate, along with the OH^- ions released from the H_2O molecules, as CH and outer product C–S–H. Migration of silicon was suggested as the probable rate-determining step; however, at low w/s ratios, reaction stops because there is no more space in which the outer product can be formed, and in this case, control presumably shifts to some process related to deposition of outer product. Barret (B68) discussed topochemical mechanisms and their relation to dissolution–precipitation processes.

Table 5.6 Numbers of atoms in $1000/N_o$ ml

Phase	Density (kg m^{-3})	Ca	Si	H	O
C_3S	3120	41.0	13.7		68.3
β-C_2S	3326	38.6	19.3		77.2
C–S–H (90% RH)	1900	14.4	8.2	66.0	63.9

Slightly modified from Ref. T26; C–S–H composition assumed $C_{1.75}SH_{4.0}$.

5.6.5 Early hydration of β-C_2S

The kinetics and mechanism of β-C_2S hydration are similar to those for C_3S, apart from the much slower rate of reaction (M53,T29,F11,O14,F22), and, as noted earlier, the products are similar apart from the much smaller content of CH. Preparations appear to be more variable in reactivity than those of C_3S; this is partly attributable to differing stabilizers (F23), but could also be due to differing amounts or natures of phases in intergranular spaces or exsolution lamellae. Preparations made at low temperatures are especially reactive (F22). Because of the low rate of heat evolution, conduction calorimetry is difficult, but shows the existence of an induction period similar to that found with C_3S (M53), which is also indicated by QXDA (O14). ESCA, SEM and solution studies show that behaviour during the early period is similar in principle to that of C_3S (M54). The ESCA data show that the connectivity of the silicate tetrahedra increases within 24 h at 50°C; this behaviour resembles that of C_3S at the end of the induction period. They show the approximate depth of the hydrated layer to be 5 nm after 15 s, and 15 nm after 6 h at 25°C. The rate of reaction is increased in the presence of C_3S (T29,O14).

6

Hydrated aluminate, ferrite and sulphate phases

6.1 AFm phases

6.1.1 Compositional and structural principles

AFm (Al_2O_3–Fe_2O_3–mono) phases are formed when the ions they contain are brought together in appropriate concentrations in aqueous systems at room temperature. Some can also be formed hydrothermally, i.e. in the presence of water under pressure above 100°C. They are among the hydration products of Portland cements. Under favourable conditions, they form platey, hexagonal crystals with excellent (0001) cleavage. Some of the AFm phase formed in Portland cement pastes is of this type, but much is poorly crystalline and intimately mixed with C–S–H. AFm phases have the general formula $[Ca_2(Al,Fe)(OH)_6] \cdot X \cdot xH_2O$, where X denotes one formula unit of a singly charged anion, or half a formula unit of a doubly charged anion. The term 'mono' relates to the single formula unit of CaX_2 in another way of writing the formula, viz. $C_3(A,F) \cdot CaX_2 \cdot yH_2O$ [or $C_4(A,F)X_2 \cdot yH_2O$], where $y = 2(x + 3)$. Many different anions can serve as X, of which the most important for Portland cement hydration are OH^-, SO_4^{2-} and CO_3^{2-}. A crystal may contain more than one species of X anion. AFm-type phases can also be prepared in which other tripositive cations, such as Cr^{3+}, replace the Al^{3+} or Fe^{3+}.

AFm phases have layer structures derived from that of CH by the ordered replacement of one Ca^{2+} ion in three by Al^{3+} or Fe^{3+} (Fig. 6.1) (A7–A9). The principal layers thus defined alternate with interlayers containing the X anions, which balance the charge, and H_2O molecules. The replacement of Ca^{2+} by the smaller Al^{3+} or Fe^{3+} ions distorts the structure of the principal layer, alternate Ca^{2+} ions moving in opposite directions from its central plane. This allows each to coordinate the oxygen atom of an interlayer H_2O molecule in addition to its six OH^- ions. The principal layer, together with the H_2O molecules thus bonded to the Ca^{2+} ions, has the composition $[Ca_2(Al,Fe)(OH)_6 \cdot 2H_2O]^+$. In the simpler AFm structures, these units are stacked in such a way as to produce octahedral cavities surrounded by three H_2O molecules from each of the adjacent layers. These cavities may contain

X anions, H_2O molecules, or both. In $C_3A \cdot CaCl_2 \cdot 10H_2O$ (or $[Ca_2Al(OH)_6 \cdot 2H_2O]Cl$), each cavity contains a Cl^- ion. In $C_3A \cdot CaSO_4 \cdot 12H_2O$, one half contain SO_4^{2-} anions and the remainder contain two H_2O molecules. In C_4AH_{13}, all contain one OH^- anion and one H_2O molecule.

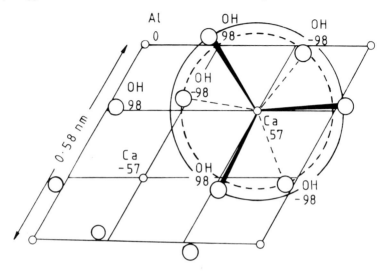

Fig. 6.1 Structure of a single principal layer of composition $[Ca_2Al(OH)_6]^+$ in an AFm phase, in ab-projection. Distances of calcium, aluminium and oxygen atoms above or below the central plane of the layer are in picometres. Hydrogen atoms are not shown. Large circles illustrate the distortion of the CaO_6 octahedra, which allows the coordination of each calcium atom to rise to 7 through the addition of an H_2O molecule (not shown) directly above or below it in this projection.

The unit cells of all AFm phases are based on hexagonal structural elements with $a = 0.57$–0.59 nm. These values are somewhat less than $\sqrt{3}$ times that of CH. The layer thickness, c', depends on the nature of the X anion and the amount of interlayer water, which can be varied by stepwise dehydration from the highest hydration state. AFm phases exist in which a complete, additional layer of H_2O molecules is present between the principal layers. There is no restriction on the separation between adjacent layers; this allows large anions, and also intercalated neutral molecules (D19,D20), to be present in interlayer sites. Many AFm phases readily undergo changes in water content and anion exchange (F24,D20), the latter including that of CO_3^{2-} for OH^-. Due care must therefore often be taken to exclude CO_2, and where necessary also to control the humidity.

Complex sequences of stacking the layers, or ordered patterns of filling of the cavities where their contents are not all the same, lead to unit cells that

6.1.2 The C_4AH_x, $C_4A\bar{C}_{0.5}H_x$ and $C_4A\bar{C}H_x$ phases

Fischer and Kuzel (F25) summarized and extended earlier work on these groups of AFm phases, which are listed in Table 6.1. In the pure CaO–Al_2O_3–H_2O system, the phase in equilibrium with solutions of appropriate composition at temperatures up to at least 50°C is C_4AH_{19}, of which two polytypes (α_1 and α_2) are known. On decreasing the RH or increasing the temperature, lower hydrates are reversibly formed, with decrease in layer thickness; a decrease in RH to 81% is sufficient to produce C_4AH_{13} (R19). C_4AH_{19} is structurally derived from C_4AH_{13} by the addition of an extra layer of H_2O molecules. A high degree of crystallinity is needed to allow this extra layer to form (D21), and it is doubtful whether it could occur in the material in a cement paste. In C_4AH_{11}, the contents of the octahedral cavities in the interlayers are reduced from (OH^- + H_2O) to OH^-, and in C_4AH_7, which is a poorly crystalline material, the H_2O molecules attached to the Ca^{2+} ions have also been lost.

The layer thicknesses indicate that the CO_3^{2-} ions are oriented perpendicular to the layers in $C_4A\bar{C}_{0.5}H_{12}$ and parallel to them in the other phases containing CO_3^{2-}. In all these phases, some octahedral cavities contain a CO_3^{2-} ion, and others varying combinations of H_2O molecules and OH^- ions. In the early literature, $C_4A\bar{C}H_{11}$ (or $C_3A·C\bar{C}·H_{11}$) was incorrectly identified as a C_3A hydrate, and $C_4A\bar{C}_{0.5}H_{12}$ as a polymorph of C_4AH_{13}. Both phases readily form on carbonation of the C_4A hydrates.

Table 6.2 includes crystal data for the C_4AH_x, $C_4A\bar{C}_{0.5}H_x$ and $C_4A\bar{C}H_x$ groups. Only data based on single crystal studies are included. Solid solutions in these groups of phases appear to be very limited, only $C_4A\bar{C}_{0.5}H_{12}$ showing some variation in layer thickness up to a maximum of about 0.825 nm associated with partial replacement of CO_3^{2-} by $2OH^-$ together with variations in H_2O content (D21,F25). The limiting composition is $C_4A\bar{C}_{0.25}H_x$. More extensive solid solution may occur in the poorly crystalline material formed in cement pastes.

6.1.3 The $C_4A\bar{S}H_x$ phases

Tables 6.2 and 6.3 give data for these phases. $C_4A\bar{S}H_{12}$ (or $C_3A·C\bar{S}·H_{12}$) is variously known as monosulphate, monosulphoaluminate, or, in the early literature, low-sulphate calcium sulphoaluminate. The water contents of the

Table 6.1 AFm phases of the C_4AH_x, $C_4A\bar{C}_{0.5}H_x$ and $C_4A\bar{C}H_x$ groups

Composition (a)	Drying conditions	Note	Layer thickness nm (a)	Interlayer contents (d) OH^-	CO_3^{2-}	H_2O
C_4AH_{19}	25°C, >88% RH	(b)	1.068	1	0	6
C_4AH_{13}	25°C, 11–81% RH	(b)	0.794	1	0	3
	40°C, 25% RH	(b)				
C_4AH_{11}	25°C, anhydrous CaCl$_2$	(b)	0.735	1	0	2
	50–90°C	(c)				
C_4AH_7	25°C, P$_2$O$_5$	(b)	0.56	1	0	0
	110–120°C	(c)				
$C_4A\bar{C}_{0.5}H_{12}$	22°C, ≥36% RH	(a)	0.8193	1/2	1/4	11/4
	40°C, 25% RH	(b)				
$C_4A\bar{C}_{0.5}H_{11.25}$	35°C,	(a)	0.763	1/2	1/4	19/8
$C_4A\bar{C}_{0.5}H_{10.5}$	25°C, anhydrous CaCl$_2$	(b)	0.726	1/2	1/4	2
	80°C	(a)				
$C_4A\bar{C}_{0.5}H_{6.5}$	25°C, P$_2$O$_5$	(b)	0.66	1/2	1/4	0
	105°C	(a)				
$C_4A\bar{C}H_{11}$	25°C, saturated CaCl$_2$	(a)	0.756	0	1/2	5/2
$C_4A\bar{C}H_8$	95°C	(a)	0.72	0	1/2	1
$C_4A\bar{C}H_6$	130°C	(a)	0.66	0	1/2	0

(a) Ref. F25. (b) Ref. R19. (c) Ref. B69. (d) Per formula unit of $Ca_2Al(OH)_6^+$.

Table 6.2 Crystal data for AFm phases

Phase	Crystal system[a]	Space group	Unit cell parameters				Z	D_x (kg m^{-3})	Ref.
			a (nm)	b (nm)	c (nm)	Angles			
α_1-C_4AH_{19}	H	$R\bar{3}c$ or $R3c$ (?)	0.577	—	6.408	$\gamma = 120°$	3	1803	A10
α_2-C_4AH_{19}	H	$P6_3/m$ or $P6_322$	0.577	—	2.137	$\gamma = 120°$	1	1802	A10
C_4AH_{13}	H	R--	0.5752	—	9.527	$\gamma = 120°$	6	2046	F25
$C_4A\bar{C}_{0.5}H_{12}$	H	$R\bar{3}c$ or $R3c$	0.5770	—	4.9159	$\gamma = 120°$	3	1984	F25
$C_4A\bar{C}H_{11}$	T	$P\bar{1}$ or $P1$	0.5781	0.5744	0.7855	[b]	1/2	2170	F25
$C_4A\bar{S}H_{12}$	H	$R\bar{3}$	0.576	—	2.679	$\gamma = 120°$	3/2	2014	A11
α-$C_3A\cdot CaCl_2\cdot H_{10}$	M	$C2/c$ or Cc	0.998	0.574	1.679	$\beta = 110.2°$	2	2065	K23
β-$C_3A\cdot CaCl_2\cdot H_{10}$	H	$R\bar{3}c$ or $R3c$	0.574	—	4.688	$\gamma = 120°$	3	2090	K23
$C_3A\cdot CaCl_2\cdot H_6$	H	$R\bar{3}c$ or $R3c$	0.5729	—	4.1233	$\gamma = 120°$	3	2080	K24
$C_2AH_{7.5}$[c]	M	$C2/c$ or Cc	0.993	0.574	4.22	$\beta = 97.0°$	8	1943	S55
C_2AH_5	H	$R\bar{3}c$	0.573	—	5.22	$\gamma = 120°$	6	2042	S55
$C_2A\bar{S}H_8$	H	$R\bar{3}$ or $R3$	0.5747	—	3.764	$\gamma = 120°$	3	1936	K25

[a] H = hexagonal or trigonal; M = monoclinic; T = triclinic.
[b] $\alpha = 92.61°$, $\beta = 101.96°$, $\gamma = 120.09°$.
[c] For C_2AH_8, only the structural element ($a = 0.574$ nm, $c = 1.07$ nm, $Z = 1$) and X-ray density (1950 kg m^{-3}) are known (S55).

Table 6.3 AFm phases with $[Ca_2Al(OH)_6]^+$ principal layers and interlayer sulphate, chloride, aluminate or aluminosilicate anions

Composition	Drying conditions	Layer thickness (nm)	Interlayer contents[a]	Ref.
$C_4A\bar{S}H_{16}$[b]	<10°C at 100% RH	1.03	1/2 SO_4^{2-} 5 H_2O[b]	D21
$C_4A\bar{S}H_{14}$[b]	>10°C at 100% RH	0.95	1/2 SO_4^{2-} 4 H_2O[b]	D21
$C_4A\bar{S}H_{12}$	>10°C at 20–95% RH	0.893	1/2 SO_4^{2-} 3 H_2O	D21
$C_4A\bar{S}H_{10}$	>10°C at <20% RH	0.815	1/2 SO_4^{2-} 2 H_2O	D21
$C_4A\bar{S}H_8$	30–50°C over P_2O_5	0.795	1/2 SO_4^{2-} 1 H_2O	D21
$C_3A \cdot CaCl_2 \cdot H_{10}$ (α)	<28°C at 35% RH	0.788	1 Cl^- 2 H_2O	K23
$C_3A \cdot CaCl_2 \cdot H_{10}$ (β)	>28°C at 35% RH	0.781	1 Cl^- 2 H_2O	K23
$C_3A \cdot CaCl_2 \cdot H_6$	120–200°C	0.687	1 Cl^-	K24
C_2AH_8	<26°C at 45% RH	1.07	1 $Al(OH)_4^-$ [c] 3 H_2O	S55
$C_2AH_{7.5}$	>26°C at 45% RH	1.04	1 $Al(OH)_4$ 5/2 H_2O	S55
C_2AH_5	Room temp. over P_2O_5	0.87	1 $Al(OH)_4^-$	S55
$C_2A\bar{S}H_8$	Room temp. at 37% RH	1.255	1 $AlSiO_3(OH)_2^-$ 4 H_2O	K25
$C_2A\bar{S}H_4$	>135°C	1.12	1 $AlSiO_3(OH)_2^-$	K25

[a] Per formula unit of $Ca_2[Al(OH)_6]^+$.
[b] Water contents uncertain; see text.
[c] Al is 6 coordinated in C_2AH_8 (G54).

two highest hydrates are uncertain, values of 16 for the 0.96-nm hydrate (T30) and of 15 for the 1.03-nm hydrate (R19) having also been reported. An alternative formula with 18 H_2O (M55) for the 0.96-nm hydrate implies an impossibly high density for the additional water. As with C_4AH_{19}, these highly hydrated forms probably cannot occur in the poorly crystalline material of cement pastes.

$C_4A\bar{S}H_{12}$ forms solid solutions in which up to one half of the SO_4^{2-} is replaced by OH^-. The substitution does not significantly affect the length of the a axis, but c' is slightly reduced, to a value of 0.877 nm at maximum replacement (R19). The higher hydrates do not form if excess CH is present. Early reports that a continuous solid solution series exists with C_4AH_x are incorrect. They were based on observations by light microscopy, which with layer structures of this type cannot distinguish between a true solid solution, in which the components are mixed at or below a nanometre level within each layer, an oriented intergrowth consisting of blocks each containing many layers of one or other component, or intermediate possibilities, such as random interstratification of layers of different types. Distinction can be made by XRD.

A phase of approximate composition $C_4A_{0.9}N_{0.5}\bar{S}_{1.1}H_{16}$ was reported to form at high concentrations of Na_2O, Al_2O_3 and SO_3 (D22). Three hydration states, with $c' = 1.00$, 0.93, and 0.81 nm, and 16, 12 and 8 moles of H_2O, respectively, were described, and it was suggested that the structure was derived from that of $C_4A\bar{S}H_x$ by the omission of some of the Al^{3+} ions, balanced by the inclusion of Na^+, in addition to anions, in the interlayer. The concentrations used in the original work were far removed from those existing in the liquid phase of cement pastes, but later work (S56) indicated that a similar phase, though probably of different composition, could form in highly alkaline media in the presence of excess CH and absence of excess Al_2O_3. This suggests that it could occur as a hydration product of cements high in alkali. Evidence that AFm phases can accommodate cations as well as anions in the interlayer is provided by the existence of a compound $[Ca_2Al(OH)_6]K\cdot(ClO_4)_2\cdot xH_2O$ (D20).

6.1.4 Other AFm phases containing aluminium

The range of anions that can occupy the X positions is very wide (A11,D20); even anions that form calcium salts of very low solubility, such as fluoride, can be introduced using special techniques. We shall consider only those most relevant to cement chemistry. Tables 6.2 and 6.3 include data. $C_3A\cdot CaCl_2\cdot 10H_2O$ or $[Ca_2Al(OH)_6]Cl\cdot 2H_2O$ (Friedel's salt) can be formed in concrete exposed to chloride solutions. Two polytypes are known; β is the higher temperature form, produced reversibly from α at 28°C. $C_3A \cdot CaCl_2 \cdot 10H_2O$ does not form solid solutions with the sulphate AFm phases, but an ordered compound $C_6A_2 \cdot C\bar{S} \cdot CaCl_2 \cdot 24H_2O$ exists, in which interlayers containing Cl^- probably alternate with ones containing SO_4^{2-} (K23).

The C_2A hydrates are AFm phases with aluminium-containing species in the interlayer. The hydrate formed in metastable equilibrium with aqueous solution at 18°C is C_2AH_8. Scheller and Kuzel (S55) found that, contrary to some earlier opinions, C_2AH_8 shows neither polymorphism nor polytypism, but that below 45% RH at 26°C it is converted into $C_2AH_{7.5}$. The transition temperature depends on the RH. An X-ray structure determination on the lower hydrate, C_2AH_5, showed that this had the constitution $[Ca_2Al(OH)_6]$-$[Al(OH)_4]$, the interlayer aluminium and some of the interlayer hydroxyl groups being statistically distributed. Because of poor crystallinity, it was not possible to obtain such detailed information about the higher hydrates, but some evidence was obtained for a similar constitution, with H_2O molecules added, in $C_2AH_{7.5}$. In contrast, a study of C_2AH_8 using ^{27}Al magic angle NMR showed that all the aluminium was octahedrally coordinated (G54), and the constitution $[Ca_2Al(OH)_6][Al(OH)_3(H_2O)_3]OH$ was proposed. The structural element of C_2AH_8 is closely similar to that of C_4AH_{19}, and the proposed constitution could be derived from that of the latter compound merely by removing some of the hydrogen atoms from interlayer H_2O molecules and introducing aluminium atoms, perhaps statistically, into octahedral sites. There is no necessary conflict between the two sets of data, as they relate to different hydrates. C_2AH_8 is reported to form solid solutions with C_4AH_{19}, but there is disagreement as to the end of the series at which these occur (J17,D20).

An early report of the formation of $C_3A \cdot C\bar{S} \cdot H_x$ was not supported by later work (C28), but three subsequent investigations have shown that such a phase can be prepared, though it is unstable at ordinary temperatures, and possibly so even in contact with its mother liquor at 5°C. The anion is probably $H_2SiO_4^{2-}$, and hydration states with $c' \simeq 1.01$ nm and 0.89 nm, presumably analogous to those of $C_4A\bar{S}H_x$, have been observed. It was obtained by anion exchange (D20), in reaction rims around C_3A grains in pastes made from C_3A, C_3S and gypsum (R20) and by reaction between C_4AH_{19} and amorphous silica in the presence of water at 5°C (V4). Because of its instability, it is unlikely to be more than a transient hydration product of cement, at least at ordinary temperatures, but the possibility of partial replacement of SO_4^{2-} by $H_2SiO_4^{2-}$ in an AFm phase cannot be excluded.

C_2ASH_8, known by its mineral name of strätlingite and also as 'gehlenite hydrate', is well established as a natural mineral, hydration product of certain types of composite cements and laboratory product. Its crystal data

(Table 6.2) show it to be an AFm phase having an aluminosilicate anion as interlayer and probable constitution $[Ca_2Al(OH)_6][AlSiO_3(OH)_2 \cdot 4H_2O]$ (K25).

Naturally occurring AFm phases, called hydrocalumite, appear to vary in composition. The original specimen was reported to have a composition near to C_4AH_{12} with some CO_3^{2-} and a monoclinic unit cell similar to that of $\alpha\text{-}C_3A \cdot CaCl_2 \cdot 10H_2O$ (T31), but others have proved to be either the α or the β polymorph of that compound or members of a $C_3A \cdot CaCl_2 \cdot 10H_2O$–$C_4AH_{13}$ solid solution series (F26).

6.1.5 AFm phases containing iron

Iron(III) analogues of many of the phases described above have been obtained. Unless the anion imparts colour, AFm phases containing Fe^{3+} are colourless, brown colours being due to iron(III) oxide or hydroxide impurity. They have a-axial lengths of approximately 0.589 nm (S57,K26).

C_4FH_{13} has been obtained by many investigators (e.g. S57,R21). It has $c' = 0.79$ nm and forms a continuous series of solid solutions with C_4AH_{13} (S57). As in the corresponding system with Al_2O_3, the phase existing in contact with solution is the 19-hydrate (R21). $C_4F\bar{S}H_x$ phases are also well established (M56). Kuzel (K26) found that $C_4F\bar{S}H_{12}$ is isostructural with its Al^{3+} analogue, and has $c' = 0.8875$ nm; a higher hydrate with $c' = 1.021$ nm and probably containing $14\,H_2O$ exists above 90% RH. $C_4F\bar{S}H_{12}$ and $C_4A\bar{S}H_{12}$ form a continuous solid solution series at 100°C, but at 25°C or 50°C, miscibility is incomplete. At 25°C, $C_4F\bar{S}H_{12}$ accommodates up to about 50 mole % $C_4A\bar{S}H_{12}$, and the latter up to about 10 mole % $C_4F\bar{S}H_{12}$, and an intermediate phase with $Fe/Al \simeq 0.5$ exists (K26). $C_3F \cdot CaCl_2 \cdot 10H_2O$ similarly shows only limited miscibility with $C_3A \cdot CaCl_2 \cdot 10H_2O$ (K26).

Schwiete et al. (S58) found no solid solution between C_4FH_{13} and $C_4F\bar{S}H_{12}$, but in preparations made from C_4AF or C_6A_2F limited miscibility occurred at the sulphate-rich end of the series as with C_4AH_{13} and $C_4A\bar{S}H_{12}$. Fe^{3+} analogues of the Al^{3+} phases containing CO_3^{2-} appear to exist (R22,S56).

6.1.6 XRD patterns, thermal behaviour, optical properties and IR spectra

Strong peaks in the XRD powder patterns of AFm phases normally include the first and second orders of the layer thickness, and the $11\bar{2}0$ and $30\bar{3}0$

reflections, which for the Al^{3+} compounds have d-spacings at or near 0.2885 nm and 0.1666 nm, respectively. In mixtures, it is sometimes difficult to distinguish between AFm phases of similar layer thickness, and in such cases, examination of the patterns of heated samples is often useful.

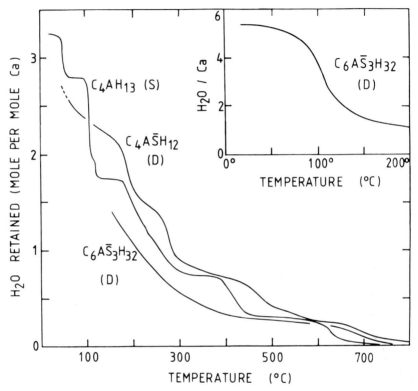

Fig. 6.2 Weight loss curves for C_4AH_{13} (static; Buttler et al. (B69)) and monosulphate and ettringite (TG; 10 deg C min^{-1}; Taylor (T16)).

Weight loss curves under either static or dynamic (TG) conditions or both have been reported for C_4AH_{13} (B69,L6), $C_4A\bar{C}H_{11}$ (T32,F25), $C_4A\bar{C}_{0.5}H_{12}$ (F25), $C_4A\bar{S}H_{12}$ (T16), $C_3A\cdot CaCl_2\cdot 10H_2O$ and other phases (A9,K24), C_2AH_8 (L6) and $C_2A\bar{S}H_8$ (K25). The curves for C_4AH_{13} and $C_4A\bar{S}H_{12}$ (Fig. 6.2) and of C_2AH_8 are all fairly similar if the quantity plotted against temperature is moles of H_2O retained per mole of CaO, and allowance is made for the effect of differing heating rates. With the carbonate, nitrate and bromide, and on static heating also with the chloride, volatiles in addition to H_2O are lost. All the curves show steps in varying degrees. Those at the lower temperatures are associated with the changes in hydration state of the AFm phase; loss of molecular water is complete by about 150°C on static, or 250°C on dynamic heating. XRD evidence showed that, for C_4AH_{13}, the plateau at about 350°C corresponds to formation of a poorly crystalline mixture of CH and $C_4A_3H_3$, and the steps at about 410°C and 630°C to decomposition of these phases to CaO and $C_{12}A_7H$ plus CaO, respectively (B69). The curve for $C_4A\bar{S}H_{12}$ shows similar features, but their significance has not been determined.

The DTA curves of C_4AH_{13} and $C_4A\bar{S}H_{12}$ are characterized by a series of endotherms below 300°C, which correspond to the steps in the TG curve (K27,M57). Fig. 6.3 includes a typical curve for $C_4A\bar{S}H_{12}$; that of C_4AH_{13} is similar, but with peak temperatures some 10 deg C lower. The peak temperatures depend on the technique and amount present, the 200°C peak being typically shifted to 180–190°C for the material present in cement pastes (B70). DTA curves have been reported for other AFm phases, including $C_4A\bar{C}H_{11}$ (T32), $C_3A \cdot CaCl_2 \cdot 10H_2O$ (A9,K24) and $C_2A\bar{S}H_8$ (K25).

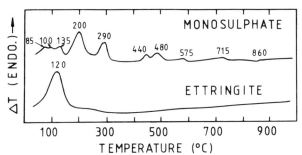

Fig. 6.3 DTA curves for monosulphate and ettringite (10 deg C min^{-1}). Taylor (T5).

As would be expected from their layer structures, AFm phases are optically negative. The Al^{3+} phases typically have $\omega = 1.50$–1.56, $\varepsilon = 1.49$–1.54; for the Fe^{3+} phases, ω is typically 1.56–1.61 and ε is 1.54–1.60.

Henning (H34) reviewed IR spectra of hydrated calcium silicates and aluminates; data for C_4AH_x and C_2AH_x phases, $C_4A\bar{S}H_{12}$, $C_4A\bar{C}H_{11}$ and $C_4F\bar{S}H_{12}$ were included. Other data for phases in the CaO–Al_2O_3–H_2O system (B71) and for $C_4A\bar{S}H_{12}$ (B72) have been reported.

6.2 AFt phases

6.2.1 Compositions and crystal structures

AFt (Al_2O_3–Fe_2O_3–tri) phases have the general constitutional formula

$[Ca_3(Al,Fe)(OH)_6 \cdot 12H_2O]_2 \cdot X_3 \cdot xH_2O$, where $x \leq 2$ and X represents one formula unit of a doubly charged, or, with reservations, two formula units of a singly charged, anion. The term AFt refers to the three units of CX in an alternative way of writing the formula, $C_3(A,F) \cdot 3CX \cdot yH_2O$ [or $C_6(A,F) \cdot X_3 \cdot yH_2O$], where $y = x + 30$. AFt phases are formed under broadly similar conditions to AFm phases, but at higher ratios of CaX to $C_3(A,F)$ and rarely above about 90°C. The range of anions that can occupy the X sites is smaller, and singly charged anions can possibly only be accommodated to a limited extent. The most important AFt phase is ettringite, $[Ca_3Al(OH)_6 \cdot 12H_2O]_2 \cdot (SO_4)_3 \cdot 2H_2O$ or $C_3A \cdot 3CaSO_4 \cdot 32H_2O$; a phase of or near this composition is formed during the early hydration of most Portland cements. In the early literature, it was often called high-sulphate calcium sulphoaluminate. Ettringite also occurs as a natural mineral. Phases of AFt type also exist in which other cations replace the Ca^{2+} or Al^{3+} or both; an important example is thaumasite, $Ca_3[Si(OH)_6 \cdot 12H_2O](CO_3)(SO_4)$.

AFt phases form hexagonal prismatic or acicular crystals. Their structures (C29,M58,M59) are based on columns in hexagonal array, running parallel to the prism (c) axis, with the X anions and, usually, H_2O molecules in the intervening channels (Fig. 6.4). The columns, which are of empirical formula $[Ca_3(Al,Fe)(OH)_6 \cdot 12H_2O]^{3+}$, are composed of $(Al,Fe)(OH)_6$ octahedra alternating with triangular groups of edge-sharing CaO_8 polyhedra, with which they share OH^- ions. Each calcium atom is also coordinated by four H_2O molecules, the hydrogen atoms of which form the nearly cylindrical surface of the column. The channels contain four sites per formula unit with six calcium atoms, of which, in ettringite, three are occupied by SO_4^{2-} ions and one by two H_2O molecules. The repeat distance along the column is approximately 1.07 nm.

Ettringite is trigonal, with $a = 1.123$ nm, $c = 2.150$ nm, $Z = 2$, $D_x = 1775$ kg m^{-3} (S59); the space group is P31c, apparent higher symmetry being due to twinning or disorder (M59). It is optically negative, with $\omega = 1.463$, $\varepsilon = 1.459$ (M60). The doubling of c is due to ordering of the SO_4^{2-} ions and H_2O molecules in the channels. Analogues with Fe^{3+}, Mn^{3+}, Cr^{3+} or Ti^{3+} in place of Al^{3+} exist (B73,B74,S59); solid solution in the Fe^{3+}–Al^{3+} series is almost continuous, but there is probably a small gap at 70–80

Fig. 6.4 Crystal structure of ettringite. (A) Part of a single column in (11$\bar{2}$0) projection; A = Al, C = Ca, H = O of an OH group, W = O of an H_2O molecule. Hydrogen atoms are omitted, as are the H_2O molecules attached to those calcium atoms lying in the central vertical line of the figure. (B) Projection on the ab-plane, showing columns (large circles) and channels (small circles); the unit cell, with $a = 1.123$ nm, is outlined. Modified from Struble (S59).

Hydrated Aluminate, Ferrite and Sulphate Phases 179

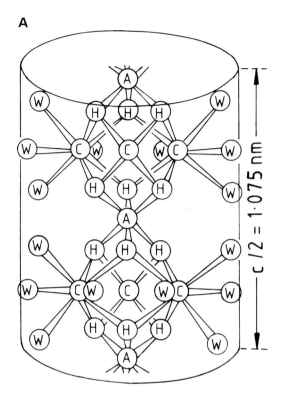

mole % Fe^{3+} (B73). This Fe^{3+} phase has $a = 1.1182$ nm, $c = 2.2008$ nm (S59). Analogues also exist with CrO_4^{2-} replacing SO_4^{2-} (B73,B75) and with Sr^{2+} replacing Ca^{2+} (B76). A carbonate analogue, $C_6A\bar{C}_3H_{32}$, is well established (C28). Its unit cell ($a = 1.0834$ nm, $c = 2.1250$ nm) (S59) and refractive indices ($\omega = 1.480$, $\varepsilon = 1.456$) are near those of ettringite, but the cell parameters differ sufficiently to permit distinction by XRD. Hollow, tubular crystals have been observed (C28).

Midgley and Rosaman (M57) concluded from DTA evidence that SO_4^{2-} can be partly replaced by OH^- in the AFt phase of cement pastes. Pöllman et al. (P22a) showed that the fully OH^--substituted phase, C_6AH_{36}, exists but is easily carbonated. Material described as a silicate analogue was later shown to have also contained CO_3^{2-} (C28) but a product identified by X-ray microanalysis as $C_6AS_3H_{31}$ has since been found in reaction rims around C_3A particles in pastes of that compound with C_3S and gypsum (R20). The compound $C_3A \cdot 3CaCl_2 \cdot 30H_2O$ has been reported to form at temperatures below $0°C$, but to be unstable at $20°C$ (S60,S61). Later attempts to prepare it failed (P22a). Several other AFt phases have been described (J18,P22a). A higher hydrate of ettringite, $C_6A\bar{S}_3H_{36}$, is reported to exist at high relative humidities (P22a).

Thaumasite, $[Ca_3Si(OH)_6 \cdot 12H_2O](SO_4)(CO_3)$ or $C_3S\bar{S}\bar{C}H_{15}$, has a structure similar to those of the AFt phases, with Si^{4+} replacing Al^{3+} and SO_4^{2-} and CO_3^{2-} groups in the channel sites (E2,E3). The octahedral coordination of the Si^{4+} was first established from IR evidence (M61). Thaumasite is hexagonal, with space group $P6_3$, $a = 1.104$ nm, $c = 1.039$ nm, $Z = 2$, $D_x = 1886$ kg m^{-3} (E2); the refractive indices are $\omega = 1.470$, $\varepsilon = 1.504$. A Mn^{4+} analogue, jouravskite, exists (G55). Thaumasite and ettringite are not completely miscible, but limited solid solutions probably occur (E2).

6.2.2 Properties

If crystals of natural ettringite are dehydrated, lattice shrinkage occurs, mainly in the ab plane, giving a product with $a \simeq 0.84$ nm, $c \simeq 1.02$ nm and probable constitution $[Ca_3Al(OH)_6 \cdot 3H_2O]_2(SO_4)_3$; this has been observed using crystals of the natural mineral heated at $110°C$ (B77) and subsequently examined by XRD, or examined by electron diffraction, dehydration then occurring in the high vacuum (G56). Most of the molecular water has been lost, and the columns have fused together. Several phases of similar structure, such as despujolsite, $[Ca_3Mn(OH)_6 \cdot 3H_2O](SO_4)_2$ (G57), occur as natural minerals. Synthetic ettringite becomes almost amorphous when heated or subjected to a high vacuum, and gross morphological changes are easily effected in the electron microscope. The differing behaviour may be related to defect concentrations.

At ordinary humidities, ettringite begins to lose water rapidly at about 50°C. Fig. 6.2 includes a TG curve, determined at 10 deg C min^{-1}. Curves obtained at lower heating rates for ettringite and its Fe^{3+} and Cr^{3+} analogues have been reported (B73). Thaumasite is thermally more stable, rapid weight loss beginning only at 110°C (B78). DTA curves have been reported for ettringite (M57,B73), its Fe^{3+} (M57,B73) and CO_3^{2-} (C28) analogues and thaumasite (B78). Fig. 6.3 includes a curve for ettringite. All show strong endotherms at 110–150°C, the temperature varying somewhat with the composition, technique and amount present. When present in cement pastes, ettringite typically gives a peak at 125–130°C (B70).

In XRD powder patterns, ettringite is readily recognized by its strong, low angle peaks at 0.973 nm (10$\bar{1}$0) and 0.561 nm (11$\bar{2}$0), which disappear on heating or intensive drying at room temperature. The corresponding peaks of thaumasite have spacings of 0.956 nm and 0.552 nm. Some investigators have found it impracticable to determine ettringite quantitatively by QXDA, because crystallinity is easily lost on grinding (M62,O15), but Crammond (C30) reported satisfactory techniques for determining ettringite, thaumasite and gypsum by this method. Ettringite has also been determined by DTA or DSC (O15). IR spectra of ettringite, thaumasite and related phases have been reported (B73-B75,M61,B78,S59,B70).

6.3 Other hydrated phases

6.3.1 Hydrogarnet phases

These phases have structures related to that of grossular or garnet ($Ca_3Al_2Si_3O_{12}$). The latter has a cubic structure, in which the silicon, aluminium and calcium atoms are in tetrahedral, octahedral and distorted cubic coordination, respectively; each atom is bonded to one silicon, one aluminium and two calcium atoms. In the hydrogarnets, this structure is modified by omission of some or all of the silicon atoms, the charge being balanced by replacing each of the oxygen atoms to which it was attached with a hydroxyl group. The Al^{3+} may be partly or wholly replaced by Fe^{3+}. Complete replacement of silicon by four hydrogen atoms in grossular thus gives $Ca_3[Al(OH)_6]_2$ or C_3AH_6, and solid solutions exist within a compositional region bounded by C_3AH_6, C_3FH_6, C_3AS_3 and C_3FS_3. In recent mineralogical nomenclature, phases in the C_3AH_6–C_3AS_3 series are collectively called hydrogrossular, and the names katoite and hibschite are used more specifically to denote those in the C_3AH_6–$C_3AS_{1.5}H_3$ and $C_3AS_{1.5}H_3$–C_3AS_3 ranges, respectively (P23).

C_3AH_6 is the only stable ternary phase in the $CaO-Al_2O_3-H_2O$ system at ordinary temperatures, but neither it nor any other hydrogarnet phase is formed as a major hydration product of typical, modern Portland cements under those conditions. Minor quantities are formed from some composite cements and, in a poorly crystalline state, from Portland cements. Larger quantities were given by some older Portland cements, and are also among the normal hydration products of autoclaved cement-based materials. C_3AH_6 is formed in the 'conversion' reaction of hydrated calcium aluminate cements (Section 10.1).

Hydrogarnet phases crystallize in various cubic forms, of which at ordinary temperatures icositetrahedra are probably the most usual. The space group is Ia3d, with Z = 8; C_3AH_6 has a = 1.25755 nm, D_x = 2527 kg m^{-3} (K28). Detailed structural studies have been reported on C_3AH_6 or C_3AD_6 (K29,F27,B79) and on phases in the $C_3AH_6-C_3AS_3$ (B80,S62) and $C_3FH_6-C_3FS_3$ (C31) series. The hydrogen atoms lie outside the tetrahedra of oxygen atoms to which they are attached, and which surround the empty silicon sites. They form neither hydrogen bonds nor, contrary to some early views, H_4 clusters.

The compositions of naturally occurring specimens indicate that the $C_3AH_6-C_3AS_3$ solid solution series is continuous (P23), but synthetic studies show that it is unlikely that all compositions in this series can be obtained at any one temperature; probably, those in the range from C_3AH_6 to about $C_3AS_2H_2$ can be obtained at relatively low temperatures, and the more siliceous ones only above about 360°C (R6). Contrary to many earlier opinions, C_3FH_6 can be prepared in the absence of SiO_2, though it is metastable and easily decomposed at ordinary temperatures (R21).

Equations relating the cell parameter and coordinates of the oxygen atoms in hydrogarnets to composition have been proposed (B80,B81), but give mediocre agreement with the observed cell parameter for C_3AH_6. The equations implicitly assume that Végard's law is obeyed over the entire $C_3AH_6-C_3AS_3$ series, but there is probably a break near the C_3AS_3 end (R6). An empirical equation giving better agreement over the more highly hydrated compositions relevant to cement chemistry ($SiO_2/CaO \leqslant 0.67$) is

$$a = 1.171 + 0.016Fe_2O_3 + 0.0144H_2O \qquad (6.1)$$

where a is the cell parameter in nanometres and Fe_2O_3 and H_2O are the values of x and y, respectively, in the formula $C_3A_{1-x}F_xS_{3-y/2}H_y$. In order to determine the composition of a phase of this general composition from the XRD powder pattern, a second quantity must be used. Zur Strassen (Z10) showed that the intensity ratio of the 022 peak to the 116 peak (d = 0.4445 nm and 0.2040 nm, respectively, for C_3AH_6) was suitable. Table 6.4

gives some values of a and of this ratio, calculated from equation 6.1 and from the crystal structure, respectively (T5).

Studies by weight loss, XRD and IR spectroscopy show that, on static heating, C_3AH_6 decomposes at 200–250°C to a mixture of $C_{12}A_7H$ and CH, of bulk composition $C_3AH_{1.43}$ (B79,B82). At 500–550°C, the CH decomposes to CaO, and on prolonged heating at 810°C, C_3A is formed (K28). The $C_{12}A_7$ and C_3AH_6 structures are closely related (B12) and the $C_{12}A_7H$ forms topotactically, with all three axes parallel to those of the C_3AH_6 (B79). The a-axis of $C_{12}A_7H$ (1.198 nm) is only slightly shorter than that of C_3AH_6 (1.258 nm), and the lattice shrinkage at 200–250°C is gradual.

Table 6.4 Calculated values of the cell parameter a and ratio of intensities of the 022 to the 116 XRD powder reflections for some hydrogarnet phases (T5)

x (moles Fe_2O_3 in formula)	$C_3Al_{1-x}F_xH_6$		$C_3Al_{1-x}F_xSH_4$		$C_3Al_{1-x}F_xS_2H_2$	
	a (nm)	I_{022}/I_{116}	a (nm)	I_{022}/I_{116}	a (nm)	I_{022}/I_{116}
1.0	1.273	2.22	1.245	1.77	1.216	1.36
0.5	1.265	1.15	1.237	0.78	1.208	0.43
0.0	1.257	0.41	1.229	0.17	1.200	0.03

Passaglia and Rinaldi (P23) discussed IR spectra and TG curves for C_3AH_6 and other hydrogarnet phases. The TG curve of C_3AH_6 shows major loss at 250–310°C and further loss at 450–550°C, but in that of a katoite specimen the two steps were barely distinguishable. Majumdar and Roy (M63) reported DTA and IR data for C_3AH_6. The refractive indices of C_3AS_3 and C_3AH_6 are 1.734 and 1.604, respectively; those of the solid solutions are linearly related to the composition (P23).

6.3.2 CAH_{10}

This phase is formed as a hydration product of calcium aluminate cements (Section 10.1). Buttler and Taylor (B83) gave tentative crystal data and described the thermal behaviour. The crystals are hexagonal prisms, but none large enough for a single crystal XRD have been obtained, and attempts to determine the unit cell by electron diffraction have failed due to loss of crystallinity in the instrument. The unit cell is possibly hexagonal, with $a = 1.644$ nm, $c = 0.831$ nm, $Z = 6$ [CAH_{10}], $D_x = 1730$ kg m^{-3}, prism axis c, and the ionic constitution is possibly $Ca_3[Al_6(OH)_{24}] \cdot 18H_2O$, with rings of six edge-sharing octahedra. Much of the water is very loosely bound; water loss begins on drying at 80% RH, and the H/A ratio is about 7 at 45% RH or 5.5 on drying over P_2O_5. TG curves at 0.5 or 5 deg min^{-1} (L25) show no definite steps; at 110°C, H/A is about 4 for the slower, or 6 for the faster heating rate. On static heating at $p(H_2O) = 6$ torr, H/A is about 6 at 50°C, 3 at 110°C and 1 at 200°C.

The products of thermal decomposition depend on how easily the water can escape (B83). If escape is easy, a nearly amorphous product retaining some of the structural features of CAH_{10} is formed, but if it is not, C_3AH_6 and gibbsite (AH_3) are formed. The latter situation can occur if a hardened paste of a calcium aluminate cement containing CAH_{10} is heated. Under conditions not giving C_3AH_6, the XRD pattern varies continuously with temperature; the crystallinity deteriorates, the cell parameters decrease slightly and relative intensities change markedly. Little change occurs up to about 70°C, but by 120°C (H/A $\simeq$ 2.5), only the 1.4 nm peak and a close doublet near 0.7 nm remain, and the relative intensity of the 1.4 nm peak is much reduced. By 200–350°C, the material is amorphous; by 1000°C, CA is formed. The marked loss in crystallinity that occurs when H/A falls below 4 can be attributed to the fact that further loss entails dehydroxylation. With some preparations small amounts of C_2AH_8 and AH_3 are also formed. The rehydration behaviour has been studied (B84).

DTA curves at 10 deg C min^{-1} show a large endotherm at 130–150°C and a smaller one at about 290°C. Since the first of these is due to loss of molecular water, its height is affected by any preliminary drying that the specimen may have undergone. DTA has been used to determine the relative amounts of CAH_{10} and AH_3, and thus indirectly also of C_3AH_6 in calcium aluminate cement concretes, but caution is needed because the CAH_{10} may have undergone partial dehydration and also because its thermal decomposition can itself yield AH_3. To some extent, this caution also applies to the determination of CAH_{10} by QXDA.

A ^{27}Al NMR study (G54) confirmed the octahedral coordination of the aluminium atom. Henning (H34) reported IR spectra for CAH_{10}, CAH_7 and CAH_4.

6.3.3 Brucite, hydrotalcite and related phases

Brucite [magnesium hydroxide; $Mg(OH)_2$] is isostructural with CH. It is formed in Portland cement concrete that has been attacked by magnesium salts, and on hydration of Portland cements high in MgO and possibly of Portland cements in general. It has $a = 0.3147$ nm, $c = 0.4769$ nm, $Z = 1$, $D_x = 2368$ kg m^{-3}, $\omega = 1.561$, $\varepsilon = 1.581$ (S63). Three polytypes of aluminium hydroxide [$Al(OH)_3$], gibbsite, bayerite and nordstrandite, contain layers essentially similar to those in brucite, but with an ordered pattern of cation vacancies and a different relationship between adjacent layers, resulting from hydrogen bonding.

There exists a range of phases structurally related to brucite as the AFm phases are to CH; that is, some of the Mg^{2+} ions are replaced by tripositive ions, typically Al^{3+} or Fe^{3+}, and the charge is balanced by anions which, together with H_2O molecules, occupy interlayer sites. Allman (A11) reviewed them. The first structure determination on this group of phases was on sjögrenite, a polytype of $[Mg_{0.75}Fe_{0.25}(OH)_2](CO_3)_{0.125}(H_2O)_{0.5}$ (A12). They are sometimes called hydrotalcite-type phases, after a polytype of the Al^{3+} analogue of sjögrenite. They are formed on hydration of slag cements, and as minor hydration products of Portland cement. These products possibly approximate in composition to meixnerite, $[Mg_{0.75}Al_{0.25}(OH)_2](OH)_{0.25}(H_2O)_{0.5}$ (K30), which, like the other phases named above, occurs as a natural mineral.

Because Mg^{2+}, Al^{3+} and Fe^{3+} are all relatively similar in radius, the ratio of di- to tripositive ions is not fixed, as in the AFm phases. Brindley and Kikkawa (B85) discussed the factors limiting it. The minimum $M^{2+}/(M^{2+} + M^{3+})$ ratio is 2/3, and could be set by a requirement to avoid the occurrence of M^{3+} ions in adjacent sites; in phases of this ratio, the cations are ordered (H35), giving unit cells with a approximately $\sqrt{3}$ times that of brucite. At higher ratios, the cations are disordered, though ordered regions may occur locally, and the upper limit of the ratio may be set by nucleation of pure brucite layers if the proportion of tripositive ions falls below a certain value. In a study of synthetic meixnerites, pure phases were obtained only at Mg/(Mg + Al) ratios between 0.67 and 0.76, suggesting an upper limit close to 0.75 (M64), and it is probably significant that this ratio is much the most usual in the natural minerals.

As with the AFm phases, anion exchange reactions occur (H35), and meixnerite readily takes up CO_2, giving a material similar to hydrotalcite. Among the natural minerals of the group, ones with interlayer CO_3^{2-} are relatively common, and ones with interlayer OH^- very rare. This may be a further indication that replacement of OH^- by CO_3^{2-} occurs easily. Differing schemes for the packing of H_2O molecules and CO_3^{2-} ions in the interlayer have been proposed (A12,I8).

Layer thicknesses are similar to those of the AFm phases, and it would be difficult to distinguish the two groups of phases by XRD if only the basal reflections could be observed. Polytypism is common; for example, sjögrenite, with a two-layer structure, has a three-layer polytype, called pyroaurite. Three-layer structures are the ones normally formed at room temperature. Meixnerite, which is a three-layer form, has $a = 0.30463$ nm, $c = 2.293$ (3 × 0.764) nm, space group $R\bar{3}m$, $Z = 3$, $D_x = 1950\,\mathrm{kg\,m^{-3}}$ (K30). Thermal dehydration (B86) and IR spectroscopic (H35) studies on Mg–Al phases of this group have been reported.

6.3.4 Sulphate phases

Table 6.5 lists these phases and gives crystal and optical data. The gypsum used in cement manufacture is usually of natural origin, but byproduct gypsums produced by the chemical industry have also been used. The most important is phosphogypsum, resulting from manufacture of phosphoric acid by the wet process. It contains impurities, especially phosphate and fluoride, that retard setting and which, if present in relatively high proportions, may have to be partially removed by appropriate pretreatments (M65). Another byproduct gypsum, which is likely to be produced in increasing quantity, is that resulting from flue gas desulphurization.

Dehydration of gypsum in air at 70–200°C gives hemihydrate ($CaSO_4 \cdot \simeq 0.5H_2O$) or γ-$CaSO_4$ ('soluble anhydrite'), the extent of dehydration depending on the temperature and duration of heating and RH of the surrounding atmosphere. γ-$CaSO_4$ is probably never completely anhydrous (B88). These products have essentially identical crystal structures, which are hexagonal or pseudohexagonal, with the H_2O molecules in channels parallel to the hexagonal c-axis. The channels can accommodate methanol (R23). Hemihydrate is known as a natural mineral, called bassanite. Above about 200°C, anhydrite (also called 'insoluble anhydrite'; Section 2.5.1) is formed.

Reported α and β forms of hemihydrate appear to differ only in degree of crystallinity or crystal size, the more crystalline α form being obtained by dehydration of gypsum in the presence of liquid water or aqueous solutions and the β form by heating in air (B89). γ-$CaSO_4$ is readily rehydrated to hemihydrate. There has been uncertainty as to the maximum H_2O content of these products and whether variation in this quantity is continuous. Kuzel (K31), studying highly crystalline preparations, found that variation was continuous from 0.53 to 0.62 H_2O, but that a miscibility gap existed at 0.03–0.53 H_2O; the space group was $P3_121$ for 0.62 H_2O and $I2$ for 0.53 H_2O. Some studies indicate a maximum H_2O content well above 0.5; thus Abriel (A13) reported a crystal structure refinement for a preparation with 0.8 H_2O, for which the space group was $P3_121$. The H_2O molecules were found to be statistically distributed, with four molecules among five positions. Abriel concluded that larger unit cells arose through ordering of the H_2O molecules, which was possible only at H_2O contents of or below 0.5. This would account for the larger cell with space group $I2$ found at H_2O contents around 0.5 and for the reversion to a smaller cell in γ-$CaSO_4$. The extent to which these structural distinctions are relevant to the less crystalline products obtained by heating gypsum in air is uncertain.

Reaction in the presence of water between K_2SO_4 and $CaSO_4$ can produce syngenite, $KC\bar{S}_2H$, data for which are included in Table 6.5.

Table 6.5 Crystal and optical data for hydrated sulphate phases

	Gypsum $CaSO_4 \cdot 2H_2O$	Hemihydrate $CaSO_4 \cdot 0.8H_2O$	Hemihydrate $CaSO_4 \cdot 0.5H_2O$	γ-$CaSO_4$ $CaSO_4 \cdot <0.05H_2O$	Syngenite $K_2Ca(SO_4)_2 \cdot H_2O$
Crystal data:					
Crystal system	Monoclinic	Trigonal	Monoclinic	Hexagonal	Monoclinic
Space group	I2/a	$P3_121$	I2	$P6_222$	$P2_1/n$
a (nm)	0.5679	0.6968	0.6930	0.69695	0.6225
b (nm)	1.5202	–	1.2062	–	0.7127
c (nm)	0.6522	0.6410	1.2660	0.63033	0.9727
Angles (°)[a]	$\beta = 118.43°$	$\gamma = 120°$	$\alpha = 90°$[b]	$\gamma = 120°$	$\beta = 104.15°$
Z	4	3	12	3	2
D_x (kg m^{-3})	2310	2783	2733	2558	2607
Reference[c]	P24	A13	L26	L26	B87
Optical data and morphology (W3):					
α	1.5205	$\simeq 1.56$	1.559	1.505	1.5010
β	1.5226	–	1.5595	–	1.5166
γ	1.5296	$\simeq 1.59$	1.5836	1.548	1.5176
Optic sign and 2V	(+)58°	+	(+)14°	+	(−)28.3°
Morphology	Tablets, (010) cleavage	Hexagonal; prisms (length c); rarely, hexagonal plates			Tablets or prisms

[a] Where not 90°.
[b] Axes chosen to clarify relation to the trigonal and hexagonal modifications; $b = 0.696\sqrt{3}$ nm, $c = 2 \times 0.633$ nm.
[c] Also to crystal structure refinements.

6.4 Equilibria and preparative methods

6.4.1 The $CaSO_4$–H_2O, $CaSO_4$–$Ca(OH)_2$–H_2O and $CaSO_4$–K_2SO_4–H_2O systems

At ordinary temperatures, the metastable solubility of hemihydrate or γ–$CaSO_4$ is considerably higher than the solubility of gypsum, which is thus precipitated on mixing either of the former materials with water (Fig. 6.5); the setting of plaster is based on this reaction. Anhydrite displaces gypsum as the stable phase above 42°C, or at somewhat lower temperatures if other solutes are present, but is not readily precipitated, and gypsum can persist metastably up to at least 98°C, above which temperature hemihydrate is stable relative to gypsum.

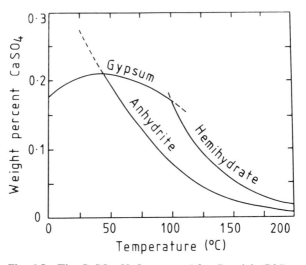

Fig. 6.5 The $CaSO_4$–H_2O system. After Posnjak (P25).

The solubility of gypsum is only slightly decreased in the presence of CH, and vice versa (Table 6.6). The effect, which could scarcely be explained if the only dissolved species were Ca^{2+}, OH^- and SO_4^{2-}, occurs because substantial proportions of the OH^- and SO_4^{2-} ions are complexed to give $CaOH^+$ (Section 5.5.2) and neutral $CaSO_4$, respectively. The formation constant of the latter species is 204 mol l^{-1} (S51).

Gartner et al. (G58) discussed equilibria governing precipitation of gypsum, CH and syngenite with reference to the early stage of cement hydration. They concluded that the equilibrium activity products at 25°C were $(2.547 + 2.258I) \times 10^{-5}$ mol^2 l^{-2} for gypsum, 8.25×10^{-6} mol^3 l^{-3}

for CH and $(13.9I - 0.3) \times 10^{-5}$ mol^5 l^{-5} for syngenite, where I is the ionic strength. As Bailey and Hampson (B90) noted earlier, it is essential to take account of both complex formation and activity coefficients.

Table 6.6 Invariant points in the system CaO–Al$_2$O$_3$–SO$_3$–H$_2$O at 20°C (D24, B91). Labelling of points as in Fig. 6.8

		Concentrations (mmol l^{-1})		
Point	Solid phases	CaO	Al$_2$O$_3$	CaSO$_4$
w	Ca(OH)$_2$	20.7	–	–
o	Hydrous alumina	–	≃0.01	–
b	Gypsum	–	–	15.2
v	Ca(OH)$_2$, C$_4$AH$_{19}$	21.5	0.0329	–
y	C$_4$AH$_{19}$, C$_2$AH$_8$	10.9	0.94	–
p	C$_2$AH$_8$, hydrous alumina	5.79	6.74	–
c	CH, gypsum	19.7	–	12.4
j	Hydrous alumina, gypsum	–	≃0.01	15.2
f	Ettringite, gypsum, Ca(OH)$_2$	20.6	0.0272	12.2
e	Ettringite, gypsum, hydrous alumina	15.2	0.103	0.151
d	Ettringite, monosulphate, Ca(OH)$_2$	21.1	0.022	0.0294
n	Ettringite, monosulphate, hydrous alumina	5.98	1.55	5.88

6.4.2 The CaO–Al$_2$O$_3$–H$_2$O, CaO–Al$_2$O$_3$–SiO$_2$–H$_2$O and CaO–Al$_2$O$_3$–SO$_3$–H$_2$O systems

Fig. 6.6 shows equilibria in the CaO–Al$_2$O$_3$–H$_2$O system mainly as found by Jones and Roberts (J17) at 25°C and reported by Jones (J19), who also reviewed earlier studies. Table 6.6 gives invariant concentrations at 20°C. The stable phases in contact with solution at 25°C are gibbsite (AH$_3$), C$_3$AH$_6$ and CH, but metastable solubility curves for AFm phases, CAH$_{10}$ and hydrous alumina can also be obtained.

The curve shown for AFm phases contains a discontinuity at Y. Jones and Roberts considered that the curve CY related to C$_4$AH$_{19}$, and the curve YT to a solid solution of C$_4$AH$_{19}$ with C$_2$AH$_8$ which had C/A = 2.0 at T and C/A = 2.4 at Y; the prolongation TT' of this latter curve related to C$_2$AH$_8$. The results of Dosch and Keller (D20) do not support the conclusion that solid solutions with C/A ratios of 2.0–2.4 exist, but the data of Jones and Roberts could probably be equally well explained by the formation of oriented intergrowths on a submicrometre scale, which in the present case would be virtually indistinguishable from solid solutions by XRD, light microscopy or solubility relations. Not all workers have observed the discontinuity at Y. The curve for the AFm phases may be slightly variable, as small changes in conditions could determine whether separate crystals of C_4AH_{19} and C_2AH_8, intergrowths or solid solutions are formed.

Data have also been obtained at other temperatures from 1°C to 50°C (J19); some are considered in Section 10.1.5. At 50°C or above, the other ternary phases are rapidly replaced by C_3AH_6.

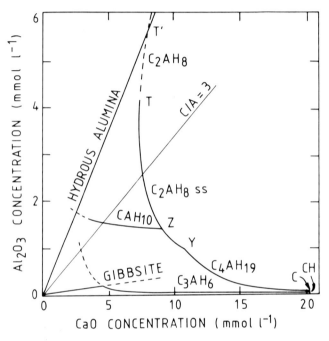

Fig. 6.6 The $CaO-Al_2O_3-H_2O$ system at 25°C, mainly after Jones and Roberts (J17); CAH_{10} solubility curve at 21°C, Percival and Taylor (P26).

No detailed studies on the $CaO-Al_2O_3-SiO_2-H_2O$ system at ordinary temperatures have been reported, but, based largely on considerations of solubility products, Dron (D23) suggested the main features of a probable metastable equilibrium diagram, a modified form of which is shown in Fig. 6.7. The diagram indicates the metastable coexistence of CH, C-S-H and C_4AH_{19}, and that strätlingite cannot coexist with CH. In cement pastes, the AFm and AFt phases appear to accommodate some silicon; the extents of these substitutions, and the related question of the scale on which the hydrated phases are mixed, are considered in Sections 7.2 and 7.3.

Equilibria in the $CaO-Al_2O_3-SO_3-H_2O$ system were studied at 25°C by Jones (J20) and at 20°C by d'Ans and Eick (D24). Following Brown (B91), the latters' results are shown on a three-dimensional diagram in which the functions plotted are the tenth roots of the concentrations (Fig. 6.8). This figure represents the metastable equilibria involving AFm phases and hydrous alumina. The stable equilibria, which involve C_3AH_6 and crystalline AH_3, were also studied but are less relevant to cement hydration chemistry. Table 6.6 gives data for the invariant points and probable AFm compositons. Seligmann and Greening (S56) considered that an AFm solid solution may have a stability field in the quaternary system. The CaO–Al_2O_3–SO_3–H_2O system modified by various additions of alkalis has also been studied (J21). In the main, the general form of the phase relations appears to be unchanged, but the invariant concentrations are considerably altered.

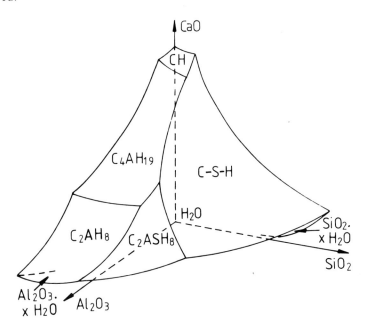

Fig. 6.7 Suggested general form of the metastable equilibria in the CaO–Al_2O_3–SiO_2–H_2O system at ordinary temperature, showing solubility surfaces. Modified from Dron (D23).

Ogawa and Roy (O16) studied the stability of ettringite in the presence of water at elevated temperatures and summarized earlier work. Ettringite appears to be stable in water up to at least 90°C at atmospheric pressures, 130°C at 6.7 MPa or 145°C at 27 MPa. The principal decomposition product is monosulphate.

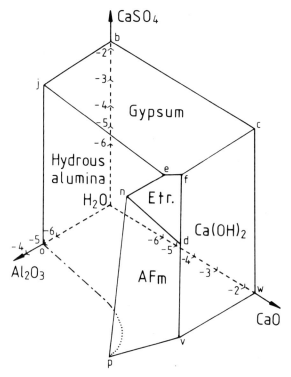

Fig. 6.8 Metastable equilibria in the CaO–Al$_2$O$_3$–SO$_3$–H$_2$O system at 20°C, showing crystallization surfaces, based largely on the data of d'Ans and Eick (D24). Concentrations are expressed as logarithms to the base 10 of values in mol l^{-1} and, following Brown (B91), are plotted on scales proportional to their tenth roots. Concentrations at the lettered invariant points are given in Table 6.6.

6.4.3 Preparative methods

Mylius (M60) described the preparation of many calcium aluminate hydrates and related compounds. As with C–S–H, it is normally essential to exclude atmospheric CO$_2$ and to avoid prolonged contact with glass apparatus. Dosch and Keller (D20) described methods for obtaining many AFm phases by anion exchange or other special procedures.

The C$_4$A and C$_2$A hydrates are most conveniently prepared by adding CaO or saturated CH solution to a supersaturated calcium aluminate solution obtained by shaking CA or white calcium aluminate cement (Section 10.1.1) with water. Such solutions typically contain up to about 1.2 g CaO l^{-1} and 1.9 g Al$_2$O$_3$ l^{-1}, these concentrations depending on the shaking time, temperature, proportioning and particle size of the starting material. The amount of CaO or CH solution added must be such as to produce the desired compound and a solution in metastable equilibrium with it, in accordance with the solubility relations (Fig. 6.6 and Table 6.6). CAH_{10} may be prepared in this way at 5°C, no added CaO or CH being added. C_3AH_6 may be prepared from metallic aluminium and CH solution at 50°C, or by hydrothermal treatment of C_3A or a mixture of CaO and AH_3 with water at 150°C.

Ettringite may be prepared by mixing solutions of CH, gypsum and $Al_2(SO_4)_3$. Mylius used 1 l of saturated CH (1.2 g CaO l^{-1}) diluted to 1.2 l (A), 530 ml of saturated gypsum (2.05 g $CaSO_4$ l^{-1}; B) and a solution of 1.78 g of $Al_2(SO_4)_3 \cdot 18H_2O$ in 770 ml of water (C). Solutions A and B were first mixed, and solution C then added, with shaking. The mixed solution was shaken at 18°C until an initially precipitated gel had been replaced by crystalline material. The molar proportions are $8CaO : 1Al_2(SO_4)_3 : 3CaSO_4$.

Monosulphate is troublesome to prepare. Mylius mixed 500 ml of a supersaturated calcium aluminate solution (649 mg CaO and 969 mg Al_2O_3 l^{-1}) at 18°C, with shaking, with 2924 ml of saturated CH (1.25 g CaO l^{-1}) and 334 ml of saturated gypsum (1.93 g $CaSO_4$ l^{-1}). After 30 min, crystallization was complete; the mixture was filtered, washed four times with a little water followed by 96% ethanol and ether, and dried over $CaCl_2$ and soda lime without evacuation. Monosulphate may also be prepared hydrothermally (A8).

Carlson and Berman (C28) described the preparation of $C_4A\bar{C}H_{11}$ and $C_6A\bar{C}_3H_{32}$. $C_4A\bar{C}H_{11}$ was obtained using a supersaturated calcium aluminate solution, saturated CH and Na_2CO_3. The most satisfactory method for preparing $C_6A\bar{C}_3H_{32}$ employed a supersaturated calcium aluminate solution, NH_4HCO_3 and a concentrated CH solution containing sucrose.

Crammond (C32) reviewed reports of the formation of thaumasite in the laboratory and in deteriorated building materials. It forms readily at 4°C from mixtures that contain the appropriate ions and also 0.4–1.0% of reactive alumina. It does not form at 25°C or above. A little Al^{3+} may be essential, and reported failures to reproduce published syntheses may be due to the use of aluminium-free materials.

6.5 Hydration reactions of the aluminate and ferrite phases

6.5.1 Reaction of C_3A with water or with water and calcium hydroxide

The reactions of C_3A and C_4AF with water, alone or in the presence of calcium hydroxide or sulphate or both, have been widely studied for the light they may cast on the mechanism of Portland cement hydration, and especially on the action of gypsum in controlling setting. The results of such studies must be interpreted with caution. The aluminate and ferrite phases in Portland cement clinker differ greatly in composition from the pure compounds. In normal clinkers, they are intimately mixed with each other and react in an environment chemically and microstructurally influenced, and perhaps dominated, by the silicate phases, and which is further affected by the presence of alkalis. The hydration of CA and $C_{12}A_7$ has similarly been studied for its bearing on that of calcium aluminate cements, and is considered in Section 10.1.4.

Many studies have shown that the reaction of C_3A with water in suspensions at ordinary temperatures gives C_2AH_8 and C_4AH_{19}, which are converted into C_3AH_6 at a rate that depends on such factors as the temperature, w/s ratio, grain size, and presence or absence of CO_2. At temperatures above about 30°C or in pastes, conversion is rapid and the AFm phases may not be observed; local temperature increase due to the strongly exothermic nature of the reaction probably has an important effect. The formation of gel as a precursor to the AFm phases, mentioned in several early reports, has been confirmed by SEM and XRD (B92) and HVTEM (S41) studies, which indicate that the initial product consists of thin foils, apparently growing on the C_3A surfaces, and not detectable by XRD. Their composition is not known. ESCA evidence (J22) suggested that the initial product has a relatively high CaO/Al_2O_3 ratio, possibly equal to 4. On the other hand, the CaO/Al_2O_3 ratio of the solution rises above 3.0 when precipitation begins, indicating that a solid product with CaO/Al_2O_3 below 3.0 has been formed (B93,G59). Another ESCA study (B94) indicated that the surface layer of the starting material is hydroxylated and deficient in calcium before coming into contact with liquid water. Inspection of Fig. 6.6 indicates that the first product to form might be C_2AH_8 or some form of hydrous alumina, the latter tending to redissolve later. In either case, this would explain the observed increase in the CaO/Al_2O_3 ratio of the solution.

In pastes, the hydration of C_3A, as followed calorimetrically, is slightly retarded if CH is present (C33). The products are essentially the same as those formed in the absence of CH, but the crystals of AFm phases are smaller (G60). In dilute suspensions, the first crystalline product to form is C_4AH_{19}; this causes the CaO/Al_2O_3 ratio in the solution to decrease (B93).

6.5.2 Reaction of C_3A with water in the presence of calcium sulphate

Studies using calorimetry, electron microscopy, XRD and DTG (S64,S65, C33) have shown that this reaction occurs in two stages, both of which are strongly exothermic. The first stage is characterized by a peak in the heat evolution curve during the first 30 min, and yields ettringite. In the second stage, which is characterized by a peak typically occurring at 24–48 h, the ettringite reacts further and AFm phases are formed. The reactions may be represented by the following equations:

$$C_3A + 3C\bar{S}H_2 + 26H_2O \rightarrow C_6A\bar{S}_3H_{32} \qquad (6.2)$$

$$2C_3A + C_6A\bar{S}_3H_{32} + 4H_2O \rightarrow 3C_4A\bar{S}H_{12} \qquad (6.3)$$

$$C_3A + CH + 12H_2O \rightarrow C_4AH_{13} \qquad (6.4)$$

where the C_4AH_{13} could occur either in a solid solution with monosulphate or in separate crystals. In ordinary Portland cements, the ratio of SO_3 to Al_2O_3 is typically about 0.6, suggesting that the final products will be $C_4A\bar{S}H_{12}$ and C_4AH_{13}, but some AFt phase nevertheless often persists.

Studies by TEM (B95) and HVTEM (S41) show that, as in the absence of gypsum, the formation of a distinctly crystalline reaction product is preceded by that of a layer of amorphous or poorly crystalline material having a foil or irregular, platey morphology on the surface of the grains. There is no direct evidence concerning the composition of this material, but Brown (B91) concluded from a consideration of the phase equilibria (Fig. 6.8) that, depending on the concentrations in the solution in contact with the C_3A when the latter began to react, the first Al_2O_3-containing phase to form might be either AH_3 or ettringite. The morphology of the product does not support the view that it is ettringite, though it might be amorphous material of similar composition.

The ettringite produced in the first stage of reaction tends to form as stubby, prismatic crystals up to 1 μm long and close to the C_3A surfaces, but it is also formed away from these surfaces, thus indicating a through solution mechanism (M66,S41). Further evidence for the latter is provided by the observation that, in pastes of C_3A and C_3S with gypsum, some of the C_3A grains form hollow shells only partly occupied by unreacted starting material or hydration products (S41). The ettringite crystals are readily damaged by the electron beam. Longer needles, of high aspect ratio, can also form. The amount of available space (M66), pH (B95) and sulphate ion concentration (S41) have been variously considered to control the morphology.

The calorimetric evidence, and other evidence discussed in Section 7.6.2, indicates that reaction of the C_3A is markedly retarded in the presence of gypsum and that gypsum and CH together are more effective than gypsum alone (C33). The effect of gypsum has been attributed to the protective action of a layer of ettringite (S65,C33), but it has been queried whether this layer is sufficiently impermeable to have such an effect (M66). Retardation has also been attributed to an underlying layer of hydrous alumina (C34) or AFm phase (P27,G61); this view is consistent with the electron microscopic evidence mentioned earlier. The retarding properties of such a layer might well depend on its composition, thus explaining the different rates of reaction observed in the absence of admixtures and with CH or gypsum or both. Retardation has also been attributed to the blocking of active sites on a modified C_3A surface by adsorbed sulphate ions (S66,F28); however, Na_2SO_4 has no significant effect on the rate of reaction (C33).

The reaction of the aluminate phase in a cement paste is influenced by the presence of the silicate phases and of alkali (Section 7.6). Ghorab and El Fetouh (G62) studied the effect of the latter on reactions in the pure system.

6.5.3 Reaction of the ferrite phase

Many investigations have shown that the hydration products of phases in the $C_2(A,F)$ series are essentially similar to those formed from C_3A under comparable conditions (C35,S57,C36,D10,C37). The experimental evidence includes XRD, DTA, DTG, TEM, SEM, IR and calorimetry. The first crystalline products to form in the absence and presence of $CaSO_4$ are AFm and AFt phases, respectively, the AFt phases being later replaced by AFm phases, as in the case of C_3A. Both types of product phase contain Fe^{3+} as well as Al^{3+}, and tend to undergo further change to give hydrogarnet phases. ESCA evidence (B94) indicates that the surface of C_4AF is hydroxylated and deficient in calcium and iron before contact with liquid water. The deficiency in iron possibly arises from zoning, considered in Section 2.3.1.

Different preparations of ferrite phase of a given composition and particle size distribution appear to vary greatly in reactivity (e.g. Ref. S41), perhaps because of zoning and differences in distribution of cations among tetrahedral and octahedral sites (Section 1.5.1). Some of the earlier studies indicated that reaction is accelerated in the presence of CH, but later work showed that, as with C_3A, reaction is retarded by CH, more strongly by gypsum and still more strongly by the two together. Under comparable conditions, C_4AF appears normally to react more slowly than C_3A, though initially at least, the reverse may sometimes be the case (S41). It appears generally agreed that the rate of reaction of the ferrite phase decreases with the Fe/Al ratio.

Several investigators have noted that the Al/Fe ratio in the product phases tends to be higher than that of the starting material, and have concluded that an iron(III) oxide or hydroxide is also formed. Teoreanu et al. (T33), using Mössbauer spectroscopy and XRD, concluded that the hydration products formed from C_2F or C_4AF at ordinary temperatures included FH_3. With C_2F above 75°C, this was replaced by hematite. Fukuhara et al. (F29) concluded from calorimetric evidence that the AFt phase formed from C_4AF in pastes at 20°C had an Al/Fe ratio of about 3 and postulated that FH_3 was also formed. Rogers and Aldridge (R21) found that in $C_2(A,F)$ preparations hydrated at 4°C the only product detectable by XRD was $C_4(A,F)H_{19}$, but SEM with X-ray microanalysis evidence showed that an amorphous iron oxide or hydroxide was also formed. In products formed at 80°C, hematite was identified by SEM with X-ray microanalysis, XRD and Mössbauer spectroscopy. Hydrogarnet phases formed from C_6A_2F or C_4AF contained only small proportions of iron as did that formed from C_6AF_2 at 150°C; those formed from C_6AF_2 at 21°C or 80°C had Fe/(Al + Fe) ratios of 0.30–0.35. Fortune and Coey (F30), using XRD and Mössbauer spectroscopy, similarly found that hydrous iron(III) hydroxide was present among the products of C_4AF hydrated at 72°C. The hydrogarnet phase that was the main product had an Fe/(Al + Fe) ratio of 0.22. If CH was added, this ratio was increased to 0.32, but the hydrogarnet formed in presence of gypsum was almost free from Fe^{3+}.

Brown (B96) concluded that in C_4AF hydration in the presence of $CaSO_4$, the early product was C_2AH_8 or hydrous alumina, and that it was followed by formation of an AFt phase containing little or no Fe^{3+}; an iron oxide gel containing some Ca^{2+} was also formed. The conclusion regarding the later products agrees substantially with that of Fukuhara et al. (F29). The tendency to form iron oxide or hydroxide may be connected with the fact that species containing Al^{3+} can migrate in pastes relatively easily, whereas those containing Fe^{3+} cannot, and are thus largely confined to the space originally occupied by the anhydrous material from which they come (T19).

6.5.4 Enthalpy changes

Table 6.7 gives standard enthalpies of formation of some relevant compounds, and Table 6.8 gives enthalpy changes for hydration reactions of C_3A calculated from them. The strongly exothermic nature of these reactions is apparent. Lack of data for the standard enthalpies of the hydration products containing iron precludes similar calculations for the ferrite phase. For the hydration of C_3A to give C_3AH_6, Lerch and Bogue (L24) obtained an experimental value of $-234\,kJ\,mol^{-1}$; other workers have obtained similar results. Experimental determinations of the enthalpy of hydration of C_4AF have given $-203\,kJ\,mol^{-1}$ (products not established; Ref. L24), and, in the presence of excess CH to give hydrogarnets, $-193\,kJ\,mol^{-1}$ for C_4AF and $-208\,kJ\,mol^{-1}$ for $C_4A_{0.68}F_{0.32}$ (B60). For C_4AF in the presence of gypsum to give ettringite and FH_3, a value of $-352\,kJ\,mol^{-1}$ has been reported (F29). Houtepen and Stein (H36) gave standard enthalpies of formation of some other AFm phases determined experimentally, and Babushkin *et al.* (B62) gave values for other compounds estimated on the basis of crystal chemistry.

Table 6.7 Standard enthalpies of formation (ΔH_f°; kJ mol^{-1}) for some compounds relevant to hydration of the aluminate and ferrite phases (25°C except where otherwise stated)

Compound	ΔH_f°	Ref.	Compound	ΔH_f°	Ref.
C_3A	−3587.8	W10	$C_6A\bar{S}_3H_{32}$	−17539	W10
				−17528	S67
C_4AF (20°C)	−5090.3^a	T8	$C_4A\bar{S}H_{12}$	−8778	W10
				−8752	S67
C_6A_2F	−8104.9	N11	C_4AH_{13}	−8318	H36
$C\bar{S}H_2$	−2022.6	W10	$C_4A\bar{C}H_{10.68}$	−8176	W10
$C\bar{S}H_{0.5}$ (β)	−1574.65	W10	C_3AH_6	−5548	W10
CH	−986.1	W10	H_2O (liquid)	−285.83	W10

^{a} Using data for oxides, from Ref. W10.

Table 6.8 Enthalpy changes for some hydration reactions of C_3A, calculated from the data in Table 6.7

Reaction	ΔH (kJ per mole of C_3A)
$C_3A + 3C\bar{S}H_2 + 26H_2O \rightarrow C_6A\bar{S}_3H_{32}$	−452
$C_3A + C\bar{S}H_2 + 10H_2O \rightarrow C_4A\bar{S}H_{12}$	−309
$C_3A + CH + 12H_2O \rightarrow C_4AH_{13}$	−314
$C_3A + 6H_2O \rightarrow C_3AH_6$	−245
$2C_3A + C_6A\bar{S}_3H_{32} + 4H_2O \rightarrow 3C_4A\bar{S}H_{12}$	−238

7

Hydration of Portland cement

7.1 General description of the hydration process and products

7.1.1 Evidence from X-ray diffraction

Unless otherwise stated, this chapter relates to ordinary Portland cements hydrated in pastes at 15–25°C and w/c ratios of 0.45–0.65. XRD powder studies on such pastes have been reported by many investigators (e.g. C38,M67). The rates of disappearance of the phases present in the unreacted cement are considered more fully in Section 7.2.1. Gypsum and other calcium sulphate phases are no longer detectable after, at most, 24 h, and the clinker phases are consumed at differing rates, alite and aluminate phase reacting more quickly than belite and ferrite. The ratio of belite to alite thus increases steadily, and after about 90 days at most, little or no alite or aluminate phase is normally detectable.

The peaks of CH begin to appear within a few hours. Within a few days they dominate the XRD pattern. A diffuse band at 0.27–0.31 nm and a sharper one at 0.182 nm develop more slowly; these are similar to the bands of C–S–H observed in C_3S or β-C_2S pastes (Fig. 5.6), but taking into account the amount of C–S–H likely to be present, they suggest an even lower degree of crystallinity than in the latter materials.

The hydrated aluminate phases that are observed depend on the cement composition and on the time and other conditions of hydration. With most ordinary Portland cements under the conditions mentioned above, ettringite peaks are detectable within a few hours and increase in intensity to a maximum at about 1 day. They then tend to weaken, and may disappear completely, though with many cements they persist indefinitely. The 0.973- and 0.561-nm peaks are especially prominent, but many other peaks can often be detected, indicating that at least some of the ettringite is highly crystalline. Peaks of AFm phases typically begin to appear after about 1 day and become stronger as those of ettringite weaken. This behaviour is similar to that observed with pastes of C_3A (Section 6.5.2) and can be explained in the same way.

The only peaks clearly attributable to AFm phases are usually those at about 0.288 nm (11.0), 0.166 nm (30.0) and the first- and second-order basal reflections corresponding to one or, usually, more layer thicknesses. Taking into account the amounts likely to be present, the XRD evidence suggests that much of the AFm material is poorly crystalline. If the specimen is kept reasonably moist (RH ≥ 80%) and no special precautions are taken to exclude CO_2 during specimen preparation and examination, the principal layer thickness is usually 0.89 nm, attributable to $C_4A\bar{S}H_{12}$ or its solid solution with C_4AH_{13}, and weaker peaks corresponding to layer thicknesses of approximately 0.82 nm, 0.79 nm, 0.76 nm, or any combination of these. The 0.82-nm spacing could be due to $C_4A\bar{C}_{0.5}H_{12}$ or $C_4A\bar{S}H_{10}$ or both, the 0.79-nm spacing most probably to C_4AH_{13} and perhaps also to hydrotalcite-type material, and the 0.76 nm spacing to $C_4A\bar{C}H_{11}$. The phases containing CO_3^{2-} form by carbonation of C_4AH_{13}, and the $C_4A\bar{S}H_{10}$ by partial dehydration of $C_4A\bar{S}H_{12}$. In all these formulae, the possibility of some replacement of Al^{3+} by Fe^{3+} is assumed. To establish the nature of the AFm phases in the untreated paste with more certainty, it will often be necessary to prepare and examine the specimen under CO_2-free conditions and at more than one relative humidity.

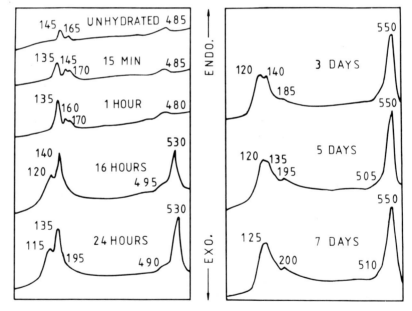

Fig. 7.1 DTA curves for pastes of a typical Portland cement. Bensted and Varma (B97).

Copeland *et al.* (C38) observed that the XRD patterns of many pastes included peaks of a hydrogarnet in addition to those of other hydrated aluminate phases, and obtained evidence that this had formed by hydration of the ferrite. The unit cell parameter was 1.24 nm. Hydrogarnet phases do not seem to form in any quantity at ordinary temperatures from modern cements. Luke and Glasser (L27) found a hydrotalcite-type phase by XRD in a residue from which the other phases had been removed by chemical extraction, and both a hydrotalcite-type phase and a poorly crystalline product probably approximating to a hydrogarnet have been found by TEM (Section 7.1.3). Both appear to be minor products. Calcite, if present, is detectable by a sharp peak superimposed on the maximum in the C–S–H band at 0.3035 nm, and if in sufficient quantity by additional peaks.

7.1.2 Evidence from differential thermal analysis and infrared spectroscopy

Bensted and Varma (B97) summarized the use of these methods for the examination of cement pastes. Both are mainly useful for following the phase changes in relatively young pastes. Fig. 7.1 shows DTA curves for pastes of a typical Portland cement at ages up to 7 days. The peaks at 145°C and 165°C given by the unhydrated cement are due to gypsum, and that at 485°C to a small amount of CH formed during storage, probably by hydration of free lime. The peak at 135–140°C that appears within 15 min is due to ettringite; its intensity passes through a maximum at 16–24 h. The peak at 115–125°C is due to C–S–H or cement gel, and that at 530–550°C to CH. The relatively large crystals of this phase that are formed on hydration in the paste decompose at a higher temperature than the smaller ones formed on hydration during storage. The peak at 185–200°C is due to AFm phases. One would expect all these peak temperatures to vary with the technique used, but that relative positions would be essentially unchanged.

In the first effective studies of hydrated cements by DTA, Kalousek and co-workers (K27,K32,K33) observed the early formation of ettringite and its subsequent replacement by what was termed a solid solution, but which was probably a mixture of AFm phases. They found no C_3AH_6 or other hydrogarnet phases. Quantitative or semiquantitative determinations of gypsum and ettringite indicated that less than half the total SO_3 present could be accounted for by these phases. The authors concluded that the solid solution was eventually replaced by a product which they termed 'Phase X', and which was possibly a gel containing all the oxide components of the original cement.

Infrared absorption spectra (B97) of unhydrated cements typically show moderate to strong bands at 525 and 925 cm^{-1} due to alite, and at 1120 and 1145 cm^{-1} from S–O stretching vibrations, and weak bands in the 1650 and 3500 cm^{-1} regions due to H_2O molecules. The early formation of ettringite on hydration is shown by a change in the sulphate absorption to a singlet centred at $1120\,\text{cm}^{-1}$, and the subsequent replacement of ettringite by monosulphate by a further return to a doublet at 1100 and $1170\,\text{cm}^{-1}$. The hydration of the silicate phases causes a shift in the broad Si–O absorption band from 925 to $970\,\text{cm}^{-1}$. After several days, the silicate absorption tends to obscure that from sulphate. Changes also occur in the H_2O bending band near $1650\,\text{cm}^{-1}$ and in the H_2O or OH stretching bands at $3100\text{--}3700\,\text{cm}^{-1}$; in the latter region, CH gives a peak at $3640\,\text{cm}^{-1}$, ettringite one at $3420\,\text{cm}^{-1}$ and a weaker one at $3635\,\text{cm}^{-1}$, and monosulphate ones at 3100, 3500 (broad), 3540 and $3675\,\text{cm}^{-1}$. The complexity and partial overlapping of these absorptions lessens their utility. The IR spectra also indicate that both the ettringite and the AFm phases formed in cement pastes are less crystalline than the pure materials, and that the AFm phases become poorer in SO_4^{2-} as hydration proceeds.

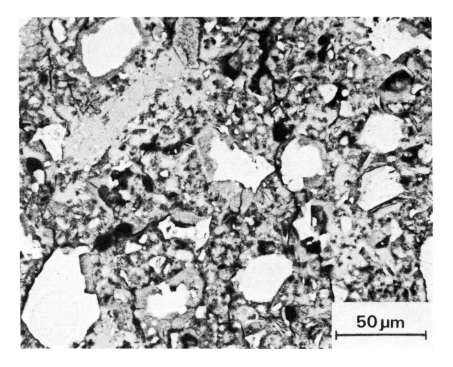

Fig. 7.2 Backscattered electron image of a mature Portland cement paste, aged 2 months. Successively darker areas are of unreacted cement grains (bright), sometimes with visible rims of hydration products, $Ca(OH)_2$, other ('undesignated') regions of hydration products, and pores (black). Scrivener and Pratt (S28).

7.1.3 Evidence from light and electron microscopy

Studies on the microstructures of cement pastes have given results broadly similar to those on calcium silicate pastes (Section 5.3.1), though with some important differences at early ages. As with calcium silicate pastes, the first effective studies were made using light microscopy of thin sections (B51), which were successively augmented by TEM of ground and redispersed material (G41,C21), SEM of fracture surfaces (C22,W18,D13), X-ray (T19,U12,U13) or backscattered electron (S28,S68) imaging of polished sections, TEM examination of ion-thinned sections (R24) and other techniques.

The development of microstructure is considered in Section 7.4. The broad microstructures of pastes more than about 28 days old are similar to those of C_3S pastes. Fig. 7.2 shows a typical backscattered electron image, obtained under conditions maximizing contrast from compositional differences. The following types of region are distinguishable by their differing grey levels and by their compositions determined by X-ray microanalysis.

(1) The brightest areas are of unreacted clinker phases. Individual cement grains are usually polymineralic and, as in a clinker (Section 4.3.1), the different phases within them can be distinguished by their differing grey levels or by X-ray microanalysis.

(2) To varying extents, the clinker phases are replaced by hydration products that have formed in situ, and which appear darker on the backscattered electron image. Most noticeably, the alite and belite are replaced by material approximating in composition to the C–S–H of calcium silicate pastes, but which tends to have a higher Ca/Si ratio and contains small amounts of other elements, especially aluminium, iron and sulphur. Initially, these hydration products are observed as rims on the unhydrated material. As the paste ages, the latter is progressively replaced, until ultimately no significant quantity remains. The in situ hydration products have often been called 'inner product', but this term may be criticized on the grounds that the apparent outlines of the cement grains may lie within the true ones, and 'late product' may be a better description (J10). Only the larger cement grains, unreacted or otherwise, can be clearly distinguished in this way.

(3) CH can be observed as areas darker than the unreacted clinker phases but brighter than the other hydration products. As in calcium silicate pastes, these appear to have grown in regions initially occupied by water. Although the areas appear discrete on two-dimensional sections, they are not necessarily so in the three-dimensional material. They can engulf small cement grains.

(4) The remaining space, which forms the matrix in which the above regions appear to be embedded, consists of material varying in grey level from material similar to the in situ product to darker regions, representing pores. Like the CH, these appear discrete in two dimensions, but are not necessarily so in three. This material is often called 'outer product', but the term is imprecise, as it must include not only products formed in what was originally water-filled space, but also that formed in space initially occupied by the interstitial material or by the smaller cement grains, the outlines of which can no longer be distinguished. The term 'undesignated product' is preferable. For most cements, the majority of individual microanalyses of the hydration product in these regions show compositions varying between those of C–S–H and either AFm phase or CH (H4). Regions large enough for accurate microanalysis on polished surfaces ($\simeq 5\,\mu m$) are rarely of pure AFm phase.

Transmission electron microscopy of ion-thinned sections provides data at higher resolution than can be obtained with polished sections. Rodger and Groves (R24) described regions which had probably formed in situ from the ferrite phase, and which consisted of C–S–H, a hydrotalcite-type phase and a poorly crystalline phase containing iron that could have been the precursor of a hydrogarnet. The particles of this last constituent were almost spherical and some 200 nm in diameter. The same investigation also showed that much of the product formed in situ from alite or belite was essentially pure calcium silicate hydrate.

7.2 Analytical data for cement pastes

7.2.1 Determination of unreacted clinker phases

The experimental considerations applying to calcium silicate pastes (Sections 5.1 and 5.2) are equally relevant to cement pastes. Of the methods so far used in attempts to determine the degrees of reaction of the individual clinker phases as a function of time, QXDA (C39,D12,T34,P28) has proved much the most satisfactory. Procedures are essentially as for the analysis of a clinker or unreacted cement (Section 4.3.2), but it is necessary to take account of overlaps with peaks from the hydration products, and especially with the C–S–H band at 0.27–0.31 nm. The water content of the sample must be known, so that the results can be referred to the weight of anhydrous material. If a sample of the unhydrated cement is available, and its quantitative phase composition has been determined, it may be used as the reference standard for the individual clinker phases in the paste.

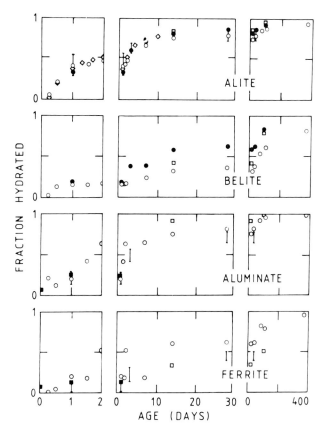

Fig. 7.3 QXDA results for the fractions of the clinker phases reacted in Portland cement pastes. Filled circles: Copeland and Kantro (C39), w/c = 0.65. Diamonds: Bezjak et al. (B98), sample C2. Vertical lines: Osbaeck and Jøns (O8), range for 7 samples. Open circles: Dalziel and Gutteridge (D12). Open squares: Patel et al. (P28), samples cured at 100% RH. Filled squares: Tang and Gartner (T34), clinker interground with gypsum.

The precision and accuracy of the technique should not be overestimated. If a cement paste contains 50% of alite initially, an error of 3% in the determination of that phase leads to one of 6% in the percentage reaction; if it contains 5% of aluminate phase initially, an error of 1% leads to one of 20% in the percentage reaction. With existing techniques, errors much smaller than these are unlikely to be achieved for untreated pastes. The precision may be greatly increased for the aluminate and ferrite by first removing the silicates by chemical extraction methods (Section 4.3.3). Fig. 7.3 shows some results from the literature. The considerable differences between the results of different investigations are no doubt partly due to the limitations of the technique, but also to differences between cements or curing conditions, which are discussed in Section 7.7.

7.2.2 Non-evaporable and bound water

As with calcium silicate pastes, the gelatinous nature of the principal reaction product renders any definition of chemically bound water somewhat arbitrary. The three definitions of water content described in Section 5.2.2 for calcium silicate pastes are relevant to cement pastes, viz. non-evaporable water, chemically bound water and water seemingly essential to the formation of the hydration products in a saturated paste.

Water retained after D-drying, known as non-evaporable water, has often been wrongly identified with chemically bound water. It excludes much of the interlayer water in C–S–H, AFm and hydrotalcite-type phases and much of the water contained in the crystal structure of AFt phases. It is often used as a measure of the fraction of the cement that has reacted, but can only be approximate in this respect, because the clinker phases react at different rates and yield products containing different amounts of non-evaporable water. Fully hydrated cement pastes typically contain about 23% of non-evaporable water, referred to the ignited weight. Copeland *et al.* (C38) determined the non-evaporable water contents of a series of mature cement pastes and carried out regression analyses on the cement composition. For pastes of w/c ratio 0.8 and aged 6.5 years, they obtained the approximate expression:

$$H_2O = CaO - 0.5SiO_2 + 5Al_2O_3 - 5Fe_2O_3 - 2SO_3 \qquad (7.1)$$

where H_2O represents the quantity of non-evaporable water and all quantities are in moles. The coefficient for SO_3 was of low reliability, and other sets of data given in the same paper suggest a value nearer to -1. For a typical cement composition (65% CaO, 21% SiO_2, 5.5% Al_2O_3, 3.0% Fe_2O_3, 2.3% SO_3), and using a coefficient of -1 for SO_3, this gives a non-evaporable water content of 20.4%.

The content of chemically bound water is approximately that retained on equilibration at 11% RH of a sample not previously dried below saturation. For fully hydrated pastes of typical cements, it is about 32%, referred to the ignited weight (F13,T35). There are no systematic data relating this quantity to cement composition. The total content of water essential for complete hydration in a saturated paste is defined as that present in such a paste having the minimum w/c ratio at which complete hydration is possible

(Section 8.2). For typical Portland cements, it is 42–44% referred to the ignited weight.

7.2.3 Thermogravimetry and determination of calcium hydroxide content

The discussion of methods for determining CH in calcium silicate pastes (Section 5.2.1) applies also to cement pastes; TG and QXDA are probably the most satisfactory methods. Fig. 7.4 shows a typical TG curve for a mature cement paste. The step at 425–550°C is due primarily to decomposition of CH, and its height, estimated as shown in Fig. 5.3, probably affords the best available method for determining this phase. It is nevertheless subject to at least two sources of error: curves for AFm phases show a step in the same range (Fig. 6.2) and the decomposition of CH is not quite complete within this range (Section 5.1.3). It is probably not practicable in the present state of knowledge to correct for these errors, and doubtful whether either TG or QXDA gives results with an accuracy better than ±1%.

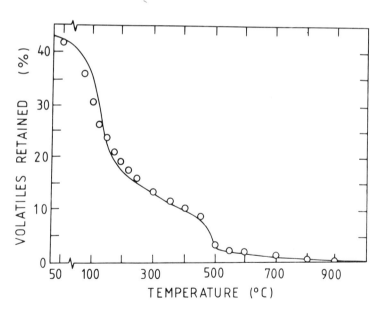

Fig. 7.4 TG curve for a Portland cement paste (w/c = 0.5), moist cured at 25°C for 14 months; full curve, with points calculated as described in the text shown by open circles. Heating rate, 10 deg C min^{-1}, flow rate of dry, CO_2-free N_2, 15 ml min^{-1}; sample weight, 50 mg. Taylor (T5).

The loss above 550°C is due partly to CO_2 and partly to the final stages of dehydration of C–S–H and the hydrated aluminate phases. It is not practicable to distinguish the contributions from TG evidence alone, and, unless evolved gas analysis is used, a separate determination of CO_2 should be made. As with calcium silicate pastes, serious errors arise if TG determinations are carried out on material that has been treated with an organic liquid, e.g. to stop hydration. Losses above 550°C of more than about 3%, referred to the ignited weight, indicate serious carbonation either from this or other causes.

The loss below the CH step is due to decomposition of C–S–H and the hydrated aluminate phases. Although the TG curves of pure AFm phases are markedly stepped in this region (Fig. 6.2), those of cement pastes normally show only slight indications of steps. Weak peaks can, however, sometimes be seen on DTG curves. The absence of steps is probably due to a combination of low crystallinity, the presence of other phases and the presence of AFm phases of different compositions in mixture or solid solution or both. For typical experimental conditions with a 50 mg sample, heating rate of 10 deg C min^{-1} and N_2 flow rate of 15 ml min^{-1}, the volatiles retained at about 150°C, after correction for CO_2, correspond to the non-evaporable water, and those retained at about 100°C to the bound or 11% RH water, but this last temperature, in particular, is very dependent on experimental conditions (T5).

For pastes of typical ordinary Portland cements cured for 3–12 months, the CH content found by thermal methods or QXDA is typically 15–25%, referred to the ignited weight (P29,R14,M37,H37,T17,D12). Pressler *et al.* (P29) found that for pastes of various ages of ordinary (US Type I) Portland cements, it was linearly related to the content of non-evaporable water, but that for cements high in belite (US Type IV), it tended to a maximum while the latter continued to increase. This is readily explained, since belite yields only a little CH on hydration. The author has noticed similar behaviour even with modern cements high in alite, and that the CH content can possibly even decrease slightly after 28–91 days (T5).

7.2.4 Determinations of hydrated aluminate and silicate phases

Opinions have differed as to the possibility of determining the hydrated aluminate phases by thermal or X-ray methods. The determination of ettringite was discussed in Section 6.2.2. Bensted (B99), who used DTA, found that for ordinary Portland cements the ettringite content increased with time during the first 2 h to maximum values of 2.2–2.8%, and that the quantity of ettringite formed at any given time increased with the w/c ratio.

There are probably no effective direct methods at present for determining either C–S–H or AFm phases in cement pastes; in both cases, this is probably attributable to the low degree of crystallinity. Odler and Abdul-Maula (O15) found that determination of AFm phase by QXDA was only semiquantitative. Postulated quantitative phase compositions of cement pastes may, however, be tested by comparing observed and calculated TG curves (Section 7.3.3).

7.2.5 Analyses of individual phases

The general considerations mentioned in Section 5.2.3 in relation to calcium silicate pastes apply equally to cement pastes. Table 7.1 gives results of X-ray microanalyses of the product formed in situ from the larger alite or belite grains, obtained by analysis of polished sections in the EPMA or SEM. Most of the investigators reported considerable scatter between the results of individual analyses. The reported mean Si/Ca ratios are 0.49–0.60 (Ca/Si 1.7–2.0). There are indications that the Mg/Ca, Al/Ca and S/Ca ratios are correlated with those of the cements, or probably more directly with those of the alite or belite that they contain (H4). Rayment and Lachowski (R29) noted a significant dependence on w/c ratio, a 5-year-old paste with w/c = 0.6 giving a unimodal distribution of Ca/Si ratio with a centroid at 1.75, whereas one with w/c = 0.3 gave a bimodal distribution with centroids at 1.75 and 1.95. The bimodularity was due to variations in silicon content, and the distribution of Ca/(Si + Al) ratios was unimodal. Apart from a possible decrease in the S/Ca ratios with age (H4), there are no clear correlations with age or temperature. Results for a paste hydrated at 45°C (U12) do not differ significantly from those for ones hydrated at lower temperatures.

Several investigators have reported X-ray microanalyses of polished sections of the 'undesignated product' formed in space formerly occupied by water, interstitial material or small cement grains (R28,R29,T19,H4). Fig 7.5 shows a plot of Al/Ca ratios against Si/Ca ratios for individual microanalyses from varying parts of the microstructures of typical Portland cement pastes. Most of the analyses clustering near the lower right-hand corner are of the material formed in situ from the larger alite or belite grains and described in the previous paragraph. The spots analysed were not chosen at random, and CH is much under-represented. The analyses intermediate between those of the in situ product and either CH or AFm phase were, in general, obtained from the undesignated product, suggesting that the latter is largely a mixture on or below a micrometre scale of C–S–H with CH or AFm phase. Rayment and co-workers (R28,R29) obtained broadly similar results, but expressed them by noting a tendency for the ratio of (Ca + Mg) to (Si + Al + Fe + S) to be more nearly constant than that of Ca to Si.

Table 7.1 Results of X-ray microanalyses by EPMA or SEM of the gel formed in situ from alite or belite in Portland cement pastes[a]

Na	Mg	Al	Si	S	K	Fe	w/c	Temp. (°C)	Age	Ref.
tr.	0.05	0.07	0.59	0.01	0.02	0.01	0.5	20	8 days	R25
n.d.	n.d.	n.d.	0.59	n.d.	n.d.	n.d.	0.5	25	28 days	R26
n.d.	0.02	0.06	0.49	0.02	n.d.	0.03	0.3	?	5 years	R27
n.d.	n.d.	0.06	0.60	0.02	n.d.	0.02	0.6	?	5 years	R28
n.d.	0.03	0.08	0.60	0.03	n.d.	0.015	0.45	25	23 years	T19
n.d.	0.05	0.10	0.49	0.01	n.d.	0.05	?	?	136 years	R27
0.01	0.02–0.08	0.05–0.07	0.50–0.53	0.01–0.03	0.01–0.02	0.01–0.02	?	20–25	3–180 days	H4[b]
n.d.	n.d.	0.12(?)	0.49	n.d.	n.d.	n.d.	0.40	25	4 years	U12
n.d.	n.d.	0.14(?)	0.53	n.d.	n.d.	n.d.	0.40	45	60 days	U12

[a] Means or other measures of central tendency, expressed as atom ratios relative to Ca. n.d. = not determined or not reported; tr. = trace.
[b] Data for 6 pastes from 3 cements.

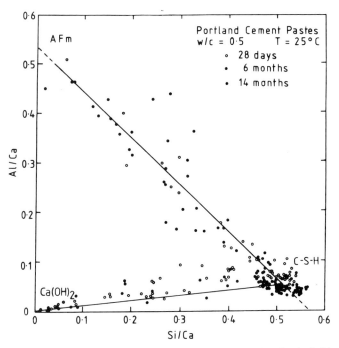

Fig. 7.5 Al/Ca atom ratios plotted against Si–Ca atom ratios for individual X-ray microanalyses of typical Portland cement pastes. Modified from Harrisson *et al.* (H4).

As noted earlier, regions of pure AFm phase of sufficient size to be analysed by EPMA or SEM microanalysis are rare in most cement pastes. Harrisson *et al.* (H4) found them to be low in iron and sulphur (Al/Ca = 0.50, Si/Ca = 0.05, S/Ca = 0.12, Fe/Ca = 0.01) and that the AFm phase mixed with C–S–H in the undesignated product tended to be higher in sulphur (mean S/Ca = 0.17). The same investigation showed no significant correlations between the Mg/Ca, Al/Ca, S/Ca and Fe/Ca ratios. The authors concluded that little or none of the iron replaces aluminium in the AFm phase, and that magnesium in excess of some quite small amount did not replace calcium in the C–S–H. X-ray microanalyses of one of the old Portland cement pastes studied earlier by Copeland *et al.* (C38) confirmed that the principal hydrated aluminate phase was a hydrogarnet and showed its composition to be near $Ca_3Al_{1.2}Fe_{0.8}SiO_{12}H_8$ (T19).

Studies of ground and redispersed cement pastes by analytical electron microscopy (L28-L30,T17) showed wide variations between individual analyses, even within single particles of micrometre dimensions, and gave mean Ca/Si ratios of 1.5–2.0 for the C–S–H. The mean ratios of minor elements relative to Ca + Mg were broadly similar to those found by X-ray microanalysis, but some particles contained virtually no aluminium or iron. Particles of AFm phases had lower Fe/Al ratios than the cements from which they were formed, and contained significant proportions of silicon. Parallel studies on calcium silicate pastes gave mean Ca/Si ratios that appear to have been low (Section 5.2.3). Later work, on ion-thinned sections (R24), indicated that the product formed in situ from the cement grains had a Ca/Si ratio of 1.5 ± 0.15. Most of the volumes analysed were found to be pure calcium silicate, but some contained very small amounts of aluminium, sulphur and potassium; no aluminate sulphate was found as a second phase. The 'outer' gel had an average Ca/Si ratio of 1.6 ± 0.2 and many areas contained aluminium, sulphur or potassium. The apparent conflicts between these results and those obtained by EPMA or SEM microanalysis have not been resolved.

7.2.6 Silicate anion structure

Silicate anion structures in Portland cement pastes have been studied by the methods described in Section 5.3.2 for calcium silicate pastes. Trimethylsilylation (TMS) studies (L20,T12,S69,T36,L31,M43,M44) show that, as with C_3S, the proportion of the silicon present as monomer decreases with age and that the hydration products contain dimer, which is later accompanied and eventually partly replaced by polymer ($\geqslant 5$ Si). Some results have indicated that fully hydrated pastes of cement differ from those of C_3S in that substantial proportions of the silicate occur as monomer (S69,L31), but the results of a study in which pastes of C_3S, β-C_2S and cement were compared (M44) suggest that the differences between the anion structures of cement and C_3S pastes are probably within the considerable experimental errors inherent in the method. The recovery of monomer from unhydrated β-C_2S was only 66% and results for cement pastes can only be considered semiquantitative.

Another study (L31,M43), in which the TMS derivatives were examined by gel permeation chromatography and other methods, gave the results shown in Table 7.2. There are no significant differences between the distributions of anion size within the polymer fractions of the cement and C_3S pastes, and, though the lower totals for the cement pastes impede comparison, probably none between the relative proportions of monomer, dimer and polymer that cannot be explained by kinetics, side reactions or other experimental inaccuracies. In all cases, the data for the molecular weights of the polymer TMS derivatives, and the mean connectivities of the polymeric anions, are consistent with the hypothesis that the latter are linear, with pentamer as the major component and little of the material in chains containing more than 11 tetrahedra.

A ^{29}Si NMR study (B100) showed that for cement, as for C_3S pastes, the content of Q^0 silicate tetrahedra decreases with time and that those of Q^1 and later of Q^2 tetrahedra increase. After 180 days, the degree of hydration, estimated from the intensities of the NMR peaks, was approximately 90%. These results are consistent with those obtained by the TMS method, and suggest that the hydration products present after 180 days contain at most only a small proportion of monomer. The possible effects on the silicate anion structure of drying, whether during hydration as a result of localized water shortage or subsequently, were considered in Section 5.3.2.

Table 7.2 Comparison of silicate anion structures in pastes of C_3S and Portland cement (M43)

Starting material:	C_3S	Cement	Cement
Curing time (months):	12.0	9.5	13.5
% of total Si recovered as:			
Monomer	14	11	11
Dimer	50	34	36
Polymer	34	34	34
Molecular weights of polymer TMS derivatives:			
Mode	1300	1300	1300
Number mean	1700	1650	1700
Weight mean	2900	2950	2700
Mean connectivity of SiO_4 tetrahedra in polymer anions:	1.8	1.8	1.8

7.3 Interpretation of analytical data

7.3.1 The nature of the cement gel

Copeland et al. (C39,C25) treated C_3S pastes, and C–S–H prepared in other ways, with various sources of Al^{3+}, Fe^{3+} or SO_4^{2-} ions. The XRD peaks of the phases providing the ions disappeared, and changes occurred in the micromorphology of the gels and in the non-evaporable water contents and fractions of the CaO extractable by an organic solvent. It was concluded that the ions were taken up by the C–S–H and that up to about one silicon atom in six could be replaced by aluminium, iron or sulphur, and that aluminium and iron could also replace calcium. These results suggested that the principal hydration product in Portland cement pastes was a substituted C–S–H, a conclusion that later appeared to be supported by the results of the

X-ray microanalyses described in Section 7.2.5. However, none of the evidence shows that the substitutions occur at the crystal chemical level postulated by Copeland *et al.*, and there are some indications that this is not the case.

Stade *et al.* (S70) examined C–S–H preparations containing Al^{3+} by ^{27}Al magic angle NMR. At a Ca/Si ratio of 0.95, the aluminium was tetrahedrally coordinated; this is also true of aluminous tobermorite (K34), which is of broadly similar composition and structure. At higher Ca/Si ratios, octahedral aluminium was also present. The fraction of the aluminium in octahedral coordination increased with Ca/Si ratio, and at Ca/Si = 1.51 reached unity. The aluminium in pastes made from alite or from C_3S with a little C_3A was also largely in octahedral coordination (S71). Electron spin resonance and Mössbauer examinations of C–S–H preparations containing Fe^{3+} showed that the iron was in octahedral coordination (S72).

These results show that the Al^{3+} and Fe^{3+} present in the cement gel do not replace Si^{4+}, except perhaps to a minor extent. This conclusion is reinforced by most of the results of the silicate anion studies described in Section 7.2.6. If there was any significant replacement of silicon at a crystal chemical level, the average anion size and SiO_4 connectivity would both be reduced. Further evidence of a similar nature is provided by the observation that the average anion size in C–S–H preparations made in suspension is not significantly affected by the presence of Al^{3+} (S73). The silicate anion evidence also makes it appear unlikely that there is significant replacement of silicon by sulphur; it is, in any case, improbable that such replacement would occur in a dimeric or chain anion, as a Si–O–S link would be readily hydrolysed.

Stade and Müller (S71) considered that C–S–H of high Ca/Si ratios contained layers of CH structure, in which some of the calcium could be replaced by aluminium or iron. Because of the large difference in size, any such replacement could probably not occur at random, but one calcium atom in three can be replaced in an ordered way, and then gives a layer of AFm structure. This suggests the hypothesis that most or all of the Al^{3+}, Fe^{3+} and SO_4^{2-} ions incorporated into the gel are not present as substituents at a crystal chemical level, but in layers of AFm phase intimately mixed with those of C–S–H. Such admixture might occur at various levels, ranging from that of single layers, as in an interstratified clay mineral, to one sufficiently coarse for the AFm phase to be detectable by XRD. This hypothesis readily accounts for the evidence from XRD and other sources that the AFm phase in cement pastes is poorly crystalline.

If this hypothesis is correct, the gel formed in situ from the alite or belite is merely a limiting case, in which the proportion of AFm layers is at a minimum. This minimum possibly depends on the contents of Al^{3+} and other foreign ions in the anhydrous phases, and the extents to which they tend to pass into the pore solution on hydration. If the C-S-H constituent of the gel is assumed to have the same Ca/Si ratio as in calcium silicate pastes, one would expect that the ratio would be about 1.9 (Si/Ca = 0.53) in a gel with Al/Ca = 0.07. This agrees with some, but not all, of the data in Table 7.1. The hypothesis might explain the observations of Rayment and Lachowski (R29) on the bimodal distribution of Ca/Si ratio in the in situ gel and the relative constancy of its Ca/(Si + Al) ratio. It probably offers the most satisfactory explanation of the existing data but needs to be further tested. Continued studies by TEM of ion-thinned sections may be expected to yield valuable data in this respect.

7.3.2 Fe_2O_3, SO_3 and minor oxide components in cement gel

Most of the studies on the hydration of C_4AF with or without gypsum (Section 6.5.3) have indicated that the hydration products are AFm, AFt or hydrogarnet phases of higher Al/Fe ratio, together with an iron(III) hydroxide gel. It has also been suggested that a calcium ferrite gel is formed (B96). There is, however, no experimental evidence that an iron hydroxide gel is formed in cement pastes, and Harchand *et al.* (H38) concluded from a Mössbauer study that it was absent and that the Fe^{3+} entered the AFm phase. The ferrite phase in cement may well behave differently from pure C_4AF, because of the large difference in composition and the effects of other constituents.

Studies of polished sections by X-ray (T19) and backscattered electron (S68) imaging indicated that the Fe^{3+} in Portland cement does not migrate through the pore solution on hydration, but remains in products formed in situ. In agreement with this, TEM of ion-thinned sections showed the presence of a poorly crystalline iron-rich phase resembling a hydrogarnet, which was probably formed as part of the in situ hydration product of the ferrite phase (R24). A hydrogarnet high in iron would approximate in composition to the calcium ferrite gel suggested on other grounds. It could accommodate in its octahedral sites not only Al^{3+} and Fe^{3+}, but also Ti^{4+} and Mn^{3+}, which are concentrated in the ferrite phase of the clinker.

In an ordinary Portland cement, only some two-thirds of the Fe^{3+} occurs in the ferrite, the rest being contained largely in the alite and aluminate (Table 4.3). On hydration, the Fe^{3+} in these other phases probably does not enter a hydrogarnet, but goes into AFm phases or layers formed in situ. This would account for the observation by analytical electron microscopy that small amounts of Fe^{3+} are present in the AFm phases.

One would expect that any relatively large crystals of periclase in a clinker would hydrate to give brucite. However, in modern clinkers, most of the Mg^{2+} is not present in this form, but in the major clinker phases, especially alite and ferrite, and, if the MgO content exceeds about 1.5%, in small crystals of periclase intimately mixed with aluminate and ferrite in the interstitial material. In all these cases, it is associated with comparable or larger amounts of Al^{3+}, and one would then expect that a hydrotalcite-type phase would be formed. TEM and XRD evidence (Sections 7.1.1 and 7.1.3) indicate that this is the case. Mg^{2+}, like Fe^{3+}, does not readily migrate through the pore solution, and the formation of a hydrotalcite-type phase is readily explained by the substantial magnesium content of the ferrite. A hydrotalcite-type phase formed from the Mg^{2+} and part of the Al^{3+} present in the alite or belite could be mixed with C–S–H in the manner suggested for AFm phase.

Sulphate present in the gel could be accommodated as interlayer anions associated with AFm layers. C–S–H can, however, take up $CaSO_4$ in the absence of Al^{3+} or Fe^{3+}, and thus of AFm layers. Odler (O17) showed that $CaSO_4$ thus bound was very readily removed, and suggested that the Ca^{2+} and SO_4^{2-} ions were adsorbed. The K^+ and Na^+ ions that are found in the cement gel by X-ray microanalysis are partly the result of deposition from the pore solution when the latter evaporates during specimen preparation or in the high vacuum of the instrument, and these, too, with equivalent amounts of OH^-, may be adsorbed; after deducting the fractions deposited from the pore solution, the K/Ca and Na/Ca ratios in the gel are typically $\simeq 0.01$ and <0.01, respectively (Section 7.5.2).

The considerations discussed in this and the preceding section suggest that the cement gel is a complex material, comprising several types of solid constituent together with the pore solution. Most of the solid constituents, viz. the C–S–H, AFm and hydrotalcite-type phases, have layer structures, and the layers are probably mixed on scales varying from the nanometre to the micrometre level. Hydrogarnet and AFt phase, neither of which has a layer structure, could also be mixed with the other constituents on or below a micrometre scale. The relative proportions of the different solid constituents vary in microstructurally different regions. If this description is correct, C–S–H describes only one, or more probably two, of the solid constituents, albeit the most important. Provided that this is recognized, the term 'C–S–H gel' may still be considered appropriate for the material; alternatively, the term 'cement gel' is possibly to be preferred.

7.3.3 The stoichiometry of cement hydration

Many attempts have been made to formulate the stoichiometry of cement hydration (e.g. P30,T17,T35). The following treatment incorporates features of those in the references mentioned. It applies to ordinary Portland cements hydrated in pastes of w/c ratio 0.4–0.6 at 15–25°C for times greater than about 1 day. Modifications would be needed for other curing conditions or other types of cement, such as sulphate-resisting cements.

The essential input data are (a) the bulk chemical composition of the cement, (b) the quantitative phase composition of the cement and the chemical compositions of its individual phases, (c) the fraction of each phase that has reacted, (d) the w/c ratio, (e) the CO_2 content of the paste and an estimate of how it is distributed among phases, and (f) the composition of each hydrated phase for the specified drying condition. If (b) is unknown, it may be estimated as described in Section 4.4, and if (c) is unknown, it may be estimated from the age as described by Parrott and Killoh (P30), or, more simply though less precisely, by using empirical equations (D12,T37). If the phase composition by volume and porosities are to be calculated, densities of phases are also required.

In the initial stages of the calculation, quantities of elements are best expressed as millimoles per 100 g ignited weight. The hydration products of the ferrite phase are first considered; it is assumed that all the magnesium released from it on hydration, with an equivalent amount of its aluminium, enters a hydrotalcite-type phase and that all the iron, silicon, titanium and manganese released from it, with some of the calcium and aluminium, enter a hydrogarnet-type phase. The remaining calcium and aluminium from the ferrite are pooled with the atoms supplied by the other cement phases. It is then assumed that the magnesium, with an equivalent amount of the octahedral cations (aluminium, titanium, manganese and iron), enters a hydrotalcite-type phase. The silicon is assumed to occur as C–S–H, and by solving two simultaneous equations using the amounts of sulphur and octahedral cations, the amounts of AFm and AFt phases are obtained. This procedure can be modified, if desired, to place some silicon in the aluminate phases, or aluminium in the C–S–H or both; three equations will then have to be solved. Calcium not present in any of the phases thus far considered, or in $CaCO_3$, is assumed to occur as CH. On this basis, a table is constructed giving the amount of each element present in each phase, expressed in millimoles per 100 g ignited weight.

For any selected drying condition, a table is now completed, similar to the previous one, but with the H_2O and CO_2 content of each phase included and the quantities of elements expressed as weight percentages of oxide components. Table 7.3 shows results for a typical, mature cement paste thus calculated. It was assumed that the C–S–H had Ca/Si = 1.7 and Al/Ca = 0.02, that the AFm and AFt phases each had Si/Ca = 0.05 and that the hydrogarnet-type phase had the composition $Ca_{2.95}Al_{0.1}Fe_{1.75}Ti_{0.1}Mn_{0.05}$-

Table 7.3 Calculated mass balance and compositions by weight and by volume for a 14-month-old cement paste (w/c = 0.5)[a]

Percentages on the ignited weight for material equilibrated at 11% RH

	Na$_2$O	MgO	Al$_2$O$_3$	SiO$_2$	SO$_3$	K$_2$O	CaO	TiO$_2$	Mn$_2$O$_3$	Fe$_2$O$_3$	Other	CO$_2$	H$_2$O	Total
Alite	0.0	0.0	0.0	0.7	0.0	0.0	1.9	0.0	0.0	0.0	0.0	0.0	0.0	2.7
Belite	0.0	0.0	0.0	0.5	0.0	0.0	1.1	0.0	0.0	0.0	0.0	0.0	0.0	1.6
Aluminate	0.0	0.0	0.0	0.0	0.0	0.0	0.1	0.0	0.0	0.0	0.0	0.0	0.0	0.1
Ferrite	0.0	0.1	0.6	0.1	0.0	0.0	1.4	0.0	0.0	0.6	0.0	0.0	0.0	2.9
C–S–H	0.0	0.0	0.5	18.6	0.0	0.0	29.6	0.0	0.0	0.0	0.0	0.0	11.4	60.2
Ca(OH)$_2$	0.0	0.0	0.0	0.0	0.0	0.0	18.5	0.0	0.0	0.0	0.0	0.0	5.9	24.4
AFm	0.0	0.0	3.2	0.4	1.1	0.0	7.9	0.0	0.0	0.7	0.0	0.0	7.8	21.4
AFt	0.0	0.0	0.6	0.1	1.5	0.0	2.4	0.0	0.0	0.1	0.0	0.0	4.0	8.7
Fe HP	0.0	0.0	0.1	0.2	0.0	0.0	1.7	0.1	0.0	1.5	0.0	0.0	1.0	4.6
Mg HP	0.0	1.1	0.4	0.0	0.0	0.0	0.0	0.0	0.0	0.1	0.0	0.0	1.1	2.6
CaCO$_3$	0.0	0.0	0.0	0.0	0.0	0.0	0.9	0.0	0.0	0.0	0.0	0.7	0.0	1.6
Other	0.2	0.0	0.1	0.2	0.0	0.4	0.0	0.1	0.0	0.0	0.2	0.0	0.0	1.2
Total	0.2	1.2	5.6	21.0	2.6	0.4	65.4	0.3	0.1	3.1	0.2	0.7	31.2	132.0

Volume percentages for paste equilibrated at 11% RH

Alite	Belite	Alum.	Ferr.	C–S–H	CH	AFm	AFt	Fe HP	Mg HP	CaCO$_3$	Other	Pores
1.0	0.6	0.0	1.0	33.4	13.2	12.7	6.2	1.8	1.7	0.7	0.6	27.0[b]

Water contents (percentages on the ignited weight)
Bound water: 31.2%   Non-evaporable water: 21.9%

Porosities[c]
Capillary, 15.5%   Free water, 27.0%   Total water, 41.9%

[a] Modified from Ref. T35. 'Other' components mainly P$_2$O$_5$ (0.2%); other phases mainly alkalis (adsorbed or in solution), insoluble residue and P$_2$O$_5$ (form of combination unknown). Fe HP = hydrogarnet-type product from ferrite phase. Mg HP = hydrotalcite-type phase. AFm and Mg HP include poorly crystalline materials intimately mixed with C–S–H. Discrepancies in totals arise from rounding.
[b] Free water porosity.
[c] Porosities defined in Section 8.3.1.

Table 7.4 TG data, water contents and densities for phases in Portland cement pastes

	Phase and mole ratio of volatiles retained at given temperature[a]						
Temperature (°C)	C–S–H H_2O/Ca	CH H_2O/Ca	AFm H_2O/Ca	AFt H_2O/Ca	Fe HP[b] H_2O/Ca	Mg HP[c] H_2O/Mg	Any CO_2/Ca
50	(2.0)	1.00	3.00	5.20	1.76	2.67	1.00
75	(1.6)	1.00	2.80	4.50	1.76	2.49	1.00
100	(1.3)	1.00	2.36	3.00	1.76	2.11	1.00
125	0.98	1.00	2.28	1.95	1.76	2.03	1.00
150	0.82	1.00	2.19	1.33	1.76	1.95	1.00
175	0.67	1.00	2.09	1.20	1.76	1.86	1.00
200	0.58	1.00	1.60	1.05	1.76	1.42	1.00
225	0.52	1.00	1.49	0.91	1.00	1.33	1.00
250	0.46	1.00	1.39	0.76	0.50	1.24	1.00
300	0.37	1.00	0.89	0.55	0.40	0.79	1.00
350	0.27	1.00	0.79	0.43	0.30	0.70	1.00
400	0.18	0.95	0.72	0.36	0.20	0.64	1.00
450	0.14	0.80	0.60	0.32	0.15	0.53	1.00
500	0.10	0.05	0.40	0.28	0.10	0.36	1.00
550	0.07	0.02	0.30	0.24	0.08	0.27	1.00
600	0.05	0.01	0.26	0.22	0.05	0.23	1.00
700	0.04	0.00	0.14	0.07	0.00	0.12	0.80
800	0.03	0.00	0.03	0.04	0.00	0.03	0.00
900	0.02	0.00	0.01	0.00	0.00	0.01	0.00

H_2O/Ca or CO_2/Ca ratios[d]

Saturated	2.30[e]	1.00	3.50	5.33	1.76	2.50	1.00
11% RH	1.20	1.00	3.00	5.30	1.76	2.50	1.00
D-dry	0.85	1.00	2.00	1.20	1.76	1.90	1.00

Densities ($kg\,m^{-3}$)

Saturated	1900[e]	2240	1900	1770	3100	2000	2710
11% RH	2180	2240	2010	1730	3100	2000	2710
D-dry	2700	2240	2400	2380	3100	2300	2710

Densities of clinker phases ($kg\,m^{-3}$)[f]

Alite, 3150   Belite, 3300   Aluminate, 3060   Ferrite, 3570

[a] TG with heating rate 10 deg C min^{-1}, 50 mg specimen, flow rate of dry N_2 15 ml min^{-1}.
[b] Hydrogarnet-type product from ferrite phase.
[c] Hydrotalcite-type phase.
[d] H_2O/Mg in case of Mg HP.
[e] Of uncertain significance for w/c > 0.4.
[f] For typical compositions.

$Si_{0.4}O_{12}H_{10.4}$. Table 7.4 gives other data assumed. By introducing also the w/c ratio and the densities of the phases, the volume percentages of phases for the specified drying condition may now be calculated, and, by difference, the porosity. Using data for the TG curves of the phases containing H_2O or CO_2 or both, one may also calculate a simulated TG curve for the paste. The results of such a calculation are given in Fig. 7.4, the observed curve in which relates to the 14-month-old paste of Table 7.3.

Results such as those in Table 7.3, and calculated TG curves such as that in Fig. 7.4, agree broadly with experimental data, but many of the data assumed are tentative and the method needs to be tested in many more cases before its accuracy can be assessed. The hydration products listed, other than CH and, in part, the AFm and AFt phases, are probably not separate phases on more than a sub-micrometre scale, but constituents of the cement gel. If the data were available, it might be more meaningful to group the volume percentages differently, by such categories as in situ products formed at the various stages of hydration from each of the clinker phases, and undesignated product, and to give the chemical and 'phase' compositions for each. AFm phase as defined in Table 7.3 represents a weighted mean for products of differing compositions and states of subdivision and crystallinity, ranging from relatively well defined crystalline material, probably low in SO_3, to material intimately mixed with C–S–H and probably higher in SO_3. A similar comment may apply in some degree to AFt phase.

7.4 Development of microstructure

7.4.1 The early period of hydration

Many features of the development of microstructure in cement pastes are similar to those observed in C_3S pastes (Section 5.3.1), but there are some important differences. As with C_3S pastes, it is convenient to consider the process in terms of early, middle and late periods of hydration, with divisions at approximately 3 and 24 h after mixing. Fig. 7.6 shows diagrammatically the sequence of changes undergone by a typical, polymineralic cement grain, as given by Scrivener, on whose conclusions (S39,S68) the following is largely based.

Studies by HVTEM show that at high dilutions a gel layer or membrane forms over the surfaces of the grains soon after mixing (D14,G43). A product of similar appearance has been observed in pastes of C_3A with gypsum (S41), and, less distinctly, in pastes of cement of normal w/c ratios (J23,S68). It is probably amorphous, colloidal and rich in alumina and silica, but also containing significant amounts of calcium and sulphate, the composition varying with that of the underlying surface. Within about 10 min,

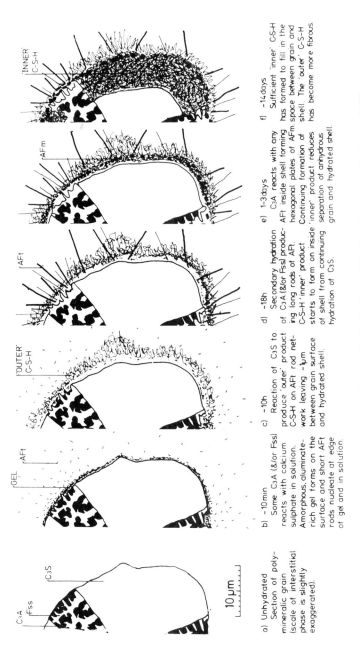

Fig. 7.6 Development of microstructure during the hydration of Portland cement. Scrivener (S39).

stubby rods of AFt phase are also seen (D25–D27). They are typically some 250 nm long and 100 nm thick. Studies using wet cells show them to occur both on the surfaces of the grains, and at some distance away (S41,S68) (Fig. 7.6b). They are probably more abundant near to the surfaces of the aluminate phase, and appear to nucleate in the solution and on the outer surface of a layer of gel. On drying, this layer shrinks, and the AFt crystals fall back onto the surfaces of the cement grains. The early products thus differ in morphology and composition from the exfoliating foils or honeycombs of C–S–H that have been observed in C_3S pastes.

The first study using a wet cell, made at high w/c ratios, showed tubular growths radiating from the cement grains, which were considered to have formed by a 'silicate garden' mechanism (D14). Later work showed that they were rich in calcium, aluminium and sulphur, and that they did not form if C_3S was substituted for cement (B101). They have not been observed in the more recent studies made at normal w/c ratios, and do not appear to be a significant feature of normal cement hydration.

7.4.2 The middle period of hydration

During this period, which begins at about 3 h and ends at about 24 h, some 30% of the cement reacts. It coincides with that of strong heat evolution and is characterized by the rapid formation of C–S–H and CH. Studies using wet cells show that the undried C–S–H has a filmy, foil-like morphology (P31,S40), which on drying changes to give fibres ('Type I C–S–H') where space is freely available or honeycombs or reticular networks ('Type II C–S–H') where it is more restricted (Section 5.3.1). The morphology is also affected by the presence of foreign ions, such as Cl^-. Spherical aggregates of fibres some 2 µm in diameter, observed in dried material, probably result from the rapid reaction of small cement grains consisting only of alite (S68). The CH forms massive crystals in the originally water-filled space. Nucleation sites appear to be relatively few in number, and the growing crystals may engulf some of the smaller cement grains. All these features of the process are similar to those observed in C_3S pastes.

The C–S–H forms a thickening layer around the cement grains (D26,P31) which engulfs and perhaps nucleates on the AFt rods (Fig. 7.6c). A significant amount has formed by 3 h, and the grains are completely covered by 4 h. The shells grow outwards; by about 12 h they are some 0.5–1.0 µm thick, and those surrounding adjacent grains are beginning to coalese. At this stage, which was called the cohesion point, fracture through the shells begins to supplant fracture between them. It coincides with the maximum rate of heat evolution and corresponds approximately to the completion of setting. The structure of interconnected shells has been considered to play an important part in determining the mechanical and other properties, which thus depend on the particle size distribution of the cement (D26).

Studies using ion-thinned sections, wet cells and backscattered electron images of polished sections show that a space develops between the shell and the anhydrous material (S40,S41,S68) (Fig. 7.6c). In this respect, the hydration of cement differs from that of C_3S, in which the C–S–H grows directly over the C_3S surfaces, without any detectable separation (S41). By 12 h, the spaces are up to 0.5 μm wide. They are likely to be filled with a highly concentrated or colloidal solution, and the shells are evidently sufficiently porous at this stage that ions can readily migrate through them (S68). The existence of spaces shows that reaction proceeds by dissolution and precipitation; further evidence for this is provided by the fact that the C–S–H also deposits on the surfaces of pfa particles, if these are mixed with the cement (D28). Some other relatively unreactive or inert admixtures behave in the same way.

Towards the end of the middle period, a renewed growth of AFt crystals takes place (D27,P31,S41) (Fig. 7.6d). They are markedly more acicular than those formed earlier; their lengths are typically 1–2 μm, but sometimes up to 10 μm. Their formation is associated with a shoulder on the heat evolution curve (Section 7.5.1). Their formation implies an increase in the rate of reaction of the aluminate, or less probably the ferrite phase, which is probably related to the reaction of the alite (S68).

7.4.3 The late period of hydration

With the decreasing permeability of the shells, C–S–H begins to deposit also on their insides, and its surface advances inwards more quickly than that of the alite retreats. Grains smaller than about 5 μm appear to react completely before the end of the middle period, and before much material has deposited inside the shells; many that originally contained aluminate phase are empty (S68). The outer product from such grains is often absorbed in the shells surrounding adjacent, larger grains. With larger grains, the spaces between shell and core fill up, and by about 7 days they have disappeared; at this stage, the shells are typically some 8 μm thick and consist mainly of material that has been deposited on their inner surfaces (Fig. 7.6e and f) (S68). The separations between shell and core seen on polished or ion-thinned sections rarely exceed about 1 μm.

Separated shells, up to at least 10 μm across and sometimes completely hollow, were originally observed on fracture surfaces (B102) and have been called 'Hadley grains'. These are not observed on ion-thinned sections, and possibly result from the core falling out during specimen preparation (S68). It would be difficult to reconcile any widespread occurrence of such cavities in a mature paste with the bulk density of the material (P31).

The concentration of SO_4^{2-} must drop rapidly inside the shells as the aluminate phase reacts, and AFm phase often forms within the shells, any AFt phase formed there initially being replaced by AFm as a result of continued reaction of the aluminate phase (S41,S68). A single specimen may show AFm phase within the shells and AFt phase outside them. The XRD evidence (Section 7.1.1) shows that significant quantities of AFt phase may persist, apparently indefinitely; this is presumably material that has been precipitated outside the shells. If the SO_4^{2-} concentration in the solution outside the shells drops before these have sufficiently isolated the anhydrous grains, relatively large ($\simeq 10\,\mu m$) crystals of AFm phase can form throughout the paste (S68).

After the spaces between shells and cores have filled up, reaction is slow and, in contrast to that occurring earlier, appears to occur by a topochemical mechanism (Section 5.6.4). In old cement pastes, three regions of C–S–H can thus be distinguished in the relicts of the larger fully reacted grains (S74), *viz.* (a) an outer layer some 1 µm thick that has formed through solution in originally water-filled space, (b) a middle layer some 8 µm thick that has been deposited, also through solution, on the inside of the shell and thus in space originally occupied by the cement grain, and (c) a central core, that has formed topochemically. As with C_3S pastes, all the C–S–H probably has a foil morphology, which is modified in varying ways by the effects of drying and in accordance with the amount of space available.

Except in pastes many years old, the hydration of belite has proved difficult to observe by electron microscopy (S68), although XRD evidence shows that considerable reaction has occurred within 14 days. In a 23-year-old paste studied by backscattered electron imaging, the belite grains had reacted completely, but the original lamellar structure remained, providing further evidence of a topochemical mechanism (S74). The belite appears to react preferentially along the exsolution lamellae. The reaction of the ferrite phase has similarly proved difficult to observe, though backscattered electron imaging showed the formation in the same sample of rims high in iron, and TEM examination showed the formation of the in situ product containing C–S–H, hydrotalcite-type material and poorly crystalline iron-rich material mentioned in Sections 7.1.3 and 7.2.5.

7.5 Calorimetry, pore solutions and energetics

7.5.1 The early and middle periods

Fig. 7.7 shows the heat evolution curve for a typical Portland cement paste, determined by conduction calorimetry. It broadly resembles those given by C_3S (Fig. 5.14) and comparison with the microstructural evidence (Section 7.4) shows that it can largely by explained in a similar way. The initial peak (1) is attributable to a combination of exothermic wetting and the early-stage reactions, which with cement give a gelatinous coating and rods of AFt phase. Rehydration of hemihydrate to give gypsum may contribute. The main peak (2) corresponds to the middle-stage reaction, in which the main products, as with C_3S, are C–S–H and CH. The gradually decreasing rate of heat evolution after 24 h corresponds to the continuing slow reactions of the late stage, which again give mainly C–S–H and CH. As with C_3S, the process may be divided into the five stages of initial reaction, induction, acceleratory and deceleratory periods, and the final period of slow reaction.

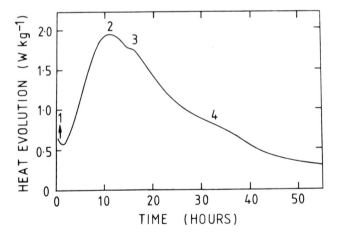

Fig. 7.7 Rate of heat evolution at 20°C for a typical Portland cement; for significance of numbered peaks, see text. After Pratt and Ghose (P31).

Many cements show a shoulder or more definite peak (3) at about 16 h. This has often been associated with the conversion of AFt into AFm phase, but comparison with the microstructural evidence shows that this is incorrect (P31); it is associated with the renewed formation of ettringite. A further, less distinct shoulder (4) has been associated with hydration of the ferrite phase (P31) or with the conversion of AFt to AFm phase (S68). It may correspond to a peak at about 50 h observed by Stein (S75). In all these reactions involving the aluminate or ferrite phases, the principal exothermic component is probably the reaction of the anydrous compound with water and not the precipitation or subsequent reactions of hydrated compounds.

Many studies of concentrations in the pore solution during the early and middle periods have been reported (L32,L33,T38,U12,G58). Fig. 7.8 shows typical results. Relatively high concentrations of Ca^{2+}, K^+, Na^+, SO_4^{2-} and OH^- are quickly reached. Between 2 and 12 h in the example given, the concentrations change relatively little, indicating an approximate balance between the continued dissolution of the cement phases and the precipitation of products. At 12–16 h, the concentrations of Ca^{2+} and SO_4^{2-} fall sharply, and the solution thereafter is essentially one of alkali hydroxides. The sharp fall at 12–16 h corresponds to the renewed growth of ettringite observed in the SEM and to shoulder (3) on the heat evolution curve.

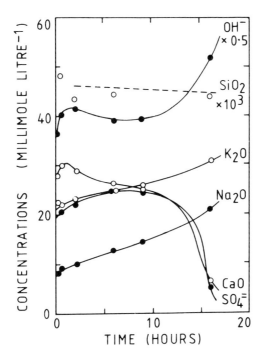

Fig. 7.8 Concentrations in the pore solution (scaled as indicated in the cases of OH^- and SiO_2) of a Portland cement paste of w/c ratio 0.5. After Lawrence (L32).

Concentrations of SiO_2 and Al_2O_3 are low throughout, values of 0.03–0.05 mmol l^{-1} for Si and below 5 ppm for Al having been reported for times from 20 s to 45 min (T38).

Any more detailed interpretation demands that both activity coefficients and the existence of complex species be taken into account (B90,G58). Gartner et al. (G58) showed that, for most cements, the solution is saturated in CH within 12 min and that the saturation factor (defined as the activity product divided by its value at saturation) reaches a maximum of between 2 and 3 within 2 h. For gypsum, saturation is reached within 6 min, but the saturation factor never exceeded 1.3 in the cases studied. With cements high in K_2O, the solution can become saturated in syngenite. Gartner et al. found that the K^+ and Na^+ concentrations increased rapidly during the first 12 min and only slowly, or not at all, during the subsequent period up to 3 h.

Other methods that have been used to follow the reactions in cement pastes during the first day include measurements of AC electrical conductivity (T39,A14) and dielectric constant (A14).

One may calculate a partial mass balance for the reaction at an age of 30 min. QXDA of a typical cement retarded with gypsum showed that, at this age, 11% of the aluminate and 8% of the ferrite phase had reacted (T34). Assuming from the bulk analysis that the cement contained 12% of aluminate and 7% of ferrite, and that these phases had the typical compositions given in Table 1.2, a total of 1.0 g of CaO and 0.5 g of Al_2O_3 are released from them per 100 g of cement. For broadly similar cements at the same age, ettringite contents of about 2% have been typically reported (B99,U13). These appear to be referred to weights of paste, and suggest values of about 3% referred to that of cement. This quantity of ettringite, ignoring any iron substitution, contains 0.8 g of CaO, 0.2 g of Al_2O_3 and 0.6 g of SO_3. It thus appears that, of the Al_2O_3 released, less than half enters the ettringite. Apart from small amounts in solution, the remainder and the CaO, SiO_2 and part of the SO_3 released from other phases presumably enter the amorphous gel that is also formed.

The evidence from microstructure, calorimetry and other sources suggests that the hydration processes of cement and C_3S are essentially similar. There are important differences in the nature of the early product and in where the C–S–H formed in the middle stage of reaction begins to deposit, but in both cases it would appear that the early reaction slows down because of the deposition of a layer of product, which either isolates parts of the anhydrous surfaces from the main solution or allows the concentrations close to those surfaces to rise to values approaching the theoretical solubilities of the anhydrous compounds. In both cases, the initiation of the main reaction and the kinetics in its acceleratory phase appear to be controlled by the nucleation and growth of C–S–H.

7.5.2 Pore solutions after the first day

Apparatus for extracting pore solution from hardened cement pastes by applying pressures of some 375 MPa was first described by Longuet et al. (L34), and many investigators have reported analyses of solutions thus obtained (L34,D29,X1; for further references, see Ref. T37). With pastes of normal Portland cements more than about 1 day old, the only ions present in concentrations above a few mmol l^{-1} are K$^+$, Na$^+$ and OH$^-$. Most studies show the concentrations of these ions to rise with time, and to approach a limit after 28–90 days. Some show concentrations passing through maxima, with subsequent decreases that are usually slight. Typical concentrations after 180 days for pastes of w/c ratio 0.5 are 0.08 mol l^{-1} for Na$^+$ and 0.24 mol l^{-1} for K$^+$ for a low-alkali cement (0.16% Na$_2$O, 0.43% K$_2$O) (X1) and 0.16 mol l^{-1} for Na$^+$ and 0.55 mol l^{-1} for K$^+$ for a high-alkali cement (0.24% Na$_2$O, 1.21% K$_2$O) (L34). The corresponding OH$^-$ concentrations are 0.32 mol l^{-1} and 0.71 mol l^{-1}, respectively. Some investigators (e.g. Ref. L35) have reported concentrations varying with the pressure used to express the liquid, suggesting the possibility that high concentration gradients exist and casting doubt on the interpretation of the results. This question does not appear to have been resolved.

The K$^+$ and Na$^+$ are present in the cement partly as sulphates and partly in the major clinker phases (Section 3.5.6). When the phases containing them react, the accompanying anions enter products of low solubility and equivalent quantities of OH$^-$ are produced. The K$^+$, Na$^+$ and OH$^-$ ions are partitioned between the pore solution and the hydration products. Early estimates of the fractions remaining in solution were too high because non-evaporable water was used as a measure of bound water and the quantity of pore solution thereby overestimated.

Taylor (T37) described a method for predicting the concentrations at any desired age after 1 day from the w/c ratio and the contents of total Na$_2$O, total K$_2$O, water-soluble Na$_2$O and water-soluble K$_2$O in the cement. It was assumed that the amount of each alkali cation taken up by the products is proportional to its concentration in the solution and to the quantity of products (C-S-H and AFm phase) taking it up. This led to the equation

$$c = m_r/[V + (b \times P)] \qquad (7.2)$$

where c is the concentration in mol l^{-1}, m_r the quantity in millimoles of alkali cation released, V the volume of pore solution in ml, P the quantity of relevant products (C-S-H + AFm phase) relative to that formed on complete hydration, and b an empirical constant, called the binding factor, equal to 20.0 ml for K$^+$ and to 31.0 ml for Na$^+$. All quantities are referred to 100 g of cement. V was taken to be the total volume of water less the volume of bound water, i.e. that retained at 11% RH. The latter quantity, and m_r and P, were estimated using empirical equations.

For a Portland cement with 0.2% total Na_2O and 0.6% total K_2O, at $w/c = 0.5$ and an age of 1 year, calculation by the above method indicated that 59% of the Na^+ released and 48% of the K^+ released were taken up by the products. The ratios of alkali cations to Ca^{2+} in the products are sufficiently low (typically < 0.01 for Na^+ and $\simeq 0.01$ for K^+) as to suggest that the ions, with an equivalent quantity of OH^-, are adsorbed. As Glasser and Marr (G63) concluded earlier, Na^+ appears to be more strongly held than K^+. As was noted in Section 7.3.2, the ratios of alkali cations to Ca^{2+} found in the cement gel by X-ray microanalysis are higher than those in the solid products, because they include the materials dissolved in the pore solution, which are deposited when the latter evaporates.

7.5.3 Energetics of cement hydration

Copeland et al. (C38) reviewed the energetics of cement hydration. After the first few days, the rate of heat liberation is too low for conduction calorimetry to be a practicable means of investigation, but the total amount of heat liberated after any desired time can be determined from the heat of solution in acid, which is compared with that of the unhydrated cement. Fig. 7.9 gives average results thus obtained for different US types of Portland cements.

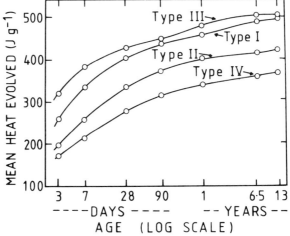

Fig. 7.9 Average cumulative heat evolution for a total of 20 Portland cements of US Types I, II, III and IV, hydrated at $w/c = 0.4$ and 21°C. After Copeland et al. (C38).

With any Portland cement, the amount of heat evolved by a given time is directly related to the amounts of the clinker phases that have reacted, and thus depends on such factors as the particle size distribution, w/c ratio and temperature and RH of curing. As the results in Fig. 7.9 show, it also depends on the cement composition. At all ages, the amount of heat liberated is greatest for the Type III (high early strength) cements and least for the Type IV (low heat) cements. The differences reflect the different average potential phase compositions, which show progressively lower contents of C_3S and C_3A, and progressively higher ones of C_2S, along the sequence from Type III to Type IV.

Using least squares regression analysis, the heat liberated at any given age was predicted from the potential phase composition by equations of the type

$$H_t = a(C_3S) + b(C_2S) + c(C_3A) + d(C_4AF) \qquad (7.3)$$

where H_t is the amount of heat evolved in $kJ\,kg^{-1}$, the formulae denote the weight fractions of the phases calculated by the Bogue equations and the coefficients a, b, c and d have the values given in Table 7.5, which also includes values for the enthalpies of hydration of the pure clinker compounds, taken from the data in Sections 5.5.3 and 6.5.4. The agreement between the least squares coefficients and the enthalpies of hydration is good for mature pastes, but there are some anomalies for younger pastes.

In principle, it should be possible to calculate the heat of hydration from the quantitative phase compositions of the unreacted mix and of the paste, using standard enthalpies of formation. The sensitivity of such calculations to small errors in the latter data probably renders this approach unsatisfactory with existing data.

7.6 Actions of calcium sulphate and of alkalis

7.6.1 Setting

Setting times are commonly defined empirically, using a Vicat needle. In this device, weighted needles of standard design are allowed to sink into the paste, and initial and final set defined as the times when the degree of penetration falls below specified levels. In the British standard, which is typical, initial set must occur not earlier than 45 min, and final set not later than 10 h.

Some workers have concluded from QXDA evidence that the rate of consumption of the aluminate phase is not significantly affected by the presence of gypsum (L33,L36,O18), and setting has been attributed largely to

Table 7.5 Coefficients in equation 7.3 for predicting the cumulative heat evolution in a cement paste of given age from the potential phase composition of the cement (for w/c = 0.4 and 21°C) (C38)

Compound	Coefficient	Value of the coefficient (kJ kg^{-1}) for age given below							Enthalpy of complete hydration[a] (kJ kg^{-1})
		3 days	7 days	28 days	90 days	1 year	6.5 years	13 years	
C_3S	a	243	222	126	435	490	490	510	−517 ± 13
β-C_2S	b	50	42	105	176	226	222	247	−262
C_3A	c	887	1556	1377	1301	1167	1372	1356	−1144[b]; −1672[c]
C_4AF	d	289	494	494	410	377	464	427	−418[d]

[a] From data in Chapters 5 and 6.
[b] Reaction with gypsum to give $C_4A\bar{S}H_{12}$.
[c] Reaction with gypsum to give ettringite.
[d] Reaction in presence of excess CH to give a hydrogarnet.

recrystallization of ettringite (L33). In contrast, Tang and Gartner (T34) found that the amounts of both aluminate and ferrite phases reacting during the first 30 min were substantially decreased in the presence of gypsum or other forms of calcium sulphate, and evidence from XRD, DTA, IR spectroscopy, conduction calorimetry and SEM shows that setting occurs during the acceleratory period and thus corresponds to a period of rapid formation of C–S–H and CH (B103). The microstructural evidence (Section 7.4.2) shows that little change occurs in the amount or morphology of the ettringite during this period, the secondary growth of ettringite occurring later. The concept of a cohesion point, at which the shells surrounding individual cement grains coalesce (D26), further indicates the importance of the relation between C–S–H formation and both setting and the initial stages of strength development. The close similarity in setting behaviour between pastes of cement and C_3S, and the relation between setting and microstructural development in C_3S pastes, provide further evidence that normal setting depends primarily on the silicate reactions.

If too little gypsum is added, or none at all, many cements undergo what is termed 'flash' or 'quick' set. This is a rapid set, with much evolution of heat; plasticity is not regained on continued mixing, and the subsequent development of strength is poor. It is associated with increased early reaction of the aluminate and ferrite phases (T34), and with the formation of plates of AFm phase throughout the paste. These can be 5–10 µm in size (L36). The meshwork of plates, which may be contrasted with the much more compact coatings of gel and small rods of AFt phase formed in the presence of an adequate supply of gypsum, accounts for the rapid stiffening, but the reason for the subsequent poor strength development is less clear. Possible explanations include the formation during the initial reaction of strongly protective layers of product around the cement grains, and the existence of a microstructure weakened by the large plates of AFm phase. Some cements low in aluminate phase do not show flash set even if no gypsum is added.

Another undesirable condition is called 'false set'. This is also a rapid set, but there is no abnormally high evolution of heat, plasticity is regained on further mixing, and subsequent strength development is not markedly affected. The most usual cause of false set is the presence of too much calcium sulphate in the form of hemihydrate, which is rehydrated to give gypsum (called 'secondary' gypsum). The setting is attributable to the interlocking of the gypsum crystals, which are tabular and can be 5–10 µm in longest dimension (L36). If the quantity of secondary gypsum is not too great, it redissolves on further mixing, and the reactions then follow their normal course. Dehydration of gypsum to hemihydrate can occur during milling; hemihydrate may also be added deliberately, if a more reactive form of calcium sulphate is required. With cements high in K_2O, syngenite can be precipitated; this also can cause false set, and, due to the removal of $CaSO_4$ from the system, even flash set (J24).

7.6.2 Optimum gypsum

The gypsum in cement affects not only the setting time, but also the strength development and the volume stability. Locher *et al.* (L33,L37), Kinare and Gartner (K35) and Tang and Gartner (T34) have discussed factors governing the optimum content and the effects of varying the source of the sulphate. The situation is complicated by the fact that, contrary to some early conclusions, the amounts needed to optimize different properties, such as strength at various ages and drying shrinkage, are not necessarily the same; also, the amount needed to optimize a given property in a concrete may not be the same as that required in a paste or mortar (K35).

During the early and middle periods of reaction in a cement paste, gypsum or other forms of calcium sulphate dissolve and react at or close to the surfaces of the clinker grains, or more specifically those of the aluminate and ferrite phases. The factor most directly influencing the course of the early reactions is not so much the relative amounts of calcium sulphate, aluminate and ferrite phases, as the rates at which the relevant ionic species are made available at the surfaces of the cement grains. The rates at which Ca^{2+} and SO_4^{2-} ions are supplied by the calcium sulphate thus depend both on the amount of the latter and on its physical and chemical nature. Hemihydrate or soluble anhydrite supply ions more quickly than does gypsum, which in turn supplies them more quickly than anhydrite. Other major factors affecting the supply of these ions are the particle size distribution of the calcium sulphate and the distribution in space of the particles; Tang and Gartner (T34) found that a given proportion of gypsum added by intergrinding was much more effective than the same proportion added by blending, all other conditions being as nearly as possible equal. Intergrinding presumably brings the gypsum particles into more intimate contact with those of the clinker. Increase in any of the factors rendering the $CaSO_4$ more readily available will tend to decrease the amount that is needed to produce a given effect in the early stage of reaction. Increase in temperature has a similar effect (C39).

Factors affecting the rate at which Al_2O_3 is supplied include its content in the clinker, its distribution among the clinker phases and especially the amount present in the aluminate, the specific surface area of the ground clinker, the reactivities of the aluminate and ferrite, and the microstructure of the clinker particles, or, more specifically, the areas of surface composed of aluminate and ferrite phases and the manner in which these phases are intergrown with each other and with the silicate phases. Any factor increasing the availability of the Al_2O_3 will tend to increase the amount of gypsum that is required to produce a given effect in the early reactions.

Sulphate ion is also supplied by the clinker, especially as alkali sulphates or calcium langbeinite, and Ca^{2+} ions are supplied by the clinker phases, including free lime. The alkali sulphates provide a highly available source of SO_4^{2-}, but the alkali cations, or more probably the OH^- ions that they produce, have additional effects (Section 7.6.3).

The minimum content of SO_3 required to control setting is typically around 2%. Increase in SO_3 content beyond the minimum required has little effect on setting unless the proportion of hemihydrate or soluble anhydrite is so high as to cause false set. It causes a progressive decrease in the shrinkage that occurs in dry atmospheres (e.g. at 50% RH) and a progressive increase in the expansion that occurs in water. Hobbs (H39) gave data for cements containing more than the normal amounts of gypsum. Mechanisms of expansion are discussed in Sections 10.2 and 12.5. If expansion is excessive, the concrete disintegrates; this limits the permissible SO_3 content, which, in typical specifications, is set at 2.5–3.0%, depending on the potential C_3A content of the clinker.

The effects on strength are complex and not well understood. In general, for any particular clinker and source of SO_3 there is an optimum content of SO_3 at early ages. In a careful study, in which strengths were corrected empirically for the effects of differing particle size distributions and contents of entrapped air, Osbaeck and Jøns (O8) found maximum 1-day strengths at 3–5% SO_3 for a number of cements having different alkali contents and clinker SO_3/alkali ratios. At later ages, somewhat variable results have been reported; in general, the optimum seems to shift to higher values but to become less pronounced. The results of Osbaeck and Jøns, and of Jelenić et al. (J25), indicate that it is more definitely upwards if the alkali content is high. In evaluating the effect of gypsum content on strength or other properties, it is necessary to ensure that other relevant factors are as far as possible kept constant. Thus, if cements with the same clinker but different gypsum contents are ground to the same specific surface area, the clinker will be less finely ground the more gypsum is present, and this will tend to decrease the rate of strength development.

Calcium sulphate additions probably affect both the quantities of products that are formed, especially at early ages, and the microstructure at all ages. Addition of $CaSO_4$ is reported to increase the rate of hydration, both of pure C_3S and of the alite in cement (C39,B104,J25,J26,L36). This could explain an increase in strength at early ages due to additions up to the optimum, but not a decrease with larger amounts. The fact that calcium sulphate additions affect the drying shrinkage also suggests that the microstructure is modified. This could occur in various ways, e.g. by changes in the micromorphology of the C–S–H or by incipient precipitation of ettringite throughout the gel. The latter effect, if it happened at an early age or to a limited extent, could augment strength through a decrease in porosity, but if it happened later or to too large an extent could decrease strength through localized expansions.

Attempts to formulate equations relating optimum gypsum content to composition and particle size distribution have proved of little practical value, and gypsum contents are normally decided on the basis of empirical strength tests. In practice, the molar ratios of total SO_3 to total Al_2O_3 in modern cements range from 0.5 to 0.9, with an average of about 0.6 (K35). In order to adjust the supply of SO_4^{2-} to that required at different stages of the hydration process, it may be desirable to use mixtures of different forms of calcium sulphate; for example, with cements high in aluminate phase, a mixture of anhydrite and hemihydrate, the latter produced by dehydrating gypsum during milling, has been recommended (L36).

7.6.3 Effects of alkalis

Alkali cations normally occur in cements either as sulphates or in the major clinker phases. In either case, the balancing anion sooner or later enters a hydration product of low solubility, and an equivalent amount of OH^- ion is released. Apart from syngenite precipitation, mentioned earlier, the effects of alkali cations on cement hydration are likely to be predominantly those of the OH^- ion. Jawed and Skalny (J24) reviewed these effects, both on the hydration process and on the properties of the hardened product. In general, early strengths are increased and late strengths decreased. Some other properties, such as drying shrinkage, are affected, and the optimum gypsum content is increased. This last effect must be taken into account when effects on strength are considered. Lerch (L38) attributed it to accelerated early reaction of the clinker phases.

The results obtained by Osbaeck and Jøns (O8) in the study mentioned in the previous section are probably typical. They show the effects of alkali sulphates on early and late strengths and on optimum gypsum contents mentioned above. The effects on strength were diminished or absent at gypsum contents above the optimum. QXDA showed that the increased early strengths were associated with increased reaction of the alite, aluminate and ferrite phases. This effect was quite marked; for example, at 1 day, the reaction of the alite increased from 26% for a cement low in soluble alkalis to 56% for an otherwise similar cement high in them. There were less definite indications of an association between the decreased strength at 28 days and decreased consumption of the clinker phases, and the authors were uncertain whether the decrease in strength was due to this cause or to the formation of a less favourable microstructure.

Bezjak *et al.* (B98) similarly found that the fractions of alite reacting at early ages were substantially increased in the presence of alkali sulphate, both in Portland cement and in mixtures of alite with C_3A. After 10 days, which was the longest time studied, the fraction reacting was still enhanced in the presence of alkali sulphate in the case of a cement low in aluminate phase, but was slightly lowered in that of cement higher in aluminate. Applying a theory of the kinetics of hydration described in Section 7.7.2, they attributed the acceleration at early ages primarily to an increase in the permeability of the layer of hydration products surrounding the alite at a stage soon after the rate of reaction has become controlled by diffusion.

Tang and Gartner (T34) studied a clinker low in alkalis and in SO_3, but high in aluminate, which they blended with varying mixtures of calcium sulphate, calcium alkali sulphate and alkali sulphate phases. They, too, found that alkali sulphates increased the early strength; without exception, the 1-day strengths obtained using blends of gypsum or hemihydrate with aphthitalite, syngenite or calcium langbeinite were greater than those obtained at the same ratio of total SO_3 to Al_2O_3 with the calcium sulphates alone, calcium langbeinite and syngenite being especially effective. The efficacy of calcium langbeinite as a set-controlling agent was earlier recognized by Moir (M7). At 28 days, in contrast to the findings of Osbaeck and Jøns, there was little significant difference in strength between the cements with and without alkali sulphates, though those of the former tended to be lower.

Jelenić *et al.* (J26) found that, for pastes of alite and C_3A with varying amounts of gypsum and with or without alkali sulphate, high strengths at 28 days were correlated with low C_3A consumption at 1 day. The late strengths were increased, not decreased, in the presence of alkali provided that the gypsum content was suitably adjusted. Tang and Gartner (T34) similarly found little correlation between the amount of aluminate phase consumed at 2 min and the compressive strength at 1 day, but there was a negative correlation between this quantity and the strength at 28 days. They concluded that a reduced initial aluminate consumption tended to improve not only the workability of the fresh paste, but also the ultimate strength of the hardened material.

7.7 Kinetics and modelling of the hydration process

7.7.1 Kinetics: experimental data

The kinetics of cement hydration are concerned with the relations between the degree of hydration (α) and the age (t), and the factors that influence them. We shall describe curves of α against t as kinetic curves. From the practical standpoint, the kinetic curve controls the way in which the physical properties develop as curing proceeds.

α may relate either to an individual clinker phase or to the cement as a whole. Because cement is a mixture of phases that react at different rates, there are problems in determining, and even in defining, α for the whole cement; we shall define it as the weight fraction of the cement that has reacted, irrespective of phase. In principle, it may be obtained by summing the amounts of the individual phases that have reacted, which in turn can be determined by QXDA or in other ways. Because of the experimental difficulty and questionable precision of this approach, less direct methods have often been used, based on the determination of such quantities as non-evaporable water, cumulative heat evolution (C38), or chemical shrinkage (K36). It has normally been assumed, with varying degrees of justification, that these quantities are proportional to α as defined above.

Major influences on the kinetic curve of a cement include the phase composition of the clinker, the particle size distribution of the cement and the RH and temperature regimes during curing. Other influences include the w/c ratio, the content and distribution of admixtures, including gypsum, the reactivities of individual clinker phases and probably others, such as the microstructures of the clinker and of the cement particles.

By definition, the kinetic curve of a cement is the weighted sum of the curves for its constituent phases as they occur in that cement. The reactivities of individual clinker phases were considered in Section 4.5 and some effects of particle size distribution, which is a particularly important variable, in Section 4.1.4. Although many data relating particle size distribution directly to strength exist, much less is known about its relation to degrees of reaction. Parrott and Killoh (P30) presented data indicating that the rate of hydration, as represented by that of heat evolution, was proportional to the specific surface area during the period of hydration in which the rate was controlled by diffusion.

The rates of reaction of the clinker phases are greatly influenced by the RH of the atmosphere in which curing occurs. For a typical Portland cement paste of w/c ratio 0.59 cured at 20°C and 100% RH, Patel et al. (P28) found the fractions of the alite, belite, aluminate and ferrite phases hydrated after 90 days to be respectively 0.94, 0.85, 1.00 and 0.51. If the RH was lowered to 80%, the corresponding values were 0.77, 0.19, 0.83 and 0.32. The hydration rate of the belite thus appears to be especially sensitive to RH. On the basis of earlier data from the literature, Parrott and Killoh (P30) concluded that the effect of RH on the hydration rate ($d\alpha/dt$) of each of the phases could be represented by a factor $\{(RH - 0.55)/0.45\}^4$.

Copeland and Kantro (C39), using QXDA, found that alite and belite in a Portland cement hydrated more rapidly at w/c = 0.65 than at w/c = 0.35. Heat of hydration data (C38) indicated that the rate of hydration in Portland cement pastes is higher at w/c = 0.6 than at w/c = 0.4 at all ages from 1 day to 6.5 years. Taplin (T40) found that at low w/c ratios the rate was significantly reduced at later ages. Based on this last result, Parrott and Killoh (P30) considered that $d\alpha/dt$ was reduced by a factor $f(w/c)$ which applied only at values of α greater than $1.333 \times w/c$:

$$f(w/c) = \{1 + 4.444(w/c) - 3.333\alpha\}^4 \qquad (7.4)$$

In contrast to these results, Locher (L39), working with pure C_3S, found that hydration was more rapid at a low w/c ratio.

Temperature has a large effect, especially in the earlier stages of hydration; for example, from QXDA, Copeland and Kantro (C39) found that in a Portland cement paste of w/c = 0.57, the fraction of the alite hydrated at 2 days was 0.28 at 5°C, 0.63 at 25°C and 0.81 at 50°C. Apparent energies of activation calculated from such data were 41 kJ mol^{-1} at $\alpha = 0.6$ and 26 kJ mol^{-1} at $\alpha = 0.7$; for belite, a value of 56 kJ mol^{-1} at $\alpha = 0.4$ was obtained. The decrease in the apparent energy of activation in the case of alite was attributed to a gradual change in rate control from a chemical process to diffusion.

7.7.2 Interpretation of kinetic data

The kinetics of cement hydration are dominated by the effects associated with the particle size distribution of the starting material, and attempts to explain them in which this is ignored can lead to very misleading results (T41,B98,B105,K37,J27,K38,K39). Even laboratory-prepared samples with close distributions (e.g. 2–5 µm) (K20) are far from monodisperse from the kinetic standpoint. Two approaches to the resulting problems of interpretation will be considered.

Bezjak, Jelenić and co-workers (B98,B105,J27) assumed that during the hydration of each particle, up to three processes were rate controlling at different stages. The first was nucleation and growth of a product. The second was a process occurring at a phase boundary. The third was a diffusion process. Each supplanted a predecessor when it became the slowest of the three, at a time that depended on the initial radius (r) of the particle; more than one rate-determining process could thus operate simultaneously. Given the rate equations and appropriate rate constants, it was possible to calculate the value of α at any desired time for a particle of given radius, and by summing the amounts of material that had hydrated for particles of all sizes, the value of α for the whole material.

In the rate equations relating α to t for a single particle, the first did not involve r. In the second, a function of α was proportional to $(t - t_o)/r$, where t_o is the time at which the process became rate determining. In the third, a function of α was proportional to $(t - t_o)/r^2$. Kinetics represented by equations of the second and third of these types are described as linear and parabolic, respectively. It was shown that the kinetic curves of a number of alite and cement pastes, some of which contained added alkali sulphates, could be satisfactorily explained (B105). For the cements, diffusion became virtually the sole rate-controlling process at values of α varying between about 30% and 60%. This appears to agree broadly with the evidence from apparent energies of activation noted in the previous section.

Knudsen (K37–K39) assumed that the rate equation relating α to t for a single particle could be either linear or parabolic. Like Bezjak *et al.*, he also assumed that the total amount of material reacted was the sum of the amounts calculated for all the individual particles, and thus, given the rate equation for a single particle, the rate constant and the particle size distribution, obtained the value of α for the whole material. In contrast to Bezjak *et al.*, he concluded that the results were highly insensitive to the precise form of the rate equation, and that all that could be learned about the latter from the kinetic curve was whether it was linear or parabolic, and the value of the rate constant. Using data from various sources, Knudsen concluded that, for the values of $α > 0.15$ to which his model was applicable, some cements followed linear kinetics for virtually the whole range of values up to $α = 1$, and others parabolic kinetics. A change from linear to parabolic kinetics could be produced by adding $CaCl_2$, and the possibility of a gradual change from linear to parabolic kinetics at high values of α was not excluded. It is not obvious how the conclusion that some cements show linear kinetics throughout can be reconciled with the evidence from apparent energies of activation.

Knudsen's model led to the prediction that, if linear kinetics were followed, the age at which 50% of the cement has hydrated is proportional to the fineness constant r_o (or x_o) in the Rosin/Rammler distribution (equation 4.1); for parabolic kinetics, it predicted that this age is proportional to r_o^2 (K40). Evidence was presented in support of this conclusion for cements considered to follow linear kinetics. The theory did not predict any relation to the breadth of the particle size distribution, which is represented by the slope of the Rosin–Rammler curve.

In neither of these approaches is it necessary to specify the rate-controlling step more precisely than is given above. The discussion on the hydration kinetics of C_3S pastes (Section 5.6) is probably substantially applicable also to cement pastes. One may speculate that in the theory of Bezjak *et al.*, the nucleation and growth of C–S–H determines the rate until the anhydrous phases cannot supply the ions sufficiently quickly. Dissolution of the anhydrous phases then becomes rate determining until the permeability of the layer of product has decreased to a level such that diffusion takes over. It would be highly desirable to examine the implications of both theories in the light of the more recent microstructural evidence (Section 7.4). It would also be desirable to determine whether Knudsen's more pessimistic conclusion regarding the information that can be extracted from the experimental data results from a more realistic appraisal of the precision of the data, or from the differences between his assumptions and those of Bezjak *et al.*

7.7.3 Mathematical modelling of the hydration process

This has the objective of quantifying knowledge of the process in a manner that permits the prediction of experimentally accessible properties from similarly accessible inputs, thus allowing the effects of changes in inputs or assumptions contained in the model to be examined. Necessary inputs comprise such data as cement composition, w/c ratio and curing conditions, and typical outputs include weight or volume fractions of phases in a paste or concrete, content of non-evaporable water, cumulative heat evolution or physical properties such as strength, porosity or permeability defined in various ways. Output might also include quantitative description of microstructural features.

The most comprehensive attempt to model hydration to date is due to Parrott (P30,P32,P33). In essence, it contains three steps, viz:

(1) The degree of hydration of each clinker phase is calculated at 56 ages. For each step, three different rate equations, each with assumed constants, were tested, and that giving the smallest increment of reaction was used. The rate equations were modified to allow for the effects of temperature, RH and w/c ratio as indicated in the previous section (P30). Two of the equations were similar to the first and third of those used by Bezjak and Jelenić (B105), the third being an empirical equation introduced to deal with the later stages of hydration. In so far as the clinker phases were treated separately, the simultaneous existence of more than one rate-determining step was assumed. Allowance was made for the effect of particle size distribution.

(2) The phase composition of the paste by weight and by volume, and the non-evaporable water content, were calculated using procedures

broadly similar to that described in Section 7.3.3. Porosities were then calculated by a method indicated in Section 8.3.1. The heat evolved was assumed to be the sum of the amounts given by each clinker phase, which in turn was assumed to be proportional to the fraction of that phase that had reacted. Extensions of the treatment permitted the effects of admixture of the cement with pfa to be considered.

(3) The compressive strength, permeability to water and methanol exchange rate were calculated, using empirical equations relating them to the porosity; such relations are discussed in Chapter 8. Examples of predictions from the model were given, and the desirability of continued updating of the procedure in the light of new knowledge of the hydration process was indicated.

Jennings and Johnson (J28) described a mathematical model simulating the development of microstructure during C_3S hydration. The anhydrous grains, treated as spheres having a realistic size distribution and randomly placed, were assumed to react forming layers of inner and outer hydrates, together with crystals of CH growing in the water-filled space, which were also assumed to be spherical. The model could clearly be extended to cement and refined in various ways, and the approach offers considerable promise as a means of exploring the microstructural consequences of differing assumptions regarding the particle size distribution and other characteristics of the unhydrated cement, the stoichiometry of the hydration reactions, and the densities and other properties of the hydration products. This type of model could perhaps be advantageously combined with that developed by Parrott.

The representation of kinetics by appropriate equations is another aspect of modelling, some aspects of which have been reviewed by Brown *et al.* (B106). A more complete understanding of the kinetic laws governing the various stages of the hydration process, especially if it took into account the more recent microstructural evidence, would contribute to the power of models of the kinds described above.

8

Structure and properties of fresh and hardened Portland cement pastes

8.1 Fresh pastes

8.1.1 Workability

Workability is a general, descriptive term that indicates the ease with which a concrete can be mixed, transported, placed and compacted to give a uniform material. There is no single measure of it, but various empirical tests that provide information on particular aspects are widely used. The most important is the slump test, in which the material is moulded by lifting away a conical container in which it was placed. The subsidence of the resulting pat provides a measure of the ability of the material to flow under its own weight. Neither this, nor any of several other empirical tests that are in use, give results that are simply related to any fundamental rheological properties. Their main value is for quality control of a given concrete mix rather than for the comparison of different mixes.

The workability of a concrete mix is by no means dependent only on the physical properties of the cement paste it contains, but an understanding of it requires one of those properties. For some specialist uses in which cement is used without an aggregate, the latter are directly relevant. The most important properties are concerned with rheology, and this section deals primarily with these properties in Portland cement pastes, free from admixtures, prior to setting. From the chemical standpoint, this period comprises that of initial reaction and induction period. From the practical standpoint, it includes those of mixing, placing and compaction.

8.1.2 Rheology

The resistance of a fluid to flow may be considered in terms of the situation existing between two parallel planes, one of which is moving in its own plane relative to the other. It is assumed that flow is confined to the single direction thus defined, and that the velocity varies linearly with distance in the direction perpendicular to the planes. Liquids of simple and stable molecular structure generally obey the Newtonian law

$$\tau = \eta \times \dot{\gamma} \qquad (8.1)$$

where τ is the shear stress (Pa), η is the viscosity (Pa s) and $\dot{\gamma}$ is the shear rate (s^{-1}), equal to the velocity of one surface relative to the other divided by the distance between them. The resistance to flow at a given temperature and pressure is thus defined by a single constant, the viscosity.

Suspensions show Newtonian behaviour only if there is no long-range structure. In varying degrees, many approximate in behaviour to the Bingham model, represented by the equation

$$\tau = \tau_o + (\mu \times \dot{\gamma}) \qquad (8.2)$$

where τ_o is the yield value and μ is the plastic viscosity. No flow occurs until the shear stress exceeds τ_o. The relation between τ and $\dot{\gamma}$ may not be linear, and may show hysteresis. Fig. 8.1 shows some of the possibilities. A curve that is concave towards the shear rate axis (Fig. 8.1C) indicates breakdown of structure due to shear. Positive hysteresis (Fig. 8.1E) shows that the structure does not reform immediately on decreasing the shear rate. Some fluids show thixotropy, that is, the structure breaks down under shear stress and reforms slowly when this is removed. A second ascending curve will lie between the previous ascending and descending curves, whereas if the structure does not reform, it will coincide with the descending curve. Thickening under stress (Fig. 8.1D) and negative hysteresis (Fig. 8.1F) are other possibilities. For various reasons, the curves obtained in practice can differ considerably in form from the idealized ones shown in Fig. 8.1, especially at low shear rates. The quantity $\tau/\dot{\gamma}$ for a non-Newtonian fluid is called the apparent viscosity. Its value varies with $\dot{\gamma}$ and, if hysteresis occurs, with the previous history of the sample.

Rheological studies on cement pastes and concrete have been reviewed in a book (T42) and several shorter articles (H40,L40,S76). The results in the literature show wide variations, many of which reflect the large effects of seemingly minor differences in experimental technique. Since cement pastes do not show Newtonian behaviour, methods giving only a single parameter are inadequate. Viscometers in which the material is studied in shear between a shallow cone and a plate, or between concentric cylinders, have generally been used.

Tattersall (T43) found that pastes of w/c ratio 0.28–0.32 and age 4.5 min followed the Bingham model at low rates of shear, but that at higher rates the structure broke down irreversibly. Several other investigators have obtained similar results, but negative hysteresis has also been observed (e.g. Ref. R30), probably due to the use of hysteresis cycles of long duration, in which the structural breakdown due to shear is outweighed by the effects of continuing hydration (B107). The rheological behaviour is strongly affected by the method of mixing, since the initial structure is broken down if this produces sufficient shear. Other major factors affecting the plastic viscosity and yield value are the w/c ratio, particle size distribution, age, presence of admixtures (Chapter 11) and, to a lesser extent, cement composition. There are severe difficulties in interpreting hysteresis loops, and in studies of the variation in shear stress with time a constant speed of rotation has been found to give more certain information (T42,L40).

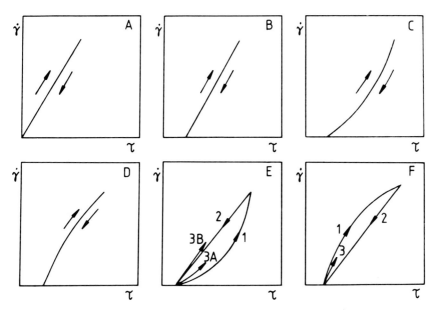

Fig. 8.1 Idealized plots of shear rate ($\dot{\gamma}$) against shear stress (τ) for fluids of various types. (A) Newtonian fluid. (B) Bingham fluid. (C) Shear thinning. (D) Shear thickening. (E) Positive hysteresis: 1, 2, 3A thixotropy; 1, 2, 3B, rheodestruction. (F) Negative hysteresis with antithixotropy.

8.1.3 Models of fresh paste structure

Helmuth (H40) considered the cement particles in a paste to constitute a single floc, in which, however, there were regions of varying solid content. On shearing, the particles became more uniformly distributed, and the static floc structure was reformed on standing. Tattersall and Banfill (T42) considered that this model did not sufficiently explain the irreversible nature of the breakdown caused by shearing, and proposed that during the initial reaction aggregates of particles already in contact were coated with a continuous membrane of gel which, if destroyed by shearing, was replaced by separate coatings around each particle, which were less effective in binding the particles together.

It is difficult to test such models directly, since studies on dilute suspensions are of doubtful relevance to pastes and neither light nor electron microscopy is readily applied, but Uchikawa et al. (U14) described a method of specimen preparation for SEM examination. The paste was mixed for 3 min and, after various times, was placed on the surface of a metal sample holder already cooled in liquid N_2. The sample holder was then further cooled by immersion in liquid N_2 and the specimen was examined at $-120°C$. For a Portland cement paste without admixtures 5 min after mixing, flocs of small particles and of large particles with adhering small ones were observed, but there was no continuous floc structure. The average distance between flocs was about 3 µm. At 2 h, C–S–H and ettringite were observed on the grain surfaces and the flocs were more definite. At 6 h, the layers of hydration products were much thicker and a three-dimensional structure was formed through the linking of the larger grains through smaller ones.

One would expect the final distribution of cement particles after mixing is complete to be reflected in that of the hardened paste, which can be more readily studied, but no studies on the relations between mixing conditions and hardened paste structure appear to have been reported.

To an extent that increases with the w/c ratio, fresh cement pastes exhibit the phenomenon of bleeding, i.e. settlement of the solid particles. The interparticle attractions are sufficiently strong that particles of all sizes settle at the same rate, typically about $2 \mu m\ s^{-1}$. Settlement also tends to increase the w/c ratio at the top and to decrease it at the bottom of the sample. It decreases with increased fineness or increased early hydration rate of the cement. In a concrete, it can produce layers of water beneath aggregate particles or reinforcing bars.

8.2 Hardened cement pastes: models of structure

8.2.1 The Powers–Brownyard model

The rest of this chapter deals with hardened Portland cement paste (hcp), which has the properties of a rigid gel. It is a relatively rigid and strong solid of high porosity and internal surface area. Variations in the relative humidity (RH) of the surrounding atmosphere cause it to gain or lose water and also produced small but important changes in volume. It is sometimes called a xerogel, i.e. a gel from which the dispersion medium (pore solution) has been removed. Powers and Brownyard (P20) described the broad structure of the material by a model based largely on evidence from total and non-evaporable water contents and water vapour sorption isotherms. Powers later modified the model in minor respects (P34).

In this model, hcp is assumed in the general case to comprise three components from the volumetric standpoint, *viz.* (a) unreacted cement, (b) hydration product and (c) capillary pores. Individual solid phases are not considered, whether in the cement or in its hydration products, which were collectively called 'cement gel'. This term may be found confusing, because it includes the CH, which forms relatively large crystals and cannot reasonably be considered part of a gel; we shall substitute the term 'hydration product'. The water present in the paste was categorized as evaporable or non-evaporable, the latter being defined in the later work (P34) as that retained on D-drying. Evaporable water, when present, was considered to reside partly in the capillary pores, and partly in so-called gel pores within the hydration product. This latter part was called gel water.

The content of non-evaporable water, relative to that in a fully hydrated paste of the same cement, was used as a measure of the degree of hydration. Portland cement paste takes up additional water during wet curing, so that its total water content in a saturated, surface dry condition exceeds the initial w/c ratio. Evidence from water vapour sorption isotherms indicated that the properties of the hydration product that were treated by the model were substantially independent of w/c and degree of hydration, and only slightly dependent on the characteristics of the individual cement. The hydration product was thus considered to have a fixed content of non-evaporable water and a fixed volume fraction, around 0.28, of gel pores.

The hydration product occupies more space than the cement from which it is formed, and the capillary pores were regarded as the remnants of the initially water-filled space. Their volume thus decreases, and that of the gel pores increases, as hydration proceeds. Evidence from water vapour sorption isotherms indicated that the hydration product was composed of solid units having a size of about 14 nm, with gel pores some 2 nm across (P34). The width of the capillary pores could not be determined from the available data, but they were considered to be generally much wider than the gel pores, though tending to become narrower as the water-filled space was used up, and thus in some regions indistinguishable from gel pores.

We shall use the following symbols:

w/c   = initial water/cement ratio, corrected for bleeding.
w/c*   = critical value of w/c below which complete hydration cannot occur; typically about 0.38.
w_t/c   = total water/cement ratio in the saturated, surface dry state.
w_n/c   = ratio of non-evaporable water to cement.

248   Cement Chemistry

w_e/c = ratio of evaporable water to cement in the saturated, surface dry state, equal to w_t/c less w_n/c.
V_c = specific volume of unhydrated cement; typically about 3.17×10^{-4} m^3 kg^{-1}.
V_g = mean specific volume of the gel water, taken here as 1.00×10^{-3} m^3 kg^{-1}.
V_n = mean specific volume of the non-evaporable water in m^3 kg^{-1}.
D_w = density of pore solution in fresh paste, taken here as 1000 kg m^{-3}.
m_g = mass of gel water per unit mass of cement in a fully hydrated paste; typically about 0.21.
m_n = mass of non-evaporable water per unit mass of cement in a fully hydrated paste, typically about 0.23.
α = mass fraction of cement hydrated.
α_{max} = maximum value of α possible at $w/c \leqslant w/c^*$.

8.2.2 Minimum water/cement ratio for complete hydration; chemical shrinkage

Powers and Brownyard (P20) determined the non-evaporable and total water contents of pastes of many cements made up at different w/c ratios and cured for different lengths of time. In Fig. 8.2 the ratios of evaporable water to cement in saturated, mature pastes of a typical cement are plotted against those of total water to cement. The initial w/c ratios are shown by the scale at the top; up to about 0.06 kg of water per kg of cement is taken up during wet curing. The data show a distinct break at w/c = 0.38, and are fitted by the equations

$$w_e/c = w_t/c - 0.227 \quad \text{for w/c} > 0.38 \tag{8.3}$$

and

$$w_e/c = 0.482 \times w_t/c \quad \text{for w/c} < 0.38 \tag{8.4}$$

The value of w_t/c at the break is 0.44.

Similar results were obtained for other Portland cements. In the following interpretation of them, the small changes in total paste volume that occur during curing are ignored, and any air entrapped or deliberately introduced during mixing is regarded as being outside the system. Because the hydration product was considered to occupy more space than the cement from which it is formed, and since the total volume of the paste scarcely changes on hydration, complete hydration cannot occur if w/c is below a certain value. For the cement under discussion, this value is 0.38, and a mature paste of this w/c ratio consists entirely of hydration product. The hydration product contains, per kg of cement, 0.227 kg of non-evaporable water and, in the saturated condition, 0.211 kg of gel water, making a total of 0.438 kg. The pastes with w/c < 0.38 consist of unreacted cement and hydration product, and thus have $(w_e/c)/(w_t/c)$ equal to 0.211/0.438, or 0.482. The pastes with w/c > 0.38 consist of hydration product and capillary pores, which in the saturated condition are filled with water, and thus have w_e/c equal to $w_t/c - w_n/c$.

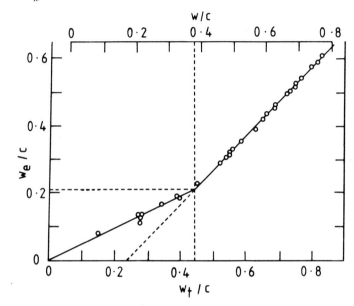

Fig. 8.2 Relations between the initial (w/c), total (w_t/c) and evaporable (w_e/c) water/cement ratios for saturated, mature pastes of a Portland cement. After Powers (P34).

Most investigators appear to have agreed with the observation that complete hydration cannot occur if w/c is below a value in the region of 0.38, but Rössler and Odler (R31) reported that hydration was complete in a paste with w/c = 0.22, and further investigation may be required. On the Powers–Brownyard theory, the value of the minimum depends on the volume ratio of hydration product to cement, which in turn depends on the gel porosity of the hydration product. It was assumed that this porosity was fixed by the inability of new particles to nucleate in pores smaller than a certain size and by that of existing particles to grow beyond a certain size.

If a paste having w/c < 0.44 is cured under sealed conditions, there will be insufficient water to fill the gel pores completely, and the capillary pores will

250   Cement Chemistry

be empty. The effective RH will be low, and hydration will become very slow or stop, even though free space remains. This phenomenon is called self-desiccation. Even at higher w/c ratios under sealed conditions, partial emptying of the capillary pores will retard hydration, and can produce irreversible changes in pore structure which reduce the final strength. It is therefore important to provide an adequate supply of water during curing. If a paste is kept in contact with an excess of water during curing, the total volume of paste and water decreases. This effect is called chemical shrinkage, and may be used to follow the course of hydration. For this purpose, it has the merits that hydration need not be stopped and that the determination can readily be automated. Knudsen and Geiker (K36) described a device for automated determination on up to 30 samples.

8.2.3 Calculation of volumetric quantities

This section gives expressions and values for some important volumetric quantities. The data for the cement discussed in the previous section are adopted as typical.

(1) The *total volume of paste* per unit mass of cement is $V_c + (w/c)/D_w$, where V_c is the specific volume of the cement and D_w the density of the pore solution in the fresh paste. Assuming a value of $3.17 \times 10^{-4} \, m^3 \, kg^{-1}$ for V_c and one of $1000 \, kg \, m^{-3}$ for D_w, this gives $3.17 \times 10^{-4} + (w/c)/1000 \, m^3 \, kg^{-1}$.

(2) The *volume of hydration product* per unit mass of cement reacted is obtained by considering the special case of a fully hydrated paste having the critical w/c ratio, w/c*, in which only hydration product is present; it is $V_c + (w/c^*)/D_w$. For the general case, it is thus $[V_c + (w/c^*)/D_w] \times \alpha$, where α is the degree of hydration. Inserting the values of V_c, w/c* and D_w, this gives $7.0 \times 10^{-4} \times \alpha \, m^3 \, kg^{-1}$.

(3) The *volume of unreacted cement* per unit mass of cement is $V_c \times (1 - \alpha)$, or $3.17 \times 10^{-4} \times (1 - \alpha) \, m^3 \, kg^{-1}$.

(4) The *specific volume of the non-evaporable water*, V_n, is obtained by taking the volume of hydration product less those of the cement from which it was formed and the gel water, and dividing by the mass of non-evaporable water. This gives $[(w/c^*)/D_w - V_g \cdot m_g]/m_n$, or $0.74 \times 10^{-3} \, m^3 \, kg^{-1}$.

(5) The *maximum degree of hydration*, α_{max}, for a paste with w/c ≤ w/c*, may be calculated by equating the total volume of paste to the sum of the volumes of unreacted cement and hydration product; it is $(w/c)/(w/c^*)$, or $(w/c)/0.38$. This assumes that an unrestricted supply of

water is available during curing. If no such water is available, the degree of hydration is limited by the amount of water present initially to a maximum value of $(w/c)/(m_n + m_g)$, or $(w/c)/0.44$.

(6) The *ratio of the volume of hydration product to that of the cement* from which it is formed may be calculated by dividing the former quantity by the volume of cement reacted per unit mass of cement, which is $\alpha \times V_c$, giving $1 + [(w/c^*)/(D_w \times V_c)]$. This gives a value of 2.20.

(7) The *porosity of the hydration product* is given by dividing the volume of gel pores per kg of reacted cement by that of the hydration product. This gives $V_g \times m_g/[V_c + (w/c^*)/D_w]$, or 0.30. The value of 0.28 given by Powers and Brownyard was obtained assuming a different value for V_g. It should be noted that this quantity is the volume of gel pores relative to that of all the hydration products, including the CH.

Fig. 8.3 shows the mass and volume relationships for a fully hydrated, saturated paste of w/c ratio 0.5, calculated using the above expressions and values. Following Powers and Brownyard, the hydrated cement is treated from a purely volumetric standpoint as a composite of reacted cement, non-evaporable water and gel water. The specific volume of the non-evaporable water was assumed to be 0.73×10^{-3} m^3 kg^{-1}, and that of the capillary and gel water to be 1.00×10^{-3} m^3 kg^{-1}. The results are approximate, for several reasons; for example, the pore solution is in reality not pure water, but an alkali hydroxide solution with a specific volume (for 0.3 M KOH) of about 0.986×10^{-3} m^3 kg^{-1}.

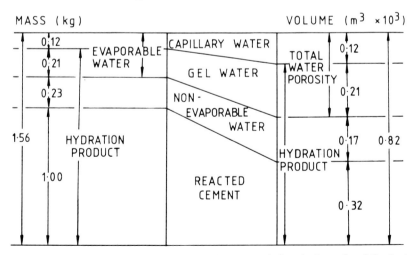

Fig. 8.3 Diagram illustrating the Powers–Brownyard description of a fully hydrated and saturated Portland cement paste of initial water/cement ratio 0.5. All quantities refer to 1 kg of cement.

8.2.4 Later models of hardened paste structure

Brunauer and co-workers (B55,B108) considered that the gel particles of the Powers–Brownyard model consisted of either two or three layers of C–S–H, which could roll into fibres. D-drying caused irreversible loss of interlayer water, and the specific surface area could be calculated from water vapour sorption isotherms, which gave values in the region of $200 \, \text{m}^2 \, \text{g}^{-1}$ for cement paste. Sorption isotherms using N_2 give lower values of the specific surface area; this was attributed to failure of this sorbate to enter all the pore spaces.

Feldman and Sereda (F31,F32) regarded the gel as a three-dimensional assemblage of C–S–H layers, which tended to form subparallel groups a few layers thick and which enclosed pores of dimensions ranging from interlayer spaces upwards (Fig. 8.4). They considered that much of the gel water of the Powers–Brownyard model was interlayer water. Unlike Brunauer, they considered that the loss of interlayer water on D-drying was reversible; consequently, sorption isotherms using N_2, not water, gave a true measure of the specific surface area. In a modification of this model, Daimon et al. (D30) considered that the gel consisted of particles having an internal structure similar to that shown in Fig. 8.4, together with pores having an equivalent radius of 1.6–100 nm. The smaller pores within the gel particles were 0.6–1.6 nm in equivalent radius, and pores of both types tended to have narrow entrances.

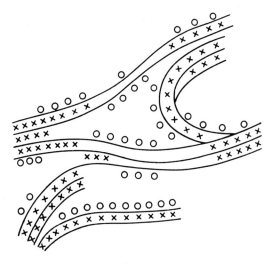

Fig. 8.4 Feldman–Sereda model of the structure of the C–S–H gel of Portland cement paste, showing C–S–H layers (lines), interlayer water molecules (crosses) and adsorbed water molecules (circles). After Ramachandran et al. (R32).

In Wittmann's (W23,W24) 'Munich model', hcp is described as a xerogel, together with the crystalline constituents. The xerogel is considered to consist of separate particles, for which no particular internal structure is assumed. Emphasis is placed on the role of water molecules adsorbed on, or lying between, these particles. Below 50% RH, the particles are considered to be in contact, but at higher humidities, intervening water molecules exert a disjoining pressure which keeps them apart to a limited extent. The model was used primarily in the development of a theory of dimensional changes.

The Powers–Brownyard model explains why complete hydration does not occur if w/c is below a certain value, and provides a partial explanation of the dependence of some important physical properties, especially compressive strength, on w/c and degree of hydration. No other satisfactory explanation of the minimum w/c needed for complete hydration appears to have been given, and the most important feature of the model is probably the conclusion that a given volume of cement can react only if an additional volume of free space approximately 1.2 times as great is available. The direct evidence that the relevant characteristics of the hydration product are independent of w/c and degree of hydration was, however, based largely on data from water vapour sorption isotherms for severely dried pastes, the interpretation of which has been strongly criticized (F13). Further tests of the correctness of the hypothesis are needed.

The Feldman–Sereda model was based on the studies of sorption properties, porosities and relations between water content and physical properties. Alone among the proposed models, it is clearly compatible with the microstructural evidence and with the probable relationships between C–S–H gel and crystalline compounds. It is incompatible with that of Brunauer, but not with the essential features of that of Powers and Brownyard in its original form if the nature of the gel porosity is reinterpreted. Calculations of bound water (Section 7.3.3) indicate that about a third of the gel porosity of the Powers–Brownyard model is interlayer space, the remainder being micro or fine meso porosity* of the kind shown in Fig. 8.4. However, as that figure illustrates, the boundary between interlayer space and micropores is ill defined.

The models differ in whether, to what extent, or on what scale the gel is regarded as being composed of separate particles. The microstructural evidence suggests that the situation differs with the region considered, the inner or late product being an essentially continuous, though porous, assemblage of layers, whereas the outer product is composed of similar material but is partly or wholly particulate on a 100-nm or micrometre scale. Powers (P34) discussed his model in relation to the meagre evidence on

*The International Union of Pure and Applied Chemistry has adopted the following definitions of pores by width: micropores, < 2 nm; mesopores, 2–50 nm; macropores, > 50 nm.

microstructure then available. He recognized that the inner and outer products might differ in microstructure and that the porosities defined in the model could be means, but considered that the hydration product would tend towards a uniform porosity. In recent years, much progress has been made both in the study of microstructure by electron microscopy and in that of pore structure by other methods, but relatively little has been done to relate the two approaches. Further understanding of the structure of cement paste requires a more complete synthesis of the existing models, the results of investigations on microstructure and on pore structure, and information of the structure of C–S–H and other hydration products at the crystal chemical level.

8.3 Pore structure

8.3.1 Porosities obtained by calculation

In the Powers–Brownyard model, the volume of capillary pores per unit mass of cement is obtained by subtracting the volumes of hydration product and of unhydrated cement from the total volume; using the symbols defined in Section 8.2.1, this gives $[(w/c) - \alpha(w/c^*)]/D_w$. The volume of gel pores per unit mass of cement is equal to $m_g \times V_g \times \alpha$. Expressed as fractions of the total volume, the capillary porosity is $[w/c - \alpha(w/c^*)]/[(D_w \times V_c) + w/c]$ and the gel porosity is $m_g V_g \times \alpha/[V_c + (w/c)/D_w]$. In all the above expressions, the maximum possible degree of hydration, α_{max}, equal to $(w/c)/(w/c^*)$, must be substituted for α if $w/c \leqslant w/c^*$.

Table 8.1 Calculated porosities, based on the Powers–Brownyard model

w/c Ratio	Fraction of cement hydrated (α)	Capillary porosity	Gel porosity	Total water porosity
0.3	0.00	0.49	0.00	0.49
0.3	0.79	0.00	0.27	0.27
0.4	0.00	0.56	0.00	0.56
0.4	1.00	0.03	0.29	0.32
0.5	0.00	0.61	0.00	0.61
0.5	1.00	0.15	0.26	0.41
0.6	0.00	0.65	0.00	0.65
0.6	1.00	0.24	0.23	0.47

Powers and Brownyard defined the total porosity as the sum of the capillary and gel porosities. Because the term 'total porosity' has been used by some later workers in other senses, we shall refer to such values as 'total water porosities'. As hydration proceeds, the capillary and total water porosities decrease and the gel porosity increases. Table 8.1 gives some values of the three types of porosity obtained from the above formulae, using the data given in Section 8.2.2 as typical.

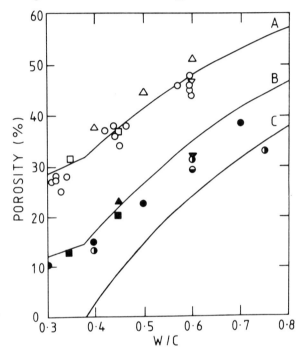

Fig. 8.5 Relations between porosities (volume percentages) and water/cement ratio for mature Portland cement pastes. The experimental data are for pastes at least 8 months old, and the calculated curves relate to a typical cement aged 18 months. Open symbols: total water porosities. Filled or half-filled symbols: mercury porosities. Curve A: total water porosity. Curve B: free water porosity. Curve C: capillary porosity. References to data: ○ (P20); □ (S77); △ (F33); ▽ (M68); ■ (S78); ▲ (F34); ● (O19); ▼ (M68); ◐ (D31); ◑ (H41). In the last two cases, porosities by volume were estimated from data referred in the original sources to masses of dried paste, assuming the latter to have contained 0.23 kg of water per kg of cement having a specific volume of $3.17 \times 10^{-4}\,\mathrm{m^3\,kg^{-1}}$.

In Section 7.3.3, a method was described for calculating the quantitative phase composition of a cement paste by weight and by volume for various drying conditions. Fig. 8.5 includes porosities thus calculated for 18-month-old pastes of a typical Portland cement at varying w/c ratios and corresponding to (A) the D-dried state, (B) a state that includes all bound water, including that present in interlayer spaces, but no water in larger pores, and (C) capillary porosities, as defined by Powers and Brownyard. Conditions (B) and (C) are possibly realized approximately by taking material not previously exposed to an atmosphere below saturation and equilibrating it at 11% RH and about 90% RH, respectively (F14,Y5). The porosities defined by condition (B) will be called free water porosities.

Parrott and co-workers (P30,P32,P35,P33) described a more sophisticated method for modelling the hydration process. The fraction of the total water porosity that was below 4 nm was calculated by multiplying the volume fraction of C–S–H by an appropriate factor, which depended on whether the C–S–H was formed from alite or belite, the temperature and the amount of space available. The constants assumed were based on experimental data obtained using a procedure based on methanol sorption (Section 8.3.4). The effect of drying was allowed for (P35) by introducing a factor of $0.7 + 1.2(RH - 0.5)^2$ for $0.5 < RH < 1$, or of 0.7 for $RH \leqslant 0.5$. These refinements allow some deviation from the Powers–Brownyard postulate of a fixed volume ratio of gel porosity to product. Typical results for the volume fractions of pores larger than 4 nm in mature pastes of a cement with an alite content of 56% were approximately 0.26, 0.16 and 0.07 for w/c ratios of 0.65, 0.50 and 0.35, respectively (P32). For the two higher w/c ratios, these results are near the capillary porosities of Powers and Brownyard, but for w/c 0.35 the latter value is zero.

8.3.2 Experimental methods: general points

The problems of defining and determining densities of finely porous solids, indicated in Section 5.3.3, apply to hcp. A determination of the density of such a material is also one of porosity, since both properties are related to the solid volume; the porosity per unit volume of material is equal to $1 - [m_s/(D_s \times V)]$, where m_s and D_s are, respectively, the mass and density of the solid and V is the total volume. The density and porosity determined by any method entailing contact with a fluid can vary with the extent to which the solid has been dried, how it has been dried and the fluid employed. Fluids may differ in their abilities to penetrate the pore system, and the pore structure may be altered both during drying and by the action of the fluid subsequently introduced.

These comments apply also to studies of pore size distribution or specific surface area, which have been widely studied using sorption isotherms or, in the former case, mercury intrusion porosimetry (MIP). Gregg and Sing

(G64) have described these methods. In both cases, the material must first be dried to water contents lower than those likely to be encountered in the normal environment of the material, and the effects can be serious and difficult to allow for. The amount of water considered to form part of the solid obviously affects the definition of porosity, but the extent and manner of drying also affect the pore size distribution.

The most widely used drying procedures have been desorption at ordinary temperatures, including D-drying, or at elevated temperatures, and solvent replacement. Freeze drying has also been used (K41). In solvent replacement, the specimen, normally in a saturated state, is immersed in an organic liquid, such as methanol. The course of replacement can be followed from the weight change; it has been claimed that all the pore water can be replaced (P36,D32). The organic liquid is then removed. Because of the lower surface tension, one might expect that less damage would occur as the liquid evaporates, and several investigators have concluded that this is the case (P36,H41,D32). In contrast, Beaudoin (B109,B45) concluded that straight chain aliphatic alcohols seriously affect the pore structure, partly through reaction with CH. Methanol and other organic liquids are retained strongly by cement or C_3S pastes, with resulting effects on the TG and DTA curves (Section 5.1.2). It does not necessarily follow that the pore structure is affected; this applies especially to the coarser part of the distribution, which is probably the more important for many of the physical properties.

Porosities have sometimes been reported as percentages by volume of the paste and sometimes as volumes per unit weight of dried paste. The first is the more meaningful, especially as in the latter case the water content of the material has not always been given. A similar comment applies to specific surface areas (L41).

8.3.3 Determination of porosities by pyknometry

Total water porosities are obtained experimentally from the loss in weight when a saturated paste is D-dried or subjected to some procedure regarded as equivalent, such as heating to constant weight at 105°C. Reabsorption of water by materials thus dried gives identical results (D32). The procedures should be carried out under CO_2-free conditions. A value for the specific volume of the evaporable water must be assumed; this has usually been 1.00×10^{-3} m^3 kg^{-1}. Fig. 8.5 includes typical values thus obtained.

Powers and Brownyard (P20) found that porosities of D-dried pastes obtained using helium as the pyknometric fluid were lower than those obtained using water. Feldman (F33) confirmed this and found that a lower value was also obtained using methanol; if a D-dried sample was exposed to helium, the latter flowed in quickly at first, and then more slowly. The porosities calculated from the initial, rapidly accepted volumes of helium agreed with those obtained with methanol. If the material was dried only by equilibration at 11% RH, the porosities obtained with helium, methanol or saturated aqueous CH were all similar to those obtained for D-dried pastes using methanol. For D-dried pastes, the porosities obtained using helium were 23.3%, 34.5%, 42.1% and 53.4% for w/c ratios of 0.4, 0.5, 0.6 and 0.8, respectively. The corresponding values obtained with water were 37.8%, 44.8%, 51.0% and 58.7%; these are included in Fig. 8.5. Day and Marsh (D32) confirmed that porosities obtained by admitting methanol or propan-2-ol to an oven-dried paste were substantially lower than total water porosities; however, if the undried paste was treated with the organic liquid, the volume exchanged was closely similar to the total water porosity.

Feldman (F33) explained his results in terms of the Feldman–Sereda model as follows. If the paste is D-dried, water is lost from the interlayer spaces. If the sample is immersed in water, this water is reabsorbed, and the total water porosity therefore includes the volume of the interlayer space. Methanol does not penetrate into this space, and helium does so only slowly, so that lower porosities are obtained. If the sample is equilibrated at 11% RH, the interlayer spaces are largely filled, and the lower value is obtained irrespective of the fluid used. The helium porosities reported by Feldman were near to the calculated free water porosities.

8.3.4 Sorption isotherms; specific surface areas

Sorption isotherms, using water, N_2 or other sorbates, have been widely used to study the pore structure of hcp. The results obtained by different investigators show considerable variations, which arise in part from differences in the conditions of preliminary drying.

Desorption of water from a saturated paste at ordinary temperature is a slow process. Fig. 8.6 gives results for a fully hydrated C_3S paste of w/c 0.5, in which each point represents the results of a separate experiment on a slab 1 mm thick (G65). Loss of water is especially slow at 30–80% RH. Parrott et al. (P37,P36) similarly found that in the initial desorption, equilibrium was reached within a few days at RH < 30%, but that at RH values of 40–80%, many weeks were needed; there were indications that larger pores emptied through smaller ones. Drying at these humidities led to a coarsening of the pore structure, which was indicated by a marked reduction in the specific surface area accessible to N_2, from over $100 \, m^2 \, g^{-1}$ at 90% RH to around $30 \, m^2 \, g^{-1}$ at 10% RH and was also associated with irreversible shrinkage (P38). On resaturation, the higher surface area was regained, but the pore size distribution did not completely revert to that originally found.

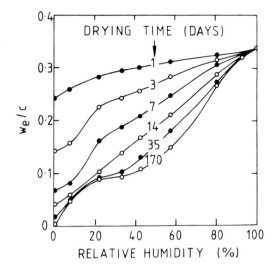

Fig. 8.6 Desorption curves for an initially saturated C_3S paste of w/c ratio 0.5, hydrated for 5.8 years at 25°C, and in the form of slab samples 1 mm thick. Each point represents the result of a separate experiment in which the sample was exposed to an atmosphere of given relative humidity for the drying time shown. After Grudemo (G65).

Water sorption isotherms for hcp show marked hysteresis. Powers and Brownyard (P20) found that, while it was difficult to obtain reproducible desorption curves, the low-pressure part of the water vapour resorption curve varied little with w/c ratio, between different Portland cements, or, if allowance was made for the contents of unreacted cement, with the degree of hydration. This was their main direct evidence for the conclusion (Section 8.2.1) that the properties of the hydration product considered in their model were essentially independent of these variables. However, the water sorption iostherms obtained by different investigators have varied considerably (e.g. Refs P20 and S79), and it is not clear to what extent the above conclusion would stand had different desorption conditions been used.

In principle, isotherms at low partial pressures of the sorbate may be used to determine specific surface areas by the Brunauer–Emmett–Teller (BET) method (G64). In this method, it is assumed that molecules of the sorbate are adsorbed on surfaces that can include the walls of pores, provided that the distance between molecules on opposing walls is large compared with molecular dimensions. From a plot derived from the isotherm, and given the effective cross-sectional area of the sorbate molecule, the specific surface area of the sorbent and the net heat of adsorption are obtained. Using water as sorbate, specific surface areas of about 200 m^2 per g of D-dried paste have typically been obtained for mature cement pastes of normal w/c ratios

(P20,H42); however, much of the water has been shown to enter the vacated interlayer spaces, thus rendering the BET equation at best only partially applicable (F31).

Using N_2 or other non-polar sorbates, widely varying and usually much lower values have been obtained. Lawrence (L41), reviewing earlier studies, noted that for well-cured pastes, the values ranged from under 10 to nearly $150 \, m^2 \, g^{-1}$, and even higher values, up to $249 \, m^2 \, g^{-1}$, have been reported (L42). In contrast to values for fully hydrated pastes obtained using water, the N_2 areas tend strongly to increase with w/c ratio, but are also much affected by the conditions under which the sample has been dried. It would appear that the value obtained using water, whatever its precise significance, is dominated by the structure at or near the individual layer level and is thus relatively insensitive to coarser features of the structure. With a possible reservation regarding material dried by solvent replacement, N_2 cannot penetrate the interlayer spaces and can therefore give information about structural features at a somewhat coarser level.

8.3.5 Pore size distributions

The hysteresis shown by sorption isotherms of hcp, whether using water, N_2, or other sorbates, indicates the presence of mesopores (G64). At partial pressures above those to which the BET theory applies, capillary condensation occurs. If a liquid forms a meniscus in a cylindrical pore, its vapour pressure is lowered to an extent that increases with decreasing pore radius. Using an equation due to Kelvin, or other procedures ultimately derived from it, the isotherm may be used to estimate pore size distributions in the approximate range of 1–30 nm and may give indications regarding the shapes of the pores. It may be necessary to allow for the fact that the meniscus forms in a pore narrowed by an adsorbed layer of sorbate. From an early study of this type on D-dried cement pastes, Mikhail et al. (M69) showed that the size distribution of the pores available to N_2 was probably unimodal and that with increase in w/c from 0.35 to 0.7 it shifted to higher values in both the low and the high parts of the distribution. It is difficult to reconcile this conclusion with the unvarying pore size distribution within the gel implied by the model of Powers and Brownyard.

Lawrence and co-workers (L41,L43,L44) concluded from sorption data using N_2 and butane that the microstructure partly collapses during normal drying. This agrees with Parrott's (P37,P36) conclusion, noted earlier. Pastes that had been rapidly dried were likely to be closest in structure to undried pastes; the sorption results indicated that their structures were dominated by platey particles or lamellae that formed slit-shaped mesopores and micropores. Slow drying produced small pores more nearly cylindrical or spherical. Methanol treatment at least partially prevented the microstructural collapse, but did not always produce an increase in the recorded surface area, though it always caused a change in the shape of the isotherm; this was possibly due to chemisorption of the methanol.

Using N_2 sorption on samples solvent-dried by methanol exchange, Hansen and Almudaiheem (H41) found the volumes of pores with diameters smaller than 4 nm and accessible to N_2 to be 0.052 ml, 0.035 ml and 0.038 ml per g of dry paste for mature pastes of w/c ratio 0.4, 0.6 and 0.75 respectively. These results correspond to volume porosities of about 9%, 5% and 4%, respectively.

Patel et al. (P28) determined the volumes of pores smaller than 4 nm or 37 nm in methanol-exchanged samples, using a methanol sorption method in which porous glass granules of known and approximately monodisperse pore size were used as a reference standard. Using also the total water porosity, the volumes in the ranges larger than 4 nm and 37 nm were also obtained. For a cement paste of w/c 0.59 cured for 90 days at 100% RH, this gave values of approximately 47% for the total water porosity, 17% for pores above 4 nm, and 12% for pores above 37 nm. For a paste of w/c ratio 0.71 cured in the same way, the corresponding values were approximately 57%, 36% and 18%. These results indicate volumes of pores below 4 nm considerably greater than those reported by Hansen and Almudaiheem.

Parrott et al. (P19) used butane sorption to determine the volume of pores smaller than 50 nm. For a 28-day-old alite paste of w/s 0.59, a volume porosity of 27% was obtained.

8.3.6 Mercury intrusion porosimetry (MIP)

This method is based on the fact that a liquid that does not wet a porous solid will enter its pores only under pressure. If the pores are assumed to be cylindrical, the pressure (p) needed to force the liquid into them is given by the Washburn equation

$$p = -4\gamma \cos \theta / d \tag{8.5}$$

where γ is the surface energy of the liquid, θ is the contact angle, and d is the pore diameter. For Hg, $\gamma = 0.483 \text{ N m}^{-1}$. For γ, values between 117° and 140° have been variously assumed. In critical review, Good (G66) concluded that, while a case for 180° could be made, it was probably best to assume a value of 130° in the absence of direct experimental data. The maximum pressure employed is typically around 400 MPa, which allows pores of nominal diameter down to about 3.5 nm to be intruded.

Typical results (Fig. 8.7) show that the distribution moves to smaller values as hydration proceeds. The observed porosity is mainly in the 3 nm to 1 μm range for young pastes, and in the 3–100 nm range for mature pastes. For mature pastes of low w/c ratio, which according to the Powers–Brownyard theory consist entirely of hydration product, nearly all the porosity is below 50 nm (S77). We shall refer to the porosities obtained using mercury at the maximum pressures employed as mercury porosities. Typical values for mature pastes (Fig. 8.5) are somewhat lower than the calculated free water porosities.

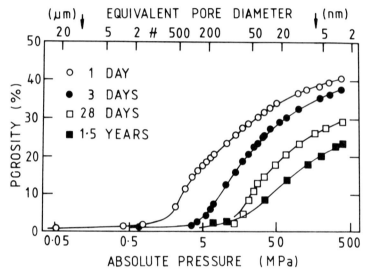

Fig. 8.7 Mercury porosimetry curves for a Portland cement paste (w/c ratio 0.47) at various ages. After Feldman (F34).

The earlier literature contains several references to 'missing' porosity, meaning the difference between mercury porosities and total water porosities. The difference clearly arises mainly from the failure of mercury to enter vacated interlayer space but has also been attributed to the encapsulation of gel by CH (D33) or to the presence of large (> 15 μm), closed pores that are not entered (A15).

MIP results do not show the sharp distinction in size between capillary and gel pores postulated by Powers and Brownyard (P20). Mikhail et al. (M69) earlier came to the same conclusion from results obtained by N_2 sorption. With increasing age, the volume of pores smaller than about 20 nm increases and that of pores larger than about 200 nm decreases, but there is no sign of two separate distributions (Fig. 8.7). The single distribution indicated by MIP corresponds to the capillary porosity and part of the gel porosity. For pure Portland cement pastes, the pore size distributions obtained using MIP are reported to be coarser for oven-dried materials than for ones dried by a solvent replacement method (M70), though D-drying is reported to give similar results to drying at 11% RH (B110).

The results for pure Portland cement pastes that have been obtained using MIP have been reasonably self-consistent, but several difficulties arise when they are compared with those given by other methods. SEM evidence (Section 7.4.3) shows the presence in significant quantity of hollow-shell grains considerably larger than 1 µm in size, which could only be entered through much smaller pores, but MIP of mature pastes shows scarcely any porosity larger than 1 µm (Fig. 8.7). The closeness of the mercury porosities intruded at maximum pressure to the calculated free water porosities implies that there is little porosity in the 0.5–3.5 nm range. This is difficult to reconcile with evidence from N_2 or methanol sorption, which indicates the existence of considerable porosity in this range. The volumes of pores wider than 4 nm found using methanol sorption (P35,P28) are much smaller than those indicated by MIP. On the other hand, Hansen (H43) found that, for solvent-exchanged pastes, the sum of the volumes of pores below and above 4 nm, obtained using N_2 sorption and MIP, respectively, agreed with the total water porosities.

Studies on other materials show that MIP determines the width distribution of pore entrances and not of the pores themselves (D34). The intrusion of mercury may also coarsen the pore structure; this need only imply that, at the higher pressures employed, some of the foils of the gel are displaced so that some pores are widened and entered while neighbouring ones are closed up. The combined result of these processes would be to produce a distribution narrower than that existing before the intrusion began, and a value for the porosity at maximum pressure that corresponded to a minimum pore width before intrusion of less than 3.5 nm. Experiments in which the mercury was removed and subsequently reintruded have indicated that the structure is usually not altered in the case of Portland cement pastes, though it is in that of pastes of composite cements (F35,D32), but cannot show whether an irreversible change occurred during the first intrusion.

8.3.7 Other methods

Parrott *et al.* (P19) impregnated alite pastes with a resin, and prepared polished surfaces which were etched with acid to dissolve the cement paste before examination by SEM. Correlations were found between the areas of pores thus determined and the volumes of pores greater than 50 nm obtained by the difference between total water porosities and the contents of small pores found using butane sorption. Subsequently, backscattered electron imaging of unetched polished surfaces, with image analysis, has provided a more direct approach, which also gave results correlating with ones obtained by a sorption method (S38,S80). Scrivener (S80) described the considerable stereological problems inherent in obtaining three-dimensional information about such features as pore connectivity from two-dimensional sections and methods for dealing with them.

The remaining methods to be described can be applied to undried material. Small angle X-ray scattering gives information on the areas of surfaces between regions of markedly different electron density. For saturated, mature cement pastes of w/c 0.4–0.6, Winslow and Diamond (W25,W26) obtained values of 700–800 m^2 per g ignited weight, decreasing to around 200 m^2 g^{-1} on D-drying. The decrease with loss of water was reversible if the specimen was not heated. Correction for CH and unreacted clinker phases would increase the values for the saturated pastes to around 1000 m^2 g^{-1}. The authors were uncertain whether the method would be sensitive to an interlayer of water molecules, but the value obtained is about what would be expected for layers of C–S–H thus separated. The decrease on drying could be attributed to the coming together of layers, and its reversibility agrees with the hypothesis that the loss of interlayer water on D-drying is reversible.

Small angle neutron scattering may be used in a somewhat similar way. It is especially sensitive to concentrations of hydrogen atoms. A study of the 3–25 nm range by this method showed a bimodal distribution of pores peaking at approximately 5 nm and 10 nm, but accounting for less than 2% of the total porosity (A16,P39). The pores were approximately spherical, and on heating the material at 105°, partial collapse of the pore structure was observed, with loss of the 10 nm peak.

Proton NMR by pulsed techniques to permit the study of relaxation effects provides information on the distribution of protons, and thus of water molecules, between different environments. An early study (S81) indicated that, at RH < 70%, the evaporable water in hcp was in an environment similar to that of the interlayer water in clay minerals or in certain crystalline hydrates. Subsequent work (B111,M71,M72,S82) has given indications of the distributions of both chemically bound and unbound water among a number of environments. The method, which can be applied most effectively only to materials very low in paramagnetic atoms (including iron), has so far been used primarily to follow the course of hydration, but it would appear that it could also provide much information on pore structure.

Low-temperature calorimetry (S83,B112) has been used to study coarse porosity. The method is based on the fact that water in pores freezes at a lower temperature than water in bulk. The ice forms through the advance of a front, analogous to the intrusion of mercury or the desorption of water. Hysteresis effects indicated the existence of necks in the pores, and the occurrence of up to three distinct peaks on curves of apparent heat capacity against temperature was interpreted as indicating maxima in the pore size distribution. Coarsening of the pore structure on drying was confirmed.

8.4 Strength

8.4.1 Empirical relations between compressive strength and porosity

The physical properties of concrete are only partly determined by those of the cement paste that it contains. We shall consider these properties only for hcp and to the extent that chemical or microstructural studies have contributed to an understanding of them. Other aspects are treated elsewhere (e.g. Ref. M73).

Factors determining the compressive strength of a cement paste include (i) the characteristics of the cement, such as clinker composition and microstructure, gypsum content and particle size distribution, (ii) the w/c ratio and the contents of air and of any admixtures present in the mix, (iii) the mixing conditions, (iv) the curing conditions, especially temperature and RH, (v) the age and (vi) the manner of testing, including the water content of the specimen. The effects of temperature and admixtures are considered mainly in Chapter 11 and some additional factors relevant to mortar or concrete are considered in Section 12.1. In various ways, the factors listed above determine the degrees of hydration of the clinker phases and the phase composition and microstructure of the hardened paste, which in turn determine its physical properties, including strength.

Many empirical relations between compressive strength and one or more of these variables have been proposed. Thus, Feret's law (1892) states that the strength is proportional to $[c/(c + w + a)]^2$, where c, w and a are the volumes of cement, water and air, respectively, and various authors have reported equations based on regression analyses relating strength to cement composition and other variables (e.g. A17,A18). The generality of such equations appears to be limited, probably because of the difficulty of taking into account all the relevant parameters.

The porosity of the hardened paste, appropriately defined, is strongly correlated with strength. As was seen in Section 8.3.1, it depends primarily on the degree of hydration and the w/c ratio. In this discussion, the following symbols will be used:

266   Cement Chemistry

σ = compressive strength
σ_o = hypothetical maximum compressive strength attainable
p = porosity, which has usually been taken to mean total water porosity
p_o = hypothetical porosity at or above which the strength is zero
A, B, C, D and E are constants

Powers (P34) found that cement pastes of various degrees of hydration and w/c ratio conformed to the relation

$$\sigma = \sigma_o X^A \tag{8.6}$$

where X is a quantity called the gel/space ratio, and equal to the volume of hydration product divided by that of hydration product plus capillary porosity, both quantities being defined as in Section 8.2.1. The value of A was about 3.0 if non-evaporable water was used as a measure of the degree of hydration, and typical values for σ_o were 90–130 MPa. This equation breaks down for mature pastes of low w/c ratio, because it implies that the strength does not then depend on the w/c ratio. In reality, strength increases with decreasing w/c ratio, even though some of the cement does not hydrate.

Several direct relationships between porosity and strength, originally found to hold for other materials, have been applied to cement pastes, *viz:*

$$\sigma = \sigma_o (1 - p)^B \text{ (B113)} \tag{8.7}$$

$$\sigma = \sigma_o \exp(-Cp) \text{ (R33)} \tag{8.8}$$

$$\sigma = D \ln (p_o/p) \text{ (S84)} \tag{8.9}$$

$$\sigma = \sigma_o (1 - Ep) \text{ (H44)} \tag{8.10}$$

Equation 8.7 is similar to that of Powers, but unreacted cement is considered equivalent to hydration product. Some of these equations break down at zero or high porosities, but for a wide range of intermediate porosities, assuming suitable values of the constants, any of them can give a reasonable fit to a given set of data. Rössler and Odler (R31) concluded that equation 8.10 was the most satisfactory. A relation identical to that of this equation, using capillary porosity, is implicit in a figure represented earlier by Verbeck and Helmuth (V5).

8.4.2 Relations between strength and microstructure or pore size distribution

There are several indications that the compressive strength does not depend solely on porosity. In general, procedures that accelerate early reaction and thus increase early strength, such as increase in temperature or addition of certain admixtures, tend to decrease later strength. It is unlikely that they significantly alter the porosity of the mature paste, but there is evidence that they affect the microstructure. If one extends consideration to a wider range of materials than normally cured Portland cement pastes, the relation between strength and porosity is markedly dependent on the broad characteristics of the microstructure. Jambor (J29) concluded that both the volume and the specific binding capacities of hydration products must be considered.

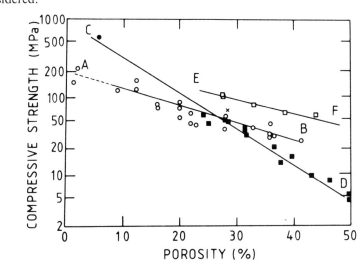

Fig. 8.8 Relations between compressive strength (log scale) and porosity; data from various sources for Portland cement pastes cured at ordinary temperatures (open circles), hot pressed Portland cement paste (filled circle), autoclaved pastes of Portland cement, sometimes with added sulphur (filled squares) and autoclaved pastes of Portland cement with 50% (open squares) or 30% (cross) of added pfa. After Feldman and Beaudoin (F36).

Feldman and Beaudoin (F36) showed that, while the strengths of Portland cement pastes could be fitted to a curve in accordance with equation 8.8, materials of a broadly similar type but composed largely of denser, crystalline particles were, for a given porosity, stronger at low porosities and weaker at high porosities (Fig. 8.8, curves AB and CD). These materials comprised hot pressed cement pastes (Section 11.9.1) and autoclaved cement pastes to which no silica had been added (Section 11.7.2). Feldman and Beaudoin considered that, while porosity was the major factor controlling strength, the morphology and density of the particles were also important. At high porosities, the better bonding properties of ill-crystallized material augmented the strength, but at low porosities, the greater intrinsic strength of the dense, crystalline particles was more important. In agreement with Bozhenov *et al.* (B114), they concluded that the highest strengths for a given porosity could be obtained from an appropriate blend of the two types of particles. Such a blend could be realized in autoclaved cement–silica materials (curve EF).

The size and shape distribution of the solid particles in a cement paste is related, though in a complex way, to that of the pores, and other workers have attempted to relate strength to pore size distribution. From studies on autoclaved materials, Mindess (M74) concluded that, for a given porosity, the strength increased with the proportion of fine pores. Jambor (J29) similarly found relations between strength and average pore size. The weaker materials, with relatively coarse pores, were ones containing dense, crystalline phases, and broadly resembled those represented by the high-porosity end of curve CD in Fig. 8.8. Odler and Rössler (O20) concluded from a study of cement pastes cured for various combinations of w/c, temperature and time that while the main factor influencing strength was porosity, pores with a radius below 10 nm were of negligible importance. This conclusion, which was based on results of MIP, was supported by the observation that the specific surface area, which depends on the content of very small pores, had very little effect on the strength. For a given porosity, a given volume of hydration products contributed more to the strength than the same volume of unreacted cement, but this effect was distinctly less than that of porosity.

These results suggest that, in attempts to relate strength to porosity, the total water porosity should not be used, the capillary or free water porosity or the volume of pores above a certain size being more appropriate. Parrott and Killoh (P30), in relation to their modelling of properties, similarly considered the volume, size and continuity of the larger pores to be the relevant quantities.

8.4.3 Mechanisms of failure

Strength cannot be explained by relating it empirically to porosity or pore size distribution; it is necessary to know what holds the material together and what happens when it fails. Many types of interatomic force exist in cement paste, and any distinction between chemical and physical forces is without obvious meaning. It has sometimes been assumed that van der Waals forces are important, but if this term is used in the restricted sense of London forces, this is unlikely. Although such forces exist between any atom and its near neighbours, in so strongly polar a material as hcp they are probably almost everywhere swamped by much stronger forces, comprising ionic–covalent bonds, ion–dipole attractions (e.g. between Ca^{2+} ions and H_2O molecules) and attractions between permanent dipoles, including hydrogen bonds. London forces are, however, probably the dominant attractive forces between the layers within a CH crystal.

Cohesion has often been attributed to the interlocking of fibrous or acicular particles. This could be important in the more porous parts of the material, but in the material as a whole, attractive forces between those parts of adjacent layers of C–S–H or other phases that are in contact are probably more important, both within particles and, in so far as the material is particulate, between them. The attractive forces could be direct, of the types mentioned above, or indirect, through interposed H_2O molecules forming ion–dipole attractions and hydrogen bonds. Even for D-dried material, analogies with crystalline tobermorite and jennite indicate that much interlayer water is still present (Section 5.4).

Various authors have discussed the application of fracture mechanics to cement paste (J29,A19,B115,P40,E4). As in other brittle materials, failure occurs through the initiation and spread of cracks, which originate in places where the local stress is high. Hardened cement pastes are about ten times as strong in uniaxial compression as in tension. It is probable that the ultimate mechanisms of failure are essentially similar, but that the cracks propagate more readily under tension. The relations between microstructure and strength must thus be sought in the features that give rise to high local stresses and in those that favour, or, conversely, that arrest the spread of cracks. This is consistent with the conclusion that porosity is a major factor but not the only one; Beaudoin and Feldman (B115) discussed some others, in addition to those noted above. The strength, however measured, is less for a saturated paste than for a dry one. In principle, this is readily explained, as the entry of water molecules between particles could weaken the attractive forces between them and thereby favour the initiation or spread of cracks. A disjoining pressure is created (Section 8.5.2).

8.5 Deformation

8.5.1 Modulus of elasticity

Curves of uniaxial compressive stress against strain for hcp are non-linear and vary somewhat with the strain rate. Caution is therefore needed in comparing values of Young's modulus obtained in different investigations, results obtained at very low stresses by the dynamic method in which a resonance frequency of vibration is determined being higher than ones given by static or slow loading methods and averaged over greater ranges of stress. Using the dynamic method, Helmuth and Turk (H45) obtained values of 20 000–30 000 MPa for saturated, mature pastes. Trends with w/c and degree of hydration, and relations with porosity, were similar to those for compressive strength. For saturated pastes, Helmuth and Turk found

$$E = E_o(1 - p_c)^3 \qquad (8.11)$$

where E is Young's modulus, E_o is its value extrapolated to zero porosity, and p_c is the capillary porosity. For two cements, Helmuth and Turk found E_o to be approximately 30 000 MPa. The same equation held equally well for total water porosity if values around 75 000 MPa were used for E_o. They also found (H46) that, for a series of pastes 6–7 months old, the irreversible shrinkage at 47% RH was linearly related to the total water porosity, whereas the reversible shrinkage was practically independent of the porosity.

Sereda *et al.* (S79) found that, as with compressive strength, the values obtained depend on the RH at which the sample is conditioned, but that the effect is in the opposite direction, saturated pastes having higher moduli than dried pastes. This was attributed to stiffening of the structure through entry of H_2O molecules into interlayer sites, possibly augmented by stronger bonding between adjacent layers (S79,F32,S85). Hysteresis occurred, and was explained by assuming that entry and exit of water begins at the edges of the layers, and that the modulus depends primarily on whether the central parts of the layers are filled. Compacts made by compressing bottle hydrated cement at 100–750 MPa gave results similar to those for pastes.

8.5.2 Drying shrinkage

Cement pastes expand slightly during wet curing; a saturated paste shrinks on being dried. For a mature paste, the linear shrinkage on strong drying is typically somewhat under 1%. The shrinkage on first drying is partly irreversible. Further wetting and drying cycles lead to some increase in the irreversible shrinkage, but after a few such cycles the variation with changes in RH is essentially reversible. A small shrinkage, called autogenous shrinkage, occurs on sealed curing. The changes in length, or more properly in volume, broadly parallel ones in water content, but the parallelism is not exact (F13). Shrinkage is complicated by the effects of moisture gradients in the specimen, and to minimize these and the resulting mechanical constraints, it is necessary to study thin slices of material. Wittmann (W24) obtained data for specimens of varying thickness and extrapolated to zero thickness.

At least four effects might contribute to reversible shrinkage:

Capillary stress

Attractive forces exist between the water in a pore and the solid surfaces; when the liquid evaporates, the tension in the meniscus is transferred to the walls, and the pore tends to shrink. The shrinkage is not necessarily reversed when all the liquid has gone, because the pore may collapse. This effect could be important between about 90% and 45% RH. Above 90% it is unlikely to be important because the pores that are being emptied are wide, and the resulting stresses are small, and below about 45%, a stable meniscus cannot form.

Surface free energy

Due to unsatisfied bonding forces, the surface of a solid particle is under tension, as in a liquid. Adsorbed molecules decrease this tension, and if they are removed, the particle tends to contract. The greatest difference occurs when the last adsorbed layer is removed, and the effect should therefore be greatest below about 20% RH. It can only occur to the extent that closer packing is possible within the solid particle, e.g. by rearrangement of smaller units. Although discussed here in relation to a particle, it could presumably also occur in a convex portion of an irregularly shaped, continuous structure.

Disjoining pressure

If two solid surfaces are in contact and the attractive forces between them are outweighed by those existing with molecules of the liquid, the latter may be drawn in between them, so forcing them apart; a disjoining pressure is said to exist. The effect occurs with many clay minerals. The main attractive forces in the latter case are probably ion–dipole forces between ions in the solid surfaces and water molecules, and the same could apply to C–S–H.

Movement of interlayer water

This is essentially the same effect as the last, but occurs between layers within a single particle. There could well be a continuum of effects, ranging from movement of water molecules between or away from the surfaces of adjacent separate particles to similar movement involving layers within a particle. An intermediate situation in such a continuum might be movement of water into or out of interlayer spaces at re-entrants at which layers of a continuous structure are splayed apart, such as those shown in Fig. 8.4.

There has been no general agreement as to the relative importance of these effects. Irreversible shrinkage probably results from an ageing process, in which particles or individual layers of the gel form permanent contacts that bond them into a more stable structure. Powers (P34) regarded disjoining pressure as the most important effect at all humidities. Feldman and Sereda (F31) considered that capillary stress and surface energy were important at the higher relative humidities, and movement of interlayer water below about 30%. Bentur *et al.* (B54) concluded from a study of calcium silicate pastes in which shrinkage and creep were related to N_2 sorption isotherms and determinations of average silicate anion size that irreversible shrinkage was due in young pastes to a reduction in pores of 4–10 nm diameter, but in mature pastes to changes in the C–S–H, but Parrott and Young (P38) found no correlation between silicate polymerization and irreversible shrinkage in mature C_3S pastes. They attributed irreversible shrinkage to capillary tension forces and reversible shrinkage to capillary tension and surface energy effects, depending on the RH. Ferraris and Wittmann (F37) considered the dominant effects to be disjoining pressure down to 40% and surface energy below 40%. Hansen (H43) concluded from a study of initial drying shrinkage in relation to weight loss and pore structure determined by MIP and N_2 sorption that the main source of stress over the entire range of RH was surface energy. Capillary tension was a subsidiary source at RH > 25%, especially at higher w/c ratios. The stresses were largely accommodated by an irreversible decrease in interlayer spacing.

8.5.3 Creep

Creep is slow deformation under load which is superimposed on the elastic strain. It can reach values several times greater than the latter, and unlike it is only partly reversed if the load is removed. The irreversible component, which is the major one, occurs on a time scale of months or longer, and the reversible component on one of days or weeks. Creep taking place at a constant RH is called basic creep. In practice, creep is likely to be accompanied by drying shrinkage, and if the loading is compressive, the resulting deformation is greater than the sum of those shown by two specimens in which the effects operate separately. The additional deformation is called drying creep. Creep occurs in tension and with other types of stress as well as in compression.

Creep decreases with age before loading and with decreasing w/c ratio (V5). If the material is heated at 110°C before loading and kept dry during test, creep almost disappears. It reappears on re-exposure to water. In contrast, increase in temperature from 10°C to 60°C during loading increases creep. Heating before loading causes the irrecoverable component of creep to decrease even if the specimen is kept saturated (P41,P42). Standing at ordinary temperature, drying at 110°C and heating under saturated conditions are all forms of ageing which produce a more stable structure. Irreversible creep may also be a form of ageing in which a similar effect is produced by stress (F38) and which occurs less readily if a more stable condition has already been reached through some other cause.

There is no general agreement on the mechanism of either reversible or irreversible creep, but the latter, at least, probably involves relative movements of particles of C-S-H or shearing movements of individual layers within particles or both (F32,B54). The presence of water appears to be essential. Bentur *et al.* (B54) suggested that interlayer Si-O-Si bonding contributes to irreversible creep, but the evidence for this is weak. Such bonding would, moreover, be unlikely to occur at ordinary temperature, and, if it did, would probably produce Q^3 tetrahedra, which are not found.

8.6 Permeability and diffusion

8.6.1 Permeability to water

The permeability of concrete and the rates at which ions and gases diffuse in it are of major importance for durability. We shall consider only the behaviour of cement paste.

Under certain experimental conditions permeability to water obeys d'Arcy's law:

$$dq/dt = K_1 \cdot A \cdot \Delta h / l \qquad (8.12)$$

where dq/dt is the flow in $m^3 s^{-1}$, K_1 is the permeability in $m s^{-1}$, A is the cross-sectional area in m^2, Δh is the head of water in metres, and l is the thickness of the specimen in metres. K_1 depends on the viscosity of the liquid, the temperature and the properties of the material. Important conditions are that the sample is completely saturated, steady flow is maintained, and osmotic effects, such as might be caused by concentration gradients in dissolved ions, are absent. Feldman (F39) gave a review, including references to experimental techniques.

Powers and co-workers (P20,P34,P43) determined values of K_1 of hcp for various combinations of w/c and age. They assumed that the capillary pores in the paste were initially continuous, but that at a certain stage in hydration they became segmented into isolated cavities, so that the water could only travel through the gel pores, causing the permeability to decrease to a very low value. They concluded that the age at which capillaries became discontinuous increased with w/c. For a typical cement, it was about 3 days at w/c = 0.4, 1 year at w/c = 0.7 and unattainable at w/c > 0.7. Reinterpreting some of these data, Verbeck and Helmuth (V5) showed that the permeability was directly related to the capillary porosity.

Subsequent studies have shown that K_1 is not related to the total water porosity but depends on the volume and connectivity of the larger pores. Mehta and Manmohan (M75) and Nyame and Illston (N16) found linear relations between log K_1 and estimates of the maximum continuous pore radius, obtained from MIP, and other quantities derived from the pore size distribution and degree of reaction. A high proportion of the flow appears to be through pores wider than about 100 nm. Typical values of log K_1 for mature pastes cured at ordinary temperatures range from around -13.4 at w/c 0.3 to around -11.8 at w/c 0.7 (M75,N16,G67,M76,H47). K_1 increases with temperature (G67,M76).

Hooton (H48) noted some pitfalls in the experimental determination of permeabilities and in the practical utilization of the results. In an experimental determination, accurate results will not be obtained unless the sample has been vacuum saturated; and, even if it is, equilibrium flow in accordance with d'Arcy's law may not occur because water is being used up to continue hydration. In practice, a concrete sample is probably not often saturated throughout; if it is not, capillary forces, as well as the pressure difference, affect the rate of flow. Applications of results obtained with pastes to concrete are further complicated by the presence in the latter of cracks, poorly compacted areas and other inhomogeneities.

8.6.2 Diffusion of ions and gases

Ushiyama and Goto (U15) studied the diffusion of ions in a Portland cement paste of w/c ratio 0.4 cured at 20°C for 28 days. Fick's second law was shown to be followed when a steady state had been reached. For Li^+, Na^+ and K^+, diffusion coefficients of 1.4–3.3×10^{-12} m^2 s^{-1} were obtained. That of Cl^- was somewhat greater, especially if the balancing cation was Ca^{2+} or Mg^{2+}. It was shown that CH and MH crystallized at the surface of the specimen, suggesting that CH within the specimen had dissolved.

Other studies on Na^+ and Cl^- (C40,P44,G68) have given results in substantial agreement with those above. At ordinary temperatures, the diffusion coefficients obtained have been 10^{-11} to 10^{-13} m^2 s^{-1} for Na^+ and 10^{-11} to 10^{-12} m^2 s^{-1} for Cl^-. The activation energies are reported to be 84 kJ mol^{-1} for Na^+ and 50 kJ mol^{-1} for Cl^- for w/c ratios of 0.35–0.45

(G68), or 42–45 kJ mol^{-1} for Cl$^-$ at w/c ratios of 0.4–0.5 and 32 kJ mol^{-1} for w/c = 0.6 (P44). These activation energies are much higher than those found for diffusion of the same ions in a dilute solution, which are typically below 20 kJ mol^{-1}. For w/c ratios of 0.35–0.45 neither the diffusion coefficient nor the activation energy varies significantly with either age or w/c ratio (G68). These observations, which contrast markedly with those on permeability, suggest that diffusion occurs by movement of the ions on the surfaces of the hydration products, or, at least, that the mechanism involves strong interactions with them (P44,G68). At w/c ratio 0.6, the lower activation energy could arise from the presence of relatively large and unconstricted capillaries not present at lower w/c ratios, which could allow the rate-limiting step to be more easily overcome (P44).

Studies on the permeability of concrete to gases, reviewed by Feldman (F39), have provided only slight information regarding the mechanism of transport through the cement paste or the relations to its pore structure. Oxgyen permeabilities of Portland cement concrete equilibrated at 55% RH are typically 10^{-14} to 10^{-17} m^2 s^{-1}, and are strongly correlated inversely with compressive strength and positively with oxygen diffusivity (L45). They increase very markedly on drying, and it cannot be assumed that their relation to pore structure is the same as that for water permeability.

9

Composite cements

9.1 Introduction

A composite cement is a hydraulic cement composed of Portland cement and one or more inorganic materials that take part in the hydration reactions and thereby make a substantial contribution to the hydration product. This definition excludes admixtures, such as $CaCl_2$, that influence the hydration process but do not themselves contribute substantially to the product. The inorganic materials will be called mineral additions. The most important are pulverized-fuel ash (fly ash; pfa), ground granulated blast furnace slag (ggbfs), natural pozzolanas and microsilica (condensed silica fume). The mineral addition may be ground together with the cement clinker and gypsum, or mixed with Portland cement when the latter is used. In British usage, these procedures are called intergrinding and blending, respectively.

Complete cements are used for various reasons. The necessity for utilizing waste materials and decreasing overall energy consumption is becoming increasingly obvious. Pfa and slag are waste materials produced in large quantities, and concretes made with them, or with natural pozzolanas, can have similar properties to ones made with pure Portland cements at lower cost per unit volume. For the cement manufacturer, intergrinding provides a degree of flexibility in adjusting the volume of cement production to changing demand. Concretes made using composite cements can have properties that are desirable for particular purposes, such as slower and decreased total heat evolution in massive structures, improved durability or, with microsilica, strengths above the normal range.

Mineral additions may be broadly categorized as pozzolanic materials or latent hydraulic cements. Neither type reacts significantly with water at ordinary temperatures in the absence of other substances. Pozzolanic materials are high in SiO_2 and often also in Al_2O_3, and low in CaO; they are sufficiently reactive that mixtures of them with water and CaO produce C–S–H at ordinary temperatures and thereby act as hydraulic cements. If they contain Al_2O_3, calcium aluminate or aluminate silicate hydrates are also formed. Because they are low in CaO, this component must be supplied in stoichiometric quantity. In a composite cement, it is provided by the Portland cement through decreased formation of CH and decreased Ca/Si ratio of the C–S–H. Pfa low in CaO, natural pozzolanas and microsilica are examples of pozzolanic materials.

Latent hydraulic cements have compositions broadly intermediate between those of pozzolanic materials and Portland cement. They act as hydraulic cements if mixed with water and a minimal amount of some other substances that serves as a catalyst or activator. Ggbfs is a latent hydraulic cement. When mixed with Portland cement, it is activated by the CH and alkali which the latter produces. As will be seen, the hydration products of the Portland cement clinker are also modified. The proportion of a latent hydraulic cement in a composite cement can be higher than that of a pozzolanic material. Some materials, such as finely ground $CaCO_3$, though used in varying proportions and having neither pozzolanic nor latent hydraulic properties, are conveniently considered along with mineral additions.

In this chapter and elsewhere, w/s denotes the ratio $w/(c + p)$ and 'percentage replacement' the quantity $100p/(c + p)$, where w, c and p are the masses of water, Portland cement and mineral addition, respectively.

9.2 Blastfurnace slag

9.2.1 Formation, treatment and use in composite cements

Blastfurnace slag is formed as a liquid at 1350–1550°C in the manufacture of iron; limestone reacts with materials rich in SiO_2 and Al_2O_3 associated with the ore or present in ash from the coke. If allowed to cool slowly, it crystallizes to give a material having virtually no cementing properties. If cooled sufficiently rapidly to below 800°C, it forms a glass which is a latent hydraulic cement. Cooling is most often effected by spraying droplets of the molten slag with high-pressure jets of water. This gives a wet, sandy material which when dried and ground is called ground granulated blastfurnace slag and often contains over 95% of glass.

In an alternative treatment, called pelletization, the molten slag is partially cooled with water and flung into the air by a rotating drum. The resulting pellets vary in size from a few millimetres to around 15 mm. The proportion of glass decreases with increasing pellet size. Pelletization has advantages to the steel manufacturer of lower capital cost, decreased emission of sulphurous gases, and formation of a drier product, which can also be sold as a lightweight aggregate, but the glass content of the size fraction used to make composite cements can be as low as 50%.

The composition of the slag must be controlled within relatively narrow limits to ensure satisfactory and economic operation of the blastfurnace, and depends on that of the ore. It therefore varies considerably between plants, but that of a given plant is unlikely to vary much unless the source of the ore is changed. Table 9.1 gives ranges and means for slags typical of those produced in western Europe. For a wider range of steel-producing countries, the contents of major components are: MgO, 0–21%; Al_2O_3, 5–33%; SiO_2, 27–42%; and CaO, 30–50% (S86).

Table 9.1 Chemical compositions of some blastfurnace slags[a] (D35)

	Mean	Minimum	Maximum		Mean	Minimum	Maximum
Na_2O	0.39	0.25	0.50	TiO_2	0.55	0.49	0.65
MgO	5.99	3.63	8.66	MnO	0.64	0.34	1.31
Al_2O_3	13.29	10.26	16.01	FeO	1.24	0.29	9.32
SiO_2[b]	33.48	31.96	37.29	S^{2-}	0.94	0.68	1.25
P_2O_5	0.13	0.00	0.34	F^-	0.16	0.06	0.31
SO_3	0.04	0.00	0.19	Cl^-	0.019	0.003	0.050
K_2O	0.70	0.44	0.98	Ign. loss	0.42	0.00	1.04
CaO	42.24	37.92	44.38	Total[c]	99.68		

[a] Means, minima and maxima for 27 slags representative of French and Luxembourg production in 1980.
[b] Total SiO_2; insoluble residue, mean, 0.41%, minimum, 0.00%, maximum, 1.32%.
[c] Corrected for O^{2-} equivalent to S^{2-}, F^- and Cl^-.

In composite cements, the granulated or pelletized slag is blended or interground with Portland cement or clinker. Ternary mixes containing also, for example, natural pozzolanas are used in some countries. Slags vary considerably in grindability, but are usually harder to grind than clinker. Intergrinding may therefore produce a cement in which the slag is too coarsely, and the clinker too finely, ground. For this and other reasons, there can be a case for separate grinding, at least with some combinations of clinker and slag (S87). The relative proportions of clinker and slag vary widely; the content of slag can be over 80%, but up to 45% is more usual. Specifications and terminology based on them vary in different countries; in the UK, Portland blastfurnace cements may contain up to 65% of slag.

In this chapter, unless otherwise stated, the term 'slag' is used to mean granulated or pelletized blastfurnace slag and 'slag cement' to mean composite cements consisting essentially of these materials with Portland cement or clinker. In these cements, the slag reacts considerably more slowly than the alite, and strength development is therefore slower to an extent that increases with the proportion of slag. For equal 28-day strengths, replacement of a Portland cement by one containing 65% of slag can lower the compressive strength by almost half at 2 days, but increase it by about 12% at 91 days (S86). Workability is similar to that obtained with Portland cements. The rate of heat evolution at early ages and the total heat evolution are both reduced. Fuller comparisons would have to take account of the proportion of slag, the particle size distributions of clinker and slag, and the fact that slags vary in reactivity.

9.2.2 Factors affecting suitability for use in a composite cement

The suitability of a slag for use in a composite cement depends primarily on its reactivity, though grindability and contents of water and of undesirable components, especially chloride, must also be considered. Reactivity most obviously depends on bulk composition, glass content and fineness of grinding, though these are probably not the only factors and the relations with composition and glass content are complex.

Several investigators have found that the presence of a small proportion of crystalline material finely dispersed in the glass improves grindability or reactivity or both (S86). Demoulian et al. (D35) found that the highest strengths at 2–28 days were given by slags containing about 5% of crystalline phases and that increase in the content of the latter up to 35% produced only small decreases in strength. They attributed the beneficial effect of small contents of crystalline phases to mechanical stress in the glass and provision of nucleation sites for hydration products, and the smallness of the decrease at high contents to a consequent change in the glass composition, which through the separation of merwinite was enriched in aluminium.

Smolczyk (S86) reviewed relations between composition and hydraulic properties. Many attempts have been made to assess slags on the basis of moduli based on bulk composition, of which $(CaO + MgO + Al_2O_3)/SiO_2$ is one of the simplest and most widely used. Minimum values for this ratio, such as 1.0 in West Germany, have been incorporated into national specifications. This and similar moduli express the fact that hydraulic activity is broadly favoured by more basic composition, but the effect of Al_2O_3 content is complex and none of the proposed moduli has proved valid for the detailed comparison of slags other than ones of relatively similar compositions produced within a given plant. A regression analysis of compressive strengths on compositions for composite cements made using a wide range of west European slags (S88) showed that increase in Al_2O_3 content above 13% tended to increase early strengths but to decrease late strengths. MgO in amounts up to 11% was quantitatively equivalent to CaO. Minor components were found to have important effects: that of MnO was always negative, but those of P_2O_5 and of alkalis were more complex.

9.2.3 X-ray diffraction and microstructure of slags

Regourd (R34) reviewed structural and other aspects of slags. XRD patterns show an asymmetric, diffuse band from the glass peaking at about 31° 2θ (CuK$_\alpha$; $d = 0.29$ nm) and extending from about 20° to about 37° and a weaker band at about 48° 2θ (0.19 nm). Crystalline phases, if present in sufficient quantity, give superimposed, sharper peaks; melilite and merwinite are the most usual. Neither periclase nor lime is found and, since it is present either in the glass or in these inert, crystalline phases, MgO is not a potential cause of expansion, as it may be in a clinker (S89).

Glass contents have been determined by light microscopy and by difference from QXDA determinations of crystalline phases (S90,R34,R35,U16). Investigators have differed as to how well the results of the two methods agree. Both will fail if the individual crystalline regions are below a certain size, which is much smaller for QXDA than for light microscopy, but light microscopy complements QXDA by providing microstructural information and does not depend on a possibly uncertain choice of reference standards. Image analysis with the SEM is a further possible method.

Individual grains of ground granulated or pelletized slag are markedly angular. The melilite and merwinite occur as inclusions in the glass, ranging in size from ones easily detectable by light microscopy to dendritic growths in which the individual dendrites are less than 100 nm wide (S90,D35). The melilite crystals may be zoned (U16). Metallic iron can occur. Crystalline inclusions and gas-filled pores can be very unevenly distributed; even in slags high in crystalline material, grains consisting only of glass may be seen using light microscopy (D35). In slags of high glass content, the composition of the glass is reported to be uniform, but near to melilite crystals it can be lower in aluminium (U16).

Some workers using small angle X-ray scattering and light and electron microscopy have found evidence of phase separation within the glass (S90,R34) but others have concluded that this does not occur (U16); slag glasses may vary in this respect.

9.2.4 Internal structures of slag glasses

The internal structures of slag glasses have been widely discussed in terms of the classical theory of network-forming and network-modifying elements. On this model, the glass structure is based on a continuous, though incompletely connected, anionic network composed of oxygen, silicon and the other relatively electronegative elements, and its charge is balanced by the calcium and other electropositive elements. Consideration of the compositions of typical slag glasses (Table 9.2) suggests that this model, which has proved valuable for more acidic glasses, may be unrealistic for slags. Some important atomic ratios, derived from the data in Table 9.2, are: O/Si = 4.4; O/(Si + Al) = 3.0; O/(Si + Al + P + Mg + Fe + Mn) = 2.5. These values indicate that nothing resembling a continuous network can be formed by the silicon, aluminium and oxygen atoms alone. Dron and Brivot's (D36) model, in which the network-forming atoms are present in straight or branched chains of various lengths, is more realistic. Studies of six slags using ^{29}Si NMR and IR spectroscopy, TMS and an alkali extraction method showed the predominant silicate species to be in all cases monomer and dimer (U16). The bulk compositions were all near that in Table 9.1.

Table 9.2 Atom ratios in a typical ground granulated blastfurnace slag

K^+	Na^+	Ca^{2+}	Mg^{2+}	Fe^{2+}	Mn^{2+}	Ti^{4+}	Al^{3+}	Si^{4+}	P^{5+}	O^{2-}	S^{2-}	F^-
0.6	0.5	30.7	6.1	0.7	0.3	0.3	10.6	22.4	0.1	98.5	1.2	0.3

For the mean of the analyses in Table 9.1; referred to 100 ($O^{2-} + S^{2-} + F^-$).

Evidence on the coordination of aluminium and magnesium in slag glasses is conflicting. From molar refractivities and other evidence, Chopra and Taneja (C41) concluded that both the aluminium and the magnesium are tetrahedrally coordinated; this is the situation in melilites. X-ray spectroscopic results have been held to favour tetrahedral coordination of the aluminium (G69) or a mixture of tetrahedral and octahedral coordination (Y7). From electron paramagnetic resonance, Royak and Chkolnik (R36) concluded that magnesium, aluminium and titanium could each be in either tetrahedral or octahedral coordination and that increased proportions in octahedral coordination enhanced hydraulic activity.

Structural defects affect reactivity; four samples of the same slag liquid, cooled at different rates, differed in reactivity, though all were shown by XRD to be completely glassy (F40). Photoelectric emission data showed that they contained different numbers of defects.

Granulated or pelletized slags are materials formed under markedly non-equilibrium conditions, and it is not surprising that their hydraulic activities are not determined solely by bulk composition and glass content. Reactivity may be presumed to depend on glass composition rather than on bulk composition. For a given combination of glass content and bulk composition, further effects varying reactivity could thus include the following:

(1) Crystalline size and morphology, through their effects on stress in the glass or nucleation sites for hydrated phases.
(2) The bulk composition of the crystalline material, through its effect on that of the glass, and the profile of compositional variation in the glass

near to the crystalline surfaces, which could produce regions of either enhanced or decreased reactivity.
(3) The degree of phase separation within the glass.
(4) The coordination numbers of the aluminium and magnesium, and the distributions of atoms and bonding patterns within the glass, e.g. variation in the number of Si–O–Al linkages compared with those of Si–O–Si and Al–O–Al.
(5) The numbers and types of electronic or structural defects.

9.2.5 Hydration chemistry of slag cements

Daimon (D37), Regourd (R37) and Uchikawa (U17) have reviewed slag cement hydration. Many XRD studies have shown that the principal hydration products are essentially similar to those given by pure Portland cements, but the quantities of CH found by this or other methods are in varying degrees lower than those which would be given by the Portland cement constituent if the slag took no part in the reaction (S91,C42,K42,R37). With percentage replacements above about 60%, the CH content may pass through a maximum and then decrease. With slags high in MgO, a hydrotalcite-type phase has been detected (K43).

The microstructures of slag cement pastes are also essentially similar to those of pure Portland cement pastes, apart from the lower CH contents (R34,U12,U17,U18,H49). As with the clinker phases, layers of in situ reaction product form at the boundaries of the slag grains and gradually extend inwards. A light microscopic study indicated that these were only 0.5 µm wide after 90 days (K42). Backscattered electron images with another slag (H49) showed them to be barely detectable at 28 days, but up to 10 µm wide at 6 months and up to 15 µm wide at 14 months. In the older pastes, fully reacted relicts of slag particles were abundant.

The quantitative determination of unreacted slag has proved difficult. Luke and Glasser (L27,L46) made a comparative study of methods described in the literature. They found the most satisfactory to be a modification of that of Demoulian et al. (D38), in which constituents other than unreacted slag were dissolved using an alkaline solution of EDTA in aqueous triethanolamine. Other extraction methods gave unacceptably high residues with pastes of Portland cements. The EDTA solution did not dissolve hydrotalcite; this is likely to make the results for unreacted slag high.

The scanty results in the literature for percentages of the slag that has reacted (Table 9.3) show wide discrepancies. The extents to which these result from differences between slags, experimental errors, proportions in the

Table 9.3 Percentages of slag reacted in slag cements

Reference	A. As percentage of initial weight of slag							B. As percentage of total ignited weight						
	L27	K42	T44	U17	R38	K42	A20	L27	K42	T44	U17	R38	K42	A20
% Slag in the cement	30	40	40	40	67	70	80	30	40	40	40	67	70	80
Age (days):														
0.25	—	—	—	—	—	—	—	—	—	—	—	—	—	3
1	—	—	—	—	—	—	4	—	—	—	—	—	—	5
3	—	17	—	10–20	19	19	6	—	7	—	4–8	8	13	6
7	—	31	—	—	—	23	8	—	12	—	—	—	16	11
28	42	32	30	30–40	13–23	29	13	13	13	12	12–16	9–15	20	15
90	40	—	—	—	16–26	—	19	12	—	—	—	11–17	—	19
180	40	—	—	50–60	16–28	—	24	12	—	—	20–24	11–19	—	20
365	64	—	—	—	19–36	—	25	19	—	—	—	13–24	—	—
540	—	—	65	—	—	—	—	—	—	26	—	—	—	—

Determined by the method of Kondo and Ohsawa (K42), with correction for other phases undissolved except for Ref. L27, for which an EDTA method was used.

cement or other causes are not clear. The data are somewhat more consistent if expressed as percentages on the total ignited weight than as ones on the initial weight of slag. This suggests that, for a given slag, the rate of reaction may be limited by the rate at which the necessary ions are supplied by the clinker or taken up by its hydration products.

Because the sulphur, iron and manganese in a slag are present in lower oxidation states, TG curves or determinations of ignition loss on slag cement pastes must be carried out in oxygen-free atmospheres. The S^{2-} ion is present in the glass, and its behaviour on hydration is not understood. Slag cements develop a transient green colour on hydration, which suggests that polysulphides are formed. The possibility of interactions between SO_4^{2-} and sulphur in lower oxidation states, either during hydration or on heating in a neutral atmosphere, cannot be overlooked. If SO_2 is formed in the latter case, it could significantly affect the interpretation of a TG or ignition loss determination.

9.2.6 X-ray microanalysis

X-ray microanalyses (R26,U18,H4,H49) and analytical electron microscopy (T44) have indicated Ca/Si ratios of 1.55–1.79 for the C–S–H surrounding the clinker grains. Some have also indicated higher contents of aluminium, potassium and sodium. These Ca/Si ratios are generally lower than those of 1.7–2.0 that are characteristic of pure Portland cement pastes. The material formed in situ from the slag is markedly lower in CaO and higher in aluminium and magnesium. Regourd et al. (R26) found C/S 0.9–1.3, C/A 4.6–4.8 and C/M 1.4–3.2 for a 28-day-old paste made with a cement containing 70% of slag and hydrated at 20°C.

Harrisson et al. (H4,H49) represented the results of X-ray microanalyses of individual spot analyses in all parts of the microstructure other than unreacted clinker grains on plots of Al/Ca ratio against Si/Ca ratio and of Mg/Ca ratio against Al/Ca ratio (Figs 9.1 and 9.2). If the analyses of the material formed in situ from the slag are excluded, the plot of Al/Ca against Si/Ca is broadly similar to those obtained for pure Portland cement pastes, and may be interpreted in the same way (Section 7.2.5).

Taken together, the results in the two figures show that the material formed from the slag has an approximately constant Si/Ca ratio of 0.62, and ratios of Mg/Ca and of Al/Ca that vary from point to point on a micrometre scale but which are related to each other by the equation shown in Fig. 9.2. This was interpreted (H49) as indicating mixtures in varying proportions of

C–S–H having Si/Ca 0.62, Al/Ca 0.09, with a hydrotalcite-type phase having Al/Mg 0.38. As with Portland cement pastes, the material described in Fig. 9.1 as C–S–H may really be an intimate mixture of the latter with AFm and, to a minor extent, hydrotalcite-type structures.

Tanaka *et al.* (T45) partially masked the surfaces of disks of slag glass with gold coatings; the disks were then embedded in cement pastes for various periods, after which they were removed and examined by SEM. This showed that two layers of in situ product were formed. The innermost one had Si/Ca 0.64, Al/Ca 0.32 and Mg/Ca 0.34; the outer one had Si/Ca 0.61, Al/Ca 0.24, Mg/Ca 0.20. The product formed immediately outside the original boundary of the slag had Si/Ca 0.48, Al/Ca 0.09, Mg/Ca 0.17. The two layers of in situ product could have been mixtures in differing proportions of C–S–H, a hydrotalcite-type phase and possibly AFm phase.

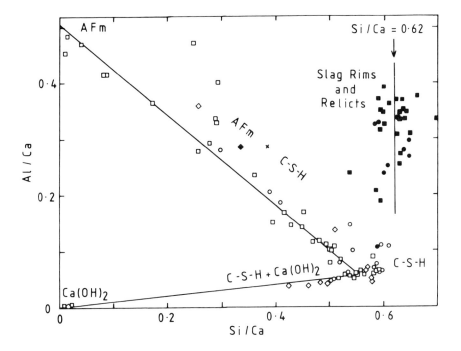

Fig. 9.1 Si/Ca and Al/Ca atom ratios from individual X-ray microanalyses of a slag cement paste (40% slag; w/s = 0.5) hydrated for various times at 25°C. Mg/Si < 0.15: ◇, 28 days; □, 6 months; ○, 14 months. Mg/Si ≥ 0.15: ◆, 28 days; ■, 6 months; ●, 14 months. Harrisson *et al.* (H49).

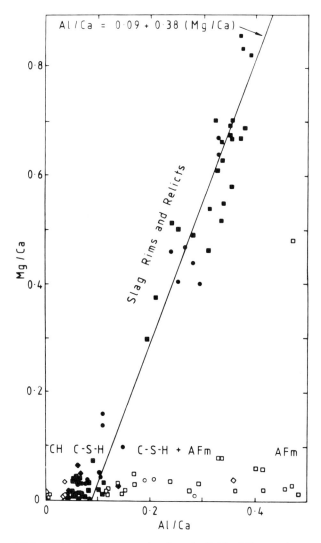

Fig. 9.2 Al/Ca and Mg/Ca atom ratios from individual X-ray microanalyses of the same slag cement paste as in Fig. 9.1 Si/Ca < 0.5: ◇, 28 days; □, 6 months; ○, 14 months. Si/Ca ≥ 0.5: ◆, 28 days; ■, 6 months; ●, 14 months. Harrisson et al. (H49).

9.2.7 Stoichiometry of slag cement hydration

Comparison of the average proportions, compositions and densities of the phases present in the material formed in situ from the slag with corresponding data for the unreacted slag suggests that little net movement of either magnesium or oxygen in or out of the latter occurs when it reacts; however, substantial fractions of the aluminium, silicon and calcium are expelled and an equivalent amount of hydrogen taken up (H49). The ions thus released contribute to the hydration products formed in the initially water-filled space, but more calcium is needed to alter the ratios to those in a mixture of C–S–H and AFm phase. This is provided partly at the expense of CH formation, and partly through decrease in the Ca/Si ratio of all the C–S–H, including that formed in situ from the alite and belite. On this basis, given also the CH content and estimates of the amounts of clinker phases and slag that have reacted, one may calculate a mass balance table by a simple extension of the procedure described in Section 7.3.3. Water contents, volume percentages of phases and, by difference, porosities, may similarly be calculated. Table 9.4 gives results of these calculations for the cement to which Figs 9.1 and 9.2 relate.

For any specified drying condition, the calculated water contents are lower and the porosities higher than those of pure Portland cement pastes, and this appears to be true in varying degrees of composite cements in general. Experimental observations support this conclusion. Non-evaporable water contents of 2-year-old pastes of w/s ratio 0.5 typically decrease with slag content from around 23% for pure Portland cements to 10–13% for cements with 90% of slag (C42). For the paste to which Table 9.4 refers, the observed non-evaporable water content was 17.7% (H49). Porosities and their relations to physical properties are discussed in Section 9.7.

9.2.8 Activation of slag glasses

If ground granulated or pelletized blastfurnace slag is placed in water alone, it dissolves to a small extent, but a protective film deficient in Ca^{2+} is quickly formed, and inhibits further reaction (K20,R39). Reaction continues if the pH is kept sufficiently high. The pore solution of a Portland cement, which is essentially one of alkali hydroxides, is a suitable medium. The supply of K^+ and Na^+ ions is limited, but these ions are only partially taken up by the hydration products, and the presence of solid CH ensures that the supply of OH^- is maintained. The slag can be rendered active by OH^- ions supplied in other ways, e.g. from calcium or sodium hydroxide, sodium silicate or sodium carbonate (R34). Although several hypotheses have been proposed, the mechanism of the attack on the slag glass is not established. If it is similar to that suggested for the clinker phases (Section 4.5.1) one would expect reactivity to increase with the fraction of the oxygen atoms bonded only to calcium and a single silicon atom or, especially, to calcium and a single aluminium atom.

Table 9.4 Calculated mass balance and compositions by weight and by volume for a 14-month-old slag cement paste

Percentages on the ignited weight for material equilibrated at 11% RH

	Na_2O	MgO	Al_2O_3	SiO_2	SO_3	K_2O	CaO	TiO_2	Mn_2O_3	Fe_2O_3	Other	CO_2	H_2O	Total
Alite	0.0	0.0	0.0	0.4	0.0	0.0	1.2	0.0	0.0	0.0	0.0	0.0	0.0	1.6
Belite	0.0	0.0	0.0	0.3	0.0	0.0	0.7	0.0	0.0	0.0	0.0	0.0	0.0	1.0
Aluminate	0.0	0.0	0.0	0.0	0.0	0.0	0.0	0.0	0.0	0.0	0.0	0.0	0.0	0.1
Ferrite	0.0	0.1	0.4	0.1	0.0	0.0	0.8	0.0	0.0	0.4	0.0	0.0	0.0	1.7
C–S–H	0.0	0.0	0.7	15.8	0.0	0.0	23.8	0.0	0.0	0.0	0.0	0.0	9.2	49.4
Ca(OH)$_2$	0.0	0.0	0.0	0.0	0.0	0.0	7.1	0.0	0.0	0.0	0.0	0.0	2.3	9.4
AFm	0.0	0.0	3.0	0.4	1.1	0.0	7.7	0.0	0.1	0.7	0.0	0.0	7.5	20.6
AFt	0.0	0.0	0.1	0.0	0.3	0.0	0.5	0.0	0.0	0.0	0.0	0.0	0.8	1.8
Slag HP	0.1	1.8	1.3	2.8	0.2	0.0	4.2	0.1	0.1	0.0	0.0	0.0	3.7	14.1
Fe HP	0.0	0.0	0.0	0.2	0.0	0.0	1.0	0.0	0.0	0.9	0.0	0.0	0.6	2.7
Mg HP	0.0	0.6	0.2	0.0	0.0	0.0	0.0	0.0	0.1	0.0	0.0	0.0	0.7	1.5
CaCO$_3$	0.0	0.0	0.0	0.0	0.0	0.0	1.5	0.0	0.0	0.0	0.0	1.2	0.0	2.7
Slag glass	0.1	1.3	1.8	5.4	0.0	0.1	6.7	0.1	0.1	0.2	0.2	0.0	0.0	16.0
Slag res.	0.0	0.2	0.0	0.7	0.0	0.0	1.0	0.0	0.0	0.0	0.0	0.0	0.0	2.0
Other	0.2	0.0	0.1	0.1	0.0	0.3	0.0	0.1	0.0	0.0	0.4	0.0	0.0	1.2
Total	0.4	4.0	7.6	26.2	1.6	0.4	56.2	0.4	0.3	2.2	0.6	1.2	24.7	126.0

Volume percentages for paste equilibrated at 11% RH

Clinker phases	C–S–H	Ca(OH)$_2$	AFm	AFt	Slag HP	Fe HP	Mg HP	CaCO$_3$	Slag glass	Slag res.	Other	Pores
1.6	27.2	5.0	12.3	1.2	8.1	1.1	0.9	1.2	6.7	0.8	0.6	33.3

Water contents (percentages on the ignited weight)
Bound water: 24.7%   Non-evaporable water: 17.7%

Porosities
Capillary, 17.7%   Free water, 33.3%   Total water, 45.8%

Ratios of water/cement/slag = 0.5:0.6:0.4. Slag glass reacted = 22%, CH content = 9.4%, both referred to the ignited weight of composite cement. Slag HP = in situ product from slag. Fe HP = hydrogarnet-type product from ferrite phase. Mg HP = hydrotalcite-type product, excluding that formed in situ from slag. Slag res. = unreacted crystalline part of slag. 'Other' components mainly P$_2$O$_5$. 'Other' phases mainly insoluble residue and alkalis present in or adsorbed on products or contained in the pore solution. Discrepancies in totals arise from rounding. Results modified from Ref. H49.

Calcium sulphate accelerates the reaction of slag glasses, probably because precipitation of ettringite provides a sink for the Ca^{2+} and $Al(OH)_4^-$ ions released from the slag. It is often described as an activator, but is not very effective unless a little alkali is also present (R37) and is perhaps more properly described as a powerful reactant. The calcium sulphate present in a slag cement thus contributes to increasing the activity of the slag. The reaction between slag and $CaSO_4$ is utilized to a much greater extent in supersulphated cements.

9.2.9 Supersulphated cements

Supersulphated cements consist of 80–85% of slag with 10–15% of anhydrite and about 5% of an activator, which is usually Portland cement clinker, and are somewhat more finely ground than Portland cements. A slag high in Al_2O_3 is preferred. XRD shows that the main hydration products are C–S–H and ettringite. The content of the latter reaches a limit in about 3 days, by which time virtually all the anhydrite has reacted (S90,K42). More water is taken up during curing than with Portland or normal slag cements, and contents of non-evaporable water are higher. TEM examination of replicas (M77) confirmed the formation of ettringite, some crystals of which were over 120 µm long.

Kondo and Ohsawa (K42) studied a cement with 80% of slag (18.5% Al_2O_3), 15% $CaSO_4$ and 5% C_3S. At 3 days, 91% of the $CaSO_4$ and 22% of the slag had reacted, and the content of non-evaporable water was 26%. To convert all the SO_3 supplied by the $CaSO_4$ into ettringite, about 19% of the slag would have to react, which agrees well with the observed value. The quantity of ettringite formed would be 43%, referred to the ignited weight. The high content of non-evaporable water is difficult to explain, as ettringite normally loses at least a half of its water on being dried under the conditions used; however, XRD indicated that the ettringite formed from the cement was particularly resistant to dehydration, and a little non-evaporable water would also have been contributed by the C–S–H. Daimon (D37) reported broadly similar results.

9.3 Pulverized fuel ash (pfa; fly ash) low in CaO

9.3.1 Properties

Pfa is ash separated from the flue gas of a power station burning pulverized coal. It must be distinguished from the coarser ash that collects at the bottom of the furnace; the most uniform and highest quality ash is likely to be that produced by efficient, base-load power stations. Ravina (R40) summarized the technology. The chemical and phase compositions depend on those of the minerals associated with the coal and on the burning conditions. In general, anthracitic or bituminous coals give ashes high in glass, SiO_2, Al_2O_3 and Fe_2O_3 and low in CaO, whereas sub-bituminous coals or lignites give ashes higher in CaO and often also in crystalline phases. The pfa used in Europe and Japan to make composite cements typically contains under 10% of CaO. The American designations of Class F and Class C fly ash, though based on contents of $SiO_2 + Al_2O_3 + Fe_2O_3$ above and below 70%, respectively, correspond approximately to low- and high-CaO ashes in the sense used here. Section 9.3 deals exclusively with pfa containing less than about 10% of CaO.

Kokubu (K44) and Regourd (R34) reviewed the major characteristics of pfa. Others (W27,T46,W28,H50) have described the pfa produced in the UK more fully. Table 9.5 gives data for a good-quality ash. SEM (e.g. Ref. R41) shows such pfa to contain a high proportion of spherical and largely glassy particles, formed by the rapid cooling of droplets of liquid. Etching with HF (R41) or partial hydration (H51) reveals inclusions of prisms of mullite and smaller crystals of quartz, which are less reactive. Hollow spheres, called cenospheres, occur; ones that contain smaller spheres are called plerospheres. K_2SO_4, $CaSO_4$, hematite (Fe_2O_3) and magnetite (Fe_3O_4) may be present as smaller particles adhering to the surfaces of the spheres. The iron oxides and quartz also occur as separate, angular particles. Carbon, if present in sufficient amounts, forms porous particles that can be either spherical or irregular in shape (R41). For the more finely divided ashes, 50% by weight is in particles smaller than about 10 μm; for the coarser ones, the corresponding size is around 40 μm.

Table 9.5 Chemical and phase compositions and density of a typical pfa low in CaO and high in glass (W27)

Chemical composition (%)

Na_2O	MgO	Al_2O_3	SiO_2	P_2O_5	SO_3	K_2O	CaO	TiO_2	Mn_2O_3	Fe_2O_3	C	H_2O	Total
1.5	1.6	27.9	48.7	0.2	1.2	4.2	2.4	0.9	tr.	9.5	1.5	0.3	99.9

Phase composition (%)

Quartz	Mullite	Hematite	Magnetite	Carbon	Glass[a]	Density (kg m^{-3})
2.8	6.5	1.6	1.9	1.5	86	2220

[a] By difference.

XRD patterns of pfa include a diffuse band from the glass, which for ashes low in CaO peaks at 22–23° 2θ (CuK$_\alpha$). This is slightly above the value for vitreous silica. Possibly in all such ashes, the glass shows phase separation on a 5–15-nm scale (Q1). The proportion of glass and other amorphous aluminosilicates, if any, may be determined by difference from QXDA on the superimposed peaks of the crystalline phases (D12); allowance should be made for the content of amorphous carbon. The same method shows that, in many ashes, the FeO and Fe_2O_3 are mainly present as magnetite and hematite. Hubbard et al. (H50) found a correlation between the K_2O/Al_2O_3 ratio and the percentage of amorphous aluminosilicates. Since it is overwhelmingly the latter that take part in the pozzolanic reaction, they suggested that this ratio, multiplied by 10, might be used as a pozzolanic potential index (PPI).

Due to the presence of hollow spheres and spongy material, the density of pfa is difficult to define. The smaller spheres are not readily broken by grinding, and values of 2100–2400 kg m^{-3}, increasing with iron oxide content, are typical (K44), but these must be influenced by the presence of carbon and voids, as Al_2O_3–SiO_2 glasses have densities of 2500–2700 kg m^{-3} and the densities of the crystalline constituents range from 2650 kg m^{-3} (quartz) to 5240 kg m^{-3} (hematite). Mather (M78) found that fine grinding of a particular pfa caused the density to increase from 2440 to 2780 kg m^{-3}. Watt and Thorne (W27) found that the weight distributions of density for five pfas all peaked at 2500–2600 kg m^{-3}, and this is probably a reasonable estimate of the true density of the reactive part of the material.

9.3.2 Factors governing suitability for use in composite cements

Important properties governing the suitability of a pfa for use in a composite cement are the content of unburned carbon, the ability to decrease the water demand of the concrete and the pozzolanic activity. The second property partly depends on the first. An excessive content of carbon also produces a discoloured concrete and interferes with the action of some concrete admixtures, especially air-entraining agents. The effect on water demand is important because the pozzolanic reaction of the pfa is slow, and to obtain a specified 28-day strength, the w/s ratio must be lower than for an otherwise similar mix containing a pure Portland cement. Early work, reviewed by Kokubu (K44), established that partial replacement of cement by a good-quality pfa increases the workability. Typically, to produce a concrete with the same 28-day strength and the same slump as one made using a pure Portland cement and the same coarse and fine aggregate, some of the cement and some of the fine aggregate are replaced by pfa and the content of water per unit volume of concrete is decreased.

The enhanced workability of concrete containing good-quality pfa has generally been attributed to the fact that the latter is largely composed of smooth, spherical particles (K44), together with the increased proportion of paste and the decreased formation at this stage of hydration products. Uchikawa (U17) reviewed rheological studies. Pfa containing a high proportion of coarse ($>$ 45 µm) material is unsuitable for blending but may be suitable for intergrinding, because the effect of the latter is largely to separate particles present in agglomerates and to break up some of the larger cenospheres; the spherical particles are not otherwise damaged (M79).

Chemical tests of pozzolanicity have proved of limited use for evaluation, because the 28-day strength depends much more on the w/s ratio than on this property, and there appears to be no effective substitute for direct testing of the relevant properties on mortars or, preferably, on concretes. The degree of pozzolanic activity is, however, important for the development of strength and decrease in permeability at later ages. For equal 28-day strengths, the strengths at 91 days or more normally exceed those of otherwise similar concretes made with pure Portland cements.

9.3.3 Rates of consumption of clinker phases and pfa, and contents of calcium hydroxide

Uchikawa (U17) reviewed the hydration chemistry of pfa and other composite cements. Pfa cements differ from pure Portland cements notably in (i) the hydration rates of the clinker phases, (ii) CH contents, which are lowered both by the dilution of the clinker by pfa and by the pozzolanic reaction, (iii) the compositions of the clinker hydration products and (iv) formation of hydration products from the pfa. The two last aspects cannot be wholly separated.

Calorimetric studies show that pfa retards the reaction of alite in the early stage of reaction (G70,J30,H51,W29), but with one exception (W29) studies on the middle stage show the alite reaction to be accelerated (K45,L47,H51,T44,D12). SEM shows that C–S–H and CH are deposited on the surfaces of pfa grains before these have started to react significantly (D28). The accelerating effect is probably due mainly or entirely to the provision of additional nucleation sites on the pfa, and occurs with other fine powders (K46). A QXDA study showed that the aluminate and ferrite react more rapidly in the presence of pfa, but with belite there was no detectable effect up to 28 days and a marked reduction in rate of consumption thereafter (D12).

Fig. 9.3 shows typical results for the contents of CH in interground cements containing various proportions of pfa. At early ages, the CH content, referred to the mass of clinker, increases with increasing proportion of pfa due to the acceleration of the alite reaction, but at later ages it decreases due to the pozzolanic reaction. The decreases would, of course, be greater if the CH contents were referred to the total ignited weight, including the pfa.

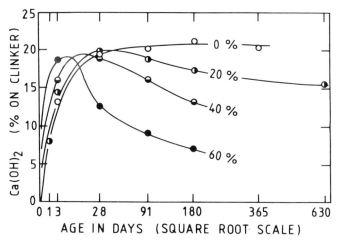

Fig. 9.3 Calcium hydroxide contents of pastes of Portland–pfa cements containing 0, 20, 40 and 60% of low-CaO pfa. Taylor et al. (T44).

SEM (H51) showed that the glass in one pfa cement was heavily etched after 7 days and that many of the particles up to 1–2 μm in size were consumed within 28 days. The crystalline phases appeared to be inert. A QXDA study of another pfa cement (D12) confirmed an earlier report (K44) that the ratio of mullite to quartz does not change on hydration, and supported the view that the crystalline phases in pfa do not react significantly within 1 year. In contrast, a TEM study (R42) of a 1-year-old specimen indicated that some, at least, of the quartz reacts at a rate comparable to that of the glass.

Quantitative data on the rate of consumption of pfa are few and somewhat variable. Those based on differences between the CH contents of pure Portland and pfa cements are suspect, because the calculation involves the effects of pfa substitution both on the rate of consumption of the clinker phases and on the compositions of the products, which are not fully understood. Unreacted pfa has been directly determined by dissolution of the other phases with HCl (C43) or with salicylic acid in methanol followed by HCl (T44), chemical separation of the residual pfa followed by QXDA determination of its content of crystalline phases (D12) and a trimethylsilylation method (U19). A method based on EDTA extraction was found unsatisfactory (L46).

Fig. 9.4 shows results from several investigations on pfa cements at or near 20°C. The tentative curves of equal mass of pfa reacted are based on the more typical of the data. Such curves are necessarily U-shaped if extended over the entire range of pfa content from 0 to 100%. The discrepancies between the results of different investigations, which are especially marked at the later ages, probably arise partly from experimental errors and partly from differences in the reactivity of the pfa or the conditions under which it was used. It is difficult, however, to reconcile the higher values shown with the amounts of CH present on any reasonable assumptions concerning the stoichiometry (Section 9.3.6).

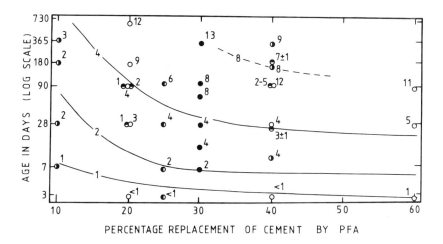

Fig. 9.4 Degrees of reaction of low-CaO pfa in pastes of Portland–pfa cements hydrated at 15–25°C. Numbers against data points denote kg of pfa reacted per 100 kg of composite cement at the age and percentage of pfa in the composite cement indicated. The curves of equal quantity of pfa reacting are based on the more typical of the results. Sources of data: ◑ (K45,K47); ○ (T44); ◐ (U19,U17); ◔ (C43); ● (D12).

Of the pfa characteristics that influence reactivity, the glass content appears to be much the most important, but specific surface area, glass composition and the effect of stress in the glass caused by the crystalline inclusions may also be relevant (U17). Of external factors, the RH, temperature (C43) and alkali content of the cement are probably the most important. Sulphate ion may also enhance reactivity by promoting the removal of Al^{3+} from the glass (U17). The rates of the pozzolanic reaction and of strength development are more sensitive to temperature than are those of hydration and strength development for pure Portland cements (e.g. Ref. H52).

9.3.4 Microstructure and compositions of the hydration products

Backscattered electron and X-ray images obtained in the SEM (O21,U17,S68) show that the microstructures of hardened pastes of cement or C_3S with pfa broadly resemble those of pure Portland cements, though, as would be expected, there is considerably less CH. Reaction rims may be seen around the pfa particles in sufficiently old pastes.

The hydrated material has been analysed by X-ray microanalysis and analytical electron microscopy. In a 3-day old paste, that formed in situ from alite or belite did not differ significantly in composition from the corresponding product in pure Portland cement pastes (H4), but at later ages Ca/Si is lower and Al/Ca higher (R25,R26,T44,U17,U18,R42). Ca/Si is typically about 1.55, but the value decreases with age and ratio of pfa to clinker. Uchikawa (U20,U17) reported a value of 1.01 for a 4-year-old paste with 40% replacement of cement by pfa. Several of the studies (R25,T44,U20,U17) showed that the C–S–H was higher in alkalis if pfa was present, but one cannot tell to what extent potassium or sodium apparently present in the C–S–H has been deposited from the pore solution on drying. For material close to the pfa particles in a 10-year-old mortar, Sato and Furuhashi (S92) found a Ca/Si ratio of 1.1–1.2.

For the microstructure as a whole, a general impression of compositions can be obtained by plotting the Al/Ca ratios of individual spot analyses against the Si/Ca ratios, as for Portland or slag cement pastes (Fig. 9.5; compare Figs 7.5 and 9.1). For the paste to which Fig. 9.5 relates, the inner product of the alite or belite had mean Si/Ca = 0.61 (Ca/Si 1.63) and mean Al/Ca = 0.10, with small contents of sodium, magnesium, sulphur, potassium and iron (H4). In reflection, the space resolution does not allow reliable analyses of the material formed in situ from the pfa to be obtained. Material near to the pfa particles was often found to include CH; where this was not present, the analyses suggested mean Si/Ca $\simeq$ 0.62 and mean Al/Ca $\simeq$ 0.18. The analyses in this group were notably high in potassium and sulphur (K/Ca 0.04–0.06, S/Ca 0.04–0.07). Some relatively large crystals of AFm phase found in a 6-month-old paste had compositions broadly intermediate between monosulphate, monosilicate and C_4AH_x.

Rodger and Groves (R42) studied ion-thinned sections of a C_3S–pfa paste by TEM, including microanalysis and diffraction. Most of the unreacted pfa particles consisted of impure alumina–silica glass, sometimes with inclusions of mullite. Some were of nearly pure silica glass, with or without quartz, or of iron oxides. In a 1-year-old paste, a typical alumina–silica particle about 12 μm in diameter had developed a reacted zone some 800 nm thick and consisting largely of radially fibrillar material described as C–S–H, with denser regions of hydrogarnet and a denser, outer rim of C–S–H. The hydrogarnet had a cell parameter of 1.24 nm and composition $C_{12}A_3FS_4H_{16}$. The C–S–H in the reacted zone had Si/Ca 0.70, Al/Ca 0.20 and Fe/Ca 0.14; that in the rim had Si/Ca 0.68, Al/Ca 0.09 and Fe/Ca 0.05. Both contained small amounts of potassium, sodium, magnesium, titanium, sulphur and other elements. Surprisingly, hydrogarnet was also observed in the reacted zones of pfa particles consisting of almost pure silica. It was suggested that these provided favourable nucleation sites, as the aluminium and iron must have come from outside. Concentric bands of differing apparent density within the reacted zones were sometimes observed, and had possibly formed by a Liesegang ring mechanism.

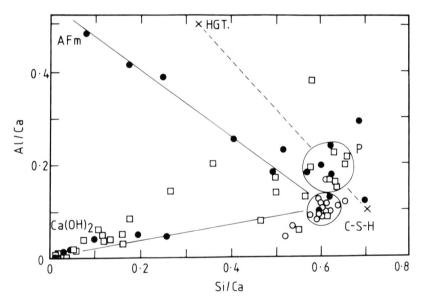

Fig. 9.5 Si/Ca and Al/Ca atom ratios from individual X-ray microanalyses of a 91-day-old paste of a Portland–pfa cement (28% of a pfa low in CaO; w/s = 0.5). Open circles, in situ product from alite or belite; open squares, regions near to pfa particles, and in some cases also to $Ca(OH)_2$; full circles, undesignated product. P: suggested composition of in situ product from pfa, possibly composed largely of C–S–H and hydrogarnet (HGT). Modified from Harrisson et al. (H4).

TMS evidence shows that silicate anion condensation occurs more rapidly in C_3S–pfa pastes than in pure C_3S pastes. Mohan and Taylor (M80) found that at 1 year, about 60% of the silicon (excluding that in unreacted pfa) was present as polymer in the former case and 40% in the latter. Uchikawa and Furuta (U9) reported data showing ratios of silicon present as polymer to that present as dimer of 0.56 for a pure C_3S paste 180 days old and of 0.80

for one containing pfa. These results agree with the evidence on C–S–H composition.

9.3.5 The nature of the pozzolanic reaction

The primary reaction of any pozzolanic material is an attack on the SiO_2 or Al_2O_3–SiO_2 framework by OH^- ions. It may be supposed that the OH^- ions attach themselves to silicon and other network-forming atoms, with consequent breaking of bonds between the latter and oxygen atoms. After this has occurred several times, the silicate or other oxy anion is detached from the framework. It may either remain in situ or pass into the solution. The charges of those that remain are balanced, partly by H^+, and partly by metal cations. Since a cement pore solution is essentially one of potassium and sodium hydroxides, the immediate product is likely to be an amorphous material with K^+ and Na^+ as the dominant cations, but the more abundant supply of Ca^{2+} and the lower solubility of C–S–H and hydrated calcium aluminate or silicoaluminate phases will ensure that this is only an intermediate product. Its presence is indicated by the relatively high potassium contents observed in or near to the reacting pfa particles.

The products formed in situ from the cement and in the initially water-filled space are modified by the removal of CaO, and the provision of additional SiO_2 and Al_2O_3. The CH content is lowered, initially because less is formed and later also by dissolution of that already present (Fig. 9.3). There is a decrease in the Ca/Si ratio of the C–S–H, even including that formed in situ from the alite, and additional C–S–H and AFm phase are formed.

The products formed in situ from the pozzolanic material are likely to reflect its composition and the situation in which SiO_2 and Al_2O_3 are readily available compared with CaO. One therefore expects to find C–S–H of low Ca/Si ratio, together with strätlingite or hydrogarnet or both. The compositions represented in Fig. 9.5 could arise from mixtures consisting largely or wholly of C–S–H and a hydrogarnet similar in composition to that found by Rodger and Groves (R42). The constituent of the in situ product described by the latter as C–S–H is unlikely from its high Ca/Si, Al/Ca and Fe/Ca ratios to have been pure C–S–H, but might have been a very close mixture of a C–S–H of lower Ca/Si ratio with other phases high in aluminium or iron.

9.3.6 Stoichiometry of pfa cement hydration

The relevant data for pfa cements are too scanty and too uncertain to allow more than a tentative approach to the hydration stoichiometry. Table 9.6 gives results calculated for the paste to which the data in Fig. 9.5 apply. The CH content and quantity of pfa glass reacted were estimates based on Figs 9.3 and 9.4, respectively, and calculation was by a modification of the method described in Section 7.3.3. It was assumed that the product formed in situ from the pfa had the composition given by the X-ray microanalyses and that its amount was such that the number of oxygen atoms per unit volume did not change when it was formed. As with pastes of pure Portland or slag cements, it is uncertain how much, if any, of the aluminium and other elements are true substituents in the C–S–H and how much are present in phases closely admixed with it. An Al/Si ratio of 0.05 for the C–S–H was arbitrarily assumed.

As with the slag cement discussed in Section 9.2.7, the calculated water contents for different humidity states are lower, and the porosities higher than for comparable Portland cement pastes. The water contents are lower because replacement of CH by C–S–H or hydrated aluminate phases causes relatively little change in the H_2O/Ca ratio, so that the water content tends to fall towards the value which would be given by the Portland cement constituent alone. Diamond and Lopez-Flores (D39) found that, for two 90-day-old pastes similar to that under discussion (30% pfa, w/s = 0.5), the non-evaporable water contents were 12.5% and 13.0%, while that of a Portland cement paste was 15.4%. The porosities are discussed in Section 9.7.

9.4 Natural pozzolanas

9.4.1 Properties

Most natural pozzolanas are of volcanic origin, though some are sedimentary. Some clays and other materials that are unsuitable for use in concrete in their natural state become usable as pozzolanas if heat treated. Both natural pozzolanas and heat-treated materials have been used with lime since ancient times, but today they are mainly used as constituents of pozzolanic cements. Several reviews are available (M81,M82,M83).

The volcanic deposits range from unconsolidated materials to ones showing varying degrees of compaction and chemical alteration. The consolidated ones are called tuffs. Not all such materials are pozzolanic; the usual active constituents are glass high in silica and often highly porous, or zeolites, or both. Italian pozzolanas, Santorin Earth (from Greece) and some rhyolites (e.g. from the USA) are examples of poorly consolidated deposits, and trass (from Germany) is one of a tuff. Some of these materials include

Table 9.6 Calculated mass balance and compositions by weight and by volume for a 91-day-old paste of a pfa cement

Percentages on the ignited weight for material equilibrated at 11% RH

	Na$_2$O	MgO	Al$_2$O$_3$	SiO$_2$	SO$_3$	K$_2$O	CaO	TiO$_2$	Mn$_2$O$_3$	Fe$_2$O$_3$	Other	CO$_2$	H$_2$O	Total
Alite	0.0	0.0	0.0	1.1	0.0	0.0	3.2	0.0	0.0	0.0	0.0	0.0	0.0	4.4
Belite	0.0	0.0	0.1	0.9	0.0	0.0	1.9	0.0	0.0	0.0	0.0	0.0	0.0	2.9
Aluminate	0.0	0.0	0.1	0.0	0.0	0.0	0.1	0.0	0.0	0.0	0.0	0.0	0.0	0.2
Ferrite	0.0	0.1	0.8	0.1	0.0	0.0	1.7	0.1	0.0	0.7	0.0	0.0	0.0	3.5
C–S–H	0.0	0.0	1.0	13.8	0.0	0.0	17.8	0.0	0.0	0.0	0.0	0.0	6.9	39.4
Ca(OH)$_2$	0.0	0.0	0.0	0.0	0.0	0.0	9.8	0.0	0.0	0.0	0.0	0.0	3.2	13.0
AFm	0.0	0.0	2.4	0.3	0.9	0.0	6.1	0.1	0.0	0.6	0.0	0.0	5.9	16.2
AFt	0.0	0.0	0.6	0.1	1.4	0.0	2.2	0.0	0.0	0.1	0.0	0.0	3.7	8.1
Pfa HP	0.0	0.0	0.3	1.2	0.1	0.1	1.9	0.0	0.0	0.1	0.0	0.0	0.7	4.6
Fe HP	0.0	0.0	0.0	0.2	0.0	0.0	1.2	0.1	0.0	1.0	0.0	0.0	0.7	3.2
Mg HP	0.0	0.9	0.3	0.0	0.0	0.0	0.0	0.0	0.0	0.1	0.0	0.0	1.0	2.4
CaCO$_3$	0.0	0.0	0.0	0.0	0.0	0.0	0.5	0.0	0.0	0.0	0.0	0.4	0.0	0.9
Pfa glass	0.2	0.3	4.4	9.4	0.0	0.8	0.2	0.2	0.0	0.9	0.0	0.0	0.0	16.5
Pfa res.	0.0	0.0	1.7	1.4	0.0	0.0	0.0	0.0	0.0	1.4	0.7	0.0	0.0	5.2
Other	0.3	0.0	0.1	0.1	0.0	0.8	0.0	0.0	0.0	0.0	0.7	0.0	0.0	2.1
Total	0.6	1.5	11.8	28.9	2.4	1.7	46.4	0.5	0.1	5.1	1.4	0.4	22.0	122.6

Volume percentages for paste equilibrated at 11% RH

Clinker phases	C–S–H	Ca(OH)$_2$	AFm	AFt	Pfa HP	Fe HP	Mg HP	CaCO$_3$	Pfa glass	Pfa res.	Other	Pores
3.8	21.1	6.8	9.4	5.5	2.3	1.2	1.4	0.4	7.7	2.3	1.0	37.0

Water contents (percentages on the ignited weight)
Bound water: 22.0%   Non-evaporable water: 14.8%

Porosities
Capillary, 29.6%   Free water, 37.0%   Total water, 48.0%

Ratios of water/cement/pfa = 0.5:0.72:0.28. pfa glass reacted = 6.0%, CH content = 13%, both referred to the ignited weight of composite cement. pfa HP = in situ product from Pfa. Fe HP = hydrogarnet-type product from ferrite phase. Mg HP = hydrotalcite-type phase, excluding any present in pfa HP. Pfa res. = unreacted non-vitreous material from Pfa. 'Other' components mainly C and P$_2$O$_5$. 'Other' phases mainly insoluble residue, and alkalis present in or adsorbed on products or contained in the pore solution. Discrepancies in totals arise from rounding (T5).

varying proportions of inactive minerals (e.g. quartz, felspars, pyroxenes, magnetite) that are harmless. Others contain constituents, such as organic substances or clays, that interfere with setting or strength development. Zeolites or minerals of related types that are found in natural pozzolanas include analcime, leucite, chabazite, phillipsite and clinoptilolite. The last of these occurs in substantial quantities virtually pure. Zeolites are readily synthesized by mild hydrothermal treatment of starting materials of appropriate composition, and their formation as alteration products of volcanic glasses during cooling is easily understandable.

Diatomaceous earth is composed of the siliceous skeletons of microorganisms. It is pozzolanic, but its use in concrete is much restricted by its very high specific surface area, which greatly increases the water demand. Some clays react significantly with lime at ordinary temperatures, but while this property can be of value for soil stabilization, their physical properties preclude their use in concrete. Many clay minerals yield poorly crystalline or amorphous decomposition products at 600–900°C (Section 3.3.2), and if the conditions of heat treatment are properly chosen, these have enhanced pozzolanic properties. Heat-treated clays, including crushed bricks or tiles, can thus be used as pozzolanas; in India, they are called surkhi. Other examples of natural rocks that have been used as pozzolanas, usually after heat treatment, include gaize (a siliceous rock containing clay minerals found in France) and moler (an impure diatomaceous earth from Denmark). The heat-treated materials are called 'artificial pozzolanas', and this term is sometimes used more widely, to include pfa.

Table 9.7 gives the chemical compositions of some natural pozzolanas. Many of the volcanic materials are not grossly different in composition from typical low-lime pfa, which they also resemble in having a glass high in SiO_2 and Al_2O_3 as a major active constituent. They differ in having zeolites as an additional active constituent and in their microstructure and physical properties. Some of the natural pozzolanas have high ignition losses, due wholly or partly to the presence of zeolites.

9.4.2 Hydration reactions

Early studies, reviewed by Malquori (M81), showed that natural pozzolanas take up CH, including that produced by Portland cement, with the formation of products similar to those formed on hydration of the latter material. They also showed that the zeolites present in many of them were at least as reactive in this respect as the glassy constituents. Zeolites are cation exchangers, but the amounts of CaO they take up are much greater than can be thus explained; moreover, cation exchange could not explain the develop-

Table 9.7 Chemical compositions of some natural pozzolanas

	Na_2O	MgO	Al_2O_3	SiO_2	K_2O	SO_3	CaO	Fe_2O_3	Loss	Total
Sacrofano (Italy)	–	–	3.05	89.22	–	2.28	0.77	4.67	–	99.99
Bacoli (Italy)	3.08	1.2	18.20	53.08	0.65	7.61	9.05	4.29	3.05	100.24
Segni (Italy)	0.85	4.42	19.59	45.47	6.35	0.16	9.27	9.91	4.03	100.05
Santorin Earth	3.8	2.0	13.0	63.8	2.5	–	4.0	5.7	4.8	99.6
Rhenish Trass	1.48	1.20	18.29	52.12	–	5.06	4.94	5.81	11.10	100.00
Rhyolite pumice	4.97	1.23	15.89	65.74	1.92	–	3.35	2.54	3.43	99.07

From a compilation in Ref. M82.

ment of strength that occurs on mixing with lime and water. It could be the initial process, but the main reaction is one in which the aluminosilicate framework of the zeolite is destroyed and C–S–H and hydrated aluminate phases are formed. The broad similarity between the chemical and phase compositions of many natural pozzolanas and Pfa's suggests that one might expect to obtain similar products, though not necessarily the same rates of reaction.

Strätlingite (C_2ASH_8; Section 6.1.4) was first synthesized by reaction between CH and metakaolin in suspensions at ordinary temperature (Z9). Depending on the composition and conditions, similar reactions of natural pozzolanas can yield, besides C–S–H, strätlingite, hydrogarnet or C_4AH_x. Sersale and Orsini (S93) studied the reactions of herschelite [a synonym for chabazite; $KNa(Si_4Al_2O_{12})\cdot 6H_2O$ approx.], analcime [$Na(AlSi_2O_6)\cdot H_2O$ approx.] and some natural pozzolanas, synthetic glasses and blastfurnace slag glasses with saturated CH solution in suspensions at ordinary temperature. In all cases, substantial amounts of CaO were removed from the solution, and XRD showed that C–S–H and one or more hydrated aluminate phases had been formed. In many cases, the presence of a 0.167-nm peak showed that the C–S–H was more specifically C–S–H(I).

Except with the slag glasses, the aluminate phase most often formed in these reactions was strätlingite. When it was not found, C_4AH_{13}, $C_4A\bar{C}H_{11}$ or hydrogarnet was usually detected. Sersale and Orsini considered that high concentrations of SiO_2 and Al_2O_3, which might arise through the presence of alkalis, favoured the formation of strätlingite and that high concentrations of CaO and Al_2O_3 favoured that of hydrogarnet; C_4AH_{13} was formed from glasses high in SiO_2 and low in Al_2O_3. The reactions of the natural pozzolanas with CH in pastes were also examined. In all cases, the products were C–S–H, strätlingite and C_4AH_{13} or $C_4A\bar{C}H_{11}$. In agreement with earlier results, the natural and synthetic glasses, tuffs and pure zeolites were found to be of comparable reactivity.

Costa and Massazza (C44) concluded from a study of natural pozzolanas of varied types that reactivity in mixtures with CH at w/s = 2 and 40°C depends during the first 28 days on the specific surface area and at later ages on the contents of SiO_2 and Al_2O_3 in the active constituents. A comparative study of five natural pozzolanas and three low-CaO pfas in pastes with cement showed that the CH contents of the pozzolanic cements were considerably lower than those of the pfa cements at 3–60 days, but virtually the same at 90 days, the pozzolanas thus appearing to react more rapidly than the pfas at early ages but more slowly later. Determinations of the unreacted mineral admixture in pastes with CH showed that at 90 days 23–30% of the natural pozzolana had reacted, compared with 11–15% for the pfas. The similarity in CH contents suggests, however, that these values may not apply to mixtures with cement.

X-ray microanalyses of C_3S-pozzolana pastes showed a gradual decrease in Ca/Si ratio on passing from regions near the unreacted C_3S to ones near the pozzolana (O21). TMS studies of pastes of C_3S, β-C_2S and cement with and without natural pozzolanas (U9,M44) showed that, in the presence of the latter, formation of polymeric anions is accelerated and their mean molecular weight is increased. TMS results and determinations of combined water showed that the hydration of β-C_2S is almost completely suppressed in the presence of a pozzolana and that, in pastes with C_3S, 16–29% of the pozzolana had reacted in 180 days (M44). The effect on β-C_2S hydration is similar to that found using QXDA for pfa (D12). Chemical extraction showed 10–45% of the pozzolanas in pastes with C_3S to have reacted in 28 days, compared with under 10% for a pfa (U9).

9.5 Microsilica (condensed silica fume)

9.5.1 Properties

Microsilica is a byproduct of the production of silicon or silicon alloys by reducing quartz in an electric furnace. Some SiO is lost as a gas and is oxidized by the air, giving a very finely particulate solid. The material may require some benficiation. The particles are spherical, typically around 100 nm in diameter, and consist largely of glass; Table 9.8 gives chemical compositions. The material formed in making elementary silicon typically contains 94–98% SiO_2, but for that formed in making the alloys 86–90% SiO_2 is more usual. Typically, the average particle size is about 100 nm and the N_2 specific surface area is 15–25 $m^2 g^{-1}$ (H53). The true density is above 2000 kg m^{-3} but the bulk density of the freshly filtered powder is only 200 kg m^{-3}. This makes the material difficult to handle unless it is agglomerated into nodules or used as a slurry.

A comparative study of microsilicas from 18 sources showed considerable variation in composition and properties, one of those examined containing as little as 23% of SiO_2 and having a specific surface area of only 7.5 $m^2 g^{-1}$ (A21). The same study showed that in most of the samples the diffuse XRD peak from the glass accounted for 98–99.5% of the total diffracted intensity and that it peaked at the value of 0.405 nm characteristic of vitreous silica. The commonest crystalline impurities detected were KCl, quartz, metallic iron and iron silicide, and pozzolanic reactivity was found to depend more on the chemical composition and nature of impurities than on the fineness or SiO_2 content. A surface layer of carbon, if present, greatly decreased reactivity.

Viewed as a mineral addition in concrete, microsilica is characterized by its small particle size and high pozzolanic activity. In the absence of a water-reducing admixture, the percentage replacement of cement is probably effectively limited by the high water demand to about 5%, but this can be increased by adding superplasticizers. By lowering the w/s ratio in such mixes to 0.2–0.3, concretes with high compressive strengths can be obtained (A22). Bleeding is also reduced. With special mixes and processing conditions, compressive strengths of up to 270 MPa have been obtained (Section 11.9).

Table 9.8 Chemical compositions of microsilica from the production of elementary silicon and 75% ferrosilicon alloy (H53)

	Si	75% Fe–Si
SiO_2	94–98	86–90
Al_2O_3	0.1–0.4	0.2–0.6
Fe_2O_3	0.02–0.15	0.3–1.0
MgO	0.3–0.9	1.0–3.5
CaO	0.08–0.3	0.2–0.6
K_2O	0.2–0.7	1.5–3.5
Na_2O	0.1–0.4	0.8–1.8
C	0.2–1.3	0.8–2.3
S	0.1–0.3	0.2–0.4
Loss	0.8–1.5	2.0–4.0

9.5.2 Hydration reactions

Traetteberg (T47) showed that microsilica used as an addition with cement has considerable pozzolanic activity, mainly in the period 7–14 days after mixing, and that the reaction product formed with CH probably had a Ca/Si ratio of about 1.1. Several subsequent studies have shown that the pozzolanic reaction is detectable within hours and also that the early reaction of the alite is accelerated (H37,H54,H55). Huang and Feldman (H54,H55) studied the hydration reactions in some detail. In pastes with 10% or 30% replacement and w/s ratios of 0.25 or 0.45, the CH content passed through maxima usually within the first day before beginning to decrease; in those with 30% replacement, it had reached zero by 14 days. Table 9.9 gives some of the results obtained for CH content and non-evaporable water in these pastes. As with pfa cements, and for the same reason, the non-evaporable water contents of mature pastes are considerably lower than those of comparable pastes of pure Portland cements.

Table 9.9 Contents of calcium hydroxide and non-evaporable water in some pastes of Portland cement with and without microsilica (percentages on the ignited weight) (H53)

Weight ratios in mix			Maximum in CH content		CH content at 180 days (%)	Non-evaporable water content at 180 days (%)
Cement	Microsilica	Water	%	Age (days)		
1.0	–	0.45	–	–	16.2	20.7
0.9	0.1	0.45	11.2	10	8.3	15.5
0.7	0.3	0.45	5	1	0.0	15.3
1.0	–	0.25	–	–	10.4	≃14.5
0.9	0.1	0.25	7.2	1	2.2	≃11.2
0.7	0.3	0.25	≃2	1	0.0	≃12.5

X-ray microanalyses show that the Ca/Si ratio of the C–S–H decreases markedly as the content of microsilica increases. For the product surrounding the clinker grains in a mortar made with a cement containing 30% of crystalline slag, Regourd *et al.* (R26) found this ratio to be 1.70; for a cement containing 25% of slag and 5% of microsilica it was 1.43. Comparable decreases were observed for cements containing granulated slag or pfa. Ca/Si ratios of 1.3 and 0.9 were reported for 200-day-old pastes with 13% and 28% replacements, respectively (R43). Uchikawa (U12) reported a value of 1.43 for Ca/Si, with substantial aluminium, 0.33% Na_2O and 0.32% K_2O, for a paste with 10% replacement cured for 60 days at 35°C.

There do not appear to be any data on the quantities of microsilica consumed, but these may be roughly estimated from CH contents and C–S–H compositions, using a method of calculation analogous to that outlined in Section 9.3.6 for low-calcium pfa. The following conclusions thus obtained are based on the data of Huang and Feldman (H55) for 200-day-old samples, and quantities are referred to 100 g ignited weight of composite cement. In the paste with 30 g of microsilica, in order to eliminate all the CH and bring the Ca/Si ratio of the C–S–H down to 0.9, about 22 g of microsilica would have to react. In the paste with 10 g of microsilica, to give a CH content of 8.3% and a Ca/Si ratio of 1.43, virtually all the microsilica would have to react. On the assumption that the Ca/Si ratio of the C–S–H does not fall below 0.8, the maximum amount of microsilica that can react is about 35 g per 100 g of cement, assuming that hydration of the latter is substantially complete.

9.6 Other mineral additions

9.6.1 Class C fly ash

Pfa formed on burning lignite or sub-bituminous coal is typically higher in CaO, MgO and SO_3, and lower in SiO_2 and Al_2O_3, than the low-CaO materials described in Section 9.3. Bulk chemical compositions vary widely (Table 9.10). Pfa with less than 70% (SiO_2 + Al_2O_3), and thus normally high in CaO, is termed Class C fly ash in the USA.

As with pfa low in CaO, fine, spherical particles consisting largely of glass predominate, but the glasses are higher in CaO and the assemblages of crystalline phases are different. In a study of 26 ashes from the western USA, McCarthy *et al.* (M84) found that quartz, lime, periclase, anhydrite and a ferrite spinel were ubiquitous or nearly so. Other crystalline phases present in some cases or in minor amounts included C_3A, merwinite, alkali sulphates, melilites, mullite, sodalite and possibly others. Due to the presence of

Table 9.10 Chemical compositions of some fly ashes high in CaO (S94)

Origin	Na$_2$O	MgO	Al$_2$O$_3$	SiO$_2$	SO$_3$	K$_2$O	CaO	Fe$_2$O$_3$	Loss	Total
Montana, USA	0.2	6.8	20.3	35.2	1.1	0.5	25.0	6.3	0.3	95.7
North Dakota, USA	7.3	7.9	12.5	30.2	9.6	0.6	23.6	4.6	1.8	98.1
North Dakota, USA	5.4	5.5	10.7	27.9	12.3	1.5	21.6	9.9	1.4	96.2
Saskatchewan, Canada	6.1	2.9	21.9	47.9	1.1	1.0	13.3	4.9	0.1	99.2

free lime, some Class C fly ashes set and harden on being mixed with water alone, and are thus true hydraulic cements. The combination of C_3A and anhydrite, if present, contributes to the hydraulic activity by forming ettringite. In some cases, the state of the lime or the content of anhydrite is such that delayed expansion may occur.

Diamond (D40) found a linear relation between the position of the glass peak in the XRD pattern and the content of CaO in the bulk analysis. This is consistent with observations on other glasses that the peak position moves to higher angles as the proportion of network modifiers increases, but the peak position is unlikely to depend only on the bulk composition because varying amounts of CaO are present in crystalline phases. In later work (D41), two fly ashes of broadly similar composition were compared. One, with 28.7% CaO, contained much lime and anhydrite, and gave a glass peak at 27° 2θ. The other, with 25.8% CaO, was low in these phases and gave a glass peak at 32° 2θ. The two ashes behaved differently on mixing with water, but these were less pronounced if cement was also present.

The particles of coal burn essentially independently of each other, and are likely to vary both in the nature and relative amounts of their inorganic constituents and in heating and cooling regime; this is likely to produce wide variations in the bulk chemical and phase compositions of individual fly ash particles (D42). Hemmings and Berry (H56) reviewed work on fly ash glasses, including microstructural, spectroscopic, DTA and chemical evidence. Two types of glass, called I and II, can be concentrated by density separation. Glass I contains minimal proportions of Ca^{2+} and other network modifiers, and tends to form relatively large, sometimes hollow, particles. Glass II contains substantial contents of network modifiers, and tends to form smaller, solid particles. The diffuse XRD peaks in general result from the overlap of contributions from these constituents. The mechanisms of formation of the particles were discussed. Evolution of H_2O and CO_2, and the varying viscosities of liquids of different compositions, are important factors affecting particle size and cenosphere formation.

Little has been reported on the hydration chemistry of high-CaO fly ashes. Diamond and Lopez-Flores (D39) found that at ages ranging from 1 h to 180 days, 30% replacements of a Portland cement by any of three different high-CaO ashes produced only small decreases in the content of non-evaporable water. This contrasts with their observations with low-CaO fly ashes (Section 9.3.6). As was noted in Section 9.3.6, the non-evaporable water content depends mainly on the amount of CaO in the hydration products, and the observation thus suggests that hydration of the high-CaO ashes occurs at a rate comparable with that of the clinker phases. Grutzeck et al. (G71) examined some pastes made using a high-CaO fly ash and a coarsely ground Portland cement by XRD and SEM. The hydration products detected at ages of 3–56 days were essentially those given by pure Portland cements at similar ages, with the addition in all cases of strätlingite and, at 3 days, a trace of C_2AH_8. The SEM showed that the particles of fly ash glass were replaced in situ by radial growths of fibrous C–S–H. These were barely detectable at 3 days but reaction was seen to be well advanced by 7 days. The formation of C_2AH_8 and strätlingite indicates substantial release of Al_2O_3 from the fly ash, even at 3 days.

9.6.2 Other pozzolanic or hydraulic additions

The ash from burning of rice husks or some other agricultural wastes is high in SiO_2 and if burning is carried out under properly controlled conditions at 350–600°C the ground material is strongly pozzolanic (M85,C45,J31). The material thus formed consists largely of amorphous silica retaining the cellular microstructure of the original material and has a very high specific surface area. Its possible utilization in conjunction with CH or cement has received attention for reasons of economics and energy conservation, and also because it can be used to make concrete that is very resistant to acid attack.

Rice husk ash has characteristics broadly similar to those of microsilica but can be even more strongly pozzolanic. A study of pastes with C_3S (K48) showed the CH content to pass through a maximum of 3% at 7 days, referred to the ignited weight; by 28 days, it had fallen to 1%. The hydration product was C–S–H with a Ca/Si ratio estimated at about 1.3 by analytical electron microscopy, or 0.9–1.2 from the contents of CH and unreacted starting materials. There were indications that an initial product with a Ca/Si ratio of 0.1–0.2 was formed.

The nature and availabilities of specific industrial waste products may be expected to vary with changes in technology, including those resulting from depletion of existing sources of energy or raw materials or from increasing recognition of the need to reduce environmental damage and pollution. As one example, both SO_2 and NO_x emissions from coal combustion may be decreased by burning at a low temperature in the presence of limestone ('primary flue gas desulphurization') (O22). A second example is provided by the Lurgi or other processes in which gaseous or liquid fuels are made from coal or lignite, which will become increasingly important as resources of petroleum and natural gas are depleted (e.g. Ref. N17). In all these cases, residues are produced which, though of limited interest at present, could become major constituents of composite cements or concrete at some future date. A number of other waste materials, such as slags from production of steel or non-ferrous metals, are also of potential interest (R34).

9.6.3 Calcium carbonate and other fillers

Composite cements may contain mineral additions other than, or as well as, ones with pozzolanic or latent hydraulic properties. Regourd (R34) reviewed the use of ground limestone, which is widely used in France in proportions of up to 27%. The limestones used consist substantially of calcite, with smaller proportions of quartz or amorphous silica and sometimes of dolomite. They must be low in clay minerals and organic matter because of the effects these have on water demand and setting, respectively. The XRD peaks of the calcite are somewhat broadened, indicating either small crystallite size or disorder or both; IR spectra confirm the occurrence of disorder.

The limestone is commonly interground with the clinker, and because of its softness becomes considerably more fine than the latter. For an overall specific surface area (Blaine) of $420 \, m^2 \, kg^{-1}$, 50% of the filler can be below 700 nm, compared with 3 µm for the clinker. The effect on 28-day strength of partial replacement of clinker by filler can be compensated by finer grinding of the clinker, and for equal 28-day strengths, the 1-day strength of a cement containing filler can be higher than that of one not containing it. This is due to the greater fineness of the clinker and the effects of the filler.

The effects of the limestone are partly physical and partly chemical. As with many other finely divided admixtures, including pfa, the hydration of the alite and aluminate phases is accelerated. Because of its fineness the material also acts as a filler between the grains of clinker, though it is unlikely to be as effective in this respect as microsilica. Chemically, it reacts with the aluminate phase, producing $C_4A\bar{C}H_{11}$, thus competing with the gypsum.

The increased rate of reaction of the alite in the presence of a limestone filler was demonstrated by light microscopic and SEM determinations of unreacted clinker, SEM observations on the thickness of the zones of hydration product surrounding the alite, and XRD observations on relative contents of CH (G72). In pastes with a high proportion of filler, the CH is concentrated into large masses, which form bridges between the grains of filler. This contrasts with the smaller and more evenly dispersed crystals formed in the absence of filler. The presence of up to 25% of fillers does not greatly affect workability, but with highly reactive or finely ground clinkers there can be some acceleration of set, probably due to formation of $C_4A\bar{C}H_{11}$ (B116).

9.7 Pore structures and their relation to physical properties

9.7.1 Porosities and pore size distributions

Studies on the pore structures of pastes of composite cements have presented difficulties, which probably arise from discontinuity of the pores. There is also strong evidence that mercury intrusion alters the structure. Feldman (F35) carried out experiments in which mercury was intruded, removed and reintruded into mature pastes. With Portland cements, the curves obtained on the second intrusion differed little from those obtained on the first, though the method cannot show whether damage occurs during the first intrusion, and there are indications that this is the case (Section 8.3.6). With slag or pfa cements the apparent pore structure became coarser. Feldman concluded that the pastes of composite cements contained relatively large but discontinuous pores, separated by walls that were broken by the mercury. Broadly similar changes on repeated intrusion were observed with microsilica cements (F41).

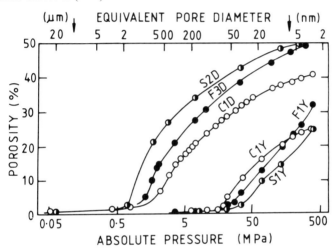

Fig. 9.6 Pore size distributions for pastes of Portland and composite cements (w/s = 0.45, curing temperature 21°C), determined using mercury intrusion porosimetry. C1D, C1Y: Portland cement pastes cured for 1 day and 1 year. S2D, S1Y: Portland–blastfurnace slag cement (70% slag) pastes, cured for 2 days and 1 year. F3D, F1Y: pfa cement pastes (13% CaO in fly ash) cured for 3 days and 1 year. After Feldman (F34).

Fig. 9.6 compares curves obtained on first intrusion for pastes of Portland and composite cements. At early ages, pfa or slag cement pastes are more porous than comparable Portland cement pastes because of the relatively slow reactions of the mineral additions, but the pore size distribution curves indicated by MIP are essentially similar in shape. For mature pastes, the apparent distributions for the composite cements indicate that there is a greater proportion of fine porosity. For the pfa cement paste, the mercury porosity obtained at maximum pressure are greater than that of the Portland cement paste. Day and Marsh (D32) agreed with Feldman that discontinuity in the pore structure produced regions that were inaccessible to mercury until the pressure was high enough for the mercury to break through the blockages, and that an inaccurate assessment of the pore size distribution resulted.

For Portland cement pastes, mercury porosities agree with those obtained using helium except at low porosities, when they are lower (B110). For pastes containing pfa or silica fume, they tend to be higher than those obtained with either helium or alcohols (F41,M86,D32), presumably because the latter fluids cannot break into the blocked regions. For pastes containing microsilica, they are lower than those obtained by equilibration with water at 11% RH (F41). This indicates that even mercury at high pressure does not completely enter all the pore space. Feldman and Huang (F41) concluded that in blends with microsilica a discontinuous pore structure was formed within 7 days.

Uchikawa et al. (U18) studied pastes of Portland and composite cements by N_2 sorption. In all cases, the pore size distributions peaked at 2 nm. Partial replacement of cement by slag increased the height of the peak, and partial replacement by pfa decreased it.

Calculations based on reaction stoichiometry and densities of phases support the conclusions from experimental observations that mature pastes of composite cements are more porous than comparable pastes of Portland cements. This is indicated by the results in Table 7.3, 9.4 and 9.6. Similar calculations for 180-day-old pastes of w/s 0.45 indicate free water porosities of about 24% for a typical Portland cement, 35% for a cement with 40% slag, 35% for one with 40% pfa and 32% for one with 30% microsilica. The calculated values are in all cases somewhat higher than observed mercury porosities (F34,F41).

9.7.2 Relations between pore structure and physical properties

Feldman (F34) plotted the logarithms of compressive strength and Young's modulus against mercury porosity for a number of pastes, and extrapolated to zero porosity. For both quantities, the intercept at zero porosity was higher for pfa cements than for Portland cements. Slag cements also gave a higher intercept than Portland cements for Young's modulus, but the intercepts for strength were virtually identical. While the precise significance of the mercury porosities is uncertain, the results suggest that pastes of composite cements are less deformable for a given porosity than those of Portland cements and that pfa cement pastes are also stronger. Feldman considered that this might be because the pastes of composite cements contained more C–S–H and less CH.

Partial replacement of Portland cement by a mineral addition can greatly decrease the permeability to water provided the age is such that sufficient reaction of the addition has occurred. For a paste with w/s 0.47 and 30% replacement of cement by pfa, cured at 20°C for 1 year, Marsh *et al.* (M87) found a permeability of $10^{-15.4}$ m s^{-1}; the corresponding paste of pure Portland cement had a permeability of $10^{-11.7}$ m s^{-1}. For the pfa cement pastes, the decreases in permeability were found to run parallel to the consumption of CH in the pozzolanic reaction. The permeability is similarly decreased through the use of microsilica (H47), slag (R44,M86) or rice husk ash (M86). Its value has been correlated with pore size distribution (M86), but it is more likely to be related to the continuity of the pores (M87).

10

Calcium aluminate, expansive and other cements

10.1 Calcium aluminate cements

10.1.1 Introduction

A cement made by fusing a mixture of limestone and bauxite or other aluminous material low in silica was patented in 1908 by Bied, of the Pavin de Lafarge Company in France. It was developed as the successful result of a search to find a cement having high resistance to chemical attack by sulphates, and became known as Ciment Fondu. Calcium aluminate cements were found to have excellent resistance to many other forms of chemical attack and to set normally but to harden rapidly; concrete can be produced that has useful strength in 6 h and a strength at 24 h equal to that given by Portland cement in 28 days. They were also found to provide good abrasion resistance in low-temperature applications. It was found that their successful use in a wide range of general and specialized applications depended on proper mix design and workmanship, and that failure to meet these requirements could lead to deterioration of the concrete.

The total quantity of heat evolved by a calcium aluminate cement on hydration is typically 325–400 kJ kg^{-1}, which is somewhat less than that for a typical Portland cement, but due to the speed of the reaction, nearly all of it is produced during the first day. Calcium aluminate cements are therefore suitable for use at low temperatures (down to $-10°C$), e.g. for winter construction in cold climates or repairs in cold stores.

Mixtures of calcium aluminate and Portland cements, sometimes also with slaked lime (CH) or additional gypsum, can be used to make rapid setting mixtures for grouting and similar applications. The strengths obtainable (e.g. 25 MPa at 5 h) (G73) are relatively high for cements of this type.

Used with appropriate refractory aggregates, calcium aluminate cements may be used to make refractory castables having applications in the steel and other heat-using industries. Cements made from bauxite may thus be used at hot face temperatures up to 1350°C. White calcium aluminate cements, which are low in iron and higher in alumina, can be used with appropriate aggregates to make castables similarly resistant up to 2000°C.

Calcium Aluminate, Expansive and Other Cements 317

Refractory castables have good resistance to thermal shock.

Calcium aluminate cements have also been termed aluminous or high-alumina cements, but the former term is less specific and the latter is more appropriately used for the white cements high in Al_2O_3 mentioned above. The cements described here are those manufactured in France and the UK. Those made elsewhere are similar in essential respects.

10.1.2 Manufacture; chemical and mineralogical compositions

Ciment Fondu is normally made by complete fusion of limestone and bauxite at 1450–1600°C. In order to produce a cement with the desired rapid-hardening properties, both raw materials must be low in SiO_2. The molten clinker is tapped off continuously from the furnace, solidifies and is typically crushed and ground to a fineness of about $300\,m^2\,kg^{-1}$. Some iron is reduced to Fe^{2+}. The colour of cements produced from bauxite can vary from yellow brown to black, but is commonly greyish black. White calcium aluminate cements are usually made by sintering calcined alumina with quicklime (calcium oxide) or high-purity limestone.

Table 10.1 Typical compositions of calcium aluminate cements (weight percentages)

Type of cement	Al_2O_3	CaO	FeO + Fe_2O_3	FeO	SiO_2	TiO_2	MgO	K_2O + Na_2O	SO_3
Ciment Fondu	38–40	37–39	15–18	3–6	3–5	2–4	<1.5	<0.4	<0.2
40% Alumina	40–45	42–48	<10	<5	5–8	≃2	<1.5	<0.4	<0.2
50% Alumina	49–55	34–39	<3.5	<1.5	4–6	≃2	≃1	<0.4	<0.3
50% Al_2O_3 (Low Fe)	50–55	36–38	<2	<1	4–6	≃2	≃1	<0.4	<0.3
70% Alumina	69–72	27–29	<0.3	<0.2	<0.8	<0.1	<0.3	<0.5	<0.3
80% Alumina	79–82	17–20	<0.25	<0.2	<0.4	<0.1	<0.2	<0.7	<0.2

Table 10.1 gives typical chemical compositions of commercially produced calcium aluminate cements. The essential compound in all of them, because it develops the main hydraulic activity and is consequently responsible for the strength development, is monocalcium aluminate, CA. In white calcium aluminate cements, it can occur with various combinations of the other binary phases of the CaO–Al_2O_3 system and corundum. The crystal structures of these compounds and the phase equilibria relating to their formation were considered in Chapter 2. In the sintering process by which these cements are made, the reaction conditions are of the utmost importance, as

the products of a given mix can vary from a mixture of CA and CA_2 to the full range of phases $CaO-C_3A-C_{12}A_7-CA-CA_2-CA_6-Al_2O_3$ (S95). Most of the commercial cements contain CA and $C_{12}A_7$; those high in Al_2O_3 may also contain CA_2, CA_6 or α-Al_2O_3.

In Ciment Fondu, most of the grains of the ground cement observable by backscattered imaging in the SEM are polymineralic (H57). Sorrentino and Glasser (S96) discussed qualitative and quantitative phase compositions. The crystalline phases may include five or even six of the following: CA, ferrite of one or more compositions, $C_{12}A_7$, β-C_2S, C_2AS, pleochroite and wüstite. Perovskite ($CaTiO_3$) may also be present, and glass if crystallization is incomplete. CA typically makes up 40–50% of the material, and is a solid solution often containing Fe^{3+} and a little silicon; the limit for Fe^{3+} is probably about 5% expressed as Fe_2O_3 (J32). The *ferrite* phase [Ca_2-$(Al_xFe_{1-x})_2O_5$; Chapter 1] constitutes 20–40%. The composition, which includes up to 3.5% of TiO_2, varies more widely than in Portland cement and depends on the manufacturing conditions (S97). X-ray microanalysis is easier than in Portland cement clinkers because there is no intergrowth with aluminate phase.

The other phases are present in amounts generally below 10% and often below 5% (S96). The C_2S (Section 1.3) contains iron and aluminium in similar proportions to the belites of Portland cement clinkers. The *gehlenite* (C_2AS; Section 2.2.5) and $C_{12}A_7$ typically contain about 9% and 5% of Fe_2O_3, respectively (S97). The behaviour of the melilite during the crystallization of the Fondu melt can be followed on the pseudo-binary phase diagram C_2AS–'C_2FS', where 'C_2FS' is a mixture of phases (S98). *Pleochroite*, of possible composition $Ca_{20}Al_{22.6}Fe^{3+}_{2.4}Mg_{3.2}Fe^{2+}_{0.3}Si_{3.5}O_{68}$, and the calcium-containing *wüstite* [(Fe,Ca)O] were described in Chapter 2. Neither the C_2S nor the wüstite is easily detectable by XRD.

Sorrentino and Glasser (S98) described phase equilibria in the CaO–Al_2O_3–Fe_2O_3–SiO_2 system. For compositions similar to those in Ciment Fondu under oxidizing conditions, the possible equilibrium assemblages of four solid phases are CA–C_2S–C_2AS–ferrite and CA–C_2S–$C_{12}A_7$–ferrite, where CA, C_2AS and ferrite represent solid solutions. Under mildly or moderately reducing conditions (S99), either pleochroite or wüstite or both may also be present, the possible equilibrium assemblages of five solid phases being CA and ferrite together with (a) C_2S, $C_{12}A_7$ and pleochroite, (b) C_2S, C_2AS and pleochroite or (c) C_2AS, pleochroite and wüstite. These results imply that ferrite is always likely to be present and that $C_{12}A_7$ cannot coexist stably with C_2AS. Caution is needed in applying them to Ciment Fondu, because equilibrium is not necessarily reached, so that phase assemblages incompatible at equilibrium may occur.

Because of the complexity and variety of the possible phase assemblages, the wide ranges of solid solutions and departures from equilibrium arising from the nature of the manufacturing process, the calculation of quantitative phase composition from bulk chemical analysis presents a much more difficult problem than for a Portland cement. Calleja (C46) obtained approximate solutions by omitting the FeO and other minor components and making suitable assumptions as to the compositions of the CA, $C_{12}A_7$ and C_2AS phases. Sorrentino and Glasser (S96) also considered the effects of FeO and MgO and consequent formation of pleochroite and wüstite.

10.1.3 Reactivities of the phases and methods of studying hydration

In commercial calcium aluminate cements, the only phases that hydrate significantly at early ages are normally CA and, if present, $C_{12}A_7$. Some sintered cements made at relatively low temperatures also contain free lime or CH, which are very active in the hydration process. The reaction of CA_2 present in many white cements high in Al_2O_3, is very slow (B117), possibly because of gel formation.

In Ciment Fondu, the ferrite phase seems to play no significant part in early hydration at 20°C, but at 30–38°C over 80% was found to have reacted by 2 months (C47). The melilite and pleochroite seem to be unreactive. When belite is present, silicate ions can be detected in the solution within a few minutes, but then disappear; it seems that precipitation occurs and further dissolution is inhibited. Among the minor oxide components, TiO_2 and MgO mainly occur in the unreactive phases. Na_2O and K_2O scarcely affect the solution equilibria at early ages, as their concentrations are very low (M88).

The reactions of the anhydrous phases with water were first studied by LeChatelier, whose conclusion that they occur by congruent dissolution followed by precipitation of hydration products has been supported by many subsequent investigations, which have established the reaction mechanisms in greater detail and the nature of the products. The methods that have given the most information have been qualitative or quantitative XRD, determinations of solution compositions, thermal methods, conduction calorimetry and studies of setting behaviour. Among recent studies may be noted those of Barret and Bertrandie (B118,B119), Cottin and George (C47), Rodger and Double (R45), Fujii *et al.* (F42), Ménétrier-Sorrentino *et al.* (M89), Edmonds and Majumdar (E5,E6) and Capmas and Ménétrier-Sorrentino (C48). Other methods have included neutron diffraction (C49), NMR (M90,R46), TEM (C50), SEM (H57), IR (B120,P45) and Mössbauer (H58) spectroscopy, determinations of electrical conductivity (G74,M88) or dielectric constant (A23), and ultrasonic wave propagation (G75). These methods have been applied, according to their various characteristics, to mortars, pastes and suspensions.

10.1.4 Hydration reactions and products

The individual hydration products and relevant phase equilibria were described in Chapter 6. The major initial products are CAH_{10} at low temperatures, C_2AH_8 and AH_3 at intermediate temperatures and C_3AH_6 and AH_3 at higher temperatures (Fig. 10.1). With the passage of time, the other products are replaced by C_3AH_6 and AH_3 at rates that depend on the temperature and other factors; this process is called conversion. An amorphous or gel phase containing Ca^{2+} has been reported to form simultaneously with the crystalline products, in a proportion relative to the latter that increases with temperature (E5,E6).

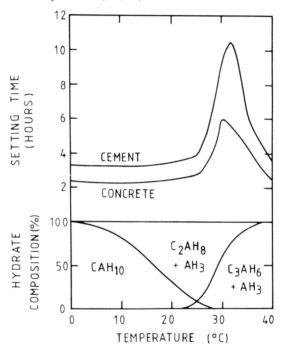

Fig. 10.1 Major hydration products and typical setting times for neat pastes and concretes of Ciment Fondu cured at various temperatures. George (G73).

Fig. 10.2 shows the changes in the solution composition that occur when a typical Ciment Fondu reacts with water at 20°C, together with the metastable solubility curves for phases in the $CaO-Al_2O_3-H_2O$ system. The CaO and Al_2O_3 concentrations increase, at an approximately constant CaO/Al_2O_3 ratio somewhat greater than 1; in the experiment shown, C_{CaO} reached 15 mmol l^{-1} in 3 min and its maximum value of 21 mmol l^{-1} in 1 h.

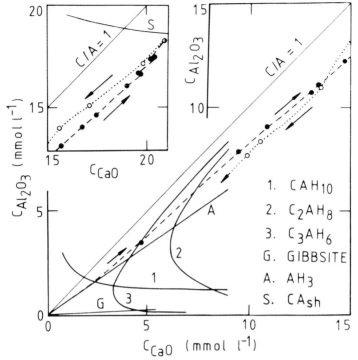

Fig. 10.2 The CaO–Al$_2$O$_3$–H$_2$O system at 20°C, showing calculated solubility curves and typical path (broken and dotted lines) followed by the composition of the solution obtained on reaction of Ciment Fondu with water in a suspension with w/s = 10. AH$_3$ = poorly crystalline hydrous alumina; CA$_{sh}$ = superficially hydroxylated CA (see text). Capmas (C51) and Ménétrier-Sorrentino (M88).

A solid product is thus being precipitated, and has a bulk CaO/Al$_2$O$_3$ ratio lower than 1. It has widely been regarded as alumina gel, but there are several indications that minor amounts of C$_2$AH$_8$ are also precipitated. SEM examination shows the presence of foils, which could well be of poorly crystalline C$_2$AH$_8$ (B121,M88) (Fig. 10.3). With some specimens of Ciment Fondu, the ascending curve shows minor arrests, during which C_{CaO} decreases (C48,M88); two such are shown in Fig. 10.2, near C_{CaO} = 20 mol l^{-1}. These arrests are also manifested by temporary reversals in the generally increasing electrical conductivity (G74,M88). Determinations of bound water show, however, that the total amount of hydrated material is very small until the maxima in C_{CaO} and $C_{Al_2O_3}$ are reached (B121,M88).

The subsequent course of the reaction, essentially as described by Capmas and Ménétrier-Sorrentino (C48), is as follows. An induction period occurs, during which the concentrations remain near their maximum values, the rates of dissolution and precipitation thus being approximately equal. The products nucleate and grow; these processes begin slowly, but accelerate. Ultimately, precipitation is massive, the concentrations drop and, in a paste, setting occurs. In the experiment shown, in Fig. 10.2, the concentrations began to decrease rapidly at about 2 h and the final point was recorded at 5 h.

Fig. 10.3 SEM secondary electron image of Ciment Fondu hydrated for a few minutes in a suspension of w/c = 10 at room temperature, showing foils, probably of C_2AH_8, and amorphous material. Ménétrier-Sorrentino (M88).

Within minutes after mixing, the solution is supersaturated in AH_3, CAH_{10}, C_2AH_8 and C_3AH_6. As noted earlier, AH_3 and C_2AH_8 begin to form while the concentrations are still rising. For a cement in which CA is the only active constituent, CAH_{10} subsequently becomes a more important product below about 15°C (Fig. 10.1). For cements containing $C_{12}A_7$, C_2AH_8 is a major product at lower temperatures.

In the conversion reaction, CAH_{10} and C_2AH_8 redissolve and C_3AH_6 is formed, together with additional AH_3, the structure of which then approximates to that of gibbsite. At 5°C, this process may take many years, but at 50°C or above, the formation of C_3AH_6 and gibbsite is virtually immediate. The only phases in the pure $CaO-Al_2O_3-H_2O$ system stable in contact with aqueous solutions up to well above 100°C are gibbsite, C_3AH_6 and CH.

10.1.5 Thermodynamic calculations

When the anhydrous phases dissolve, the species present in solution in significant concentrations are Ca^{2+}, $Al(OH)_4^-$ and OH^-; $CaOH^+$ may be ignored. Taking C_2AH_8 as an example, the equilibrium with solution is thus represented by the equation

$$Ca_2Al_2(OH)_{10} \cdot 3H_2O \rightleftharpoons 2Ca^{2+} + 2Al(OH)_4^- + 2OH^- + 3H_2O \quad (10.1)$$

The solubility product is equal to $\{Ca^{2+}\}^2 \times \{Al(OH)_4^-\}^2 \times \{OH\}^2$, where curly brackets denote species activities and $[OH^-]$ may be replaced by $2[Ca^{2+}] - [Al(OH)_4^-]$. As the concentrations are low, activity coefficients may be calculated from simplified Debye–Hückel theory. Solubility products may thus be obtained from experimental data (N18,B118,B119,C48) (Table 10.2). The variations in solubility products with temperature may be represented by empirical equations of the form

$$\log_{10} K = a + bT \quad (10.2)$$

where T is in °C (C48), and values of the constants a and b are included in Table 10.2. Standard free energies of formation may also be calculated (N18), but calculations of solubility products from them are numerically unreliable (C48). From the solubility products, again using activity coefficients calculated using simplified Debye–Hückel theory, one may obtain calculated solubility curves (B118,B119,C48), some examples of which are shown in Figs 10.2 and 10.4. It is thus possible to calculate entire curves, such as that for CAH_{10} at 40°C, or portions of curves, such as those for C_2AH_8 and C_3AH_6 at high Al_2O_3 concentrations, that are difficult or impossible to determine by experiment.

Degrees of supersaturation relative to specific hydrates may also be calculated. The solubility curve of CAH_{10} is more markedly dependent on temperature than are those of C_2AH_8 or C_3AH_6. At 5°C, a solution obtained from CA rapidly becomes highly supersaturated in CAH_{10}, but much less so in C_2AH_8 or AH_3, but with increase in temperature this situation changes; thus CAH_{10} is the major product at 5°C, but is undetectable above 30°C (C48). At 50°C, supersaturation is high only in C_3AH_6, which is rapidly formed.

Barret and Bertrandie (B118,B119,B124) explained the maximum concentrations reached in the CA–water reaction and the path by which they were attained. On contact with water, CA was assumed to develop a hydroxylated surface layer, termed CA_{sh}, of composition $Ca[Al(OH)_4]_2$, which dissolves congruently and is continuously regenerated as new areas of surface are exposed. The maximum values reached by the CaO and Al_2O_3 concentrations were considered to represent the solubility of CA_{sh}, from which a solubility product of 6.58×10^{-6} mol³ kg⁻³ was calculated. A solubility curve was thus obtained for CA_{sh}, and is included in Fig. 10.2.

Table 10.2 Equilibrium constants in the CaO–Al_2O_3–CO_2–H_2O system[a]

Reaction (H_2O molecules in solution omitted)			log K	Ref.
1. AH_3 (gibbsite)	$\rightleftharpoons$	$Al^{3+} + 3OH^-$	−33.5	S51
2. AH_3 (amorph.) $+ OH^-$	$\rightleftharpoons$	$Al(OH)_4^-$	+0.2	B118
3. CAH_{10}	$\rightleftharpoons$	$Ca^{2+} + 2Al(OH)_4^-$	−7.6	B118
4. C_2AH_8	$\rightleftharpoons$	$2Ca^{2+} + 2Al(OH)_4^- + 2OH^-$	−13.8	B118
5. C_3AH_6	$\rightleftharpoons$	$3Ca^{2+} + 2Al(OH)_4^- + 4OH^-$	−22.3	B118
6. C_4AH_{19}	$\rightleftharpoons$	$4Ca^{2+} + 2Al(OH)_4^- + 6OH^-$	−25.2	B118
7. $CaCO_3$ (calcite)	$\rightleftharpoons$	$Ca^{2+} + CO_3^{2-}$	−8.4	S51
8. $C_4A\bar{C}H_{11}$	$\rightleftharpoons$	$4Ca^{2+} + 2Al(OH)_4^- + 4OH^- + CO_3^{2-}$	−29.3	B122
9. $Ca^{2+} + OH^-$	$\rightleftharpoons$	$CaOH^+$	+1.3	S51
10. $Al^{3+} + 4OH^-$	$\rightleftharpoons$	$Al(OH)_4^-$	+32.4	B123

Constants a and b in equation 10.2 (C48)[b]

	Reaction				
	2	3	4	5	6
a	+0.05	−9.000	−14.469	−23.168	−27.837
b	+0.0174	+0.0604	+0.0192	+0.0136	+0.1073

[a] 25°C and ionic strength (I) = 0 except for No. 10, which is for $I = 0.1$ and 20°C.
[b] Recalculated to give K as log to base 10.

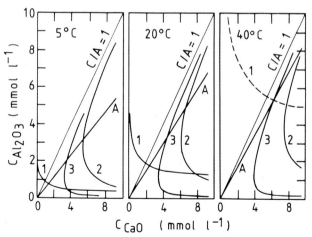

Fig. 10.4 Calculated solubility curves for (A) poorly crystalline AH_3, (1) CAH_{10}, (2) C_2AH_8 and (3) C_3AH_6 in the CaO–Al_2O_3–H_2O system at 5°C, 20°C and 40°C. Capmas (C51).

The path of increasing concentrations was shown to pass along a narrow corridor on the C_{CaO}–$C_{Al_2O_3}$ diagram, enclosing compositions for which neither AH_3 nor C_2AH_8 is rapidly precipitated. These phases were immediately precipitated on passing beyond the low-CaO and high-CaO boundaries of the corridor, respectively. The degree of supersaturation with respect to each of the hydrated phases was calculated for various combinations of C_{CaO} and $C_{Al_2O_3}$, and free energy changes for precipitation were thereby obtained. The path, which was termed a curve of minimum instability, was shown to correspond to a series of compositions for which the conflicting tendencies to precipitate AH_3 and C_2AH_8 were as far as possible minimized. The hydration of $C_{12}A_7$ was considered in a similar way.

10.1.6 Setting times; mixing and placing

The setting times of most calcium aluminate cements pass through a maximum at 25–30°C (Fig. 10.1). The effect has been observed with Ciment Fondu and with white calcium aluminate cements, and thus does not depend on the presence of phases other than CA (B124). The setting time usually shortens between 0°C and about 20°C, but at 28–30°C it can be up to eight times as long as at 20°C. Above 30°C, it again shortens very rapidly. The origin of the maximum is not clearly understood. It has been attributed to difficult nucleation of C_2AH_8 (R45,B125), but may be rather associated with that of CAH_{10} (C51).

The length of the induction period decreases with the C/A ratio in the solution, which is in turn affected by the presence of $C_{12}A_7$ (B118). For ratios up to 1.06, given by pure CA, it can be as long as 6–12 h, but at ratios above 1.20, immediate setting can occur. The mechanism and products of hydration are essentially similar to those for CA, but, as might be expected from the higher C/A ratios in the solution, only a little CAH_{10} is formed, even at 4°C (E6).

Increase in energy or time of mixing markedly accelerates the hydration of calcium aluminate cements, and if excessive can even cause quick setting in the mixer. Once mixing stops, progressive thickening occurs. Samples of a given paste mixed for different times can behave very differently during placing. This factor is often forgotten, and can lead to the addition of more water during placing, with a consequent increase in porosity and decrease in strength. The rheological properties of the fresh pastes (B126,C52) and ζ-potentials of suspensions (Z11) are broadly similar to those found for Portland cements (Section 11.4.4). As with the latter, the water demand can be decreased by adjusting the particle size distribution. Factors affecting choice of w/c ratio are discussed in Section 10.1.8.

10.1.7 Microstructural development

Fig. 10.5 shows portions of fracture surfaces of some pastes of Ciment Fondu examined by secondary electron imaging in the SEM. At 22 h, in pastes hydrated at 20°C, fibrous material is abundant, together with plates of C_2AH_8. The fibrous material is probably partially dehydrated CAH_{10}.

Cottin (C50) earlier studied pastes of Ciment Fondu, largely by the TEM replica method. The microstructures of pastes of low w/c ratios hydrated for 7 days at 12°C were too compact to show more than occasional detail, but with pastes of w/c = 0.7 much detail could be observed in cavities exposed by fracture, which were lined with small crystals of CAH_{10} and sometimes contained plates of C_2AH_8. The crystals of CAH_{10}, which were hexagonal prisms up to 1 µm long and 0.2 µm wide, were present in rounded aggregates. Hydrous alumina was seen in a paste of w/c = 1.0 hydrated at 30°C; it occurred in the lining of a cavity, mainly as rounded masses of randomly oriented platelets.

Halse and Pratt (H57) reported SEM observations on pastes hydrated at various temperatures. In those hydrated at 8°C or 23°C, the main feature was fibrous material that was considered to be hydrous alumina, but which could also have been partly dehydrated CAH_{10}. The hydrating grains of cement were surrounded by shells of hydration products, from which they tended to become separated in a manner similar to that observed with Portland cement pastes (Section 7.4.2) though the authors recognized that this could have been partly due to dehydration. Two-day-old pastes hydrated at 40°C showed spheroidal particles of C_3AH_6 and thin, flaky plates of gibbsite. In pastes mixed with sea water, hydration took place more slowly, but no other effects on microstructural development were observed.

10.1.8 Hardening; effects of conversion

Calcium aluminate cements harden rapidly as soon as the massive precipitation of hydrates begins. This may be attributed to the fact that, unlike those of Portland cement, the major hydration products are crystalline. Relatively high proportions of water are taken up in the hydration reactions, the theoretical w/c ratios needed for complete hydration of CA being 1.14, 0.63 and 0.46 for the formation of CAH_{10}, $C_2AH_8 + AH_3$ and $C_3AH_6 + 2AH_3$, respectively. For this reason, and also because of the rapid heat

Fig. 10.5 SEM secondary electron images of the fracture surface of a paste of Ciment Fondu (w/c = 0.4), hydrated for 22 h at room temperature, showing (A) CAH_{10}, and (B) plates of C_2AH_8. Ménétrier-Sorrentino (M88).

Calcium Aluminate, Expansive and Other Cements 327

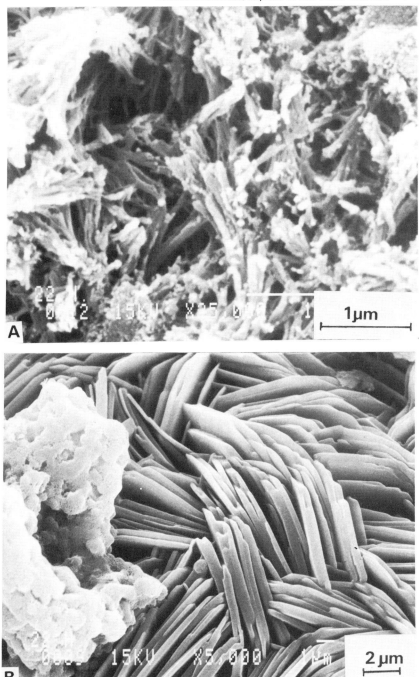

evolution, care is needed to avoid surface drying and even destruction of the concrete through the development of high internal vapour pressure. In blocks of concrete more than 300 mm thick, the temperature rise may be sufficient to cause rapid formation of C_3AH_6 during hardening under normal conditions. Strong bonds are formed with limestone aggregates (F43,B127), apparently due to formation of $C_4A\bar{C}H_{11}$, and with synthetic aggregates high in Al_2O_3, which give high abrasion and corrosion resistance (G76).

The rate of conversion at a given temperature is reduced if the RH falls below saturation. The temperatures during setting and curing are also of major importance; if they are high, subsequent conversion occurs more rapidly, even though the temperature is then low (C47). This is probably due to the presence of nuclei of C_3AH_6. Conversion causes a decrease in the solid volume and thereby increases the porosity of the concrete, which results in at least a temporary decrease in strength. Calculation based on the densities of the hydrated phases (CAH_{10}, 1730 kg m^{-3}; C_2AH_8, 1950 kg m^{-3}; C_3AH_6, 2530 kg m^{-3}; AH_3, 2440 kg m^{-3}) shows that, for pure CA, the total volume of the hydration products decreases to 47% of its original value if the initial product is CAH_{10} and to 75% if it is a mixture of C_2AH_8 and AH_3. The importance of this effect depends on the w/c ratio. At a low w/c ratio, the space available is insufficient to allow all the CA to hydrate to CAH_{10} or to C_2AH_8 and AH_3. When conversion occurs, much water is released, so that even if the pores contain no water, further hydration can occur. In addition to the aluminate hydrates, C_2ASH_8 may be formed (M91). At a high w/c ratio, more of the cement can hydrate initially, and the effects of conversion on the porosity and strength are correspondingly greater; carbonation also occurs more readily. These effects are illustrated by the curves in Fig. 10.6.

The strength after conversion is affected by the rate at which that process occurs, probably because of the effect on C_3AH_6 morphology (M91). Strength may also be affected by crystallization of AH_3, which could reduce both its pore-filling capacity and its efficacy as a binder. The size of the C_3AH_6 crystals increases with the temperature at which they are formed; this also may decrease strength. It has sometimes been said that C_3AH_6 is an intrinsically weak binder, but this is not correct. As with relatively dense, crystalline substances in general (F36), it gives lower strengths than less dense substances at high porosities, but higher ones at low porosities. This may be seen from the crossing of the curves in Fig. 10.6. Hot pressed pastes of calcium aluminate cements, in which the binder is C_3AH_6 and the porosity has been reduced to about 4%, show strengths of up to 480 MPa (Section 11.9.1).

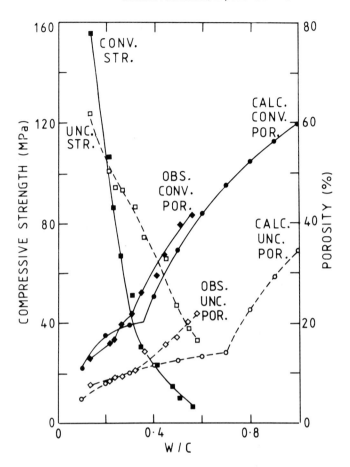

Fig. 10.6 Compressive strengths and observed and calculated porosities for pastes of Ciment Fondu of varying w/c ratios. The unconverted paste, which contained CAH_{10}, was cured for 7 days at 10°C. The converted paste, which contained C_3AH_6 and AH_3, was cured for 7 days at 70°C. After George (G73).

Conversion and its practical implications have been widely studied but not always well understood. Thus, concretes made with inadequate cement contents and excessive w/c ratios have failed in structural applications. An extensive investigation in the UK into calcium aluminate concrete in existing buildings (B128) showed that a high level of conversion was generally reached in a few years, causing a very variable loss of strength, but that this loss was in the majority of cases insufficient to endanger the structure. No adverse effects of corrosion on steel tendons removed from prestressed concrete beams in normal buildings were detected. The use of calcium aluminate cements in structural work is nevertheless not deemed to satisfy the current (1989) UK Building Regulations. In France, prohibitive restrictions were for a time placed on its use in the public sector of the construction industry, but in 1968, a new set of rules was issued after lengthy studies. Current French regulations (M92) specify that the total w/c ratio (i.e. that including any water sorbed by the aggregate) must not exceed 0.4 and that the cement content must be at least 400 kg m^{-3}.

The evidence supports the view (B128,G76) that design should be based on the assumption that conversion will occur, tests using accelerated curing being carried out to determine the converted strength. Fig. 10.7 shows the results of some long-term tests; for design purposes, a rapid increase levelling off at the minimum converted strength should be assumed.

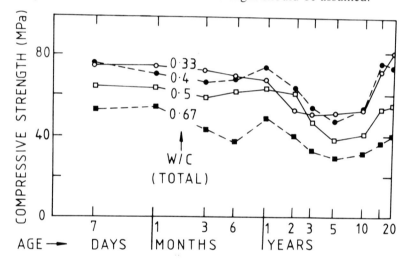

Fig. 10.7 Long-term compressive strengths of Ciment Fondu concretes cured for 24 h in moist air at 18°C and then stored outdoors in southern France. After George (G76).

Conversion is of only minor importance for the use of calcium aluminate cements in refractory applications and in mixes with Portland cement such as those described in Section 10.1.10.

10.1.9 Chemical admixtures

Many inorganic salts affect the setting times of calcium aluminate cements and are effective in concentrations of 0.5% (L6,C53,R45). The effects on setting times differ greatly both in direction and in relative magnitudes from those observed with Portland cements (Section 11.5.1). The results of different investigators show apparent discrepancies, and the effects probably depend greatly on concentration, which may even reverse their direction, but there is general agreement that lithium salts and alkali or calcium hydroxides are strong accelerators and that many common inorganic salts retard. As with Portland cement, hydroxylic organic compounds, such as sugars or citric, tartaric or gluconic acids, are powerful retarders.

Rodger and Double (R45) studied the effects of several admixtures, particularly LiCl and citric acid, by XRD, conduction calorimetry and solution analyses. The samples were slurries of a Ciment Fondu hydrated at 22°C and in the absence of admixtures gave CAH_{10} as the main product. In the mixes with LiCl, C_2AH_8 was the main product and a little $Li_2O \cdot 2Al_2O_3 \cdot 11H_2O$ was detected. The authors considered that this compound acted as a nucleation substrate for C_2AH_8. The very low concentrations in which the organic compounds are effective suggests that, as with Portland cement, they act by poisoning the nuclei of the hydration products, but the authors found this hypothesis incompatible with the observed solution compositions.

An experimental study of the $CaO-Al_2O_3-H_2O$ system in the presence of NaOH showed that at a concentration of 1 mmol l^{-1} the latter had negligible effect on the equilibria, but that at 10 mmol l^{-1} the solubility curves were displaced to lower CaO and higher Al_2O_3 concentrations. At 100 mmol l^{-1}, the solutions were very low in CaO, and precipitation did not occur within 28 days (P46). The effect of OH^- ion on the solubility curves can also be calculated by an extension of the procedure mentioned in Section 10.1.5 (B119). The results showed large shifts in the curves at an NaOH concentration of 40 mmol l^{-1}, the minimum in CaO concentration on the C_2AH_8 curve at 21°C being shifted from $C_{CaO} = 6.8 \text{ mmol kg}^{-1}$, $C_{Al_2O_3} = 2.3 \text{ mmol kg}^{-1}$ to $C_{CaO} = 1.1 \text{ mmol kg}^{-1}$, $C_{Al_2O_3} \simeq 13 \text{ mmol kg}^{-1}$. An experimental study on the effects of adding sodium phosphates of various types has been reported (B129).

Other types of admixtures used with calcium aluminate cements include water reducers and superplasticizers (Section 11.4), which also act as retarders, and thickening agents, such as carboxymethylcellulose. Complex formulations may be used for special purposes; for example, a ready-mix mortar for high-performance road repair might contain 55% Ciment Fondu and 45% sand, with glass fibre, aluminium powder, Li_2CO_3, sodium gluconate and methyl ethyl cellulose (M93).

10.1.10 Mixtures with calcite, slag, gypsum or Portland cement

Cussino and Negro (C54) reported that loss of strength in calcium aluminate cement mortars could be prevented by using calcareous aggregates or by adding finely ground calcite to the cement. They showed that $C_4A\bar{C}H_{11}$ was formed and concluded that this limited or prevented conversion. Fentiman (F44) confirmed the formation of $C_4A\bar{C}H_{11}$ but found that at 25°C or below CAH_{10} was still the main product and that conversion was delayed but not prevented. $C_4A\bar{C}H_{11}$ seems to form at the expense of C_2AH_8 and C_3AH_6 at 25–60°C but to be unstable at or above 60°C (M94,F44). Strong concentration gradients and other local inhomogeneities (B130) may explain some of these apparently contradictory results. Equilibria involving $C_4A\bar{C}H_{11}$ have also been approached theoretically (B122,B131,S100). Barret et al. (B122,B131) concluded that $CaCO_3$ would dissolve giving $C_4A\bar{C}H_{11}$ only at partial CO_2 pressures above a certain value, but that equilibrium with atmospheric CO_2 would not be reached within the pores of a concrete.

Majumdar et al. (M95) studied concretes containing equal weights of Ciment Fondu and ggbfs. On water curing at either 20°C or 38°C, the compressive strengths continuously increased over a period of 1 year, whereas those of control samples made without the slag showed the normal decreases associated with conversion. Parallel studies on pastes showed that the superior behaviour of the mixes with slag were associated with formation of strätlingite.

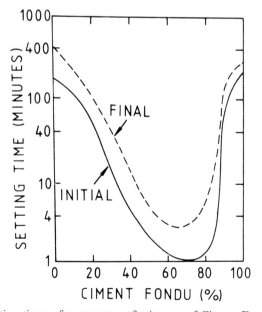

Fig. 10.8 Setting times of neat pastes of mixtures of Ciment Fondu and Portland cement. These are only general indications, as the times depend on Portland cement composition, mixing time, temperature and other factors. After Robson (R47).

Various combinations of calcium aluminate cements with calcium sulphate, Portland cement and in some cases calcium hydroxide are commonly used to make cements that set and harden rapidly. Some such mixtures also show expansive behaviour (Section 10.2.1). A simple, rapidly setting mix might contain 15–20% ordinary Portland cement, 10–30% Ciment Fondu, 10–25% CH and 40–60% sand. More complex formulations, containing various admixtures, are used to produce self-levelling mortars. The setting times of all such mixtures depend greatly on the relative proportions, the characteristics of the individual cements and the mixing time and temperature; only laboratory tests can predict the precise behaviour of a specific mix. Fig. 10.8 gives a general indication of the behaviour of neat pastes made from calcium aluminate and Portland cements. The rapid setting of the mixes high in Portland cement is possibly caused by preferential reaction of the gypsum with the calcium aluminate cement, and that of those high in calcium aluminate cement by the increase in pH due to the presence of the Portland cement. In addition to C–S–H and CH, the hydration products can include C_4AH_{13} and strätlingite (C_2ASH_8). Mixes of CA with sufficiently high proportions of calcium sulphate yield ettringite (C49).

10.1.11 Reactions of calcium aluminate concrete with external agents

Concrete made with calcium aluminate cement at a properly low w/c ratio is highly resistant to sulphate solutions, sea water or dilute acid solutions with pH > 4, including natural waters in which CO_2 is the only significant solute. Resistance may extend to pH $\simeq$ 3 if the salt formed is of sufficiently low solubility. Midgley (M96) showed that, for fully converted material exposed to a sulphate ground water for 18 years, penetration with formation of a substituted ettringite was limited to a depth of 5 mm. These properties are consistent with Lea's (L6) view that the resistance is due to the formation of a protective coating of alumina gel, coupled with the absence of CH. No fundamental studies, e.g. on microstructural effects, appear to have been reported.

In contrast, resistance to attack by alkali hydroxide solutions is low, probably because the hydrous alumina is dissolved. Attack is accentuated if CO_2 is also present, as can occur if the calcium aluminate concrete is in contact with Portland cement concrete and exposed to atmospheric CO_2 (R48). For CAH_{10}, the following reactions occur:

$$CaAl_2(OH)_8 \cdot 6H_2O + CO_3^{2-} \rightarrow CaCO_3 + 2Al(OH)_4^- + 6H_2O \quad (10.3)$$

$$2Al(OH)_4^- + CO_2 \rightarrow 2Al(OH)_3 + CO_3^{2-} + H_2O \quad (10.4)$$

The OH^- ion thus acts as a carrier for atmospheric CO_2 and is effectively regenerated; the CAH_{10} is decomposed giving $CaCO_3$ and hydrous alumina. This effect is known as alkaline hydrolysis. A similar process occurs with C_3AH_6 or calcium aluminate concretes in which conversion has occurred (B132).

In contrast to the above, carbonation in the absence of alkali increases the strength of calcium aluminate cement concrete (R49,P47). In sufficiently porous material in which conversion has occurred, some of the C_3AH_6 reacts to give calcite and hydrous alumina (P45), but in dense concrete the extent of penetration by CO_2 is small, probably because the pores do not contain the water necessary to promote the reaction (R49) and because they become blocked by $CaCO_3$ and alumina gel (B132). These observations support the general conclusion that the high resistance of dense calcium aluminate concretes to various forms of attack is at least in part due to effects that impede or prevent the ingress of the attacking substance. In addition to the blocking of pores by reaction products, these effects could include the rapid removal of incoming water by continued hydration.

10.1.12 Refractory castables

Refractory castables can be categorized as cement castables (15–25% calcium aluminate cement), low cement castables (5–10% cement) and ultra low cement castables having proprietary formulations that include under 5% of cement, microsilica and other admixtures. Depending on the service temperature and application, aggregates include exfoliated vermiculite, chamotte, sintered bauxite, brown or white fused alumina, chromite, etc. A very simple formula for a wall more than 50 mm thick and for use at a hot face temperature up to 1650°C might contain 14% of '70% Al_2O_3 cement', with 86% of a suitably graded aggregate of brown fused alumina and 0.2% of trisodium citrate on the weight of cement; the w/c ratio must be determined by an empirical test, as for a rapidly hardening mix, but would be $\leqslant 0.4$.

When the concrete is heated, the hydration products undergo conversion and dehydration. The strength decreases to a minimum value at a temperature usually between 900°C and 1100°C, though if the cement content is sufficiently high it is still enough for ordinary applications. At higher temperatures, sintering reactions occur between the cement and aggregate, and the ceramic bonding that results causes the strength and resistance to abrasion to increase. The compounds formed depend on the compositions of the cement and aggregate, but typically include such high melting point phases as CA_2, CA_6 and spinel (MA).

10.2 Expansive cements

10.2.1 General

A normal Portland cement concrete expands during moist curing, but this effect is outweighed by the drying shrinkage that occurs on subsequent exposure to an atmosphere of normal humidity. If the concrete is restrained, the resulting tensile stress may be sufficient to crack it. In an expansive cement, the expansion is increased, so as to balance or outweigh the subsequent contraction. Expansive cements, which were first described by Lossier (L48), may be categorized as shrinkage-compensated or self-stressing cements.

In a shrinkage-compensated cement, the object is to balance the drying shrinkage to prevent cracking. In practice, the expansion is restrained by reinforcement and a small compressive stress of 0.2–0.7 MPa is created in the concrete, which should preferably remain under slight compression after the normal drying shrinkage has occurred. Practical uses of shrinkage-compensated cements have included such structures as multistorey car parks or water storage tanks, in which it is desirable or essential to eliminate water leaks.

In a self-stressing cement, a larger expansion is produced, with the object of achieving greater stresses in the reinforcement and the concrete to permit the production of thin, strong articles or structures, as in mechanical prestressing. Estimates of the maximum stress that can be safely achieved have ranged from 3.5 MPa (M97) to 8 MPa (W30). Because the concrete expands triaxially, careful design and choice of curing conditions, including if necessary triaxial restraint, are essential. Self-stressing cements have been used both in the manufacture of precast pipes and other units and for in situ applications.

In any expansive cement, the amount of expansion and the period during which the expansive reaction occurs are critical. The reaction must take place after the concrete has developed some strength, but before exposure to a reduced RH places it in tension, which could cause irreversible damage.

10.2.2 Types of expansive cement

Probably a majority of practical expansive cements have depended on the modification of a Portland cement in such a way as to increase the formation of ettringite. Single expansive cement clinkers can be made, but it has been more usual to produce admixtures that are blended or interground with a normal Portland cement or clinker. Blending has the advantage that the fineness of the expansive admixture can be optimized, usually by grinding it more coarsely so as to delay the expansive reaction until the cement has developed the necessary strength. It also allows a single admixture to be used in differing proportions to make both shrinkage-compensated and self-stressing cements.

In a widely used US terminology, expansive cements based on Portland cement and calcium sulphate are categorized as Types K, M or S, according to the source of the additional Al_2O_3 that is required. Type K cements, developed by Klein and co-workers (K49), contain $C_4A_3\bar{S}$ as an essential ingredient. They are usually produced by intergrinding or blending an ordinary Portland cement clinker with additional gypsum or anhydrite and an expansive clinker containing $C_4A_3\bar{S}$. The composition of the expansive clinker can be varied considerably to utilize the raw materials most economically available; the clinkers made in the USA typically contain 8–50% of $C_4A_3\bar{S}$ with some anhydrite and usually some free lime (M97). The proportions of expansive clinker and of gypsum or anhydrite or both that are added to the Portland cement depend on the composition of the expansive clinker and the degree of expansion required. Mehta and Polivka (M97) found that Type K shrinkage-compensated cements made with Portland cement clinkers of low or moderate potential C_3A content and containing 6% or more of SO_3 resisted chemical attack by sulphates, but that Types M and S cement did not. Table 10.3 includes some analyses of Type K cements.

Table 10.3 Compositions of Type K expansive cements produced in the USA (M97)

Application	MgO	Al_2O_3	SiO_2	SO_3	CaO	Fe_2O_3	Loss
Shrinkage compensation	0.7	4.8	21.6	6.2	62.4	2.8	1.1
Shrinkage compensation	3.4	5.7	18.8	6.5	61.4	1.9	1.5
Self-stressing	2.7	8.4	14.6	12.4	57.6	1.6	1.9

Type M cements contain Portland cement, a calcium aluminate cement and additional calcium sulphate. They have been extensively studied in the USSR. Mikhailov (M98) described a self-stressing cement composed of Portland cement, calcium aluminate cement and gypsum typically in the proportions 66 : 20 : 14 and used to produce precast units. An initial cure in air followed by hot water allowed strength to be developed without formation of ettringite, which occurred during a subsequent cure in cold water. Mixtures using alumina slag as the source of the additional Al_2O_3 have also been used, and require only curing in air followed by cold water (M97,M99). Mixtures of calcium aluminate cement with gypsum (X2) or with gypsum and C_4AH_{13}, the latter made from calcium aluminate cement and lime (M98), have also been employed.

Type S cements are Portland cements high in C_3A and with suitable contents of calcium sulphate; they have found little favour as they are too difficult to control. The Al_2O_3 has also been supplied in forms other than those mentioned above. Impure alunite [$KAl_3(SO_4)_2(OH)_6$], which occurs as a natural rock, has been used, either after calcination (V6) or uncalcined (W30). In the latter case it was mixed with Portland cement clinker, anhydrite and pfa or slag and was found to dissolve relatively slowly, thus suitably delaying the expansive reaction.

The possibilities of using free lime or periclase as expansive agents have also been investigated. Kawano *et al.* (K50) and Long (L49) described clinkers in which substantial proportions of free lime are present as inclusions in alite. In both cases, the raw mix included anhydrite. Long investigated and described the processes by which this microstructure was formed.

10.2.3 Mechanism of expansion in Type K cements

In a Type K cement, the formation of ettringite from $C_4A_3\bar{S}$ may be represented by the equation

$$C_4A_3\bar{S} + 8C\bar{S}H_2 + 6CH + 74H_2O \rightarrow 3C_6A\bar{S}_3H_{32} \qquad (10.5)$$

Consideration of the relevant densities ($C_4A_3\bar{S}$, $2610\,\text{kg m}^{-3}$; $C\bar{S}H_2$, $2310\,\text{kg m}^{-3}$; CH, $2240\,\text{kg m}^{-3}$; $C_6A\bar{S}_3H_{32}$, $1780\,\text{kg m}^{-3}$) shows that the solid volume is approximately doubled. Water could be drawn in from the outside, and its volume is irrelevant. As the case of the normal hydration of Portland cement shows, an increase in solid volume is not in itself a sufficient cause of expansion. If sufficient pore space is available in which the additional product can be deposited and if the deposition occurs in that pore space without disturbing the surrounding matrix, no expansion will occur. Conversely, a small increase in solid volume may produce a large expansion if by forcing parts of the surrounding microstructure apart it causes the porosity to increase. There is thus no simple relation between the change in solid volume and the expansion of a porous, or potentially porous, material.

Any attempt to understand the expansive process must thus consider the manner in which the deposition of the products affects the microstructure. The production of ettringite can cause expansion, but does not necessarily do so; in a supersulphated cement (Section 9.2.9), much ettringite is formed, but there is no marked expansion. Comparable amounts of ettringite are reported to be formed when ordinary or sulphate-resisting Portland cements are placed in sodium sulphate solutions (H59), but the expansion is much greater with the ordinary Portland cement. For expansive cements, Nakamura et al. (N19), Mehta (M100) and Bentur and Ish-Shalom (B133,I9, I10,B134) all recognized that the mode of formation and morphology of the ettringite must be important.

In suspensions of $C_4A_3\bar{S}$, $C\bar{S}$ and CH at high CaO concentrations, the ettringite is formed as a fine-grained coating on the $C_4A_3\bar{S}$ particles, but in the absence of CH it forms long, thin needles (N19,O23,M100). In contrast, a study in which two different Portland clinkers were mixed with gypsum and an expansive clinker low in free lime showed that increase in the content of free lime in the cement increased the rate of ettringite formation but had little effect on ettringite morphology (K51). In pastes containing $C_4A_3\bar{S}$, expansion occurs if CH is present, but not if it is absent, even though ettringite is still formed (N19,O23). Expansion also requires an adequate supply of water; sealed curing either limits (N19) or prevents (M100) it, even though the w/c ratio is sufficiently high to allow a substantial quantity of ettringite to be formed.

Bentur and Ish-Shalom (B133,I9,I10,B134) studied mixtures of $C_4A_3\bar{S}$, $C\bar{S}$ and CH in pastes, alone or mixed with C_3S. They found that expansion does not begin until a certain critical degree of reaction, which was about 50% at 20°C, had been reached. Using SEM, light microscopy and mercury intrusion porosimetry, they confirmed that the ettringite was formed as porous coatings around the grains of $C_4A_3\bar{S}$, which were thus effectively caused to expand; expansion began when the expanding particles came into contact and exerted pressure on each other. In the absence of restraint, these particles could be treated as spheres, but under restraint they expanded in the directions in which the restraining forces were least. Essentially the same mechanism was described by Ogawa and Roy (O16,O24), who found that the ettringite was formed initially as very small, unoriented crystals. After the degree of reaction (defined by the quantity of ettringite formed) had reached about 50%, the ettringite developed as radial growths of longer crystals. Contact between the ettringite layers surrounding the $C_4A_3\bar{S}$ particles, and the associated expansion, began only after this stage had been reached. The change in ettringite morphology was considered to be an important first step in causing expansion.

Cohen (C55) reviewed theories of the expansion mechanism. Probably a majority of workers (e.g. Refs I9,I10,O16,O24,B133) have attributed expansion to forces exerted by the growth of the ettringite crystals. Of other theories, the most significant is that proposed by Mehta (M100), who attributed it to imbibition of water by the gelatinous layer of colloidal ettringite. On this hypothesis, expansion does not occur with the larger crystals formed at low CaO concentrations because these do not form aggregates that imbibe water. While this mechanism may contribute to the expansion, Ogawa and Roy's observations suggest that it is not the sole or even the major cause.

10.3 Other cements

10.3.1 Very rapidly hardening cements

Various methods have been used to obtain cements that set and harden rapidly. They include the use of Portland cement with admixtures and of mixtures containing both Portland and calcium aluminate cements, described in Sections 11.5 and 10.1.10, respectively. Another approach has been the manufacture of clinkers containing either $C_{11}A_7 \cdot CaF_2$ or $C_4A_3\bar{S}$, both of which hydrate rapidly under appropriate conditions with the formation of ettringite.

Regulated-set cement and jet cement are modified Portland cements in which the normal aluminate phase is replaced by $C_{11}A_7 \cdot CaF_2$ through the use of a raw mix containing CaF_2. Uchikawa and Tsukiyama (U21) gave chemical (Table 10.4) and phase compositions of two jet cements. Both contained approximately 60% of alite, 20% of $C_{11}A_7 \cdot CaF_2$, 1% of belite and 5% of ferrite. Admixtures are required to control the rate of reaction of the $C_{11}A_7 \cdot CaF_2$ and the nature of the products. One of the cements included a proprietary retarder based on citric acid, together with 2% of $CaCO_3$. The other contained 2.5% of hemihydrate. In each case, Na_2SO_4 (1%) and anhydrite were also present. The specific surface areas were around 550 m² kg^{-1}.

Table 10.4 Chemical and phase compositions of rapidly hardening materials containing $C_{11}A_7 \cdot CaF_2$ or $C_4A_3\bar{S}$

Material	Na_2O	MgO	Al_2O_3	SiO_2	SO_3	K_2O	CaO	Fe_2O_3	F	Loss	Ref.
Jet cement	0.6	0.8	11.0	13.9	11.1	0.4	58.9	1.7	1.0	0.2	U21
High strength (clinker)	–	–	28.1	11.2	7.1	–	52.0	1.6	–	–	S101

Uchikawa and co-workers (U21–U23) studied the hydration reactions of $C_{11}A_7 \cdot CaF_2$ and jet cements, including the effects of various admixtures. A wide variety of methods was used. For the two cements mentioned above, hydrated in pastes at 20°C, the main product at 3 h was ettringite. At this stage, 40–45% of the $C_{11}A_7 \cdot CaF_2$ had reacted. SEM showed needles of ettringite, 0.5–1.0 µm thick, growing from the particles of clinker and anhydrite to form a closely knit, interlocking structure. Comparable mortars of the same age had compressive strengths of around 10 MPa. The quantity of ettringite formed was such that a high proportion of the anhydrite must have reacted. Subsequently, the content of ettringite tended to decrease slightly and some monosulphate was formed; the ettringite recrystallized, giving larger crystals. The alite hydrated to give C–S–H; calculations based on assumed reaction stoichiometry indicated that the reaction was much faster than in a Portland cement, with about 65% reacting in 1 day. The compressive and bending strengths increased rather slowly between 3 h and 24 h, but thereafter more quickly, and by 7 days the former were 32–44 MPa. The type of retarder (citric acid or hemihydrate) considerably affected the hydration reactions and the pattern of strength development. The fluoride ion was believed to enter an aluminium hydroxide fluoride.

Cements that set and harden rapidly can also be made in which the principal constituents are $C_4A_3\bar{S}$ and belite. Table 10.4 includes the chemical composition of a clinker described by Sudoh *et al.* (S101). The clinker, made from limestone, byproduct gypsum, aluminous materials and clay, was burned at 1300°C. The presence of a limited amount of free lime was found to be necessary to achieve rapid strength development, and conditions such that all the SO_3 was bound as $C_4A_3\bar{S}$ were required to avoid atmospheric pollution. The phase composition was given as: $C_4A_3\bar{S}$, 54%; C_2S, 32%; C_4AF, 5%; free lime, 9%. The clinker was interground with byproduct anhydrite to a specific surface area of 400 $m^2\ kg^{-1}$; the optimum molar ratio of total SO_3 to Al_2O_3 was found to be 1.3–1.9 (20–25% SO_3). XRD and SEM of 1-day-old pastes showed the main product to be ettringite, the needles of which were 1–2 μm long and formed an interlocking network. A mortar made with the cement containing 20% SO_3 developed compressive strengths of 27 MPa at 3 h, 49 MPa at 1 day and 59 MPa at 28 days. The strength development at later ages was attributed to hydration of the C_2S.

For supporting and sealing the walls of roadways in coal mines, a material is required that can be transported as a slurry which remains stable for long periods while stationary in a pipeline but which when placed develops a modest compressive strength within a few hours. This can be achieved by employing two slurries that react and harden on being mixed. In one such system, one slurry contains a clinker containing $C_4A_3\bar{S}$ and the other contains anhydrite, CH, bentonite and alkali metal salts (L50). The $C_4A_3\bar{S}$ clinker, which can also be used as an expansive admixture with Portland cement, hydrates only slowly in the absence of CH. In the second slurry, the bentonite keeps the other solids in suspension and the alkali metal salts act as accelerators.

In most of the cements described in this section, ettringite is an important hydration product, but, as in supersulphated cements, its formation does not cause significant expansion. It is presumably formed either at a stage when the matrix of other phases is readily deformable, or in crystals which grow in a manner that allows them to be accommodated in an existing pore structure without causing disruption.

10.3.2 Low-energy cements

Environmental pollution, including the emission of CO_2, and the depletion of resources of fossil fuels and raw materials, are among the most serious problems confronting humanity today. As Boldyrev (B135) and others have stressed, the cement industry is a major user of both energy and raw materials, and can make very considerable contributions to the amelioration of these problems, especially through increased energy efficiency and by the use of industrial wastes in place of natural raw materials. Major improvements in the energy efficiency of Portland cement production have already been made, but more should be possible and the problems of raw materials and of the utilization of wastes produced by other industries remain.

Locher (L51) reviewed ways of decreasing the energy consumption in cement manufacture to values substantially below those characteristic of modern Portland cement production. They included the use of fluxes to lower the burning temperature, the production of cements containing more reactive forms of belite or based on belite together with highly reactive phases, replacement of alite by the chloride-containing phase, alinite, and the use of composite cements.

The burning temperature for production of Portland cement clinker can be decreased by about $150°C$ through the use of fluxes, but opinions have differed as to the energy saving thereby obtainable. Klemm and Skalny (K52), who reviewed the subject, estimated it at $630\,kJ\,kg^{-1}$. Christensen and Johansen (C56) considered that this figure, while possibly realistic for an inefficient, wet process kiln, was unlikely to be so for a modern, precalciner–preheater kiln, in which heat recovery is efficient. They considered a value of $105\,kJ\,kg^{-1}$ more realistic.

The major component of the enthalpy change in clinker formation arises from the decomposition of calcite, and a large saving could be achieved if the proportion of belite in the clinker could be increased. This is only of practical value if the belite can be rendered more reactive or if its low reactivity is compensated by including also some highly reactive phase, such as $C_4A_3\bar{S}$. Reactive forms of belite have been obtained in the laboratory by low-temperature methods (R50,V7) but it is not clear how the latter could be effectively utilized in practice. Rapid cooling of a clinker (e.g. at 1000 deg C min^{-1}) increases the reactivity of belite, at least in part, by preserving an α'

polymorph (G77,S102) and may offer a more practical approach. It is apparently difficult to obtain adequate early strength, but clinkers containing around 48% of α'-C_2S and giving compressive strengths of up to about 20 MPa at 3 days have been made on a pilot plant scale (S102). The clinkers were quenched in water, and in calculating the heat balance it would be necessary to take into account the extent to which the heat contained in them is recoverable.

The properties of one of the rapid-hardening cements described in the previous section show that good strength development can be obtained from cements in which the major constituents are belite and $C_4A_3\bar{S}$. Cements of a similar nature but with much higher contents of belite have been described. An alumina–belite cement (Z12) was obtained by burning at 1300°C of a mix intermediate in composition between those of Portland and calcium aluminate cements, with some gypsum. It contained 62–64% of belite, with CA, $C_{12}A_7$, $C_4A_3\bar{S}$, and $C_5S_2\bar{S}$, and was claimed to give good strength development and resistance to sulphate attack. Porsal cement (V8) is essentially similar and was made using CaF_2 as a mineralizer. Cements high in Fe_2O_3 have been described, containing, for example, 30% C_2S, 30% ferrite, 20% $C_4A_3\bar{S}$ and 20% $C\bar{S}$, and were obtained by burning suitable mixtures at 1200°C (M101). The ferrite formed at this temperature is likely to be reactive, and the clinker is probably easy to grind.

Alinite cements are discussed in the next section. Some other approaches and aspects may be considered. The partial substitution for limestone of raw materials that do not require decarbonation was noted in Section 3.2.1. The compositions of most of the available materials, such as slags, that might be thus used limit the extent to which this is possible in the production of a normal Portland cement clinker, but a wider range or higher content of waste materials might be usable in making other types of clinker. Mehta (M101) noted the possibility of using byproduct gypsum, red mud (from aluminium extraction), slag and pfa in producing high-iron cements such as those mentioned above. Some limestones unsuitable for making a normal Portland cement clinker could be used to make clinkers higher in belite. Burning at a reduced temperature might be expected to decrease the formation of NO_x.

In appraising the energy savings obtainable with any new type of cement, it is essential to consider not only fuel consumption, but also the total energy requirements, including, for example, those of winning and transporting raw materials and of grinding raw materials and clinker. The comparison should also be with modern, dry process plants. Locher (L51) emphasized the further need to consider the properties of the product, including not only the 7- or 28-day strength of the concrete, but also the durability that will be obtained given the curing conditions likely to exist in practice. If the early strength is low, the surface regions are liable to dry out before the cement in them has hydrated sufficiently, resulting in high porosity and poor durability. He also concluded that, with the low-energy cements so far investigated, the energy savings are probably no greater than can be achieved by use of composite cements containing slag, pfa, natural pozzolanas or fillers, the behaviour of which is now well understood.

10.3.3 Alinite cements

A cement in which the alite of Portland cement is replaced by alinite, a structurally related phase containing essential Mg^{2+}, Al^{3+} and Cl^-, was developed in the USSR by Nudelman and others and patented in 1977. An otherwise essentially normal raw mix is modified by adding $CaCl_2$, which allows the maximum temperature in the kiln to be reduced to 1000–1100°C. Considerable energy savings are claimed, due to lower heat losses, a lower enthalpy of reaction and easier grinding of the clinker. Bikbaou (B136) gave the phase composition as alinite, 60–80%, belite, 10–30%, $C_{11}A_7 \cdot CaCl_2$, 5–10%, ferrite, 2–10%, and discussed the process of clinker formation. $Ca_2Cl_2(CO_3)$ and spurrite were reported to form as intermediates and the alinite crystallizes from a $CaCl_2$ melt. The process of formation has been further discussed (B137,N20,A24). Locher (L51) gave a survey of more general information, based partly on the patent data. The preparation of the pure compound has been described (M102,V9).

The crystal structure of alinite has been determined (I11,V9). The constitutional formula was found to be $Ca_{9.9}Mg_{0.8}\square_{0.3}(SiO_4)_{3.4}(AlO_4)_{0.6}O_{1.9}Cl$, where $\square$ denotes a vacancy; limited variations in the Mg/Ca and Al/Si ratios are possible (V9). Alinite is tetragonal, space group $I\bar{4}2m$, with $a = 1.0451$ nm, $c = 0.8582$ nm, $Z = 2$, $D_x \simeq 3010$ kg m^{-3} (V9), and is isostructural with the silicate sulphide mineral, jasmundite, $Ca_{11}(SiO_4)_4O_2S$ (D43). A Br-analogue has been prepared (K53).

Table 10.5 Compositions of alinite clinkers (A24)

MgO	2.0–2.2	CaO	62.5–63.9	Cl (free)	0.04–0.12
Al_2O_3	4.4–4.6	Fe_2O_3	3.6–4.8	Ignition loss	0.7–1.0
SiO_2	22.7–23.3	Cl (total)	1.0–2.9	Free lime	<0.5

Agarwal et al. (A24) examined two alinite clinkers and further studied the process of their formation. The compositional range (Table 10.5) shows that, compared with typical Portland clinkers, they are of low LSF and low AR. The microstructures resembled those of typical Portland clinkers. XRD confirmed the presence of $C_{11}A_7 \cdot CaCl_2$ and showed that a calcium chloride orthosilicate was also present. On heating the clinker to 1400°C, the alinite decomposed, but much of the chlorine was retained. Von Lampe *et al.* (V9) also showed that alinite decomposes slowly at 800°C and rapidly at 1420°C.

The calcium chloride silicate most likely to be formed, in addition to alinite, is $Ca_3SiO_4Cl_2$. A study of its hydration behaviour (K54) showed it to be relatively highly reactive, C-S-H and CH being detectable by XRD in little over 1 h. $Ca_2Cl_2(OH)_2 \cdot H_2O$ was also formed at w/s = 1, though not at w/s = 10, and might thus be expected to be a product in a paste.

The hydration reactions of alinite cements do not appear to have been reported, but it may be surmised that they are similar to those of Portland cements, with the probable addition to the products of $C_3A \cdot CaCl_2 \cdot 10H_2O$ and possibly also of $Ca_2Cl_2(OH)_2 \cdot H_2O$. The Cl^- is, however, not wholly and permanently combined in these or other hydration products. In tests on reinforced concrete, serious corrosion had occurred within 5 years, showing the cement to be unsuitable for such use (L51).

11

Admixtures and special uses of cements

11.1 Introduction

Admixtures for concrete are defined as materials other than hydraulic cements, water or aggregates that are added immediately before or during mixing. Materials such as grinding aids that are added to cement during manufacture are called additives. The most important admixtures are ones added to accelerate or retard setting or hardening, to decrease the quantity of water needed to obtain a given degree of workability, or to entrain air in order to increase the resistance of the concrete to damage from freezing. Mineral additions (Chapter 10) are not included here in either category.

Many materials have been used as admixtures and the potential of many more has been examined. The literature, including the patent literature and other descriptions of proprietary materials, is voluminous; a useful review published as long ago as 1980 includes over 600 references (M103). We shall only attempt to describe the effects of the most important classes of admixtures in general terms and to discuss the chemistry underlying those effects at a fundamental level. Grinding aids are similarly considered.

The rest of the chapter deals with the hydration chemistry of Portland and composite cements at temperatures outside the range 15–25°C, including that of autoclave processes, and with specialized uses of cements in casing oil wells and in making very high strength materials.

11.2 Organic retarders and accelerators

11.2.1 Retarders

Admixtures that retard setting are of value for concreting in hot weather and other purposes, including oil well cementing (Section 11.8). Many organic materials have this property; sucrose, calcium citrate and calcium lignosulphonate are examples. They are effective in low concentrations, suggesting that they act by adsorption. Taplin (T48) noted that most contain one or more groups in which oxygen atoms are attached to adjacent carbon atoms, such as —CO.C(OH)=, or are otherwise able to approach each other closely. He also noted that adsorption on either a clinker phase or a hydration product might cause retardation, in the latter case through interfering with growth. Young (Y8), reviewing the evidence, concluded that adsorption on products, especially CH, was the prime cause of retardation. He considered that adsorption on anhydrous phases also occurred and that, with pure C_3A or C_3S, it caused an initial acceleration by favouring dissolution.

Ramachandran et al. (R32) discussed factors affecting the efficacy of a retarder with cement. They included the ratio of retarder to cement, the time at which the retarder is added, the temperature and the compositions of the cement and of the mix. The time of initial set increases with the content of retarder and generally decreases with temperature and cement content. In a typical case, addition of 0.1% of sucrose on the weight of cement might increase the time of initial set from 4 h to 14 h, while a 0.25% addition might delay it to 6 days. Sufficiently high additions can delay setting indefinitely. Retarders are more effective with cements low in aluminate, because the latter, or their hydration products, consume disproportionate amounts of retarder. Some are more effective with cements low in alkali, perhaps because the latter destroys them. Retarders are most effective if added 2–4 min after mixing, because the aluminate has by that time reacted to some extent with the gypsum and consumes less retarder. The interactions between these and other factors are sufficiently complex that the effect of a specific retarder on a given mix should be pretested before it is used on a job. This applies to admixtures in general.

11.2.2 Mechanism of retardation

Thomas and Birchall (T49) obtained strong support for the hypothesis that retardation arises from adsorption on a hydration product. They determined analytical concentrations of calcium, silicon, aluminium, iron and OH^- in the solution phases of slurries of cement (w/c = 2) with and without additions of sucrose or other sugars, at ages up to 7 h (Fig. 11.1A and B). In the absence of a retarder, the concentrations of silicon, aluminium and iron were all very low (Si = 0.04 ± 0.01 mmol l^{-1}, Al $\leqslant 0.01$ mmol l^{-1}, Fe $\leqslant 0.01$ mmol l^{-1}), and Ca/Si ratios were in the order of 1000. In the presence of sucrose at 50 mmol l^{-1}, the calcium and OH^- concentrations were somewhat increased, but those of silicon, aluminium and iron were increased by factors of up to 500, so that the Ca/Si ratio at 2 h had fallen to 6. The results were explained (T49,B138,T50) by supposing that poisoning of the surfaces of the hydration products allowed the Ca^{2+} ions to coexist in solution with the silicate, hydroxoaluminate and hydroxoferrite ions at concentrations that were impossible if the solution was in equilibrium with

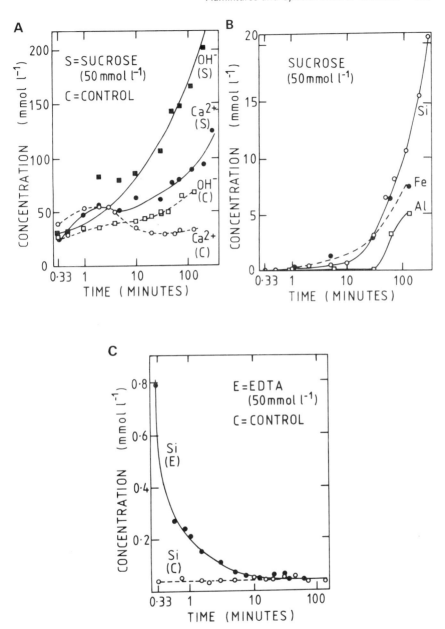

Fig. 11.1 Concentrations in the aqueous phase of Portland cement hydrated at w/c = 2.0 in water or in sucrose or EDTA solutions. After Thomas and Birchall (T49).

the unmodified hydration products. NMR studies showed that Si–sucrose complexes were not formed.

The results of a similar study in which EDTA was used in place of sucrose further supported this conclusion (T51). The silicon concentration in this case increased rapidly during the first 20 s, but soon decreased to a value similar to that in the control (Fig. 11.1C). EDTA is a strong complexing agent for Ca^{2+} in alkaline solutions; it therefore effectively removes the Ca^{2+} from the system, and allows the silicate ion concentration to increase. Because it acts in the solution and not on a surface, a stoichiometric quantity (1 EDTA anion per Ca^{2+} ion) is required, and the effect ceases when this is used up. Sucrose, in contrast, acts by adsorption, and much less is required. In accordance with these conclusions, EDTA, though an excellent complexing agent for Ca^{2+}, has a negligible effect as a retarder when used in concentrations typical for such admixtures (T48).

Thomas and Birchall (T49) summarized and extended data on the comparative retarding effects of different sugars. They distinguished three categories. Trehalose and α-methyl glucoside are ineffective as retarders; both are non-reducing sugars containing only 6-membered rings. Glucose, maltose, lactose and cellobiose, which are reducing sugars containing only 6-membered rings, are good retarders. Sucrose and raffinose, which are outstanding retarders, are non-reducing sugars containing 5- and 6-membered rings. The ability to solubilize CH was shown to be highest for maltose, lactose and cellobiose and lowest for trehalose and α-methyl glucoside, and thus does not run parallel to retarding ability. This is shown also by the behaviour of EDTA, noted above.

Experiments in which calcium oxide or hydroxide was treated with sucrose solutions (B138) showed that a 1:1 Ca–sucrose complex was formed. It was concluded (T49,B138) that the retarding species was R–O–Ca–OH, where R represents the sucrose anion, which was incorporated into the surface of a growing particle of CH or C-S-H, thereby inhibiting growth. The calcium was considered to be attached to the fructose (5-membered ring) part of the molecule. Trehalose and α-methyl glucoside were considered to be ineffective because their ability to complex calcium is very weak. The reducing sugars were considered to be less effective than sucrose or trehalose because opening of the ring allowed them to chelate calcium, giving complexes that were less readily adsorbed.

Birchall and Thomas (B138) explained the inhibition of growth of C-S-H by assuming that the CaO_2 layers are the first parts of this material to be formed, and could thus be poisoned by the retarder. Taylor (T52) earlier proposed a similar mechanism for the growth of C-S-H: starting from mononuclear complexes, such as $[Ca(H_2O)_5(OH)]^+$, repeated condensation through sharing of edges and elimination of H_2O molecules leads to the formation of a CaO_2 layer, which serves as a matrix for the formation of Si_2O_7 groups, and ultimately of dreierketten. He considered it unnecessary for a complete CaO_2 layer to be formed before silicate ions can begin to condense with it. It is most unlikely that the Si–O parts of the structure could form before the Ca–O parts because, although a column or layer of linked Ca–O polyhedra can exist without dreierketten, the reverse is not true. The manner in which the silicate tetrahedra condense is governed by the coordination requirements of the metal cations, and is entirely different if the Ca^{2+} is replaced, for example, by Mg^{2+}.

Jennings *et al.* (J33) found that with C_3S, the retarding effect of sucrose was greatest if it was dissolved in the mixing water. This contrasts with the situation for cement. They concluded that the retarder was incorporated into the initial product and impeded its transformation into a second product. This explanation is compatible with that discussed above if incorporation is taken to mean adsorption on nuclei or growing crystals of the second product. Although there may well be a phase transformation, this evidence does not seem to demand one. Even without one, the retarder would be more effective the sooner it was added, because a smaller area would need to be poisoned. The situation with cement differs due to the competing effect of the aluminate phase.

The broad conclusion indicated by the evidence considered in this section is that organic retarders are substances that are readily adsorbed on to the surfaces of growing particles of hydration products, especially C–S–H, and which do so in preference to complexing Ca^{2+} ions in aqueous solution. The monodentate mode of attachment proposed by Thomas and Birchall (T49) explains the differences in behaviour among sugars, but its applicability in this respect to a wider range of substances has yet to be examined. It does not appear to account for the observation that the effective retarders are compounds in which two oxygen atoms are located on adjacent carbon atoms, or are otherwise able to approach each other closely.

11.2.3 Practical retarders

The retarders most widely used in practice appear to be hydroxy carboxylic acids or their salts, such as citrates or heptonates, and lignosulphonates. Because of the low concentrations required, they are commonly added as solutions. Calcium or sodium lignosulphonates are waste materials from the manufacture of wood pulp, when they are formed by the breakdown of lignins. Their molecules are random, three-dimensional polymers, with molecular weights of up to 50 000, probably spherical and with many $-SO_3^-$ groups on their surfaces. The molecular structures are not completely

350   Cement Chemistry

known, and the element shown in Fig. 11.2 is purely illustrative of the types of local grouping that have been identified. Unpurified commercial lignosulphonates contain substantial proportions of sugars and related salts of sugar acids, such as gluconates, which are at least partly responsible for their retarding properties. Differing views have been expressed concerning the extent of the contribution from the lignosulphonate molecules themselves. Milestone (M104) and others have concluded that the retarding effect with C_3A is largely due to the sugar acids, but Ramachandran (R51) found that, in low concentrations with cement, purified lignosulphonates were almost as effective as the unpurified material.

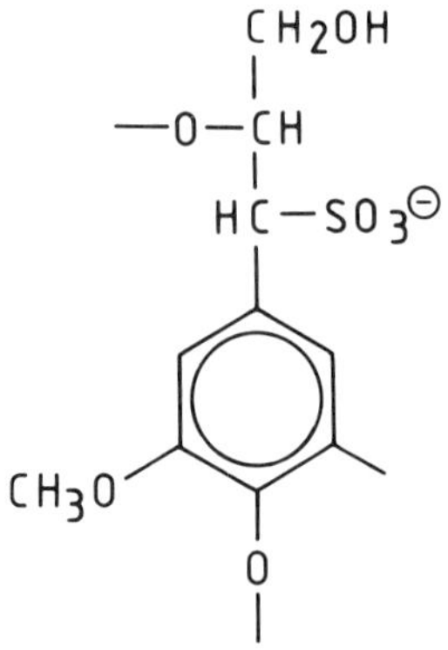

$$\begin{array}{c} CH_2OH \\ | \\ -O-CH \\ | \\ HC-SO_3^{\ominus} \end{array}$$

(with aromatic ring bearing CH_3O- and $-O-$ substituents)

Fig. 11.2   Typical element of structure from a lignosulphonate anion.

11.2.4   Organic accelerators

Calcium formate accelerates setting and hardening in a manner similar to that of many inorganic salts (Section 11.5). Oxalic acid, added to C_3S, increases the extent of the initial reaction, but has no significant effect on the length of the induction period or the subsequent major stage of reaction (O13). Calcium oxalate is precipitated, and the concentration or chemical potential of Ca^{2+} ions close to the surface of the C_3S is thereby kept low until all the oxalic acid has been used up. Triethanolamine is added to some admixture formulations to decrease the retarding effect of the water reducer. The chemistry of its action is not well understood. Ramachandran (R52) found that, added as the sole admixture in amounts greater than 0.1%, it accelerates the reaction of the aluminate phase but markedly retards that of the silicates. At 0.5%, initial set occurs within a few minutes, but final set is delayed to over 24 h. This behaviour appears to be qualitatively similar to that of glucose, which also accelerates the initial reaction of the aluminate phase, though it delays the subsequent replacement of ettringite by monosulphate and greatly retards the hydration of the silicates.

11.3 Air-entraining agents and grinding aids

11.3.1 Air-entraining agents

The ability of concrete to resist damage from freezing of the pore solution is increased by introducing voids into which the latter can expand. Air is entrapped on normal mixing, but is of limited value because the resulting voids are too large and too widely spaced. By adding a suitable surfactant, additional air is entrained and produces a stable foam, which persists in the hardened paste in the form of closely spaced, spherical pores. Air entrainment also improves workability. To avoid undue loss of strength, the total volume of pores should not be larger than necessary; the pores should therefore be small. The pores should not readily fill with water if the paste is saturated; small, isolated pores are most likely to meet this requirement. Klieger (K55) discussed the separations and total volumes of pores needed to provide adequate frost resistance. Typically, the pores are 10–250 μm in diameter and the average maximum distance from any point in the paste to the nearest void, called the void spacing factor, is about 150 μm.

Air-entraining agents are characteristically long chain molecules with a polar group at one end, and which therefore become concentrated at the air–liquid interface with their polar groups in the liquid and their non-polar parts out of it. Bubbles are formed by entrapment of air, but in the absence of the air-entraining agent are mostly destroyed because of the surface tension of the liquid. The air-entraining agent lowers the surface tension and stabilizes them. The inner surface of the bubble is thus composed of hydrophobic material formed by the non-polar parts of the molecules which acts as a barrier to the entry of water during mixing or placing and on any subsequent occasion when the paste is saturated.

Air-entraining agents may be simple compounds, such as sodium salts of fatty or alkyl aryl sulphonic acids, but some are more complex (K55). The quantity of the active ingredient used is typically of the order of 0.05% of the weight of cement. As noted earlier, the property of air entrainment overlaps in varying degrees with those of retardation and water reduction. Air entrainment increases the workability of an otherwise similar concrete; by allowing the w/c ratio to be decreased, this can partly or wholly offset the loss in strength arising from the presence of the air voids.

Admixtures are known that decrease air entrainment; butyl phosphate is an example.

11.3.2 Grinding aids

Grinding aids are sometimes used in clinker grinding, to decrease the energy required to achieve a given fineness of grinding or throughput. Massazza and Testolin (M103) reviewed their use. They are surfactants of various kinds, of which amines and polyhydric alcohols are possibly the most effective, and the amounts added are normally 0.01–0.1% on the weight of clinker. They appear to act mainly by decreasing agglomeration, though it has been suggested that they also aid fracture by preventing incipient microcracks from healing. There are conflicting reports on their effects on setting times, early strength and other properties. Grinding aids are clearly taken up by the anhydrous phases, and it is probably significant that the more effective are of different chemical types from retarders and water reducers, which are taken up mainly or entirely by the hydration products.

11.4 Water reducers and superplasticizers

11.4.1 Water reducers

Water-reducing agents, also called plasticizers, allow a given degree of workability to be achieved at a lower w/c ratio. Conventional water reducers allow w/c to be decreased by 5–15%. To achieve greater decreases, they would have to be used at concentrations that would also cause excessive retardation, excessive air entrainment or flash setting. More powerful water reducers, called superplasticizers, allow w/c to be decreased by up to about 30%; these are considered in Section 11.4.2.

Water reducers act by adsorption at the solid–liquid interface, in this respect resembling retarders. There is a wide overlap between the two properties, many retarders being in varying degrees water reducers and vice versa. Both properties also overlap to some extent with that of air entrainment, which is promoted by admixtures that act at the air–liquid interface. Conventional water reducers are typically added as solutions in concentrations of up to 0.2% on the weight of cement. Calcium and sodium lignosulphonates are widely used both as retarders and as water reducers. Their retarding effect can be decreased by treatment to lower the content of sugars and sugar acids, and can be further decreased or reversed by blending with an accelerator, or enhanced by blending with a stronger retarder. Compounds that decrease air entrainment may also be added. Other materials used as water reducers include salts of hydroxy carboxylic acids, hydrolysed carbohydrates, and hydrolysed proteins.

In general, water reducers are effective with Portland, composite and calcium aluminate cements. Many of the characteristics of retarders, described in Section 11.2.1, apply also to water reducers, and for the same reasons; thus, they are most effective if added a few minutes after mixing, and, in some cases at least, with cements low in aluminate phase and alkali.

The workability of any fresh concrete decreases with time after mixing, but this effect, which is called slump loss, is more marked if a water reducer is used. The slump nevertheless remains higher than if the latter was absent. Slump loss is caused by the slow commencement of the hydration reactions, and its increased magnitude in concrete containing water reducers is probably due to the gradual absorption of the admixture by the hydration products. Delay in adding the admixture until a few minutes after mixing minimizes it.

11.4.2 Superplasticizers

Superplasticizers are also called high-range water reducers. The marked lowering of w/c ratio that they allow makes it possible to produce high-strength concretes, especially if the mix also includes microsilica. Alternatively, the use of superplasticizers at normal w/c ratios allows the production of 'flowing' concrete that is self-levelling and can readily be placed by such methods as pumping or continuous gravity feed through a vertical pipe. The greater effect compared with conventional water reducers is due to the fact that they can be used in higher concentrations, which can be over 1% on the weight of cement, without causing excessive retardation or air entrainment. If used in similar concentrations to conventional water reducers, the degrees of water reduction obtained are also similar.

Three principal types of superplasticizer are in common use, *viz.* salts of sulphonated melamine formaldehyde polymers (SMF), salts of sulphonated naphthalene formaldehyde polymers (SNF), and modified lignosulphonate materials. Most of the available information relates to the first two types, which will alone be considered here. Both are linear anionic polymers with sulphonate groups at regular intervals (Fig. 11.3). The SMF materials are high molecular weight polymers, their most effective components possibly having molecular weights of about 30 000. The SNF materials are less highly polymerized, and have chains some 10 units in length or less. Both are commonly added as solutions. Commercial formulations often contain other substances added to alter setting behaviour or for other reasons.

Most of the characteristics of conventional water reducers are shown also by superplasticizers, at least qualitatively; thus, they are more effective if added a few minutes after mixing, retard setting and increase slump loss. Retardation of set is slight with SMF but more marked with SNF. Slump loss is considerable with both types, and is accelerated by increase in temperature. At 15°C, the enhanced fluidity typically persists for 30–60 min. Slump loss is lessened if the superplasticizer is added a few minutes after mixing, and can also be partly countered by including a retarder or, if practicable, by adding a second dose of superplasticizer before placing. Unlike conventional water reducers, superplasticizers used in normal concentrations do not cause significant air entrainment, and may even decrease the amount of entrapped air because of the greater fluidity of the mix. Air-entraining agents can be added, and must be used in higher than normal concentrations.

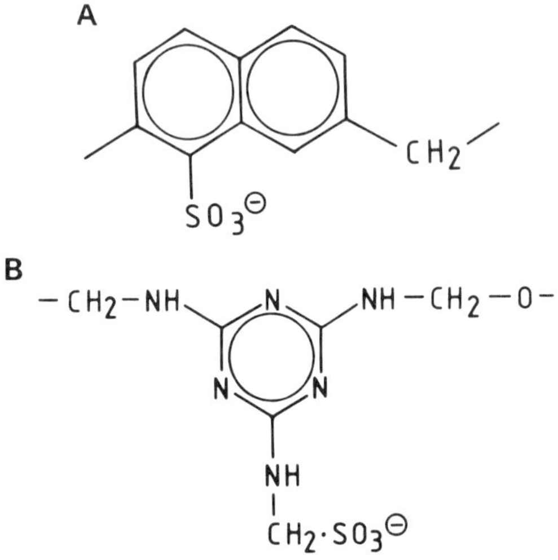

Fig. 11.3 Repeating units of the structures of superplasticizer anions: (A) naphthalene formaldehyde condensate; (B) melamine formaldehyde condensate.

11.4.3 Mode of action of water reducers and superplasticizers

The fact that the water-reducing effects of superplasticizers are similar to those of conventional water reducers used in similar concentrations suggests that the modes of action are similar, and that the essential difference is that the properties that limit the concentrations in which conventional water reducers can be used are weaker or absent in superplastizers. These properties comprise retardation, air entrainment and in some cases flash setting.

It is widely agreed that water reduction is effected through improved dispersion of the cement grains in the mixing water; flocculation is decreased or prevented, and the water otherwise immobilized within the flocs is added

Admixtures and Special Uses of Cements   355

to that in which the particles can move. With superplasticizers, this can be seen with the light microscope (D44,R32), or in greater detail by SEM of rapidly frozen samples (U14) (Section 8.1.3). With some water reducers, retardation of hydration and increased air entrainment may contribute to the increase in fluidity at any given age, though with lignosulphonates, the fluidity is increased even if air entrainment is avoided through addition of tributyl phosphate (B139), and superplasticizers in normal concentrations do not entrain air.

Adsorption of the admixture on the hydrating cement grains could decrease flocculation in at least three ways (D44). The first is an increase in the magnitude of the ζ-potential; if all the particles carry a surface charge of the same sign and sufficient magnitude, they will repel each other. The second is an increase in solid–liquid affinity; if the particles are more strongly attracted to the liquid than to each other, they will tend to disperse. The third is steric hindrance; the oriented adsorption of a non-ionic polymer can weaken the attraction between solid particles.

All water reducers, including superplasticizers, appear to contain more than one polar group in the molecule, and many are polymeric. Some simple compounds that are good retarders, such as glucose, are not water reducers. Kondo *et al.* (K56) considered that polyelectrolytes were adsorbed on the solid surfaces in the ways shown in Fig. 11.4, so that some of their polar groups bind them to the solid, while others point towards the solution.

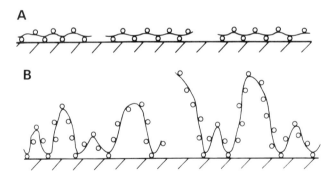

Fig. 11.4 Possible modes of attachment of polyelectrolyte anions to the surface of a particle of cement: (A) train and (B) loop modes. Circles represent negatively charged groups. After Kondo *et al.* (K56).

11.4.4 Zeta potential, rheology and nature of the sorbent phases

Ernsberger and France (E7) showed that addition of calcium lignosulphonate causes cement grains to develop a negative ζ-potential. Daimon and

Roy (D44,D45) studied the action of superplasticizers by determinations of ζ-potential and quantities adsorbed, which they related to the flow behaviour. In the absence of admixture, the charges on the grains were too small to permit reliable determinations of the ζ-potential, but with increasing contents of admixture the latter became increasingly negative, and tended towards a limit of -30 to -40 mV. Up to about 1% of SNF could be adsorbed, and the increase in ζ-potential, amount adsorbed, dispersion of the grains and increase in fluidity were all positively correlated. Daimon and Roy concluded that the increase in ζ-potential was the major cause of the improved dispersion. Further rheological studies (A25,R53) showed that superplasticizers decrease both the plastic viscosity and the yield stress. The major effect was on the yield stress, which could decrease almost to zero. Zelwer (Z11) found that, in the absence of admixtures, C_3S and cement developed weak negative ζ-potentials. That of C_3A was difficult to study due to flocculation, but appeared to be positive, becoming negative on hydration.

Kondo et al. (K56) considered that too high a molecular weight was undesirable, as the molecules could then form bridges between adjacent particles. In contrast, Anderson et al. (A26) found that for sulphonated polystyrenes, the highest ζ-potentials were obtained with material of high molecular weight, although this was not as strongly adsorbed as that of lower molecular weight. They concluded that the high molecular weight material was more likely to show loop as opposed to train adsorption (Fig. 11.4), and thus to place more of its negative charges into the diffuse double layer.

The fact that both conventional water reducers and superplasticizers are more effective if added some time after mixing provides a strong indication that adsorption probably occurs at least in part on the hydrated phases, as the anhydrous surfaces have by that time become covered with hydration products. Chiocchio et al. (C57) found that the optimum time for addition was at the start of the induction period. More of the admixture seems to be taken up by the early hydration products, especially of the aluminate phase, if it is added before the early reaction has subsided.

Costa et al. (C58) showed that superplasticizers increase the fluidity of C_3S pastes much as they do that of cement pastes. Studies on individual anhydrous and hydrated compounds in aqueous and non-aqueous media indicate that calcium lignosulphonate and superplasticizers are adsorbed by C–S–H, AFm phases or CH but not by C_3S, C_3A or C_3AH_6 (R54, R55,R56,C58,M105), though they appear to be taken up by unhydrated β-C_2S (C59). The admixtures also enter interlayer sites of C_4AH_x and perhaps also of C–S–H (R55). Intercalation of organic molecules in C_4AH_x is a well-established effect (Section 6.1.1).

Banfill (B140) noted that the amounts of superplasticizers taken up by cement were sufficient to form a layer on the anhydrous grains some 60 nm thick. He concluded that multilayer adsorption occurred and that steric hindrance was the major effect. This argument is weakened by the evidence that uptake is largely by the hydration products and that significant amounts of material are absorbed as well as adsorbed. The bulk of the evidence indicates that increase in ζ-potential is the major effect.

11.4.5 Reasons for the enhanced dispersing power of superplasticizers

As noted earlier, the difference between superplasticizers and conventional water reducers probably lies in the weaker ability of the former to act as retarders, air-entraining agents, or causes of quick setting, which allows them to be used in higher concentrations. The low air-entraining ability can reasonably be attributed to the repeating pattern of polar groups, which provides the molecule with no suitable hydrophobic regions (K56). The reason for the weakness of the retarding power is less obvious. Weak retarding power implies that the hydration products can grow despite the presence of the sorbed material. This might happen because the latter can be assimilated into them, or because the individual bonds between sorbent and sorbate are sufficiently weak that the latter can be displaced by ions adding to the product. Equilibrium with the solution would ensure that the sorbate was readsorbed on the added material.

11.5 Inorganic accelerators and retarders

11.5.1 Accelerators of setting and hardening

Calcium chloride has long been known to accelerate both the setting and hardening of Portland cement concrete. Typically, for concrete cured at 20°C, 2% of $CaCl_2 \cdot 2H_2O$ by weight of cement might shorten the time of initial set from 3 h to 1 h and double the 1-day compressive strength. The effect on strength decreases with time, and the final strength can be reduced. Some properties related to the microstructure of the paste, such as resistance to sulphate attack, are also adversely affected. The accelerating effects are greater at low temperatures. They increase with the amount of $CaCl_2$ used, and at 4% very rapid setting may occur; 2% is probably a reasonable limit. This gives a concentration of about $0.3 \, \text{mol} \, l^{-1}$ in the mixing water for $w/c = 0.5$.

Calcium chloride is typically added as $CaCl_2 \cdot 2H_2O$. Chloride ion promotes corrosion of reinforcement, and in many countries, including the UK, $CaCl_2$ is no longer used in reinforced concrete. A number of other compounds, such as calcium formate, have been found to have similar properties to $CaCl_2$, but all are both more expensive and less effective.

11.5.2 Mode of action

The addition of $CaCl_2$ increases heat evolution at early ages. Supported by other evidence, considered shortly, this indicates that the effects on setting and hardening are due to acceleration of the hydration reactions. Early studies, summarized by Murakami and Tanaka (M106), showed that a great many salts have effects qualitatively similar to those of $CaCl_2$. With some reservations, these effects occur with C_3S pastes as well as with cement pastes, and are thus seen to be due to faster hydration of the alite.

Fig. 11.5 shows heat evolution curves, obtained by conduction calorimetry, for a C_3S paste with and without addition of $CaCl_2$. When the latter is present, the main heat evolution peak begins earlier and rises and falls more steeply; its maximum is reached earlier. The rate of heat evolution at the maximum is positively correlated with the reciprocal of the time at which the maximum occurs (D18). The linear relation extends to organic retarders. As these probably act by hindering the growth of C–S–H (Section 11.2.2), this evidence suggests that the accelerators act by promoting it.

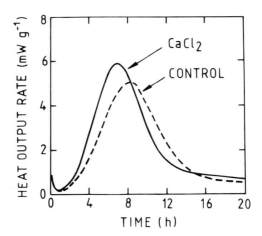

Fig. 11.5 Rates of heat output from C_3S pastes hydrated at 25°C and w/c ratio 0.6, with and without $CaCl_2$ in a concentration of 0.0204 mol l^{-1}. After Brown *et al.* (B64).

Many studies have shown that both cations and anions contribute to the accelerating effect and that an approximate rank order can be established for each type of ion (E8,M106,C60,K57,K58,D18,W31). For the present we shall exclude from consideration such ions as Mg^{2+} or F^- for which precipitation of additional phases complicates the situation. The sequences found by different investigators vary somewhat, and are also influenced by concentration and possibly by the nature of the counter ion. Kantro's (K57) results for the cumulative heat evolution at 12 h for a 3% addition of $CaCl_2 \cdot 2H_2O$ and equivalent concentrations of the other chlorides or calcium salts are reasonably typical:

$$Ca^{2+} > Sr^{2+} > Ba^{2+} > Li^+ > K^+ > Na^+ \simeq Cs^+ > Rb^+$$

$$Br^- \simeq Cl^- > SCN^- > I^- > NO_3^- > ClO_4^-$$

While there are some anomalies, the effect tends to increase with increasing charge and decreasing size of the ion. Ca^{2+} is considerably more effective than any other cation, suggesting that a specific effect is superimposed on a general one. At very high concentrations, some salts that accelerate in lower concentrations, such as NaCl, act as retarders.

At normal concentrations (e.g. $\leqslant 1$ mol l^{-1}), the accelerating effect of ions is sufficiently widespread as to suggest that some general property of electrolytes is involved. The ranking sequences are broadly similar to the Hoffmeister series, which relates to the flocculation of hydrophilic colloids; flocculation is promoted by the action of ions that compete for the H_2O molecules that otherwise keep the particles apart. A similar effect may influence the rate at which complexes can be transferred from the solution to the growing particles of C–S–H. Anomalies in relations to size and charge might be explained through differing degrees of solvation or complexation. It would indeed be surprising if the rate at which the C–S–H grows was not influenced by the nature and concentration of ions in the surrounding solution, and if at least one of the ions entering the solid did not have a specific and positive effect.

The hydration of hemihydrate or anhydrite shows interesting similarities to and differences from that of C_3S (G49). Many foreign ions accelerate or retard hydration, but the sequences of effectiveness differ markedly from those found for C_3S. Excess SO_4^{2-}, but not excess of Ca^{2+}, is a particularly strong accelerator. Murat et al. (M107) found that, for anhydrite, the effectiveness of an added cation decreased with its ionic potential, z/r, and considered that the cations mainly affected the rate of nucleation. The relation to z/r is almost the reverse of what is found with C_3S hydration.

In other approaches to the problem, attempts have been made to relate the accelerating effects either to diffusion of ions (K58,D18), the pH of the bulk solution or the solubility of CH. $CaCl_2$ and other salts decrease the pH of the solution (L52,M106,B64), but there is no apparent correlation between the magnitude of the decrease and the accelerating effect (M106) and NaOH has no significant effect on the rate of C_3S hydration in the stage of the reaction under consideration (B64,G49). This latter observation also tells against the view (S103) that acceleration is due to increased CH solubility, since the latter is decreased by addition of NaOH.

11.5.3 Effects on the compositions and structures of the hydration products

In any attempt to understand the effects of $CaCl_2$ or other accelerators on the structure or properties of the hydration products it is essential to compare pastes at equal degrees of hydration and not at equal ages. Several early studies (e.g. Refs T12 and R57) indicated that the addition of $CaCl_2$ affects the Ca/Si ratio of the C–S–H or the degree of condensation of the silicate anions, but later work does not support either conclusion. Determinations of unreacted C_3S, CH and combined water confirmed that substantially more C_3S reacts during the first day if $CaCl_2$ is present, but showed the latter to have no significant effect on either the Ca/Si or the H_2O/Si ratios of the C–S–H (L19). SEM of fracture surfaces of both young and mature samples shows that the C–S–H morphology is altered (R57,L19). At early ages, it changes from the fibrillar Type I to the honeycomb-like Type II variety. A study using TG and ^{29}Si NMR showed that admixture of C_3S with $CdCl_2$, $Pb(NO_3)_2$ or malic acid had no significant effect on the Ca/Si ratio of the C–S–H or, if equal degrees of hydration were considered, on the ratio of Q_1 to Q_2 silicate tetrahedra, and thus on the relative amounts of dimer and larger silicate ions (C60a). $CdCl_2$ is an accelerator, but $Pb(NO_3)_2$ and malic acid are retarders.

Ramachandran and Feldman (R57) determined helium porosities and densities for cement pastes containing various quantities of $CaCl_2$. They concluded that for a given degree of hydration, the effect of the latter was to increase the absolute density and thus also the porosity. At non-evaporable water contents of 12–16%, 1% or 2% additions of $CaCl_2$ slightly increased the compressive strength, but 3.5% additions decreased it.

The situation of the Cl^- ion in the hardened paste is important in relation to the possibility of promoting corrosion. A study on the hydration of C_3A and C_3S in the presence of gypsum and $CaCl_2$ showed that $C_3A \cdot CaCl_2 \cdot 10H_2O$ was formed after all the SO_4^{2-} had been used up (T53). No phase containing Cl^- was detected in the mixes with C_3S except at a

20% addition of $CaCl_2$. Ramachandran (R58) concluded from thermal and other evidence that, in cement pastes made with $CaCl_2$, much of the Cl^- was chemisorbed or present in interlayer spaces, and that about 20% of it was not leachable by water. In contrast, analysis of pore solutions expressed from 35- or 70-day-old pastes of cement with NaCl additions showed that 85–98% of the Cl^- was present in the pore solution (P48). This proportion was substantially lower for microsilica cements.

11.5.4 Precipitation effects; inorganic retarders and setting accelerators

Many salts precipitate additional phases when added to C_3S or cement pastes. If the hydroxide of an added cation is less soluble than CH, either it or a basic or complex salt is precipitated. Examples of such precipitates are $Mg(OH)_2$ or AFm phases (K57), $Zn(OH)_2$ or $CaZn_2(OH)_6 \cdot 2H_2O$ (A27) and basic lead nitrate or sulphate (T54). Precipitation will leave in solution all or part of the added anion and an equivalent amount of Ca^{2+}. The calcium salt of an added anion, if of sufficiently low solubility, is similarly precipitated; examples are provided by SO_4^{2-}, CO_3^{2-}, PO_4^{3-}, F^-, silicate, aluminate and borate. In this case, the added cation remains in solution with an equivalent amount of OH^-.

The salts in these categories have widely varying effects on setting and hardening. Some, such as lead or zinc salts, phosphates or borates, are strong retarders. Others, such as magnesium salts, behave as setting and hardening accelerators in much the same way as ones not precipitating additional phases. Yet others, such as carbonates, have effects ranging from flash set to retardation of set depending on the concentrations in which they are added. Evidently, more than one effect is involved, and these effects can be delicately balanced.

Retardation is probably caused by the formation of protective layers over the cement grains; this has been demonstrated for lead (T54) and zinc (A27) salts and carbonates (U24). In the latter case, it was shown that the layer formed at low concentrations, which caused retardation, was more compact than that formed at higher concentrations, when acceleration occurred.

The chemistry underlying accelerating effects produced by salts that yield precipitates is not well understood, and more than one mechanism probably operates. In those cases where a cation is precipitated (e.g. with magnesium salts), the Ca^{2+} released and the anion supplied would be expected to behave like the corresponding calcium salt added at the same stage. $CaCl_2$ is a less effective accelerator if added 12 h after mixing than if added immediately (C60), perhaps because by this time growth of C–S–H is becoming increasingly unimportant as a factor controlling the rate of reaction (Section 7.7.2);

this may be expected to reduce the effectiveness of the liberated Ca^{2+}. The accelerating effect of carbonates in suitable concentrations appears to be confined to the initial stage of reaction. It occurs with alite as well as with cement (U24), and is thus associated with the behaviour of that phase. It may be due to increased permeability of a protective layer, as suggested above, but might also be due to removal of Ca^{2+} from the solution, as has been postulated for oxalic acid (Section 11.2.4).

Admixtures that accelerate the early reaction are used in shotcreting or gunniting, in which a concrete mix is sprayed from a nozzle onto a vertical or steeply sloping surface, and for other applications, such as plugging leaks against pressure, in which rapid setting is essential. Those commonly used include sodium carbonate, silicate and aluminate. Setting can thus be caused to begin within minutes, or almost immediately, but subsequent hardening may be delayed and final strengths greatly reduced.

11.6 Effects of high or low temperatures at atmospheric pressure

11.6.1 Hydration at 25–100°C

The hydration chemistry of Portland cement at 25–100°C does not differ essentially from that at lower temperatures, but the reactions are accelerated and there are some differences in the microstructures and, at temperatures approaching 100°C, in the phase composition of the products. Early work suggested that strengths could be predicted by the use of maturity functions, in which curing time and temperature were regarded as equivalent. In one such approach, the compressive strength of a given concrete mix is considered to be a function of the product of the curing time and $T + 10$, where T is the temperature in °C (B141). This gives a general indication of the effect of temperature, but such expressions can only be of limited validity, and there is no general agreement on their utility.

In low-pressure steam curing, concrete is cured in saturated steam at atmospheric pressure. This is used, especially in cold climates, for the rapid production of precast units. Several reviews exist (B142,B143,I12). Various cycles are employed, including ones in which the steam treatment is followed by a period of moist curing at ordinary temperature. In all cases, it is necessary for satisfactory strength development to cure for several hours at ordinary temperature before raising the temperature and to restrict the rate of temperature increase to 10–20 deg C h^{-1}. The rate of cooling is probably less critical, but some restriction is generally considered necessary to reduce thermal stresses. The rates of increase of strength and elasticity increase with temperature, but the final values decrease, especially above about 70°C.

11.6.2 Effects on kinetics and on the ultimate extent of hydration

Several studies have shown that the content of non-evaporable water and the chemical shrinkage follow similar trends to the compressive strength; they increase more quickly at first, but the ultimate values are lower (I12,S104). The inference that the hydration reactions do not proceed as far as at lower temperatures is probably correct, but caution is needed because it depends on the assumption that the stoichiometry and compositions of the products are unchanged. QXDA studies of the hydration of pure C_3S or of the alite in cements have given varied results, some (K21,C39,C61,B53) showing no clear evidence that the ultimate degree of hydration depends on the temperature and others showing moderate (V5) or large (A28) decreases with temperature.

Verbeck and Helmuth (V5) suggested that rapid hydration leads to encapsulation of the cement grains by a product layer of low porosity, which retards or prevents further hydration. Because the remaining product would necessarily be more porous, the material would be weaker than one of the same w/c ratio and degree of hydration in which the product was uniformly distributed. An SEM examination (A28) did not support this conclusion, but it is doubtful whether the technique used (examination of fracture surfaces) would have given conclusive information, and no examinations of polished sections appear yet to have been made. Verbeck and Helmuth's hypothesis would explain the importance of precuring and slow heating and the variability in reported XRD results, since the temperature regime during the early period of hydration would be likely to affect the extent to which a protective layer is formed. An XRD examination of a cement paste that had been moist cured at 90°C for 2 months with no precuring and a rapid heating rate showed considerable quantities of unhydrated aluminate phase and hydrogarnet or C_3AH_6 (T5). This tends to support Verbeck and Helmuth's conclusion.

The reactions of pfa, ggbfs and other pozzolanic or latent hydraulic materials are accelerated by increasing temperature to a greater extent than those of the clinker phases. With pfa, substantial reaction occurs within 1–2 days at 40°C or above, which may well be reached within a large pour of concrete. Under conditions simulating the temperature regime at a depth of 3 m, and using a composite cement containing 25% of pfa, about 7 kg of pfa had reacted per 100 kg of composite cement in 3 days and 10 kg in 100 days (C43). The increased rate of reaction is also shown by accelerated consumption of CH (H52).

11.6.3 Effects on hydration products

DTA (K32,O25) and XRD (A29) studies indicate that cement pastes cured at temperatures of or approaching 100°C contain little or no detectable hydrated aluminate phases, the observation of hydrogarnet noted above thus being unusual. Kalousek and Adams (K32) considered that the Al^{3+}, SO_4^{2-} and other ions were incorporated into a 'Phase X', which possibly included all the oxide components of the cement. In pastes cured normally, much of the AFm phase is probably poorly crystalline and intimately mixed with the C-S-H (Section 7.3.1). This tendency may be increased if hydration takes place more rapidly.

When heated above about 50°C in air, synthetic ettringite yields an almost amorphous product. Buck (B144) found that, for a cement paste initially cured for 24 h at 23°C, the ettringite was destroyed on subsequent curing in sealed vials at 75-100°C. No decomposition product was detected by XRD, and the ettringite was not reformed on prolonged curing in water at room temperature. The decomposition of the ettringite did not appear to affect the compressive strength or volume stability. This effect may be a further reason for the non-occurrence of detectable crystalline hydrated aluminate phases.

The effect of temperature on the Ca/Si ratio of the C-S-H has been examined in several investigations on pastes of C_3S or β-C_2S by calculation from the contents of CH and unreacted starting material (K21,C39,O26, A28,B53,O25). The results are conflicting, but tend to indicate increases in the Ca/Si ratio with increased temperature. Some of the earlier results show marked variations in Ca/Si ratio with time, but for low degrees of reaction, results obtained by this method are very sensitive to small experimental errors. A later study on laboratory-prepared cements (O25) showed no variation with time and a significant decrease in the ratio of CH to C_3S reacting at 95°C but not at 75°C or below. It also indicated a steady decrease in bound water (defined as the ignition loss after drying with acetone and ether) from 35% on the weight of C_3S reacting at 25°C to 25% at 95°C. The decrease in bound water content could thus be due not only to decreased reaction, but also to decreased interlayer water in the C-S-H, and at the higher temperatures to replacement of ettringite by products of lower water content. X-ray microanalyses (U12) for the C-S-H of cement pastes gave Ca/Si ratios of 2.03 and 1.90 for pastes cured at 20°C and 45°C, respectively.

Studies on calcium silicate pastes show that the distribution of silicate anion size is shifted significantly upwards with a rise in curing temperature (Section 5.3.2). XRD gives no definite indication that the C-S-H formed at temperatures up to 100°C is more crystalline than that formed at ordinary temperatures, though the product of the 2-month treatment at 90°C mentioned earlier (T5) contained a little α-C_2SH, a crystalline phase which is structurally unrelated to C–S–H and readily formed at higher temperatures (Section 11.7.2).

The H_2O and N_2 specific surface areas and the volumes of pores smaller than 25–35 nm determined by N_2 sorption all decrease with curing temperature, if allowance is made for differing degrees of hydration (C62,S104, A28,B53,O20). Bentur *et al.* (B53) also found an increase in the volume of pores larger than about 4 nm, which they estimated from the amount of water lost on drying at 53% RH. They concluded that increase in temperature causes the hydration products to become more compact. Such an effect is readily explained, as one would expect the amount of interlayer water in C–S–H to decrease with temperature of formation. It is also compatible with Verbeck and Helmuth's (V5) view that the hydration products are less uniformly distributed. Pore size distributions determined by mercury intrusion porosimetry appear to show little or no variation with curing temperature (O20,B144). It is not clear whether this can be reconciled with the above observations.

11.6.4 Low temperatures

Mironov (M108) summarized the results of studies on concreting at subzero temperatures in the USSR. Hydration continues to the extent that liquid water is present; this in turn depends on the ambient conditions and on the pore structure, and is increased by precuring at ordinary temperature. At $-5°C$, with 24-h precuring, hydration continues at a useful rate, but at or below $-10°C$, it is extremely slow. Additions of $CaCl_2$ and other salts act both as accelerators and antifreezing agents (M108,M109). These conclusions were subtantially confirmed by Regourd *et al.* (R59), who examined a concrete containing 9% of $CaCl_2$ on the weight of cement that had been mixed with hot water and cast and cured for 1 year within a mine at $-10°C$. The long-term compressive strength and microstructure were not greatly different from those of a similar concrete cured at ordinary temperature.

Calcium aluminate cements behave well at low temperatures (Section 10.1).

11.7 High-pressure steam curing

11.7.1 General

The hardening of concrete or other materials based on calcium silicates may be further accelerated by the use of saturated steam under pressure in an autoclave. Typically in such processes, a mix containing Portland cement and finely ground quartz, in addition to aggregate, is cured for 2–4 h in saturated steam at 175°C (800 kPa approx.). As with steam curing below 100°C, it is necessary to precure for a few hours at ordinary temperature, and to restrict the rate of heating, and to a lesser extent that of cooling. In this way, concrete can be produced within 24 h that has a compressive strength at least as high as that obtainable in 28 days at ordinary temperature.

Autoclave processes are used to produce a wide variety of materials, ranging from dense concretes and fibre-reinforced materials, through aerated concrete with a bulk density of 400–800 kg m^{-3} to very low density thermal insulation materials. The cement provides strength to facilitate handling prior to autoclaving, but otherwise serves essentially as a source of CaO and SiO_2, and may be partly or wholly replaced by other starting materials of suitable composition. Calcium silicate bricks (P49) are made from lime and sand, and are pressed into shape when moist and autoclaved. The chemistry of the process is essentially similar to that of the cement-based materials. Aerated concretes (A30) are typically made by incorporating aluminium powder, which reacts during the initial curing below 100°C, generating H_2. In autoclave processes in general, both natural and waste materials of various kinds may be used as partial sources of the CaO or SiO_2 or both.

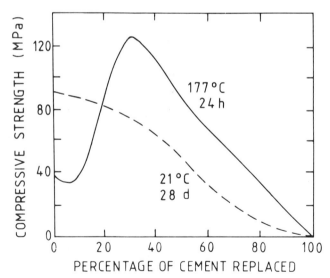

Fig. 11.6 Effects of partial replacement of cement by finely ground silica on the compressive strengths of mixes cured normally and in the autoclave. After Menzel (M110).

Early experience in making autoclaved concrete showed that the presence of siliceous material of a sufficiently reactive nature was essential for attainment of satisfactory strength. Menzel (M110) showed that the ratio of such material to cement was critical. Fig. 11.6 shows his results for mixes containing Portland cement and finely ground quartz. The highest strengths were obtained at a quartz/cement weight ratio of about 30 : 70. The optimum ratio depends on the nature and fineness of the siliceous material.

11.7.2 Basic chemistry of autoclave processes

Understanding of the chemistry of autoclave processes is due primarily to the work of Kalousek and co-workers (K32,K59–K62). Above about 150°C, for the time scales of a few hours that are used in practice, two features of cement hydration chemistry are added to those relevant at lower temperatures. Firstly, the hydration products tend to crystallize; in the absence of reactive silica, C–S–H tends to be replaced by a structurally unrelated, crystalline phase, α-C_2S hydrate. Secondly, the range of siliceous materials having effective pozzolanic properties is widened, and includes quartz and various other crystalline minerals, if sufficiently finely ground.

Menzel's results were thus explained. Addition of a small amount of quartz brings the bulk Ca/Si ratio to 2.0, which is that of α-C_2S hydrate. This phase is relatively dense and crystallizes as rectangular tablets; the product is porous and weak. With larger amounts of quartz, pozzolanic reaction occurs; the CH is consumed, formation of α-C_2S hydrate is avoided, and a C–S–H of low Ca/Si ratio is formed. The porosity is reduced, and a strong material results. The optimum addition of quartz is the maximum that can be taken up; further additions act only as a diluent, and the strength decreases. Table 11.1 gives crystal data for α-C_2S hydrate and other phases.

The optimum addition yields C–S–H with a Ca/Si ratio of 0.9–1.0. C–S–H of this composition tends to crystallize to give 1.1-nm tobermorite (Section 5.4.1). Opinions have differed as to the extent to which crystallization is desirable (K60). It has been widely found that optimum times and temperatures of autoclaving exist, which if exceeded lead to decreases in strength. As the main effect of extended time or increased temperature is likely to be an increase in the crystallinity of the binder, this suggest that, from the standpoint of strength, crystallization is either undesirable or some optimum degree of crystallinity exists. The latter view (B114,F36) appears the more probable. It has received direct experimental support in the case of aerated concrete (A30) and can also account for results obtained with a series of denser lime–quartz materials (C63). It is generally agreed that crystallization

Table 11.1 Crystal data for some synthetic, hydrothermally produced phases

Name	1.1-nm tobermorite[a]	1.1-nm tobermorite[b]	α-Dicalcium silicate hydrate
Formula	$[Ca_4(Si_5O_{16}H_2)] \cdot Ca \cdot 4H_2O$	$[Ca_4(Si_{5.5}Al_{0.5}O_{17}H_2)] \cdot Ca_{0.2} \cdot Na_{0.1} \cdot 4H_2O$	$Ca_2(HSiO_4)(OH)$
Lattice parameters			
a (nm)	0.564	0.563	0.9476
b (nm)	0.368	0.369	0.9198
c (nm)	2.26	2.28	1.0648
Crystal system	Orthorhombic	Orthorhombic	Orthorhombic
Space group	I[c]	I[c]	$P2_12_12_1$
Z	1	1	8
X-ray density (kg m^{-3})	2380	2400	2723
References			
Structure	M50, H31	M50, H31	H60, U25
Lattice parameters	H61	H61	P1[d]

[a] Typical data for normal tobermorite. Data relate to pseudocell
[b] Typical data for anomalous tobermorite with Al and Na substitution. Data relate to pseudocell.
[c] Structure not fully ordered.
[d] Card No. 29-373.

decreases drying shrinkage and improves resistance to chemical attack (K60,P49,A30).

It is doubtful whether XRD can distinguish a mixture of tobermorite and C–S–H from a uniform material of intermediate crystallinity; the situation may lie between these extremes (A30). This question is discussed further in Section 11.7.4. Crystallization is probably favoured by low bulk density; its extent is apparently minimal in calcium silicate bricks (P49), but considerable in aerated concretes (A30). In cement–silica materials, substantially all the Al_2O_3 appears to enter the C–S–H, which as its Ca/Si ratio decreases can accommodate increasing amounts of tetrahedrally coordinated aluminium (S70). NMR results (K34) support an early conclusion (K62) that 1.1-nm tobermorite, too, can accommodate aluminium in tetrahedral sites. Small amounts of hydrogarnet have sometimes been detected, especially in products made from raw materials high in Al_2O_3, such as pfa or slag. Minor amounts of tricalcium silicate hydrate (jaffeite; $C_6S_2H_3$) have sometimes been detected (A29,K61).

11.7.3 Mechanisms of reaction and equilibria

Quartz reacts relatively slowly, and substantial proportions of C–S–H having a Ca/Si ratio of about 1.75 are formed as intermediate products, which gradually react with it to give C–S–H of lower Ca/Si ratio. This process competes with the formation of α-C_2S hydrate. The conditions should be such as to minimize the latter process, since subsequent reaction of α-C_2S hydrate with quartz is slow and may not be completed in the time available. Fine grinding of the quartz and adequate mixing with the cement are therefore essential.

Several reviews of phases and equilibria in the CaO–SiO_2–H_2O system have been given (R60,T21,T55). Over 20 crystalline ternary phases were listed in the 1980 review (T55), and two more have since been reported (G78,S105). Neither 1.1-nm tobermorite nor α-C_2S hydrate is an equilibrium product in the pure system at 180°C under saturated steam pressures, the stable ternary phases under these conditions being possibly truscottite ($C_7S_{12}H_3$), gyrolite ($C_2S_3H_2$), xonotlite (C_6S_6H), hillebrandite (C_2SH) and tricalcium silicate hydrate, according to the composition (T21). Of these phases, only xonotlite forms readily under normal autoclaving conditions from mixtures in which the SiO_2 is supplied partly as quartz. It is rapidly formed from such mixtures at 200°C, and because of its good thermal stability in air is used in some autoclaved thermal insulation materials. If the SiO_2 in an autoclaved material is supplied in a highly reactive form, such as diatomaceous earth, the reaction takes a different course; gyrolite, or, at higher temperatures, truscottite, is formed as an early product, and 1.1-nm tobermorite, if formed at all, results from reaction of CaO with phases of lower Ca/Si ratio (B145).

11.7.4 Characteristics of hydrothermally formed C–S–H and tobermorite

Mitsuda *et al.* (M111) described the XRD patterns, morphology, IR spectra, thermal behaviour and other properties of hydrothermal preparations of C–S–H. The XRD patterns resembled those of C–S–H(I) (Section 5.4.3). TEM showed that all contained crumpled foils, but some also contained fibrous aggregates. The latter gave electron diffraction patterns corresponding to poorly crystalline tobermorite, in some cases probably intergrown at or near a single layer level with xonotlite or foshagite (C_4S_3H). These results show that the distinctions between C–S–H(I), tobermorite and other crystalline phases are not sharp; materials of intermediate crystallinity can exist on the nanometre scale.

1.1-nm tobermorites differ in their behaviour on heating. All lose their interlayer water at 100–300°C, but this may or may not be accompanied by shrinkage of the layer thickness to 0.93–0.94 nm (G79). Specimens that show lattice shrinkage, called normal tobermorites, crystallize as thin, pseudo-hexagonal plates with (001) cleavage and a tendency to elongation parallel to *b*. Those that do not, called anomalous tobermorites, tend to form smaller, more elongated crystals with major (001) and minor (100) cleavages. Apart from small differences in lattice parameters, which can be obscured by effects of ionic substitutions (H61), the XRD patterns are identical.

The structure of the individual layers in tobermorites was described in Section 5.4.2. In 1.1-nm tobermorites, these layers are stacked in such a way that oxygen atoms of the bridging tetrahedra (Fig. 5.7) not attached to two silicon atoms can closely approach those of the adjacent layer. A suggestion (T56,M51) that interlayer Si–O–Si links are present in anomalous tobermorites was confirmed by chemical (W32) and ^{29}Si NMR (W21,K16) studies of silicate anion type. The compositions and densities of naturally occurring tobermorites show that the contents of the pseudocell (Table 11.1) always deviate markedly from the idealized formula $C_5S_6H_5$ (M51). There are at least three effects:

(1) The silicate chains can have missing tetrahedra, as in C–S–H(I) (Section 5.4.5). This, the sole major effect in unsubstituted normal tobermorites, is compositionally equivalent to loss of the elements of SiO_2 and thus increases the Ca/Si ratio and lowers the density. In the

highly crystalline natural materials, the limiting composition is probably about $[Ca_4Si_{5.5}O_{17}H_2]Ca \cdot 4H_2O$, where square brackets exclude the interlayer material.

(2) Interlayer Si–O–Si links are formed. This effect is associated with reduction in interlayer calcium content and is probably a replacement of $(2O^- + Ca^{2+})$ by $-O-$, and is thus compositionally equivalent to loss of CaO. It decreases both the Ca/Si ratio and the density. The theoretical limiting composition is $[Ca_4Si_6O_{17}H_2]4H_2O$, but may not be realizable.

(3) Substitution of Si^{4+} by Al^{3+} is balanced by an increase in interlayer Ca^{2+} or introduction of interlayer alkali cations or both. All known naturally occurring anomalous tobermorites show effects (2) and (3).

1.1-nm tobermorite is readily synthesized using CH and finely ground quartz at 180°C. It is much less readily formed if amorphous silica is used. Synthetic studies show that normal tobermorite is an intermediate in the transition from C–S–H(I) to anomalous tobermorite (E9,H61). Its formation in preference to anomalous tobermorite is favoured by short time, low temperature, high Ca/Si ratio, and presence of Al^{3+} in the absence of alkali. The presence of Al^{3+} together with alkali favours the formation of anomalous tobermorite (E9). The presence of alkali without Al^{3+} has been variously found to impede crystallization (E9) or to favour formation of anomalous tobermorite (H61). Most of these results are readily explainable in terms of the structural differences described above. Al^{3+} is also reported to accelerate the formation of tobermorite and to retard its replacement by xonotlite at temperatures around 175°C (K62). Mitsuda and Chan (M112) found the tobermorite in some aerated concretes to be anomalous. Aluminium-substituted tobermorites have cation exchange properties (K63).

Hara and Inoue (H61) studied the lattice parameters of tobermorites. They confirmed an early observation (K62) that c increases with aluminium substitution, but found that several other variables affect the parameters to smaller extents. The a-axial lengths of highly crystalline, natural tobermorites appear to be lower (0.560–0.562 nm for the pseudocell) than those of the synthetic materials (M51).

11.8 Oil well cementing

11.8.1 General

In oil well cementing (R61,D46,S106) a cement slurry is pumped down the steel casing of the well and up the annular space between it and the surrounding rock. The main objectives are to restrict movement of fluids between formations at different levels and to support and protect the casing. More specialized operations comprise squeeze cementing, in which the slurry is forced through a hole in the casing into a void or porous rock, and plugging, in which the casing is temporarily or permanently blocked at a specified depth.

With current technology, oil wells are typically up to 6000 m deep. The temperature of the rock at the bottom of the well ('bottom hole static temperature') at that depth is 100–250°C. The maximum temperature of the slurry during pumping ('bottom hole circulating temperature') is in general lower but may still be as high as 180°C. The pressure experienced by the slurry during pumping is equal to the hydrostatic load plus the pumping pressure, and may be as much as 150 MPa. The entire depth of a deep well would not be cemented in a single operation, but even so, pumping can take several hours. In geothermal wells, the maximum temperatures encountered may exceed 300°C.

11.8.2 Types of cement and of admixtures

The slurry must remain sufficiently mobile for the pumping operation to be completed and must provide adequate strength, resistance to flow of liquid or gas and resistance to chemical attack after it has been placed. By using a high-temperature, high-pressure consistometer, in which the regime of temperature and pressure during pumping is simulated, it is possible to monitor the changes in consistency and to predict the 'thickening time' during which pumping is practicable. The thickening time is akin to final set under the simulated well conditions. The American Petroleum Institute has defined various classes of Portland cement, of which G and H are widely used. The specifications of both are typically met by sulphate-resisting Portland cements, coarsely ground to 280–340 $m^2 kg^{-1}$ (Blaine) for Class G or to 200–260 $m^2 kg^{-1}$ for Class H (S106,B146). Free lime is minimized to permit good response to admixtures. Both classes are intended to be used as basic cements at depths down to about 2500 m, and at greater depths with suitable admixtures. In practice, admixtures of many kinds are used at most depths.

Retarders and dispersants (water reducers) are widely employed, especially in the deeper wells and also to counteract the effects of other admixtures that have incidental accelerating effects. Lignosulphonates, modified lignosulphonates, cellulose derivatives and saturated NaCl are among those used. NaCl is effective up to about 130°C, and modified lignosulphonates to at least 150°C. Superplasticizers may be used up to 150°C. Accelerators, such as $CaCl_2$ (2–4% on the weight of cement) or NaCl (2.0–3.5%) are also used. In marine locations, sea water is often used for mixing, and acts as an accelerator. Sodium chloride may be added to fresh water; in addition to its accelerating or retarding properties, it reduces damage to salt and shale strata and causes the hardened paste to expand.

Various admixtures are used to modify physical properties of the slurry, though some have incidental chemical effects that are countered by the use of further admixtures. It may be necessary to adjust the bulk density, e.g. by adding hematite to raise it, or bentonite (sodium montmorillonite) or sodium silicate to lower it. Bentonite greatly increases the water demand, but may be used together with calcium or sodium lignosulphonate and sometimes NaCl to obtain a desired combination of density, fluidity, w/c ratio and thickening time. The optimum addition of untreated bentonite is 8%, but this is reduced to 2% if the bentonite has been prehydrated, i.e. mixed with water and allowed to swell before being used. Granular, lamellar or fibrous materials are occasionally added to prevent slurry from being lost in rock fissures when lightweight or thixotropic slurries are insufficiently effective. Expanded perlite (a heat-treated volcanic material), walnut shells, coal, cellophane and nylon are examples. Nylon or other fibres have been used to increase shear, impact and tensile strength but sometimes present logistical problems. Cellulose derivatives, water reducers and latex admixtures are among admixtures that reduce loss of solution from the slurry into porous strata.

More specialized admixtures include radioactive tracers, which may be detected by devices lowered down the hole to trace the movement of the slurry. Dyes or pigments may similarly be used to check its emergence. Some of the chemicals added to drilling muds are strong retarders for cement. Proprietary 'spacer fluids' are commonly pumped ahead of the cement slurry to counteract contamination. Admixtures of paraformaldehyde and sodium chromate are sometimes also used. Addition of 5–10% of gypsum produces a thixotropic slurry, which can be pumped but which gels rapidly when stationary; this can be used to help the slurry to pass permeable formations. Gypsum also causes expansion, and it or other expansive admixtures may be used to improve the seal with the rock or casing.

11.8.3 Effects of temperature and pressure

At the high temperatures encountered in deep wells, pozzolanic admixtures are essential to prevent strength retrogression, as in high-pressure steam curing (Section 11.7). Silica flour (finely ground quartz) and silica sand are the most commonly used. There are few data on the effects of prolonged exposure of cement–silica mixes to hydrothermal conditions; pastes cured for 2.5–4 months at 140–170°C and saturated steam pressures are reported to have similar strengths to those so treated for shorter times, though some replacement of tobermorite by gyrolite and xonotlite was observed (I13). In a deep well, the pressure may be much above that of saturated steam; this alters the equilibria (R60,T21) and possibly also the kinetics. At 110°C or 200°C and 7–68 MPa, mixtures of cement and quartz were shown to give C–S–H(I), which began to change into tobermorite within 32–64 h (O27). In hot, geothermal wells, many phases can form in addition to or in place of those found in autoclaved cement materials (L53).

Special problems arise in cementing wells in the Arctic, where permafrost may exist to a depth of 1000 m. It is necessary that the cement should set at low temperatures, that the surrounding ground should not be disturbed by melting or erosion during drilling or cementing or in the subsequent life of the well, and that freezable liquids should not be left in the annular space. Mixtures based on Portland cement and gypsum have been used (S106).

11.9 Very high strength cement-based materials

11.9.1 General

This section deals with cement-based materials having compressive strengths much above 100 MPa or comparable uprating of other mechanical properties. Relatively modest improvements yield materials with potential specialist uses in construction. Larger ones, especially in tensile or flexural strength and fracture toughness, offer the possibility of making low-volume, high-technology materials.

The compressive strength can be increased by lowering the w/c ratio. As the latter decreases, the particle size distribution of the starting material becomes increasingly important. Brunauer *et al.* (B147) described the preparation and properties of cement pastes with compressive strengths up to 250 MPa. Portland cement clinkers were ground to 600–900 m^2 kg^{-1} using grinding aids and subsequently mixed at w/c 0.2 with admixtures of calcium lignosulphonate and K_2CO_3.

Very high strength concretes have since been obtained using Portland cements with superplasticizers and microsilica. In so-called DSP (densified systems containing homogeneously arranged ultrafine particles) materials, the use of low w/s ratios (0.12–0.22), special aggregates, including fibres, and special processing conditions allows compressive strengths of up to 270 MPa to be obtained, with good resistance to abrasion and chemical attack

(H53,H62). The properties were attributed to a combination of effects. The particles of microsilica, being much finer than those of the cement, partially fill the spaces between the cement grains, and this, together with the superplasticizer, allows the latter to pack more uniformly. They also provide nucleation sites for hydration products, undergo pozzolanic reaction and probably improve the paste–aggregate bond.

Lime–quartz materials with compressive strengths of up to 250 MPa can be made by moulding the starting materials under a pressure of 138 MPa before autoclaving (T57,C64). Portland cement pastes have similarly been pressure moulded to allow use of w/c ratios down to 0.06 and development of 28-day strengths up to 330 MPa (R62). Roy and co-workers obtained still higher strengths by hot pressing Portland (R62,R63) or calcium aluminate (G80) cement pastes, followed by normal curing in water. A Portland cement paste pressed for 1 h at 250°C and 345 MPa and then cured for 28 days had a strength of 650 MPa. The compressive strengths and total water porosities (1.8% in the case mentioned) obeyed the Schiller relation (Section 8.4.1).

Impregnation with an organic polymer of a concrete that has already developed some strength is another way of reducing porosity and increasing strength. The concrete is first dried, and the monomer is then introduced and polymerized in situ by γ-irradiation or by including a catalyst and subsequently heating at 70–90°C (I14). Polymethylmethacrylate (PMMA) is possibly the most effective polymer. For maximum uptake of polymer, it is necessary to dry strongly (e.g. by heating at 150°C) and to evacuate, but useful amounts can be introduced without evacuation. Compressive strengths can be approximately doubled, and values around 220 MPa have been reported for cement–sand mortars. Vacuum impregnation of cement pastes and autoclaved materials with molten sulphur at 128°C similarly increased microhardness and elastic modulus by factors of up to 6 and 4, respectively (F45).

The tensile or flexural strengths of the materials described above were, in general, about one-tenth of the compressive strengths, as in normal cement pastes or concretes. In so-called MDF (macro-defect-free) cements, Birchall and co-workers (B148,K64) obtained higher relative values of flexural strength and other mechanical properties. The materials were made by incorporating a water-soluble polymer, such as poly(vinylalcohol/acetate), into a Portland or calcium aluminate cement mix to produce a dough that could be processed by methods of conventional plastic technology, such as pressure moulding, extrusion or calendering. They had flexural strengths of 40–150 MPa, Young's moduli of 35–50 GPa, compressive strengths of 100–300 MPa and fracture energies of 40–200 Jm^{-2}.

11.9.2 Phase compositions, microstructures and causes of high strength

In general, XRD and other studies show that the phase compositions of the very high strength materials described above are similar to those of weaker ones of similar types, the proportion of unreacted clinker phases increasing with decrease in porosity. The major product detected in hot pressed calcium aluminate cements was C_3AH_6 (G80); this is consistent with Feldman and Beaudoin's (F36) view that this and similar dense, crystalline phases produce high strengths if sufficiently closely welded together in materials of low porosity. The same applies to unreacted clinker phases. The microstrutures of hot pressed cements (R62) and MDF cements (S107) are dense and compact, and CH crystals, where present, are smaller and more uniformly distributed than at lower porosities. In MDF cements, CH can occur as lamellae only 10 nm thick (G37).

Pastes inpregnated with PMMA or sulphur are still sufficiently permeable to water that expansion occurs on long exposure (F46). In polymer-impregnated (S108) and MDF (R64) cement pastes, there is evidence of interaction between Ca^{2+} ions and carboxylate and possibly other groups of the polymer. In MDF pastes made with calcium aluminate cement, the polymer (PVA) was found to inhibit the normal hydration reactions of the cement, but to react with Ca^{2+} and Al^{3+} to give an ionically cross-linked polymer and calcium acetate. TEM showed the material to be essentially a dispersion of grains of clinker or hydration products in a continuous polymer matrix.

Kendall et al. (K64) considered that MDF cements owed their high strengths primarily to the absence of large flaws, which they held to be the reason for the relative weakness of normal cement pastes. This view has been challenged (B115,K65,E4). Beaudoin and Feldman (B115) discussed ways in which stress concentrations were reduced in very strong cement pastes, and both they and Eden and Bailey (E4) considered that in MDF pastes the polymer plays a major role.

12

Concrete chemistry

12.1 Cement paste in concrete

12.1.1 Portland cement mortars and concretes

Concrete cannot properly be described as a composite of coarse and fine aggregate in a matrix of cement paste otherwise identical with the aggregate free material. The microstructure of the paste close to the aggregate differs from that of cement paste in bulk, and much of the paste in a concrete or mortar is in this category (D47). Because the strength of concrete depends in part on that of the cement–aggregate bond, many of the studies on concrete microstructure have concentrated on this aspect. Several reviews exist (S109,M113,D48,M114).

Using light microscopy to examine thin sections of mortars impregnated with a coloured resin, Farran (F43) demonstrated the presence in some cases of an interfacial zone or aureole of increased porosity and presumably lower strength. Microhardness measurements confirmed this (L54). Farran et al. (F47) studied double replicas of abraded surfaces of mortars by TEM, and thereby confirmed the existence of interfacial zones up to some 35 μm wide and of distinctive microstructure. Javelas et al. (J34) examined ion-thinned sections of mortars by TEM. The surfaces of calcite or quartz grains in mortars were closely covered with layers of poorly crystalline material, which gave a diffuse diffraction ring with a spacing of 0.3 nm. This could not have been due to CH, and the material was probably C–S–H.

Several studies have been made on specimens prepared by casting cement paste against relatively large, flat and often polished surfaces of aggregate and then split apart near to the interface. Barnes et al. (B149) made SEM examinations of the interfacial zone in pastes cast against glass slides. They reported the occurrence of duplex films, comprising a layer of CH crystals in contact with the glass and oriented with their c-axes normal to its surface, and a layer of C–S–H further out into the paste. The entire duplex film was typically 1.0–1.5 μm thick, and formed part of a much thicker interfacial zone, prominent features of which included hollow-shell grains and CH crystals with c-axes both parallel and perpendicular to the glass surface. They observed similar effects in fracture surfaces of mortars (B150).

Grandet and Ollivier (G81) made XRD examinations of layers of paste at different distances away from various aggregates, exposed by progressive abrasion. CH close to the interface showed strong preferred orientation, with c normal to the interface; the degree of orientation decreased with the distance, becoming zero at about 40 µm. Monteiro et al. (M115) and Detwiler et al. (D49) confirmed this observation, which differs from those of Barnes et al. in showing no tendency to preferred orientation with c parallel to the interface. The growth of CH crystals with c normal to the interface has been described as epitactic, but this is incorrect, as there is no preferred crystallographic orientation relative to the crystal structure of the substrate, which can indeed be amorphous. Struble and Mindess (S110) observed a layer of well-oriented CH adjacent to the aggregate surface, but no duplex film, and noted that the interfacial zone appeared to be of higher w/c ratio than the bulk paste.

Wu et al. (W33) and Monteiro and co-workers (M116,M115) observed that, within the interfacial zone, the content of ettringite increases as the aggregate surface is approached.

12.1.2. Backscattered electron imaging of the interfacial zone

Studies of prepared surfaces have yielded interesting information, but their limitations are becoming increasingly apparent. As Scrivener and Gartner (S111) have noted, the aggregate surfaces in a real concrete or mortar are not polished, flat and widely separated, but in varying degrees irregular and close together. During mixing, the aggregate particles are in vigorous motion relative to the surrounding paste, and during setting, localized bleeding below aggregate particles or reinforcement can occur. Studies of fracture surfaces suffer from the further serious limitations that they are not representative of the bulk material, but of its weaker regions, and that X-ray microanalyses on them are at best only semiquantitative. The high resolution provided by secondary electron images aggravates the problem of obtaining statistically representative information. Backscattered electron images (BEI) of polished sections (S111–S114,S68) do not suffer from these limitations, and are readily combined with quantitative X-ray microanalysis and with image analysis to provide further types of quantitative information.

From an SEM study of mortars using both fracture surfaces and BEI of polished sections, Scrivener and Pratt (S112) concluded that a layer of C–S–H about 0.5 µm thick was formed over the surfaces of the sand grains during the first day. This layer resembled that formed on pfa particles. They confirmed the tendency to increased content of ettringite in the interfacial region. Large crystals of CH were sometimes formed adjacent to the sand grains, and tended to predominate on fracture surfaces because of their good cleavage; their orientation relative to the aggregate surface varied considerably. Particularly large crystals or extensive deposits of CH were found in places where local bleeding was considered to have been likely. Regions poorer in cement particles sometimes occurred on one side only of a sand grain and possibly arose from the sweeping action of the latter during mixing.

In any attempt to estimate distances within an interfacial zone from BEI of polished sections, allowance must be made for the fact that the interfaces examined are in general not perpendicular to the section. This problem may be handled by examining a sufficient number of grains (S114) or, less satisfactorily, by examining specimens in which the paste has been vibrated in contact with a single piece of aggregate with one surface in a known orientation (S111). Studies by BEI with image analysis by the former method showed that as the aggregate surface is approached within the interfacial zone, the average content of unreacted clinker phases decreases and the porosity increases (Fig 12.1). The average CH content, in contrast, appears to show only a slight increase (S111). The region within about 10 µm from the surface was found to be deficient in the usual relicts of large clinker grains, indicating that it is difficult for the latter to pack close to the interface. The grains smaller than about 5 µm had often hydrated completely, giving hollow-shell grains, which were thus relatively abundant in the interfacial region.

These results gave no indication of the presence either of duplex films or of continuous layers of CH at the aggregate interfaces in normal concretes or mortars. The phase most often in contact with the aggregate surface appears to be C-S-H; this is consistent with the early results of Javelas *et al.* (J34). Relatively large crystals or polycrystalline masses of CH may occur locally close to the aggregate surface, but in some cases at least, this results from major increases in local porosity caused by the sweeping action of the aggregate particle during mixing or from bleeding. Observations of a high concentration of CH at the interface in fracture surfaces may also result from preferred fracture through weaker regions in which this phase is present (S115).

12.1.3 The nature of the paste–aggregate bond

The attractive forces between paste and aggregate are in principle likely to be similar to those existing within the paste (Section 8.4.3). Chemical reactions between paste and aggregate can strengthen the bond by corroding the aggregate surface and thereby increasing the area of contact (F43,G82), but

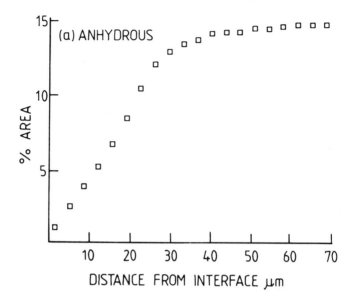

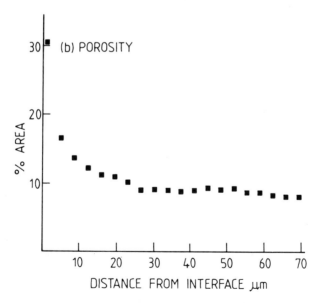

Fig. 12.1 Microstructural gradients in the interfacial region of a concrete: (a) unreacted clinker phases, (b) porosity, determined using SEM with image analysis. Each point represents a mean from 50 determinations, and the standard errors were approximately 10% relative. Scrivener *et al.* (S114).

they can also weaken the material by creating expansive forces too strong for it to withstand.

Struble *et al.* (S109) reviewed early work on the paste–aggregate bond. Postulated superficial pozzolanic reactions at ordinary temperatures between paste and quartz or other common siliceous aggregates appear to be unsupported by experimental evidence, except for the deleterious alkali–silica reaction (Section 12.4). On the other hand, there is considerable evidence that superficial reaction occurs with calcite aggregate and that it strengthens the bond with the paste (F43,B151,G82,M117). From XRD and SEM studies on composites of pastes cast against marble surfaces and on pastes made from cement mixed with finely ground calcite, Grandet and Ollivier (G82) concluded that $C_4A\bar{C}_{0.5}H_{12}$ was formed initially and that it was later replaced by $C_4A\bar{C}H_{11}$. The calcite surfaces were strongly pitted. This reaction, earlier postulated by Lyubimova and Pinus (L54), was noted in Section 9.6.3 in connection with the use of calcite as a filler.

Monteiro and Mehta (M117) found that the calcite was similarly pitted if alite was substituted for cement and reported the appearance of an XRD peak at 0.79 nm. They considered that the reaction product, either with alite or with cement, was not a CO_3^{2-}-containing AFm phase but a basic calcium carbonate. Such phases have been reported as synthetic products (S116) and as a natural mineral, defernite (P50).

12.1.4. Composite cements and other topics

Partial replacement of cement by pfa can either increase or decrease the width of the interfacial zone, depending on the characteristics of the pfa; slag has relatively little effect (C65). In concretes made with microsilica and superplasticizers, the paste is dense right up to the aggregate surfaces (R65,B152,S111). Scrivener and Gartner (S111) found some increase in porosity, but this could be almost entirely attributed to the presence of hollow-shell grains. With pastes vibrated against a single piece of aggregate, in contrast, the replacement had no significant effect on the distribution of porosity within the interfacial zone, and some clumping of the microsilica was observed. Clumping was found to occur in pastes even when a superplasticizer was used (S115), suggesting that the coarse aggregate breaks up the clumps by crushing or shearing during mixing. For a given w/s ratio, partial replacement of cement by microsilica increases the strength in concrete but not in pastes (B152). It also increases the calculated porosity (R66) (Section 9.7.1), and the strength increase in concrete of a given w/s ratio is thus due to the improvement in the paste–aggregate bond (R66,B152,S111).

Interfaces of calcium aluminate cements with aggregates were discussed in Chapter 10. The interfacial zone between Portland cement paste and steel has been studied by SEM, using various techniques of specimen preparation (e.g. A31,B153). The major features observed have been substantial deposits of CH and, further away from the interface, a relatively weak, porous zone. Bentur *et al.* (B135) also reported the occurrence of a duplex film. Interfaces of cement or C_3S pastes with zinc (A32,T58) or copper (T58) have been studied.

12.1.5 Effects at exposed surfaces

The microstructures of the surface regions of a concrete differ from that of the bulk material, and are important because they have a major effect on durability. The differences have several origins. Settlement and constraints imposed by the formwork may affect aggregate distribution. Moisture gradients can be created if a surface is exposed to a drying atmosphere, retarding or even stopping hydration in the surface layers, altering the pore structure and facilitating carbonation. Water can be lost by absorption in or leakage through formwork.

Patel *et al.* (P51) moist-cured blocks of cement pastes of w/c ratio 0.59 for 7 days and then sealed the prism surfaces of each and exposed the ends to air at 20°C and 65% RH. Methanol sorption data obtained using porous glass as a reference standard (Section 8.3.5) showed that, near to the exposed surfaces, the pore structure was markedly coarser and the diffusion time lower. TG evidence showed that less hydration had occurred and that carbonation was increased. These effects were detectable even at a distance of 50 mm from the exposed surface. Further work confirmed the marked effect of RH on hydration rates (P28) (Section 7.7.1).

Crumbie *et al.* (C66) moist-cured concrete specimens with w/c 0.4 and 0.6 in steel moulds for 24 h and then cured them in saturated CH solution for 28 days. Slices from the top, bottom, sides and bulk of each specimen were examined by methanol exchange, quantitative determination of coarse and fine aggregate and backscattered electron imaging in the SEM. The largest variations in the aggregate distribution were in the contents of coarse aggregate in the top 5 mm, which were much reduced, especially in the mix with w/c 0.6, for which the content was almost zero. The methanol exchange data showed that the porosity and permeability increased near to all the surfaces. SEM examination of a section cut perpendicular to the top surface confirmed the increase in porosity, which was about twice as high in the outermost 200 µm as in the bulk. The surface region was also depleted in CH, but the content of anhydrous material was little affected. Despite the fact that the specimens had been placed in saturated CH, some leaching appeared to have occurred.

12.2 Durability—general aspects

If it is properly designed and produced for the environment in which it has to serve, concrete is an extremely durable material. If the design or production is inadequate, it may deteriorate. Some environments exist in which no concrete is durable.

Deterioration may occur through a variety of chemical or physical processes. The rest of this chapter deals with these, with emphasis on chemical processes in which the cement paste is involved. For convenience, the various forms of attack are considered separately, but examination of deteriorated concretes often shows that more than one form of attack has been operative. This is often because one form of attack renders the concrete more susceptible to damage by another. Portland and composite cement concretes are considered; calcium aluminate cements were discussed in Section 10.1.11.

The concept of a design life for concrete structures is becoming increasingly recognized (S117), failure then being defined as deterioration of properties to an unacceptable level at an age lower than the design life. To design a concrete for a specified minimum life, it is necessary to understand the processes that cause deterioration, including the rates at which these will occur under the conditions to which it will be subjected. In this last respect, a wide range of climatic, chemical and physical factors must be considered.

12.3 Carbonation, chloride penetration and corrosion of reinforcement

12.3.1 General

Corrosion of reinforcement is probably the most widespread cause of deterioration in concrete. The expansion produced by rust formation causes the surrounding concrete to crack and spall. In a sound concrete, rusting is prevented by the high pH of the pore solution, which stabilizes an oxide film on the steel that inhibits further attack. This film is unstable at lower pH values, which can result from carbonation or leaching, or in the presence of Cl^-. Sources of the latter include sea water or salt spray, de-icing salts used on roads, certain aggregates, especially those available in desert climates, and $CaCl_2$ used as an accelerator. Though now widely prohibited, this can still affect older structures. Tensile stress can damage the protective film, and particular care must be taken to minimize ingress of CO_2 and Cl^- in prestressed structures.

Carbonation begins at exposed surfaces, and spreads inwards at a rate proportional to the square root of time (K66). The thickness of the affected layer can be approximately determined by testing a section with phenolphthalein. Chloride ion from an external source similarly penetrates inwards. The age at which corrosion is liable to begin thus depends on the minimum thickness of concrete covering the reinforcement, the resistance to penetration and the conditions to which the concrete is subjected. The depth of cover needed to provide protection over a given period can be calculated (S117,B154).

The resistance to penetration depends on the permeability of the concrete, which is affected by cement content, w/c ratio, aggregate grading, degree of compaction and adequacy of curing. Lack of bonding between the steel and concrete decreases this resistance and also destroys the protection afforded by the alkaline pore solution. Microcracks are in some circumstances closed up by the products of carbonation or by ingress of dust, but this may not occur, especially under fluctuating loads, which may thus accelerate the onset of corrosion. With a dense, well-made concrete, 15 mm of cover may suffice to provide virtually indefinite protection, but with a porous or poorly made concrete, it may allow corrosion to begin within a few years (B154). Important environmental factors, in addition to stress and availability of CO_2, O_2 and Cl^-, include RH and temperature. At low humidities, gases penetrate easily through empty pores, but lack of water may inhibit both carbonation and corrosion; at very high humidities, these effects are reversed. Chloride penetration requires the presence of water. Attack is thus most serious at intermediate humidities.

Roy (R67) reviewed carbonation and transport and reactions of chlorides. Hansson (H63) and Turriziani (T59) reviewed corrosion.

12.3.2 Carbonation

Carbon dioxide dissolves in the pore solution of cement paste, producing CO_3^{2-}, which reacts with Ca^{2+} to produce $CaCO_3$. The OH^- and Ca^{2+} ions required by these reactions are obtained by the dissolution of CH and decomposition of the hydrated silicate and aluminate phases. C_4AH_x is quickly converted into $C_4A\bar{C}H_{11}$ and ultimately into $CaCO_3$ and hydrous alumina; monosulphate and ettringite yield $CaCO_3$, hydrous alumina and gypsum. C–S–H is decalcified, initially by lowering of its Ca/Si ratio, and ultimately by conversion into a highly porous, hydrous form of silica. The reactions involving the CH and C–S–H may be represented by the following equations:

$$CO_2 + 2OH^- \rightarrow CO_3^{2-} + H_2O \quad (12.1)$$

$$Ca^{2+} + CO_3^{2-} \rightarrow CaCO_3 \quad (12.2)$$

$$Ca(OH)_2 \rightarrow Ca^{2+} + 2OH^- \quad (12.3)$$

$$xCaO.SiO_2(aq) + zH_2O \rightarrow yCa^{2+} + 2yOH^- + (x-y)CaO.SiO_2(aq) \quad (12.4)$$

SEM by backscattered electron imaging shows that, in the early stage of carbonation, a dense rim consisting largely of $CaCO_3$ is formed at the surface (Fig. 12.2). Within this zone, CH is largely absent and C–S–H, including that formed in situ from the alite or belite, is in varying degrees decalcified. The $CaCO_3$ can form relatively large, continuous masses. The microstructural aspects of the reaction have not been systematically studied, but it appears likely that extensive reorganization occurs.

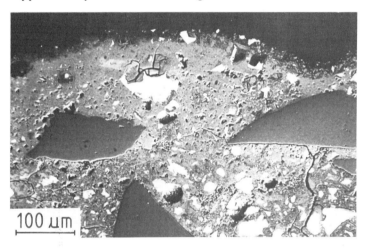

Fig. 12.2 SEM (backscattered electron image) of a concrete showing a carbonated surface layer 50–100 μm thick, forming a dense rim within which C–S–H has been decalcified to give a silica gel low in CaO. Bright regions are of unhydrated clinker and dark, angular regions are of aggregate. Courtesy WHD Microanalysis Consultants Ltd., Ipswich, UK (H64).

The results of the phenolphthalein test show that the pH falls to 8.5 or below in the carbonated zone, but no more detailed studies on the composition of the pore solution have been reported. Andrade et al. (A33) concluded that a value as low as 8.5 could only be explained if alkali cations were removed from the solution by leaching or uptake by the hydrous silica

386 Cement Chemistry

formed by decomposition of the C–S–H. In the pure $(K,Na)_2O$–CO_2–H_2O system, a pH of 8.5 corresponds approximately to formation of HCO_3^-. Suzuki et al. (S118) studied the mechanism of carbonation of C–S–H in suspensions. Barret et al. (B122) discussed equilibria relevant to the carbonation of C_4AH_x.

Apart from the effect on reinforcement corrosion, carbonation has both good and bad effects on concrete. It markedly increases the compressive and tensile strengths of Portland cement mortars and concretes, sometimes by as much as 100%, in the regions affected. The strengths also increase if the binder is a composite cement made with 40% of slag, but decrease if this proportion is increased or with supersulphated cements (M118). The changes in strength are paralleled by ones in weight for a specified drying condition, indicating that gain in strength occurs if the uptake of CO_2 exceeds the loss of bound water. Carbonation also makes a specific and substantial contribution to irreversible shrinkage; as Lea (L6) pointed out, this indicates that the C–S–H is attacked. It may cause superficial crazing, and can contribute significantly to the drying shrinkage measurements on small specimens. The reduction in permeability due to carbonation of the surface layers results in a decrease in the reversible volume changes with subsequent variations in RH.

12.3.3 Transport and reactions of chlorides

The risk of chloride-induced corrosion increases with ease of migration of Cl^- through the cement paste and with the ratio of Cl^- to OH^- in the pore solution (P52). The diffusion of Cl^- through cement pastes was discussed in Section 8.6.2. In concrete, transport can also occur through regions of enhanced permeability or microcracks and diffusion is supplemented by capillary action except in a completely saturated material. Factors affecting the Cl^-/OH^- ratio include the OH^- concentration and the extent to which Cl^- is taken up by solid phases. The OH^- concentration depends on the alkali content of the cement and will often be lower for a composite than for a pure Portland cement. Chloride is partly bound in $C_3A \cdot CaCl_2 \cdot 10H_2O$ but much remains in the pore solution (P48,M119,R67,R68). With very concentrated chloride solutions, other solid phases may be formed (Section 12.6.3). Formation of $C_3A \cdot CaCl_2 \cdot 10H_2O$ may be decreased in the presence of sulphates, due to preferential formation of monosulphate or ettringite, or by carbonation, and possibly depends on the source of the Cl^- (P52).

Page et al. (P52) determined the Cl^-/OH^- ratios in the pore solutions, Cl^- diffusivities and corrosion rates of embedded steel in pastes of cements of several types. With Portland cements, the Cl^-/OH^- ratios decreased with potential C_3A content; cements containing 30% pfa or 65% ggbfs gave lower ratios than any of the Portland cements. The OH^- concentrations were lower for the composite cements, and the low Cl^-/OH^- ratios were presumably due to greater uptake of Cl^- by the hydration products. The Cl^- diffusivities increased in the sequence slag cement < pfa cement < ordinary Portland cements < sulphate-resisting Portland cement. In conformity with these results, the corrosion rates were slightly lower for the composite cements than for an ordinary Portland cement, and considerably lower than that for the sulphate-resisting Portland cement. The authors stressed the need to study a wider range of cements and chloride solutions before any general conclusions could be made. The results nevertheless indicate that binding of Cl^-, as well as diffusivity, plays an important part in limiting chloride-induced corrosion. A microstructural study (M120) showed greater corrosion with a microsilica cement than with a pure Portland cement; replacement of cement by microsilica may be expected both to lower the OH^- concentration and to decrease the content of AFm phase able to take up Cl^-.

12.3.4 Corrosion

The corrosion of steel in concrete is an electrochemical process. Due to local compositional or structural variations, some areas become positively and others negatively charged, and electrical cells are set up. The anodic reaction and the principal cathodic reaction are respectively typified by the equations

$$3Fe + 8OH^- \rightarrow Fe_3O_4 + 4H_2O + 8e^- \quad (12.5)$$

and

$$4e^- + 2H_2O + O_2 \rightarrow 4OH^- \quad (12.6)$$

At the anodes, iron is dissolved and an oxide deposited. Electrons travel from anode to cathode within the metal, and OH^- ions travel from cathode to anode through the solution with which it is in contact. For these processes to occur, a continuing source of oxygen is needed, and the surface of the metal must remain wet. If pH is above about 11.5 (P48) and Cl^- is absent, the oxide is deposited as a thin protective film which is virtually continuous, and the rate of attack is so low as to be insignificant. The iron is said to be in a passive condition. At a lower pH, an oxide or oxyhydroxide is deposited in an incoherent form, and corrosion is rapid.

Chloride ions cause local breakdown of the passive film, even at high pH. The regions of metal thus exposed become anodes and the unaffected areas become cathodes. Since the areas of breakdown are small, high current densities can develop at the anodes, causing pitting and localized decrease in pH, which aggravates the attack. High local ratios of Cl^- to OH^- in the solution favour pitting (P52).

As was noted in Section 12.3.1, corrosion may be inhibited if the permeability of the concrete to water or oxygen is low enough, and thus does not necessarily occur even if the concrete in contact with the reinforcement is carbonated. The maintenance of the passive film requires the presence of some oxygen, and if the supply of oxygen is very restricted, as in concrete buried under moist ground or in deep water, it can all be used up. In theory, corrosion can then occur through reduction of water to H_2, but in practice, corrosion rates are little greater than when a passive film exists (H63).

Several approaches to the prevention of corrosion have been followed, in addition to that of making concrete that is less permeable and that gives adequate cover to the steel (T59). They include:

(1) Cathodic protection, in which a voltage is applied between the reinforcement and a conducting paint applied to the outer surface of the concrete.
(2) Sacrificial anodes, in which an electropositive metal or alloy is embedded in the concrete and connected electrically to the reinforcement.
(3) Corrosion inhibitors, e.g. $Ca(NO_2)_2$, which appear to act by stabilizing or otherwise altering the oxide film or by scavenging oxygen.
(4) Inert coatings of organic materials applied to the steel.
(5) Galvanizing, which, however, has been considered to retard hydration and to weaken the paste–metal bond (A32,T59).

12.4 Alkali–silica reaction

12.4.1 General

The general features of alkali–silica reaction (ASR) were first described by Stanton (S119). Hydroxide ions in the pore solution react with certain types of silica that can occur in the aggregate, resulting in internal stresses that can cause expansion and cracking. Failure may occur within days or only after years. On the surface of unrestrained concrete, it typically produces a random network of fine cracks with some large cracks ('map cracking'). In reinforced concrete, the cracks tend to form parallel to the reinforcement. Thin sections show cracks that can pass through the aggregate; a characteristic gel is present and can occur in the cracks, as rims around aggregate particles, or elsewhere in the paste (Fig. 12.3). The gel can exude from the concrete.

The necessary conditions for ASR in a Portland cement concrete are a sufficiently high content of alkali oxides in the cement, a reactive constituent in the aggregate and a supply of water. The K^+ and Na^+ are present in the cement as sulphates and in the silicate and aluminate phases. When the compounds containing these ions react, their anions enter products of low solubility, such as ettringite, C-S-H or AFm phases, and equivalent amounts of OH^- are formed. The K^+ or Na^+ are involved at this stage only in the negative sense that, since their hydroxides are soluble, they allow the OH^- to enter the pore solution. ASR is unlikely to occur in concrete made with Portland (not composite) cements if the content of equivalent Na_2O ($Na_2O_e = Na_2O + 0.66K_2O$) in the concrete is below $4\,kg\,m^{-3}$, and a practical limit of $3\,kg\,m^{-3}$ has been proposed to allow for day to day variations in cement composition (H65). An alternative criterion based on cement composition ($Na_2O_e < 0.6\%$) does not allow for varying cement content in the concrete. Alkali cations may also be supplied from external sources such as solutions of Na_2SO_4 (T60,P53) or NaCl (N21), mineral admixtures (Section 12.4.4) or aggregates. In all cases they will yield equivalent amounts of OH^- except insofar as they enter solid phases or are balanced by other anions that stay in solution.

Fig. 12.3 SEM (backscattered electron image) of a concrete showing gel (dark patches) formed by ASR, present in the cement paste in places far removed from the reactive aggregate. Bright regions are of unhydrated clinker. Courtesy WHD Microanalysis Consultants Ltd, Ipswich, UK (H64).

Diamond (D50) described the types of silica that can take part in ASR. They include quartz if sufficiently strained or microcrystalline, tridymite, cristobalite and glass or other amorphous forms, which occur in varying combinations in opals, flints, cherts and other rock types. Opals are especially reactive. Macroscopic, unstrained crystals of quartz appear to be unreactive but are possibly not completely inert. Some silicate minerals and volcanic glasses may undergo reactions similar to ASR.

If all other variables are fixed, a curve of expansion against the percentage of the reactive constituent in the aggregate often passes through a maximum at a 'pessimum' composition. For opals and other highly reactive constituents in mortars, this is typically under 10% (Fig. 12.4) (H66) but for less reactive constituents it can be much higher and may even occur at 100%. The expansion also depends on the particle size of the reactive constituent. The results of different investigators vary considerably (H67), probably on account of differences between the materials used, but expansion appears always to be greatest for material in the 0.1–1.0 mm range and is possibly zero or negligible if the particle size is below about 10 µm. At least some specimens of opal, if sufficiently finely ground, act as pozzolanas (B155).

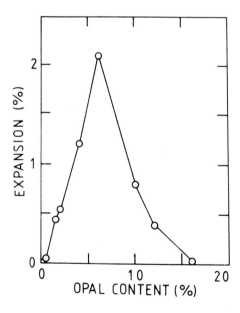

Fig. 12.4 Relation between the expansion of a mortar (w/c = 0.41, aggregate/cement = 2, age = 200 days) and the content of Beltane opal, expressed as a weight percentage on the total aggregate. Hobbs (H66).

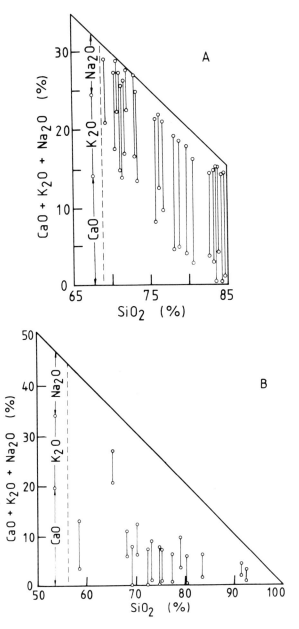

Fig. 12.5 Compositions of gels produced by ASR (A) in cracks or otherwise deposited within a concrete and (B) exuded to the exterior. In each case, the data are given as percentages on the ignited weight, and points to the left of the broken line merely illustrate how the results are expressed. Knudsen and Thaulow (K67).

12.4.2 Chemistry of alkali–silica reaction

Knudsen and Thaulow (K67) made X-ray microanalyses of the gel present in cracks in a 30-year-old concrete that had undergone ASR and compared them with typical analyses of gel exuded from the surface, reported by Idorn (I15). The gel in the cracks was, in general, much higher in CaO and lower in alkalis than the exuded gel (Fig. 12.5), and its CaO content increased on moving along a crack away from its source. The authors concluded that an alkali silicate gel formed initially reacted with Ca^{2+} ions from the cement paste, with which the exuded material had scarcely been in contact. The compositions of the gels were analogous to those of ones in the Na_2O–CaO–SiO_2–H_2O system studied by Kalousek (K19). Regourd et al. (R69) made similar observations. Diamond et al. (D51) observed that K^+, but not Ca^{2+}, penetrated deeply into opal grains, and Bhatty (B156) found that the gel formed from C_3S, opal, pfa and NaOH was essentially a calcium silicate gel and gave an XRD pattern of C–S–H(I).

These results suggest that the chemistry of ASR is essentially that of pozzolanic reaction (Section 9.3.5), the different effect on concrete properties arising mainly from the difference in particle size of the siliceous material. The following explanation of the process is essentially that given by Powers and Steinour (P54) and, in more detail, by Dent Glasser and Kataoka (D52,D53). Hydroxide ions attack the silica, causing Si–O–Si bridges to be replaced by pairs of SiO^- groups. This process, many times repeated, ultimately fragments the three-dimensional silica framework into separate silicate anions, which in a concentrated solution are of varying sizes. The negative charges, whether on oxygen atoms still forming part of the framework or on ones present in separate silicate ions, are balanced by the most readily available cations, which are K^+ and Na^+. The damaged framework is sufficiently deformable that, for some compositions at least, the material can imbibe H_2O molecules and expand; if sufficiently fragmented, it can be highly mobile. In these respects, it differs from C–S–H, which has a relatively rigid structure based on Ca–O layers. It is unstable in the presence of Ca^{2+}, which reacts with it to form C–S–H. This regenerates OH^- in the solution, e.g. from CH if the latter is the source of the Ca^{2+}. The alkali silicate constituent of the gel can possibly itself accommodate some Ca^{2+}, but a point is probably soon reached at which particles of C–S–H begin to form as a separate constituent.

In pozzolanic reaction, the alkali silicate gel is formed in an environment rich in Ca^{2+} and, except in a narrow zone close to the reacting surface, is quickly converted into C–S–H. In ASR, it is formed in an environment poor in Ca^{2+}, and massive outflows of gel may result. The cement paste cannot supply Ca^{2+} fast enough to prevent much of this gel from persisting for long periods. This situation is especially marked if the alkali silicate gel forms within the aggregate particles, as is the case with opal.

The K^+, Na^+ and OH^- concentrations in the pore solutions of mortars containing reactive aggregates are lowered due to uptake by the gel (D51). Kollek et al. (K68) showed that destructive expansion was unlikely to occur in a particular series of mortars made with pure Portland cements if the OH^- concentrations in otherwise similar mortars not containing the reactive constituent had not reached $0.6\,\text{mol}\,l^{-1}$ by 180 days. This is a correlation and not a casual relationship, and it would be wrong to regard any concentration determined in the absence of reactive aggregate as a 'trigger' value at which the aggregate begins to react.

Cole et al. (C67,C68) showed from XRD, X-ray microanalysis, IR, DTA and thermal weight loss data that the rims around sandstone and siltstone aggregates in a 30-year-old concrete contained an alkali-substituted okenite ($C_5S_9H_9$) (M121) and a precursor phase characterized by a 1.22-nm XRD spacing. This latter phase was more highly hydrated than okenite, had formed in a wet environment and on being dried changed into okenite. A possible similarity to rhodesite $[(Ca,K,Na)_9Si_{16}O_{40}\cdot 11H_2O]$ was suggested, but the latter does not change into okenite on dehydration (G83). Okenite has a structure based on silicate sheets and layers (M121), and a higher hydrate of similar structure could well exist. Similar products have been observed by other workers (B157,D54) and had previously been tentatively identified as zeolites (C68), or jennite or tricalcium silicate hydrate (B157). These crystalline phases appear to be minor, though possibly equilibrium, products of ASR.

A 'rapid chemical test' for the susceptibility of aggregates to ASR has been described (M122). The chemistry on which it is based is confused (D55).

12.4.3 The expansion process

Expansion due to ASR is generally attributed to imbibition of pore solution by the alkali silicate gel (V10); the resulting swelling pressure is not simply related to the expansion in absence of restraint (D52). In accordance with this hypothesis, many dried sodium silicate gels swell markedly on wetting and can develop high swelling pressures (S120), whereas the equivalent volume changes for C–S–H are small. Vivian (V10) found that reactive mortars stored for about 40 days in dry air did not expand, but if then placed in water expanded very rapidly. Expansion can also occur in a concrete that is continuously saturated; Hobbs (H65) has argued that this process is one in which a solid reactant yields a more voluminous solid product but which cannot properly be called imbibition. Osmosis has been postulated (H68) but there is no substantial evidence that a semipermeable membrane exists.

Three principal hypotheses have been proposed to account for the decreased expansion observed if the reactive material is more finely ground or present in quantities above the pessimum:

(1) The gel is more thinly distributed, and Ca^{2+} ions are therefore more readily available to it, accelerating its conversion into the virtually non-expansive C-S-H (P54). The thinner distribution of the gel might also in itself decrease the expansive stresses.

(2) The gel has a lower ratio of alkali cations to SiO_2, which renders it less expansive. Dent Glasser and Kataoka (D56) found the expansion of the solid phases on placing silica gel in NaOH solutions to pass through a maximum at SiO_2/Na_2O ratios that might reasonably have corresponded to pessimum compositions. Struble and Diamond (S120,S121) obtained broadly similar results for swelling pressures and amounts of unrestrained swelling for dried sodium silicate gels placed in water, but various anomalies led them to conclude that the swelling characteristics were probably more directly related to the anion structure of the gel. This depends not only on SiO_2/Na_2O ratio, but also on drying, ageing, temperature, type of silica, carbonation and possibly other factors.

(3) The gel is formed before the paste has hardened sufficiently for damage to occur (O28,H65).

These hypotheses are not mutually exclusive, though the general slowness of pozzolanic reactions renders (3) unlikely. The evidence is insufficient to determine their relative importance, which could probably be assessed from a combination of expansion data, pore solution analyses and SEM of polished sections with accompanying X-ray microanalyses made in parallel on a series of mortars of suitable ages and compositions.

Curves of expansion against time typically have the form shown in Fig. 12.6. Cracking usually begins at an expansion of 0.05-0.1%. It is unlikely that the start of the reaction is delayed, and thus probable that a certain degree of reaction can occur without producing expansion, which presumably begins when the gel can no longer be accommodated in the pore structure of the paste without producing stresses sufficient to disrupt it.

12.4.4 ASR in mortars or concretes made with composite cements

The relations between expansion and the content and particle size of the reactive aggregate early suggested that expansion might be decreased by the use of pozzolanic admixtures, and provided that the latter do not themselves supply too much alkali this has been found to be the case. Hobbs (H69) reviewed and extended data on the effects of replacements by pfa or ggbfs on expansion in mortars and concretes. In the great majority of cases, the replacements decreased expansion, though not always sufficiently to prevent failure. In a few cases in which the admixture was itself especially high in alkalis, expansion increased. Failure sometimes occurred at lower concrete Na_2O_e contents than would have been expected if these where calculated from the contribution of the Portland cement alone. Hobbs concluded that, contrary to some earlier views, allowance must be made for the alkalis supplied by the admixture, and suggested that, failing a direct test of the effectiveness of a particular pfa or slag in preventing expansion, the Na_2O_e content should be calculated by adding one-sixth of the total alkali content of the pfa, or one-half of that of the slag, to the contribution from the cement.

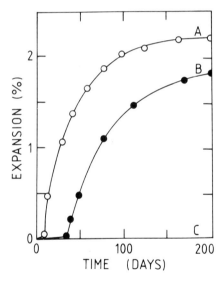

Fig. 12.6 Typical relations between expansion produced by ASR and age; data are for mortars with w/c = 0.41 and aggregate/cement = 2, (A) with 6.7% of Beltane opal in the aggregate and a cement with Na_2O_e = 1.24%, and (B) with 4.5% of Beltane opal in the aggregate and Na_2O_e = 0.8%. After Hobbs (H66).

Kollek *et al.* (K68) extended the parallel studies on expansion and pore solution compositions mentioned in Section 12.4.2 to mortars containing pfa, slag or natural pozzolanas. For each mix, several contents of the reactive aggregate (Beltane opal) were used to ensure that pessimum compositions were being compared. With a high alkali cement ($Na_2O_e = 0.92\%$), failure occurred in the absence of an admixture, but was avoided at sufficiently high replacement levels of pfa, slag or one of the natural pozzolanas used. With a cement lower in alkali ($Na_2O_e = 0.68\%$), failure did not occur in the absence of an admixture, but took place in mixes with either of the pfas, most of the natural pozzolanas and one of the slags tested.

The pore solution compositions were consistent with these results. They could not be explained by assuming that the admixtures acted as inert diluents in regard to the alkali contents of the solutions. In the mixes with pfa, failure could be correlated with an OH^- concentration greater than $0.3\,mol\,l^{-1}$ at 180 days in an otherwise similar mix not containing the reactive aggregate. The corresponding values for mixes with slags or natural pozzolanas were $0.5\,mol\,l^{-1}$ and $0.4\,mol\,l^{-1}$, respectively. The relations between the OH^- concentration in the pore solution of a mix not containing reactive aggregate and the likelihood of a failure in a mix in which one is present thus differ for the various types of cements.

The alkali cations in pfa normally occur almost entirely in the glass, and when the latter reacts may be presumed to enter the alkali-rich silicate that appears to be the initial product. When this phase is decomposed by reaction with Ca^{2+}, they will be distributed, like alkali cations from any other source, between the solution and the solid hydration products, on which they are probably adsorbed (Section 7.3.2). The C–S–H tends to take them up more strongly as its Ca/Si ratio decreases (B158,G63); consequently, the alkali cations released from the pfa are less effective in raising the OH^- concentration of the pore solution than are those released from the cement. The method outlined in Section 7.5.2 for calculating the OH^- concentration in the pore solution of Portland cement mix was extended to cover Portland–pfa cement mixes taking this into account (T37).

12.5 Sulphate attack

12.5.1 General

Portland cement concrete is attacked by solutions containing sulphate, such as some natural or polluted ground waters. Attack can lead to strength loss, expansion, cracking and, ultimately, disintegration. A 1 : 10 mortar placed in 1.8% Na_2SO_4 solution can show a linear expansion of 0.5% within a few weeks (T60). Expansion is not always an adequate measure of strength loss. The effects are minimized in a dense concrete of low permeability and by using a sulphate-resisting cement, in which there is little or no aluminate phase.

Sulphate attack has often been discussed in terms of reaction between solid phases in the cement paste and dissolved compounds, such as Na_2SO_4 or $MgSO_4$, in the attacking solution. This obscures the fact that the reactions of the cations and anions in that solution are essentially independent; for example, a solution of Na_2SO_4 may cause both sulphate attack and ASR (T60,P53), and one of $MgSO_4$ causes sulphate attack and reactions forming brucite.

XRD shows that sulphate attack leads to formation of ettringite and often also of gypsum (H59). SEM shows cracks associated with massive growths of ettringite around aggregate particles (Fig. 12.7). Gypsum can form in similar locations. Both ettringite and gypsum can also form in pores and, in the regions that have been attacked, CH disappears and the Ca/Si ratio of the C–S–H formed in situ from the alite or belite decreases, ultimately converting it into hydrous silica (C66,H64). The effects on the CH and C–S–H are thus similar to those produced by carbonation (Section 12.3.2). The expansion has generally been attributed mainly to the formation of ettringite and to a lesser extent to that of gypsum.

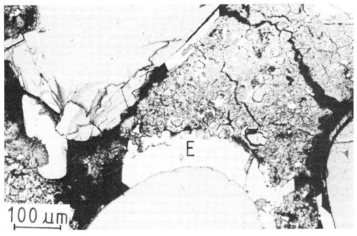

Fig. 12.7 SEM (backscattered electron image) of a concrete showing massive deposits of ettringite (one marked 'E'), which through sulphate attack has grown on aggregate surfaces, with associated cracking. Courtesy WHD Microanalysis Consultants Ltd, Ipswich, UK (H64).

The observed microstructures show that both ettringite and gypsum are deposited from solution, crystallization thus taking place in accordance with the following equations:

$$6Ca^{2+} + 2Al(OH)_4^- + 4OH^- + 3SO_4^{2-} + 26H_2O \rightarrow C_6A\bar{S}_3H_{32} \qquad (12.7)$$

$$Ca^{2+} + SO_4^{2-} + 2H_2O \rightarrow C\bar{S}H_2 \qquad (12.8)$$

The $Al(OH)_4^-$ and OH^-, and some of the Ca^{2+}, needed to form ettringite could be supplied by unreacted aluminate or ferrite phase, but the main source of $Al(OH)_4^-$ in a hardened concrete is likely to be AFm phase, represented here as monosulphate:

$$C_4A\bar{S}H_{12} \rightarrow 4Ca^{2+} + 2Al(OH)_4^- + 4OH^- + SO_4^{2-} + 6H_2O \qquad (12.9)$$

Whichever phase provides the $Al(OH)_4^-$ and OH^-, additional Ca^{2+} is needed to form either ettringite or gypsum. Reaction of cement pastes with $CaSO_4$ solutions is a special case, considered later. In other cases, CH dissolves and C–S–H is decalcified, in accordance with equations 12.3 and 12.4, respectively. These latter processes release a quantity of OH^- equivalent to the SO_4^{2-} taken up in ettringite and gypsum. If the only cations in the attacking solution are Na^+ or K^+, this OH^- remains in the solution, where it augments that already present. This can cause ASR if the aggregate is reactive. In this respect, the situation is similar to that existing if the concrete is placed in a solution of alkali hydroxide, though the local concentrations of the OH^- could be higher if this ion is formed as a byproduct of sulphate attack.

Gypsum that has been formed by the reaction shown in equation 12.8 may dissolve with precipitation of ettringite. The dissolution is represented by the reversal of the same equation, and the additional ions required could be provided by dissolution of monosulphate, represented by equation 12.9.

C_3S pastes are slowly attacked by 0.15 M Na_2SO_4 solution; more concentrated Na_2SO_4 solutions also attack β-C_2S pastes (T60). The action is possibly due to formation of gypsum. Lea (L6) stated that Na_2SO_4 solutions do not attack the C–S–H in cement pastes, but the SEM evidence does not support this view. While the observed decalcification of the C–S–H might be due to carbonation, which may take place more readily in material affected by sulphate attack, it could also be a direct consequence of the latter. C–S–H must be attacked in accordance with equation 12.4 if the Ca^{2+} concentration in the pore solution with which it is in contact falls sufficiently, and the formation of ettringite might produce such a decrease.

For cement pastes containing high proportions of finely ground gypsum, Odler and Gasser (O29) found a correlation between expansion and the amount of ettringite formed, but for mortars immersed in Na_2SO_4 solutions no such relation appears to exist (T60,H59). As with Type K expansive cements, the morphology of ettringite and the time and place of its formation are probably of prime importance. The discussion on the mechanism of expansion in these cements (Section 10.2.3) is probably largely applicable also to sulphate attack, though in the latter case the ettringite is formed around aggregate particles and elsewhere in the cement paste, and not around particles of $C_4A_3\bar{S}$.

12.5.2 Calcium sulphate

Some aggregates, notably in the Middle East, contain gypsum, which can cause so-called internal sulphate attack. Light microscopy shows that the gypsum is replaced in situ by CH, while ettringite is formed in the cement paste and can cause expansion (C69). The process entails counter migration of SO_4^{2-} and OH^- between paste and aggregate; within the paste, the reactions are presumably the same as in attack by SO_4^{2-} from an external source. The effect is lessened by using either a sulphate-resisting or a rapid-hardening Portland cement, in the latter case because the paste can become less permeable before much SO_4^{2-} has been released.

Despite its relatively low solubility, $CaSO_4$ present in solution can also attack concrete. In this case, the solution can provide all the ions that are needed to produce ettringite from monosulphate, and little or no dissolution of CH nor decalcification of C–S–H need occur. The saturated solution also slowly attacks C_3S pastes (T60); the explanation of this is not clear. The use of abnormally large quantities of finely divided gypsum interground or blended with the cement also produces expansion (O29), but little appears to be known about the microstructures that result.

12.5.3 Magnesium sulphate

Some ground waters contain significant concentrations of both SO_4^{2-} and Mg^{2+}, and are more aggressive towards concrete than SO_4^{2-} solutions of similar concentration in which the only cations are Na^+ or K^+. The action of the SO_4^{2-} is augmented by that of the Mg^{2+}, which is a particular case of the more general effect of cations forming hydroxides or basic salts of low solubility, considered further in Section 12.6.3. The Mg^{2+} reacts with OH^- ions from the pore solution, giving brucite:

$$Mg^{2+} + 2OH^- \rightarrow Mg(OH)_2 \quad (12.10)$$

The brucite forms a hard, dense skin on the mortar or concrete, which tends to hinder further attack (L6). The OH^- present in the pore solution would soon be consumed were it not replenished by dissolution of CH and decalcification of C–S–H, by the reactions shown in equations 12.3 and 12.4. In the reaction of a cement paste with an $MgSO_4$ solution, the CH and C–S–H thus serve as sources of both Ca^{2+} and OH^-.

According to Lea (L6), the ettringite formed by the reaction of SO_4^{2-} is itself ultimately decomposed by Mg^{2+}, leaving a residue of gypsum and hydrous alumina. This was attributed to a decrease in the pH of the pore solution to that of saturated $Mg(OH)_2$, which is about 10.5. Such a decrease could only occur if the alkali cations are lost from the pore solution, possibly by leaching, or if the SO_4^{2-} concentration in the solution increases. There does not appear to be any experimental evidence with which these hypotheses may be tested.

Under some circumstances, the Mg^{2+} enters sparingly soluble phases other than brucite. Roy *et al.* (R70) observed the formation of a hydrotalcite-type phase in a slag cement paste that had been treated with $MgSO_4$ solution. Its formation was possibly favoured by the enhanced availability of $Al(OH)_4^-$ provided by the slag. Cole (C70) reported the formation of a hydrated magnesium silicate in a deteriorated concrete seawall.

12.5.4 Effects of concrete properties and cement type

The superior resistance of dense, well-compacted concrete to sulphate attack indicates that the ability of the sulphate solution to diffuse into the concrete is a major factor affecting the rate of attack. The superior resistance of calcium aluminate cement concrete appears to be due to blocking of pores by alumina gel, together with absence of CH (Section 10.1.11). Autoclaved cement–silica materials are also highly resistant, probably through a combination of absence of CH and AFm phases and the formation of a more crystalline form of C–S–H (B159); hydrogarnets containing even small proportions of SiO_2 are resistant to attack (M123). Supersulphated cements show high resistance, probably because of the absence of CH and because much of the Al_2O_3 in the hydration products is already present in ettringite.

The mechanism whereby sulphate-resisting Portland cements provide better resistance is not known. Heller and Ben-Yair (H59) found that substitution of such a cement for an ordinary Portland cement greatly reduced expansion but did not significantly alter the amount of ettringite formed under given conditions. In contrast, Locher (L55) found that a sulphate-resisting Portland cement mortar, unlike one of an ordinary Portland cement, gave little ettringite when stored in a Na_2SO_4 solution. Odler and Gasser (O29) found that less ettringite was formed with a sulphate-resisting cement, but that the difference was insufficient to account for that in expansion. In agreement with Regourd *et al.* (R71), they considered that the morphology of the ettringite was different; AFt phase prepared from C_3A and gypsum consisted of well-formed needles, but that formed from C_4AF and gypsum appeared almost amorphous in SEM. They also suggested that slower crystallization of ettringite in the iron-rich cements might allow the formation of a matrix better able to resist stresses arising from that crystallization.

12.5.5 Composite cements and sulphate attack

Partial replacement of cement by granulated blastfurnace slag often improves resistance, but the effectiveness depends on the properties and proportions of the slag. Locher (L55) found that composite cements containing at least 65% of slag were resistant to sulphate attack irrespective of the Al_2O_3 content of the slag (11 or 17.7%) or the potential C_3A content of the cement (0–11%), but that those made with lower contents of slag were resistant only if the cement was low in Al_2O_3. Finer grinding of the slag further improved sulphate resistance if the content of slag was over 65%, but worsened it at lower contents of slag. With a zero-C_3A cement, 20–50% replacements by the slag containing 11% of Al_2O_3 further improved resistance, but similar replacements by that with 17.7% Al_2O_3 worsened it.

Locher concluded that resistance in slag cements arises from the type of pore structure, which at high slag contents was either less permeable or such that ettringite cannot nucleate. The effects of finer grinding and of varying the Al_2O_3 content of the slag suggest that enhanced resistance from this cause is countered by the relatively high Al_2O_3 contents of granulated blastfurnace slags, and that to obtain it requires a proper balance between the proportion, fineness, Al_2O_3 content and intrinsic reactivity of the slag. High sulphate resistance in a slag cement can probably only be obtained at the cost of some reduction in early strength.

Other mineral additions, such as pfa, can also increase sulphate resistance (R71), but with composite cements in general, the effect is likely to depend markedly on the degree of curing before exposure to the sulphate solution begins, because of the slow reaction of the mineral addition. Hughes (H70) found that 30% replacement of cement by pfa in a paste greatly increased the time to cracking if the paste had been cured in saturated CH solution for 12 weeks before being transferred to a Na_2SO_4 solution, but if the preliminary curing period was restricted to 1 week, expansion took place more quickly in the pfa–cement paste.

12.5.6 Reactions involving sulphate and carbonate

Thaumasite [$Ca_3Si(OH)_6(SO_4)(CO_3)\cdot 12H_2O$; Section 6.2.1] can form through a combination of sulphate attack and carbonation. Its crystal structure and morphology resemble those of ettringite, and its formation can cause severe softening or cracking of concrete or mortar. It can easily be misidentified as ettringite. Crammond (C32), who observed it in failed mortars and renderings on brickwork, described the conditions conducive to its formation as a constantly high RH, temperature around 4°C, adequate supply of SO_4^{2-} and CO_3^{2-} ions, and presence of reactive Al_2O_3. The prior formation of ettringite seems to be needed, probably as a nucleating agent; this would explain the need for a source of Al_2O_3. The quantity of ettringite that can be formed from a cement is limited by the amount of Al_2O_3 available, but, assuming a continuing source of SO_4^{2-}, thaumasite formation is limited only by the available CaO and SiO_2. It can therefore form in large quantities, reducing a mortar into a mush. Brickwork high in SO_4^{2-} covered with a cement-based rendering that has cracked through drying shrinkage, and existing in a cold, damp climate, provides a typical situation for thaumasite formation. Other cases in which this has been observed include the concrete lining of a hydroelectric tunnel (B160), masonry in historic buildings constructed or repaired at different times using gypsum- and cement-based materials (L56) and cement bricks made with an aggregate that contained a sulphide mineral (O30). This last case occurred in a region (Transvaal) of relatively warm climate.

Precast concrete products cured at 80°C have occasionally cracked and lost strength after fairly long subsequent exposure to the weather at ordinary temperature. Strohbauch and Kuzel (S122) found the following sequence of reactions to occur. During the initial curing at 80°C, monosulphate was formed. On exposure at 22°C, anion exchange occurred, giving $C_4A\bar{C}_{0.5}H_{12}$ and $C_4A\bar{C}H_{11}$, together with gypsum, which reacted with unaltered monosulphate to give ettringite. Thaumasite was not formed. The conditions under which CO_3^{2-} can replace SO_4^{2-} in monosulphate are not clear.

12.6 Other forms of attack

12.6.1 Physical attack

Concrete may be damaged by frost, and especially by repeated cycles of freezing and thawing. Damage usually begins with flaking at the surface, and gradually extends inwards, though deep cracks may occur. Damage associated with freezing does not occur unless a sufficient quantity of water is present in the pores, and is minimal in dense concrete of low w/c ratio and low permeability. It can be much reduced by air entrainment (Section 11.3.1). It is especially likely if freezing occurs before adequate hardening has occurred.

Various explanations have been given and may not be mutually exclusive. Early theories associated damage directly with the volume expansion that occurs on freezing; this could cause disruption if the water that freezes is sufficiently confined, e.g. between regions already frozen. The damage could also be caused by the growth of ice crystals, resulting from the migration of water as a result of differences in vapour pressure. Collins (C71) postulated a mechanism, similar to that believed to cause frost heaving in soils, in which the growth of ice lenses led to the formation of weak layers parallel to an exposed surface. Other theories (P55,P56,L57) attribute the damage to hydraulic pressure caused by the movement of water through capillaries. Powers (P55) considered that the expansion associated with freezing in the pores near the surface forced water inwards through small pores, where it could produce stresses sufficient to rupture the surrounding material. This theory led to the successful development of air entrainment. Later work showed that a properly air-entrained concrete *contracts* on freezing. This was attributed to migration of water out of the gel to the ice crystals, occurring because the vapour pressure of supercooled water is higher than that of ice at the same temperature (P56).

Studies using low-temperature calorimetry appear to provide clear evidence of the formation of an ice front that nucleates at the surface and moves inwards (B112,B161). They also showed that the pattern of ice formation is much affected by the previous drying history.

Some other forms of physical attack will be briefly mentioned. Soluble salts may be drawn into a sufficiently permeable concrete by capillary action and crystallize within the pore structure when evaporation occurs (e.g. Ref. S123). Mechanical or thermal stress or drying shrinkage can all cause concrete to crack. Thermal stresses can cause damage at low temperatures even if the pores contain little or no water. The use of de-icing salts can cause surface damage, known as salt scaling; suggested causes have included rapid cooling caused by the melting of ice, osmotic effects associated with the presence of high concentrations of de-icing chemicals in the surface layers of the concrete and chemical attack (Section 12.6.3). Surface damage from abrasion depends markedly on w/c ratio and on the quality of the surface layer of the concrete; at high w/c ratios it depends also on the properties of the aggregate. Abrasion resistance can be increased by surface treatments, e.g. with sodium silicate.

12.6.2 Leaching

Leaching of concrete by percolating or flowing water has sometimes caused severe damage, e.g. in dams, pipes or conduits, and is potentially important for the long-term storage of nuclear wastes. Pure water may be expected to remove alkali hydroxides, dissolve CH and decompose the hydrated silicate and aluminate phases. Reference to the equilibria discussed in Chapters 5 and 6 indicates that, for practical purposes, the ultimate residue will consist essentially of hydrous forms of silica, alumina and iron oxide, all the CaO having been lost. By this stage, the cement paste will have disintegrated. The equilibria also suggest that CH will be dissolved before the other phases are attacked, but, in practice, attack is likely to be simultaneous, because of the greater specific surface areas of the hydrated silicate and aluminate phases.

The rate of attack depends on the quality and shape of the concrete, the rate at which the water percolates through or flows over it, the temperature and the concentrations of solutes in the water. Attack is most likely to be serious with soft water. The partial pressure of CO_2 in unpolluted air at sea level is 32 Pa ($10^{-3.5}$ atm). Calculation based on the data in Table 12.1 shows that, for water in equilibrium with such air and not containing any other solutes, the significant species present are CO_2, HCO_3^- and H^+; $[CO_2]$ is 0.012 mmol l^{-1} and the pH is 5.6. In underground waters, $[CO_2]$ may be much higher and the pH correspondingly lower.

Table 12.1 Equilibrium constants in the system $CaO-CO_2-H_2O$

Reaction	Definition of K	log K
CO_2 (gas) $= CO_2$	$[CO_2]/p_{CO_2}$	-1.42
$CO_2 + H_2O = H_2CO_3$	$[H_2CO_3]/[CO_2]$	-2.8
$H_2CO_3 = H^+ + HCO_3^-$	$\{H^+\}[HCO_3^-]/[H_2CO_3]$	-3.5
$HCO_3^- = H^+ + CO_3^{2-}$	$\{H^+\}[CO_3^{2-}]/[HCO_3^-]$	-10.25
$CaCO_3$ (calcite) $= Ca^{2+} + CO_3^{2-}$	$[Ca^{2+}][CO_3^{2-}]$	-8.35

From Ref. P57. Species are in solution except where stated. Square brackets denote species concentrations; curly brackets denote activities. Concentrations are in mol l^{-1}; p_{CO_2} is in atm (1 atm = 101 kPa).

A solution of CO_2 can dissolve $CaCO_3$ with formation of additional HCO_3^-

$$CaCO_3 + CO_2 + H_2O \rightarrow Ca^{2+} + 2HCO_3^- \qquad (12.11)$$

and can similarly dissolve CH or Ca^{2+} and OH^- ions from C-S-H or hydrated calcium aluminate phases. In all cases, the amount of Ca^{2+} that a given volume of solution can dissolve is limited by the equilibria involving CO_3^{2-} (Table 12.1). The term 'aggressive CO_2' is used to express this amount and the rate at which attack is likely to occur. It is properly defined as the quantity of CO_2 in unit volume of solution that can react with $CaCO_3$ according to equation 12.11. As Rogers (R72) has noted, it has sometimes been incorrectly defined and calculated. It is smaller than the concentration of CO_2 present as such or as H_2CO_3, because when the solution dissolves $CaCO_3$ the concentrations of all the species change in accordance with the equilibria.

Rogers critically reviewed methods for determining total and aggressive CO_2. Natural waters may contain other anions, such as SO_4^{2-}, and in the earlier stages of leaching, at least, the situation is further complicated by the presence of alkali hydroxides from the cement pore solution. In any aqueous system of the type considered, the equilibrium composition of the solution and the nature of the solid phases can be calculated by setting up and solving equations for the various acid–base, precipitation and complexation equilibria and for charge and mass balance. The general principles of such calculations, and references to computer programs for solving the equations, may be found in works on natural water chemistry (e.g. Ref. P57).

12.6.3 Miscellaneous forms of chemical attack

This section summarizes observations on the actions on concrete made using Portland or composite cements not considered earlier. There have been few systematic or detailed chemical studies, and the information given is based largely on results reported by Lea (L6) and Kuenning (K69).

Acids have an action similar in principle to that of water, but greatly intensified; CH is dissolved and the hydrated silicate and aluminate phases are decomposed, with removal of Ca^{2+}. Those giving insoluble products, such as H_2SO_4 or H_3PO_4, tend to be less aggressive for a given concentration than those that do not, such as HCl or HNO_3. Fattuhi and Hughes (F48) made a detailed study of the action of H_2SO_4. Acid attack is produced not only by Brønsted acids that are neutral molecules, such as the above, but also by ones that are cations or anions such as NH_4^+, $Al(H_2O)_6^{3+}$, $H_2PO_4^-$ or HSO_4^-. NH_4^+ salts of strong acids, and aluminium salts of strong bases, are thus highly aggressive. Acids attack limestone aggregates as well as cement paste. Alkalis have relatively little effect, but concentrated solutions of alkali hydroxides (e.g. 5 mol l^{-1}) attack concrete, probably by decomposing the hydrated aluminate phases.

Metal cations that produce hydroxides or basic salts of low solubility attack cement paste. The behaviour of Mg^{2+} in $MgSO_4$ (Section 12.5.3) is typical, and is shown by other magnesium salts. Concentrated (3 mol l^{-1}) solutions of $CaCl_2$ or $MgCl_2$ attack Portland cement concretes by reacting with the CH to form basic salts (S124). The reactions are expansive and may contribute to salt scaling (Section 12.6.1). With saturated NaCl, $C_3A \cdot CaCl_2 \cdot 10H_2O$ is formed, and there is some loss in strength, but the effect is much less than with $CaCl_2$ or $MgCl_2$. A composite cement containing 75% of slag was resistant to all three solutions.

Some organic compounds, such as sugar solutions, attack concrete, presumably through complexing of Ca^{2+} and consequent dissolution of CH and the hydrated silicate and aluminate phases.

12.6.4 Sea water attack

Sea water (Atlantic Ocean) contains about 11 g of Na^+, 20 g of Cl^-, 2.9 g of SO_4^{2-} and 1.4 g of Mg^{2+} per litre, with smaller amounts of K^+, Ca^{2+}, Br^- and HCO_3^- (0.08 g l^{-1}). The pH is usually 7.8–8.3. Massazza (M124) gave a general account of its action on concrete, and Buenfeld and Newman (B162) discussed microstructural aspects and equilibria more fully. Attack is more serious in and somewhat above the tidal zone than on concrete that is permanently submerged. In the latter case, the combined action of Mg^{2+} and HCO_3^- produces a surface skin, typically consisting of a layer of brucite about 30 µm thick, overlain by a more slowly developing layer of aragonite. $CaCO_3$ is precipitated as aragonite rather than as calcite because of the presence of Mg^{2+} ions. This skin tends to protect a well-produced, dense concrete from further attack. To the extent that the latter occurs, the processes are those of leaching, $Mg^{2+} + SO_4^{2-}$ attack, carbonation, and Cl^- penetration, with the attendant possibility of corrosion of reinforcement, all of which are discussed elsewhere in this chapter. The effects of the Mg^{2+} and SO_4^{2-} are considerably milder than those observed with $MgSO_4$ solutions of similar concentration. The danger of corrosion decreases at depth due to the lack of oxygen.

In and above the tidal zone, cycles of saturation and evaporation tend to occur, the water being drawn in by capillary action or entering as spray, and damage may occur through salt crystallization. The protective layer of brucite and aragonite is liable to be destroyed. Mechanical damage may result from erosion due to waves, solid debris, sand or ice. Oxygen is readily available, permitting corrosion to occur.

As with concrete for use in other severe environments, the most important conditions for ensuring satisfactory performance are that it should be of high cement content, of low w/c ratio, of high strength and properly made, with adequate cover for the reinforcement, and wet cured before being exposed to sea water. Massazza (M124) concluded that pozzolanic or slag cements (>60% slag) were more durable than pure Portland cements, and noted that lime–pozzolana mortars used by the Romans in sea works are still in good condition.

12.6.5 Bacterial attack

Normal sewage is alkaline and does not significantly attack Portland cement concrete directly, but severe damage can arise through formation of H_2SO_4 by processes in which bacterial action plays a part. Anaerobic bacteria in the sewage decompose inorganic or organic sulphur compounds, giving

H_2S. This passes into the air above the liquid and dissolves in the film of water on the walls above the water level, where under the influence of aerobic bacteria, oxygen oxidizes it to H_2SO_4. The cement paste softens and the aggregate falls out.

Preventive treatments have been of several kinds (L6). The sewage may be treated, e.g. with chlorine to oxidize the sulphides, with lime to raise the pH to above 10, which decreases the activity of the anaerobic bacteria, or with appropriate salts to precipitate the sulphide. Removal of slime and silt, in which the sulphide-forming reactions appear to occur, increase in flow and design to avoid turbulence, have been found useful. The service life of the concrete is much increased by using limestone as opposed to siliceous aggregates. Various surface treatments have been used, of which one of the most effective appears to be with SiF_4 gas.

Other situations in which attack by H_2SO_4 formed through bacterial action has been reported include cooling towers (L6) and concrete floors laid on rocks containing pyrite (FeS_2) (P58).

12.6.6 Miscellaneous paste–aggregate reactions

Destructive expansion occurs in concretes made with some aggregates containing dolomite, which reacts with OH^- ions in the so-called dedolomitization reaction:

$$CaMg(CO_3)_2 + 2OH^- \rightarrow CaCO_3 + Mg(OH)_2 + CO_3^{2-} \quad (12.12)$$

The effect was first observed with certain Canadian aggregates, which also contained calcite and clay minerals. Gillott and Swenson (G84), reviewing theories of the expansion mechanism, concluded that the expansion was due to uptake of water by the clay minerals, which were exposed as a result of the dedolomitization reaction. In contrast, Tang et al. (T61) concluded from a STEM study that the clay minerals were present in the matrix material of the aggregate and not as inclusions in the dolomite, and that expansion was mainly caused by the growth of brucite crystals around the surfaces of the dolomite grains; the clay minerals provided pathways through which the alkaline solution could penetrate. Destructive reactions associated with dedolomitization have also been observed with dolomitic aggregates from Bahrein (F49), in this case possibly complicated by the presence also of gypsum.

Greening (G85) suggested that, under some conditions, ettringite could be formed by reaction between monosulphate and calcite:

$$3C_4A\bar{S}H_{12} + 2C\bar{C} + 18H_2O \rightarrow 2C_4A\bar{C}H_{11} + C_6A\bar{S}_3H_{32} \quad (12.13)$$

but it is not clear under what conditions, if any, this reaction takes place or can cause expansion in concrete.

Van Aardt and Visser (V11,V12) showed that felspars can react with CH and water at ordinary temperatures to give C_4AH_x, alkalis and other products, and that the C_4AH_x could react further with $CaSO_4$ to form ettringite. They concluded that some felspathic rocks, such as greywackes, and some shales and sandstones, might be suspect as regards reactivity towards cement paste. Way and Cole (W34) confirmed the occurrence of felspar reactions, but found no evidence of deleterious reactions with granite or basalt aggregates.

Reinforcement of a cement paste or mortar with glass fibre much increases the tensile strength and toughness, but these properties deteriorate with time, especially in a moist environment. The deterioration is retarded, but not eliminated, by using a special glass high in ZrO_2. The glass is attacked by the alkaline pore solution, but the loss of strength and toughness may not be a direct consequence of this. Changes in the microstructure at the interface between glass and cement paste take place, and may be due both to attack on the glass and to processes occurring within the cement paste (M125,S125). Growth of products, especially CH, between the filaments of which the fibres are composed alters the mode of fracture and has been considered the dominant cause of deterioration, at least during the first year (B163).

12.6.7 Fire damage

Because of its relatively low thermal conductivity and high specific heat, concrete provides good protection to steel against fire, though it may itself be extensively damaged. At low temperatures, cement paste expands on heating, but by 300°C, contraction, associated with the water loss, occurs. Normal aggregates continue to expand, and the resultant stresses lead to loss of strength, spalling and cracking. Quartz expands sharply at 573°C due to polymorphic transformation, and calcite begins to contract at 900°C due to decomposition. Blastfurnace slag, used as aggregate, is highly resistant (P59,L6).

Piasta et al. (P60) described the thermal effects on the cement paste. Below 500°C, major effects include carbonation and coarsening of the pore structure. Decomposition of CH at 450–550°C, and of $CaCO_3$ above 600°C, yield CaO, which rehydrates on cooling. The resulting expansion may disrupt a concrete that has withstood a fire without actual disintegration (L6). Above 500°C, the pore structure continues to coarsen, and incipient crystallization of β-C_2S occurs. Extensive carbonation occurs in fire-damaged buildings that have been left standing for a few years (H71).

References

Abbreviations used in the reference list are as follows:

2nd SCC = *Symposium on the Chemistry of Cements.* Stockholm 1938, Ingeniörsvetenskapsakademien, Stockholm (1939).
3rd ISCC = *Proceedings of the Third International Symposium on the Chemistry of Cement.* London 1952, Cement and Concrete Association, London (1954).
4th ISCC = *Chemistry of Cement. Proceedings of the Fourth International Symposium.* Washington 1960 (National Bureau of Standards [now National Institute of Standards and Technology] Monograph 43), 2 vols, US Department of Commerce, Washington (1962).
5th ISCC = *Proceedings of the Fifth International Symposium on the Chemistry of Cement.* Tokyo 1968, 4 vols, Cement Association of Japan, Tokyo (1969).
6th ICCC = *Sixth International Congress on the Chemistry of Cement,* Moscow, 1974, Vols 1, 2(1), 2(2) and (3), Russian with English preprints, Stroyizdat, Moscow (1976).
7th ICCC = *7th International Congress on the Chemistry of Cement,* Paris, 1980, 4 vols, Editions Septima, Paris (1980, 1981).
8th ICCC = *8th International Congress on the Chemistry of Cement,* Rio de Janeiro, 1986, 6 vols, Abla Gráfica e Editora, Rio de Janeiro (1986).
ICCem. Microsc. = *Proceedings of the International Conference on Cement Microscopy,* International Cement Microscopy Association, Duncanville, TX, USA.

A1	Aggarwal, P. S., Gard, J. A., Glasser, F. P. and Biggar, G. M. (1972). *Cem. Concr. Res.* **2**, 291.
A2	Amafuji, M. and Tsumagari, A., in *5th ISCC*, Vol. 1, p. 136 (1969).
A3	Allen, T. *Particle Size Measurement* (2nd Ed.), 454 pp., Chapman and Hall, London (1975).
A4	Anderegg, F. O. and Hubbell, D. S. (1929). *Proc. Am. Soc. Testing Mater.* **29**, 554.
A5	Anderegg, F. O. and Hubbell, D. S. (1930). *Proc. Am. Soc. Testing Mater.* **30**, 572.
A6	Aldridge, L. P. (1982). *Cem. Concr. Res.* **12**, 381.
A7	Ahmed, S. J. and Taylor, H. F. W. (1967). *Nature* **215**, 622.
A8	Allmann, R. (1968). *Neues Jahrb. Mineral. Monatsh.* 140.
A9	Ahmed, S. J., Dent Glasser, L. S. and Taylor, H. F. W., in *5th ISCC*, Vol. 2, p. 118 (1969).
A10	Aruja, E. (1960). *Acta Cryst.* **14**, 1213.
A11	Allmann, R. (1970). *Chimia* **24**, 99.
A12	Allmann, R. and Lohse, H.-H. (1966). *Neues Jahrb. Mineral. Monatsh.* 161.
A13	Abriel, W. (1983). *Acta Cryst.* **C39**, 956.
A14	Afshar, A. B. and McCarter, W. J., in *8th ICCC*, Vol. 3, p. 193 (1986).
A15	Alford, N. McN. and Rahman, A. A. (1981). *J. Mater. Sci.* **16**, 3105.
A16	Allen, A. J., Windsor, C. G., Rainey, V., Pearson, D., Double, D. D. and Alford, N. McN. (1982). *J. Phys.* **D15**, 1817.
A17	Alexander, K. M., Taplin, J. H. and Wardlaw, J., in *5th ISCC*, Vol. 3, p. 152 (1969).
A18	Aldridge, L. P., in *7th ICCC*, Vol. 3, p. VI–83 (1980).

A19 Alford, N. McN., Birchall, J. D., Howard, A. J. and Kendall, K. (1985). *J. Mater. Sci.* **20**, 1134.
A20 Abo-El-Enein, S. A., Daimon, M., Ohsawa, S. and Kondo, R. (1974). *Cem. Concr. Res.* **4**, 299.
A21 Aitcin, P. C., Carles-Gibergues, A., Oudjit, M. N. and Vaquier, A., in *8th ICCC*, Vol. 4, p. 22 (1986).
A22 Aitcin, P.-C., Sarkar, S. L. and Diatta, Y. (1987). *Mater. Res. Soc. Symp. Proc.* **85**, 261.
A23 Afshar, A. B. and McCarter, W. J. (1985). *J. Mater. Sci. Lett.* **4**, 851.
A24 Agarwal, R. K., Paralkar, S. V. and Chatterjee, A. K., in *8th ICCC*, Vol. 2, p. 327 (1986).
A25 Asaga, K. and Roy, D. M. (1980). *Cem. Concr. Res.* **10**, 287.
A26 Andersen, P. J., Roy, D. M. and Gaidis, J. M. (1988). *Cem. Concr. Res.* **18**, 980.
A27 Arliguie, G., Ollivier, J. P. and Grandet, J. (1982). *Cem. Concr. Res.* **12**, 79.
A28 Alunno-Rossetti, V., Chiocchio, G. and Collepardi, M. (1974). *Cem. Concr. Res.* **4**, 279.
A29 Aitken, A. and Taylor, H. F. W., in *4th ISCC*, Vol. 1, p. 285 (1962).
A30 Alexanderson, J. (1979). *Cem. Concr. Res.* **9**, 507.
A31 Al-Khalaf, N. N. and Page, C. L. (1979). *Cem. Concr. Res.* **9**, 197.
A32 Arliguie, G., Grandet, J. and Duval, R., in *7th ICCC*, Vol 3, p. VII–22 (1980).
A33 Andrade, C., Alonso, C., Santos, P. and Macías, A., in *8th ICCC*, Vol. 5, p. 256 (1986).

B1 Bigaré, M., Guinier, A., Mazières, C., Regourd, M., Yannaquis, N., Eysel, W., Hahn, T. and Woermann, E. (1967). *J. Am. Ceram. Soc.* **50**, 609.
B2 Boikova, A. I., Esayan, A. and Lazukin, V., in *7th ICCC*, Vol. 4, p. 183 (1981).
B3 Boikova, A. I., Fomicheva, O. I. and Zubekhin, A. P., in *8th ICCC*, Vol. 2, p. 243 (1986).
B4 Boikova, A. I., in *8th ICCC*, Vol. 1, p. 19 (1986).
B5 Burke, E., in *3rd ISCC*, p. 50 (1954).
B6 Barnes, P., Fentiman, C. H. and Jeffery, J. W. (1977). *Acta Cryst.* **A36**, 353.
B7 Bredig, M. A. (1950). *J. Am. Ceram. Soc.* **33**, 188.
B8 Barbier, J. and Hyde, B. G. (1985). *Acta Cryst.* **B41**, 383.
B9 Biggar, G. M. (1971). *Cem. Concr. Res.* **1**, 493.
B10 Banda, R. M. H. and Glasser, F. P. (1978). *Cem. Concr. Res.* **8**, 665.
B11 Berggren, J. (1971). *Acta Chem. Scand.* **25**, 3616.
B12 Büssem, W., in *2nd SCC*, p. 141 (1939).
B13 Bertaut, E. F., Blum, P. and Sagnières, A. (1959). *Acta Cryst.* **12**, 149.
B14 Bachiorrini, A. (1985). *Cem. Concr. Res.* **15**, 167.
B15 Boyko, E. R. and Wisnyi, L. G. (1958). *Acta Cryst.* **11**, 444.
B16 Büssem, W. and Eitel, W. (1936). *Zeit. Krist.* **95**, 175.
B17 Bartl, H. and Scheller, T. (1970). *Neues Jahrb. Mineral. Monatsh.* 547.
B18 Brisi, C. and Borlera, M. L. (1983). *Cemento* **80**, 3.
B19 Borlera, M. L. and Brisi, C. (1984). *Cemento* **81**, 13.
B20 Brisi, C., Borlera, M. L., Montanaro, L. and Negro, A. (1986). *Cem. Concr. Res.* **16**, 156.
B21 Brotherton, P.D., Epstein, J. M., Pryce, M. W. and White, A. H. (1974). *Austral. J. Chem.* **27**, 657.

B22	Bellanca, A. (1942). *Periodico Mineral.* **13**, 21.
B23	Bereczky, E. (1964). *Épitőanyag* **16**, 441.
B24	Bogue, R. H. (1929). *Ind. Engnrg Chem. (Anal.)* **1**, 192.
B25	Bye, G. C. *Portland Cement. Composition, Production and Properties,* 149 pp., Pergamon Press, Oxford (1983).
B26	Bucchi, R. (1978). *Cemento* **75**, 411.
B27	Bucchi, R., in *7th ICCC,* Vol. 1, p. I-1/3 (1980).
B28	Bucchi, R. (1981). *World Cem. Technol.* **12**, 258.
B29	Bogue, R. H. *The Chemistry of Portland Cement* (2nd Ed.), 793 pp., Reinhold, New York (1955).
B30	Banda, R. M. H. and Glasser, F. P. (1978). *Cem. Concr. Res.* **8**, 319.
B31	Butt, Yu. M., Timashev, V. V. and Osokin, A. P., in *6th ICCC,* Vol. 1, p. 132 (1976).
B32	Barlow, J. F., in *7th ICCem. Microsc.,* p. 243 (1985).
B33	Bucchi, R. (1981). *World Cem. Technol.* **12**, 210.
B34	Brisi, C. and Appendino, P. (1965). *Ann. Chim. (Rome)* **55**, 1213.
B35	Butt, Ju. M., Kolbasov, V. M. and Melniekij, G. A. (1974). *Zem.-Kalk-Gips* **27**, 27.
B36	Blaine, R.L. (1943). *ASTM Bull.* **123**, 51.
B37	Berger, R. L., Frohnsdorff, G. J. C., Harris, P. H. and Johnson, P. D., in *Symposium on Structure of Portland Cement Paste and Concrete* (Sp. Rpt 90), p. 234, Highway Research Board, Washington (1966).
B38	Bensted, J. (1980). *Cemento* **77**, 169.
B39	Bensted, J. and Varma, S.P. (1974). *Cem. Technol.* **5**, 378.
B40	Bensted, J. (1976). *J. Am. Ceram. Soc.* **59**, 140.
B41	Barret, P., Ménétrier, D. and Bertrandie, D. (1983). *Cem. Concr. Res.* **13**, 728.
B42	Boikova, A. I., Degen, M. G., Paramonova, V. A. and Sud'ina, V. V. (1978). *Tsement* 3.
B43	Boikova, A. I., Domansky, A. I., Paramonova, V. A., Stavitskaja, G.P. and Nikushchenko, V.M. (1977). *Cem. Concr. Res.* **7**, 483.
B44	Boikova, A. I., Grishchenko, L. V. and Domansky, A. I., in *7th ICCC,* Vol. 4, p. 460 (1981).
B45	Beaudoin, J. J. (1987). *Matér. Constr. (Paris)* **20**, 27.
B46	Bernal, J. D. and Megaw, H. D. (1935). *Proc. R. Soc. London* **A151**, 384.
B47	Berger, R. L. and McGregor, J. D. (1972). *Cem. Concr. Res.* **2**, 43.
B48	Barker, A. P. (1984). *World Cem.* **15**, 25.
B49	Bassett, H. (1934). *J. Chem. Soc.* 1970.
B50	Bentur, A. and Berger, R. L. (1979). *J. Am. Ceram. Soc.* **62**, 117.
B51	Brown, L. S. and Carlson, R. W. (1936). *Proc. Am. Soc. Testing Mater.* **36**, 332.
B52	Bailey, J. E. and Stewart, H. R. (1984). *J. Mater. Sci. Lett.* **3**, 411.
B53	Bentur, A., Berger, R. L., Kung, J. H., Milestone, N. B. and Young, J. F. (1979). *J. Am. Ceram. Soc.* **62**, 362.
B54	Bentur, A., Berger, R. L., Lawrence, F. V., Milestone, N. B., Mindess, S. and Young, J. F. (1979). *Cem. Concr. Res.* **9**, 83.
B55	Brunauer, S. and Greenberg, S. A., in *4th ISCC,* Vol. 1, p. 135 (1962).
B56	Brown, P. W., Franz, E., Frohnsdorff, G. and Taylor, H. F. W. (1984). *Cem. Concr. Res.* **14**, 257.
B57	Barret, P. and Bertrandie, D. (1986). *J. Chim. Phys.* **83**, 765.

B58 Barret, P. and Bertrandie, D. (1988). *J. Am. Ceram. Soc.* **71**, C-113.
B59 Barby, D., Griffiths, T., Jacques, A. R. and Pawson, D., in *The Modern Inorganic Chemicals Industry* (ed. R. Thompson), p. 320, Chemical Society, London (1977).
B60 Brisi, C. (1954). *Ricerca Sci.* **24**, 1436.
B61 Brunauer, S., Hayes, J. C. and Hass, W. E. (1954). *J. Phys. Chem.* **58**, 279.
B62 Babushkin, W. I., Matveev, G. M. and Mchedlov-Petrossian, O. P., *Thermodynamics of Silicates* (3rd Ed.), 350 pp., Publishers of Construction Literature, Moscow (1972).
B63 Barret, P., Bertrandie, D. and Ménétrier, D., in *7th ICCC*, Vol. 2, p. 261 (1980).
B64 Brown, P. W., Harner, C. L. and Prosen, E. J. (1986). *Cem. Concr. Res.* **16**, 17.
B65 Ball, M. C., Simmons, R. E. and Sutherland, I. (1984). *Br. Ceram. Proc.* **35**, 1.
B66 Brown, P. W., Pommersheim, J. and Frohnsdorff, G. (1985). *Cem. Concr. Res.* **15**, 35.
B67 Brown, P. W., Galuk, K. and Frohnsdorff, G. (1984). *Cem. Concr. Res.* **14**, 843.
B68 Barret, P., in *8th ICCC*, Vol. 3, p. 86 (1986).
B69 Buttler, F. G., Dent Glasser, L. S. and Taylor, H. F. W. (1959). *J. Am. Ceram. Soc.* **42**, 121.
B70 Bensted, J. (1979). *Cemento* **76**, 117.
B71 Bensted, J. (1976). *Tonind. Zeit.* **100**, 365.
B72 Bensted, J. and Varma, S. P. (1973). *Cem. Technol.* **4**, 112.
B73 Buhlert, R. and Kuzel, H.-J. (1971). *Zem.-Kalk-Gips* **24**, 83.
B74 Bensted, J. and Varma, S. P. (1971). *Cem. Technol.* **2**, 73, 100.
B75 Bensted, J. and Varma, S. P. (1972). *Silicates Ind.* **37**, 315.
B76 Bensted, J. and Varma, S. P. (1972). *Cem. Technol.* **3**, 185.
B77 Bannister, F. A., Bernal, J. D. and Hey, M. H. (1936). *Mineral. Mag.* **24**, 324.
B78 Bensted, J. and Varma, S. P. (1974). *Silicates Ind.* **39**, 11.
B79 Bartl, H. (1969). *Neues Jahrb. Mineral. Monatsh.* 404.
B80 Basso, R., Della Giusta, A. and Zefiro, L. (1983). *Neues Jahrb. Mineral. Monatsh.* 251.
B81 Basso, R. (1985). *Neues Jahrb. Mineral. Monatsh.* 108.
B82 Ball, M. C. (1976). *Cem. Concr. Res.* **6**, 419.
B83 Buttler, F. P. and Taylor, H. F. W. (1978). *Cemento* **75**, 147.
B84 Barret, P. and Dufour, P., in *7th ICCC*, Vol. 3, p. V-124 (1980).
B85 Brindley, G. W. and Kikkawa, S. (1979). *Am. Mineral.* **64**, 836.
B86 Brindley, G. W. and Kikkawa, S. (1980). *Clays and Clay Minerals* **28**, 87.
B87 Bokii, G. B., Pal'chik, N. A. and Antipin, M. Yu. (1978). *Sov. Phys. Cryst.* **23**, 141 (translated from *Kristallografiya* **23**, 257).
B88 Bensted, J. (1975). *Cemento* **72**, 139.
B89 Bensted, J. and Varma, S. P. (1971). *Nature Phys. Sci.* **232**, 174.
B90 Bailey, J. E. and Hampson, C. J. (1982). *Cem. Concr. Res.* **12**, 227.
B91 Brown, P. W., in *8th ICCC*, Vol. 3, p. 231 (1986).
B92 Breval, E. (1976). *Cem. Concr. Res.* **6**, 129.
B93 Barret, P. and Bertrandie, D., in *7th ICCC*, Vol. 4, p. 443 (1981).
B94 Ball, M. C., Simmons, R. E. and Sutherland, I. (1987). *J. Mater. Sci.* **22**, 1975.

B95 Bailey, J. E. and Hampson, C. J. (1983). *Phil. Trans R. Soc. London* **A310**, 105.
B96 Brown, P. W. (1987). *J. Am. Ceram. Soc.* **70**, 493.
B97 Bensted, J. and Varma, S.P. (1974). *Cem. Technol.* **5**, 440.
B98 Bezjak, A., Jelenić, I., Mlakar, V. and Panović, A., in *7th ICCC*, Vol. 2, p. II-111 (1980).
B99 Bensted, J. (1983). *Cem. Concr. Res.* **13**, 493.
B100 Barnes, J. R., Clague, A. D. H., Clayden, N. J., Dobson, C. M., Hayes, C. J., Groves, G. W. and Rodger, S. A. (1985). *J. Mater. Sci. Lett.* **4**, 1293.
B101 Barnes, P., Ghose, A. and Mackay, A. L. (1980). *Cem. Concr. Res.* **10**, 639.
B102 Barnes, B. D., Diamond, S. and Dolch, W. L. (1978). *Cem. Concr. Res.* **8**, 263.
B103 Bensted, J., in *Characterization and Performance Prediction of Cement and Concrete*, p. 69, Engineering Foundation, New York (1983).
B104 Bentur, A. (1976). *J. Am. Ceram. Soc.* **59**, 210.
B105 Bezjak, A. and Jelenić, I. (1980). *Cem. Concr. Res.* **10**, 553.
B106 Brown, P. W., Pommersheim, J. M. and Frohnsdorff, G. (1984). *Cem. Res. Prog.* [for] 1983, 245.
B107 Banfill, P. F. G. and Saunders, D. C. (1981). *Cem. Concr. Res.* **11**, 363.
B108 Brunauer, S. (1962). *Am. Scientist* **50**, 210.
B109 Beaudoin, J. J. (1986). *Cemento* **83**, 199.
B110 Beaudoin, J. J. (1979). *Cem. Concr. Res.* **9**, 771.
B111 Barbič, L., Kocuvan, I., Blinc, R., Lahajnar, G., Marljak, P. and Zupančič, I. (1982). *J. Am. Ceram. Soc.* **65**, 25.
B112 Bager, D. H. and Sellevold, E. J. (1986). *Cem. Concr. Res.* **16**, 709, 835.
B113 Bal'shin, M. Y. (1949). *Dokl. Akad. Nauk SSSR* **67**, 831.
B114 Bozhenov, P. I., Kavalerova, V. I., Salnikova, V. S. and Suvorova, G. F., in *4th ISCC*, Vol. 1, p. 327 (1962).
B115 Beaudoin, J. J. and Feldman, R. F. (1985). *Cem. Concr. Res.* **15**, 105.
B116 Bombled, J. P., in 8th *ICCC*, Vol. 4, p. 190 (1986).
B117 Buttler, F. G. and Taylor, H. F. W. (1959). *J. Appl. Chem.* **9**, 616.
B118 Barret, P. and Bertrandie, D., in *7th ICCC*, Vol. 3, p. V-134 (1980).
B119 Barret, P. and Bertrandie, D., in *International Seminary on Calcium Aluminates* (eds. M. Murat *et al.*), Politecnico di Torino, Turin (1982).
B120 Bensted, J. (1982). *World Cem.* **13**, 85.
B121 Bertrandie, D. (1977). Doctorate thesis, University of Dijon, France.
B122 Barret, P., Bertrandie, D. and Beau, D. (1983). *Cem. Concr. Res.* **13**, 789.
B123 Baes, E. F. and Mesmer, R. E. *The Hydrolysis of Cations*, 489 pp., Wiley, New York (1976).
B124 Bertrandie, D. and Barret, P., in *8th ICCC*, Vol. 3, p. 79 (1986).
B125 Bushnell-Watson, S. M. and Sharp, J. H. (1986). *Cem. Concr. Res.* **16**, 875.
B126 Banfill, P. F. G. and Gill, S. M., in *8th ICCC*, Vol. 6, p. 223 (1986).
B127 Bachiorrini, A. and Cussino, L. (1980). *Cemento* **77**, 183.
B128 Building Research Establishment, 'High alumina cement concrete in buildings (CP 34/75)', in *Practical Studies from the Building Research Establishment*, Vol. 1, *Concrete*, p. 228, Construction Press, Lancaster, UK (1978).
B129 Barret, P., Benes, C., Bertrandie, D. and Moisset, J., in *7th ICCC*, Vol. 3, p. V-175 (1980).
B130 Bachiorrini, A., Guilhot, B., Murat, M., Negro, A., Soustelle, M. and Fournier, A. A., in *8th ICCC*, Vol. 4, p. 376 (1986).

B131 Barret, P., Beau, D. and Bertrandie, D. (1986). *Cem. Concr. Res.* **16**, 785.
B132 Blenkinsop, R. D., Currell, B. R., Midgley, H. G. and Parsonage, J. R. (1985). *Cem. Concr. Res.* **15**, 276.
B133 Bentur, A. and Ish-Shalom, M. (1974). *Cem. Concr. Res.* **4**, 709.
B134 Bentur, A. and Ish-Shalom, M. (1975). *Cem. Concr. Res.* **5**, 597.
B135 Boldyrev, A. S., in *7th ICCC*, Vol. 1, p. V-3/1 (1980).
B136 Bikbaou, M., in *7th ICCC*, Vol. 4, p. 371 (1981).
B137 Bikbaou, M. Y., in *8th ICCC*, Vol. 2, p. 352 (1986).
B138 Birchall, J. D. and Thomas, N. L. (1984). *Br. Ceram. Proc.* **35**, 305.
B139 Bruere, G. M., in *Symposium on Structure of Portland Cement Paste and Concrete* (Sp. Rpt 90), p. 26, Highway Research Board, Washington (1966).
B140 Banfill, P. F. G. (1979). *Cem. Concr. Res.* **9**, 795.
B141 Bergstrom, S. G. (1953). *Mag. Concr. Res.* (14) 61.
B142 Budnikov, P. P. and Erschler, E. Ya., in *Symposium on Structure of Portland Cement Paste and Concrete* (Sp. Rpt 90), p. 431, Highway Research Board, Washington (1966).
B143 Butt, Yu. M., Kolbasov, V. M. and Timashev, V. V., in *5th ISCC*, Vol. 3, p. 437 (1969).
B144 Buck, A. D., in *8th ICCC*, Vol. 3, p. 417 (1986).
B145 Buckner, D. A., Roy, D. M. and Roy, R. (1960). *Am. J. Sci.* **258**, 132.
B146 Bensted, J. (1987). *World Cem.* **18**, 72.
B147 Brunauer, S., Skalny, J., Odler, I. and Yudenfreund, M. (1973). *Cem. Concr. Res.* **3**, 279.
B148 Birchall, J. D. (1983). *Phil. Trans R. Soc. Lond.* **A310**, 31.
B149 Barnes, B. D., Diamond, S. and Dolch, W. L. (1978). *Cem. Concr. Res.* **8**, 233.
B150 Barnes, B. D., Diamond, S. and Dolch, W. L. (1979). *J. Am. Ceram. Soc.* **62**, 21.
B151 Buck, A. D. and Dolch, W. L. (1966). *J. Am. Concr. Inst. Proc.* **63**, 755.
B152 Bentur, A., Goldman, A. and Cohen, M. D. (1988). *Mater. Res. Soc. Symp. Proc.* **114**, 97.
B153 Bentur, A., Diamond, S. and Cohen, M. D. (1985). *J. Mater. Sci.* **20**, 3610.
B154 Browne, R. (1986). *Chem. Ind.* 837.
B155 Bennett, I. C. and Vivian, H. E. (1955). *Austral. J. Appl. Sci.* **6**, 88.
B156 Bhatty, M. S. Y. (1985). *Cem. Concr. Aggr.* **7**, 69.
B157 Buck, A. D. and Mather, K., in *Effects of Alkalies on Cement and Concrete* (ed. S. Diamond), Publ. CE-MAT-1-78, p. 73, Purdue University, W. Lafayette, IN, USA (1978).
B158 Bhatty, M. S. Y. and Greening, N., in *Effects of Alkalies on Cement and Concrete* (ed. S. Diamond), Publ. CE-MAT-1-78, p. 87, Purdue University, W. Lafayette, IN, USA (1978).
B159 Blair, L. R. and Yang, J. C.-S., in *4th ISCC*, Vol. 2, p. 849 (1962).
B160 Baronio, G. and Berra, M. (1986). *Cemento* **83**, 169.
B161 Bager, D. H. and Sellevold, E. J. (1987). *Cem. Concr. Res.* **17**, 1.
B162 Buenfeld, N. R. and Newman, J. B. (1986). *Cem. Concr. Res.* **16**, 721.
B163 Bentur, A., Ben-Bassat, M. and Schneider, D. (1985). *J. Am. Ceram. Soc.* **68**, 203.
C1 Chan, C. J., Kriven, W. M. and Young, J. F. (1988), *J. Am. Ceram. Soc.* **71**, 713.

C2	Colville, A. A. and Geller, S. (1972). *Acta Cryst.* **B28**, 3196.
C3	Colville, A. A. and Geller, S. (1971). *Acta Cryst.* **B27**, 2311.
C4	Colville, A. A. (1970). *Acta Cryst.* **B26**, 1469.
C5	Chatterjee, A. K. and Zhmoidin, G. I. (1972). *J. Mater. Sci.* **7**, 93.
C6	Conjeaud, M. (1975). Unpublished data, quoted in Refs G13 and M24.
C7	Chatterjee, A. K. and Zhmoidin, G. I. (1972). *Inorg. Mater.* **8**, 769 (translated from *Neorg. Mater.* **8**, 886).
C8	Castanet, R. and Sorrentino, F. P., in *8th ISCC*, Vol. 2, p. 36 (1986).
C9	Chromý, S., in *6th ICCC*, Vol. 3, p. 268 (1976).
C10	Cottin, B., Rouanet, A. and Conjeaud, M., in *8th ICCC*, Vol. 2, p. 31 (1986).
C11	Chromý, S. (1982). *Zem.-Kalk-Gips* **35**, 204 (partial English translation, p. 145).
C12	Chromý, S. and Weber, M. (1981). *Zem.-Kalk-Gips* **34**, 453 (partial English translation, p. 244).
C13	Christensen, N. H. (1979). *Cem. Concr. Res.* **9**, 219.
C14	Christensen, N. H. (1979). *Cem. Concr. Res.* **9**, 285.
C15	Chromý, S. (1982). *Zem.-Kalk-Gips* **35**, 555 (partial English translation, p. 303).
C16	Chromý, S. and Hrabě, Z. (1982). *Zem.-Kalk-Gips* **35**, 368 (partial English translation, p. 221).
C17	Campbell, D. H. *Microscopical Examination and Interpretation of Portland Cement and Clinker*, 128 pp., Construction Technology Laboratories, Skokie, IL, USA (1986).
C18	Commission chimique du CETIC (1978), *Ciments, Bétons, Plâtres, Chaux* (713) 205.
C19	Conjeaud, M. and Boyer, H. (1980). *Cem. Concr. Res.* **10**, 61.
C20	Copeland, L. E. and Hayes, J. C. (1953). *ASTM Bull.* **194**, 70.
C21	Copeland, L. E. and Schultz, E. G. (1962). *J. PCA Res. Dev. Labs* **4** (1), 2.
C22	Chatterji, S. and Jeffery, J. W. (1966). *Nature* **209**, 1233.
C23	Currell, B. R., Midgley, H. G., Montecinos, M. and Parsonage, J. R. (1985). *Cem. Concr. Res.* **15**, 889.
C24	Clayden, N. J., Dobson, C. M., Hayes, C. J. and Rodger, S. A. (1984). *J. Chem. Soc. Chem. Commun.* 1396.
C25	Copeland, L. E., Bodor, E., Chang, T. N. and Weise, C. H. (1967). *J. PCA Res. Dev. Labs* **9** (1), 61.
C26	Carpenter, A. B., Chalmers, R. A., Gard, J. A., Speakman, K. and Taylor, H. F. W. (1966). *Am. Mineral.* **51**, 56.
C27	Cliff, G., Gard, J. A., Lorimer, G. W. and Taylor, H. F. W. (1975). *Mineral. Mag.* **40**, 113.
C28	Carlson, E. T. and Berman, H. A. (1960). *J. Res. Natl. Bur. Stand.* **64A**, 333.
C29	Courtois, A., Dusausoy, Y., Laffaille, A. and Protas, J. (1968). *CR Acad. Sci. Paris* **D266**, 1911.
C30	Crammond, N. J. (1985). *Cem. Concr. Res.* **15**, 431.
C31	Cohen-Addad, C. (1971). *Acta Cryst.* **A27**, 68.
C32	Crammond, N. J. (1985). *Cem. Concr. Res.* **15**, 1039.
C33	Collepardi, M., Baldini, G., Pauri, M. and Corradi, M. (1978). *Cem. Concr. Res.* **8**, 571.
C34	Corstanje, W. A., Stein, H. N. and Stevels, J. M. (1973). *Cem. Concr. Res.* **3**, 791.

C35 Chatterji, S. and Jeffery, J. W. (1962). *J. Am. Ceram. Soc.* **45**, 536.
C36 Carlson, E. T. (1964). *J. Res. Natl. Bur. Stand.* **68A**, 453.
C37 Collepardi, M., Monosi, S., Moriconi, G. and Corradi, M. (1979). *Cem. Concr. Res.* **9**, 431.
C38 Copeland, L. E., Kantro, D. L. and Verbeck, G., in *4th ISCC*, Vol. 1, p. 429 (1960).
C39 Copeland, L. E. and Kantro, D. L., in *5th ISCC*, Vol. 2, p. 387 (1969).
C40 Collepardi, M., Marcialis, A. and Turriziani, R. (1972). *J. Am. Ceram. Soc.* **55**, 534.
C41 Chopra, S. K. and Taneja, C. A., in *5th ISCC*, Vol. 4, p. 228 (1969).
C42 Cesareni, C. and Frigione, G., in *5th ISCC*, Vol. 4, p. 237 (1969).
C43 Coole, M. J. (1984). *Br. Ceram. Proc.* **35**, 385.
C44 Costa, U. and Massazza, F., in *Effects of Flyash Incorporation in Cement and Concrete* (ed. S. Diamond), p. 134, Mater. Res. Soc., University Park, PA, USA (1981).
C45 Cook, D. J. and Suwanvitaya, P., in *Fly Ash, Silica Fume, Slag and Other Mineral By-Products in Concrete* (ed. V. M. Malhotra), Sp. Publ. SP79, Vol. 2, p. 831, American Concrete Institute, Detroit (1983).
C46 Calleja, J., in *7th ICCC*, Vol. 3, p. V-102 (1980).
C47 Cottin, B. and George, C. M., in *International Seminary on Calcium Aluminates* (eds. M. Murat *et al.*), Politecnico di Torino, Turin (1982).
C48 Capmas, A. and Ménétrier-Sorrentino, D., in *UNITECR '89* (Proc. Conf. Global Advances in Refractories, ed. L. J. Trostel), Vol. 2, p. 1157, American Ceramic Society, Westerville, OH, USA, (1989).
C49 Christensen, A. N., Fjellvåg, H. and Lehmann, M. S. (1986). *Acta Chem. Scand.* **A40**, 126.
C50 Cottin, B. (1971). *Cem. Concr. Res.* **1**, 177.
C51 Capmas, A. (1989). Private communication.
C52 Chappuis, J., Bayoux, J. P. and Capmas, A., in *UNITECR '89* (Proc. Conf. Global Advances in Refractories, ed. L. J. Trostel), Vol. 2, p. 1171, American Ceramic Society, Westerville, OH, USA, (1989).
C53 Currell, B. R., Grzeskowiak, R., Midgley, H. G. and Parsonage, J. R. (1987). *Cem. Concr. Res.* **17**, 420.
C54 Cussino, L. and Negro, A., in *7th ICCC*, Vol. 3, p. V-62 (1980).
C55 Cohen, M. D. (1983). *Cem. Concr. Res.* **13**, 809.
C56 Christensen, N. H. and Johansen, V., in *Cement Production and Use*, p. 55, Engineering Foundation, New York (1979).
C57 Chiocchio, G., Mangialardi, T. and Paolini, A. E. (1986). *Cemento* **83**, 69.
C58 Costa, U., Massazza, F. and Barrilà, A. (1982). *Cemento* **79**, 323.
C59 Costa, U. and Massazza, F. (1984). *Cemento* **81**, 127.
C60 Collepardi, M., Rossi, G. and Spiga, M. C. (1971). *Ann. Chim. (Rome)* **61**, 137.
C60a Clayden, N. J., Dobson, C. M., Groves, G. W. and Rodger, S. A., in *8th ICCC*, Vol. 3, p. 51 (1986).
C61 Chiocchio, G. and Turriziani, R. (1978). *Cemento* **75**, 165.
C62 Collepardi, M., Massidda, L. and Usai, G. (1971). *Cemento* **68**, 3.
C63 Crennan, J. M., El-Hemaly, S. A. S. and Taylor, H. F. W. (1977). *Cem. Concr. Res.* **7**, 493.
C64 Cole, W. F. and Moorehead, D. R., in *Autoclaved Calcium Silicate Building Products*, p. 134, Society of Chemical Industry, London (1967).
C65 Carles-Gibergues, A., Grandet, J. and Ollivier, J. P., in *Liaisons Pâtes de Ciment Matériaux Associés*, (Proc. RILEM Colloq.), p. D8, Laboratoire de Génie Civil, Toulouse (1982).

C66 Crumbie, A. K., Scrivener, K. L. and Pratt, P. L. (1989). *Mater. Res. Soc. Symp. Proc.* **137**, 279.
C67 Cole, W. F. and Lancucki, C. J. (1983). *Cem. Concr. Res.* **13**, 611.
C68 Cole, W. F., Lancucki, C. J. and Sandy, M. J. (1981). *Cem. Concr. Res.* **11**, 443.
C69 Crammond, N. J. (1984). *Cem. Concr. Res.* **14**, 225.
C70 Cole, W. F. (1953). *Nature* **171**, 354.
C71 Collins, A. R. (1944). *J. Inst. Civ. Engnrs.* **23**, 29.

D1 Day, A. L., Shepherd, E. S. and Wright, F. E. (1906). *Am. J. Sci.* (4th series) **22**, 266.
D2 Dahl, L. A., quoted by R. H. Bogue in *2nd SCC*, p. 138 (1939).
D3 Duda, W. H. *Cement Data Book*, Vol.1: *International Process Engineering in the Cement Industry* (3rd Ed.), 636 pp., Bauverlag, Wiesbaden and Berlin (1985).
D4 Davis, P., Stringer, J. A. and Watson, D., in *Making the Most of Materials—Cement*, Chapter 7, Science Research Council, Swindon, UK (1979).
D5 Davis, P. and Longman, P. A., in *International Cement Seminar* (Proc., 19th), p. 25, Rock Products, Chicago (1984).
D6 De Keyser, W. L. (1951). *Bull. Soc. Chim. Belg.* **60**, 516.
D7 De Keyser, W. L. (1954). *Bull. Soc. Chim. Belg.* **63**, 40.
D8 De Keyser, W. L. (1955). *Bull. Soc. Chim. Belg.* **64**, 395.
D9 Danilov, V. V., in *6th ICCC*, Vol. 2, Part 1, p. 73 (1976).
D10 De Keyser, W. L. and Tenoutasse, N., in *5th ISCC*, Vol. 2, p. 379 (1969).
D11 Day, R. L. (1981). *Cem. Concr. Res.* **11**, 341.
D12 Dalziel, J. A. and Gutteridge, W. A. (1986). *The Influence of Pulverized-Fuel Ash upon the Hydration Characteristics and Certain Physical Properties of a Portland Cement Paste* (Techn. Rpt 560), 28 pp., Cement and Concrete Association, Slough, UK (1986).
D13 Diamond, S., in *Hydraulic Cement Pastes: Their Structure and Properties*, p. 2, Cement and Concrete Association, Slough, UK (1976).
D14 Double, D. D., Hellawell, A. and Perry, S. J. (1978). *Proc. R. Soc. London* **A359**, 435.
D15 Dalgleish, B. J. and Ibe, K. (1981). *Cem. Concr. Res.* **11**, 729.
D16 Dent Glasser, L. S., Lachowski, E. E., Qureshi, M. Y., Calhoun, H. P., Embree, D. J., Jamieson, W. D. and Masson, C. R., (1981). *Cem. Concr. Res.* **11**, 775.
D17 Davis, R. W. and Young, J. F. (1975). *J. Am. Ceram. Soc.* **58**, 67.
D18 Double, D. D. (1983). *Phil. Trans R. Soc. London* **A310**, 53.
D19 Dosch, W. (1967). *Neues Jahrb. Mineral. Abhandl.* **106**, 200.
D20 Dosch, W. and Keller, H., in *6th ICCC*, Vol. 3, p. 141 (1976).
D21 Dosch, W., Keller, H. and zur Strassen, H., in *5th ISCC*, Vol. 2, p. 72 (1969).
D22 Dosch, W. and zur Strassen, H. (1967). *Zem.-Kalk-Gips* **20**, 392.
D23 Dron, R., in *6th ICCC*, Vol. 2, Part 1, p. 208 (1976).
D24 D'Ans, J. and Eick, H. (1953). *Zem.-Kalk-Gips* **6**, 197, 302.
D25 Dalgleish, B. J., Ghose, A., Jennings, H. M. and Pratt, P. L. (1982). *Sci. Ceram.* **11**, 297.
D26 Dalgleish, B. J., Ghose, A., Jennings, H. M. and Pratt, P. L., in *International Conference on Concrete of Early Ages* (Proc. RILEM), Vol. 1, p. 137, Editions Anciens ANPC, Paris (1982).
D27 Dalgleish, B. J., Pratt, P. L. and Toulson, E. (1983). *J. Mater. Sci.* **17**, 2199.

D28 Diamond, S., Ravina, D. and Lovell, J. (1980). *Cem. Concr. Res.* **10**, 297.
D29 Diamond, S. (1981). *Cem. Concr. Res.* **11**, 383.
D30 Daimon, M., Abo-El-Enein, S. A., Hosaka, G., Goto, S. and Kondo, R. (1977). *J. Am. Ceram. Soc.* **60**, 110.
D31 Diamond, S. and Dolch, W. L. (1972). *J. Coll. Interface Sci.* **38**, 234.
D32 Day, R. L. and Marsh, B. K. (1988). *Cem. Concr. Res.* **18**, 63.
D33 Diamond, S. (1971). *Cem. Concr. Res.* **1**, 531.
D34 Dullien, F. A. L. *Porous Media—Fluid Transport and Pore Structure*, 396 pp., Academic Press, New York (1979).
D35 Demoulian, E., Gourdin, P., Hawthorn, F. and Vernet, C., in *7th ICCC*, Vol. 2, p. III-89 (1980).
D36 Dron, R. and Brivot, F., in *7th ICCC*, Vol. 2, p. III-134 (1980).
D37 Daimon, M., in *7th ICCC*, Vol. 1, p. III-2/1 (1980).
D38 Demoulian, E., Vernet, C., Hawthorn, F. and Gourdin, P., in *7th ICCC*, Vol. 2, p. III-151 (1980).
D39 Diamond, S. and Lopez-Flores, F., in *Effects of Flyash Incorporation in Cement and Concrete* (ed. S. Diamond), p. 112, Mater. Res. Soc., University Park, PA, USA (1981).
D40 Diamond, S. (1983). *Cem. Concr. Res.* **13**, 459.
D41 Diamond, S. and Olek, J. (1987). *Mater. Res. Soc. Symp. Proc.* **86**, 315.
D42 Diamond, S., in *Effects of Flyash Incorporation in Cement and Concrete* (ed. S. Diamond), p. 12, Mater. Res. Soc., University Park, PA, USA (1981).
D43 Dent Glasser, L. S. and Lee, C. K. (1981). *Acta Cryst.* **B37**, 803.
D44 Daimon, M. and Roy, D. M. (1978). *Cem. Concr. Res.* **8**, 753.
D45 Daimon, M. and Roy, D. M. (1979). *Cem. Concr. Res.* **9**, 103.
D46 Dowell Schlumberger. *Cementing Technology*, Nova Communications, London (1984).
D47 Diamond, S., Mindess, S. and Lovell, J., in *Liaisons Pâtes de Ciment Matériaux Associés* (Proc. RILEM Colloq.), p. C42, Laboratoire de Génie Civil, Toulouse (1982).
D48 Diamond, S., in *8th ICCC*, Vol. 1, p. 122 (1986).
D49 Detwiler, R. J., Monteiro, P. J. M., Wenk, H.-R. and Zhong, Z. (1988). *Cem. Concr. Res.* **18**, 823.
D50 Diamond, S. (1976). *Cem. Concr. Res.* **6**, 549.
D51 Diamond, S., Barneyback, R. S. and Struble, L. J., in *Alkali–Aggregate Reaction in Concrete*, Paper S252/22, CSIR, Pretoria (1981).
D52 Dent Glasser, L. S. and Kataoka, N., in *Alkali–Aggregate Reaction in Concrete*, Paper S252/23, CSIR, Pretoria (1981).
D53 Dent Glasser, L. S. and Kataoka, N. (1982). *Cem. Concr. Res.* **12**, 321.
D54 Davies, G. and Oberholster, R. E., in *8th ICCC*, Vol. 5, p. 249 (1986).
D55 Dent Glasser, L. S. and Kataoka, N. (1981). *Cem. Concr. Res.* **11**, 191.
D56 Dent Glasser, L. S. and Kataoka, N. (1981). *Cem. Concr. Res.* **11**, 1.

E1 Eysel, W. (1973). *Am. Mineral.* **58**, 736.
E2 Edge, R. and Taylor, H. F. W. (1971). *Acta Cryst.* **B27**, 594.
E3 Effenberger, H., Kirfel, A., Will, G. and Zobetz, E. (1983). *Neues Jahrb. Mineral. Monatsh.* 60.
E4 Eden, N. B. and Bailey, J. E., in *8th ICCC*, Vol. 4, p. 163 (1986).
E5 Edmonds, R. N. and Majumdar, A. J. (1988). *Cem. Concr. Res.* **18**, 311.
E6 Edmonds, R. N. and Majumdar, A. J. (1988). *Cem. Concr. Res.* **18**, 473.
E7 Ernsberger, F. M. and France, W. G. (1945). *Ind. Engnrg Chem.* **37**, 598.

E8   Edwards, G. G. and Angstadt, R. L. (1966). *J. Appl. Chem.* **16**, 166.
E9   El-Hemaly, S. A. S., Mitsuda, T. and Taylor, H. F. W. (1977). *Cem. Concr. Res.* **7**, 429.
F1   Fukuda, K. and Maki, I. (1989). *Cem. Concr. Res.*, **19**, 913.
F2   Fletcher, K. E., Midgley, H. G. and Moore, A. E. (1965). *Mag. Concr. Res.* **17**, 171.
F3   Feng, Q. L., Glasser, F. P., Howie, R. A. and Lachowski, E. (1988). *Acta Cryst.* **C44**, 589.
F4   Fundal, E., in *8th ICCC*, Vol. 2, p. 139 (1986).
F5   Flörke, H. (1952). *Neues Jahrb. Mineral. Abhandl.* **84**, 189.
F6   Fayos, J., Glasser, F. P., Howie, R. A., Lachowski, E. E. and Pérèz-Mendez, M. (1985). *Acta Cryst.* **C41**, 814.
F7   Fundal, E. (1979). *World Cem. Technol.* **10**, 195.
F8   Franke, B. (1941). *Zeit. Anorg. Allg. Chem.* **247**, 180.
F9   Fierens, P. and Verhaegen, J.-P. (1972). *J. Am. Ceram. Soc.* **55**, 309.
F10 Fierens, P., Tirlocq, J. and Verhaegen, J. P. (1973). *Cem. Concr. Res.* **3**, 549.
F11 Fujii, K. and Kondo, W. (1979). *J. Am. Ceram. Soc.* **62**, 161.
F12 Feldman, R. F. and Ramachandran, V. S. (1971). *Cem. Concr. Res.* **1**, 607.
F13 Feldman, R. F., in *5th ISCC*, Vol. 3, p. 53 (1969).
F14 Feldman, R. F. and Ramachandran, V. S. (1974). *Cem. Concr. Res.* **4**, 155.
F15 Fujii, K. and Kondo, W., in *5th ISCC*, Vol. 2, p. 362 (1969).
F16 Feldman, R. F. (1972). *Cem. Concr. Res.* **2**, 123.
F17 Farmer, V. C., Jeevaratnam, J., Speakman, K. and Taylor, H. F. W., in *Symposium on Structure of Portland Cement Paste and Concrete* (Sp. Rpt 90), p. 291, Highway Research Board, Washington (1966).
F18 Fujii, K. and Kondo, W. (1983). *J. Am. Ceram. Soc.* **66**, C220.
F19 Fujii, K. and Kondo, W. (1981). *J. Chem. Soc. Dalton Trans* 645.
F20 Fierens, P. and Verhaegen, J. P. (1976). *Cem. Concr. Res.* **6**, 337.
F21 Funk, H., in *4th ISCC*, Vol. 1, p. 291 (1962).
F22 Fierens, P. and Tirlocq, J. (1983). *Cem. Concr. Res.* **13**, 41.
F23 Fierens, P. and Tirlocq, J. (1983). *Cem. Concr. Res.* **13**, 267.
F24 Feitknecht, W. and Buser, H. W. (1951). *Helv. Chim. Acta* **34**, 128.
F25 Fischer, R. and Kuzel, H.-J. (1982). *Cem. Concr. Res.* **12**, 517.
F26 Fischer, R., Kuzel, H.-J. and Schellhorn, H. (1980). *Neues Jahrb. Mineral. Monatsh.* 322.
F27 Foreman, D. W. (1968). *J. Chem. Phys.* **48**, 3037.
F28 Feldman, R. F. and Ramachandran, V. S. (1966). *Mag. Concr. Res.* **18**, 185.
F29 Fukuhara, M., Goto, S., Asaga, K., Daimon, M. and Kondo, R. (1981). *Cem. Concr. Res.* **11**, 407.
F30 Fortune, J. M. and Coey, J. M. D. (1983). *Cem. Concr. Res.* **13**, 696.
F31 Feldman, R. F. and Sereda, P. J., in *5th ISCC*, Vol. 3, p. 36 (1969).
F32 Feldman, R. F. and Sereda, P. J. (1970). *Engnrg J.* **53**, 53.
F33 Feldman, R. F. (1972). *Cem. Technol.* **3**, 5.
F34 Feldman, R. F., in *Effects of Fly Ash Incorporation in Cement and Concrete* (ed. S. Diamond), p. 124, Mater. Res. Soc., University Park, PA, USA (1981).
F35 Feldman, R. F. (1984). *J. Am. Ceram. Soc.* **67**, 30.
F36 Feldman, R. F. and Beaudoin, J. J. (1976). *Cem. Concr. Res.* **6**, 389.
F37 Ferraris, C. F. and Wittmann, F. H. (1987). *Cem. Concr. Res.* **17**, 453.
F38 Feldman, R. F. (1972). *Cem. Concr. Res.* **2**, 521.

F39 Feldman, R. F., in *8th ICCC*, Vol. 1, p. 336 (1986).
F40 Fierens, P. and Poswick, P., in *7th ICCC*, Vol. 2, p. III-112 (1980).
F41 Feldman, R. F. and Huang, C.-Y. (1985). *Cem. Concr. Res.* **15**, 765.
F42 Fujii, K., Kondo, W. and Ueno, H. (1986). *J. Am. Ceram. Soc.* **69**, 361.
F43 Farran, J. (1956). *Rev. Matér. Constr.* (490–491) 155; (492) 191.
F44 Fentiman, C. H. (1985). *Cem. Concr. Res.* **15**, 622.
F45 Feldman, R. F. and Beaudoin, J. J. (1977). *Cem. Concr. Res.* **7**, 19.
F46 Feldman, R. F. and Beaudoin, J. J. (1984). *Cem. Concr. Res.* **14**, 785.
F47 Farran, J., Javelas, R., Maso, J.-C. and Perrin, B. (1972). *CR Acad. Sci. Paris* **D275**, 1467.
F48 Fattuhi, N. I. and Hughes, B. P. (1988). *Cem. Concr. Res.* **18**, 545.
F49 French, W. J. and Poole, A. B. (1974). *Cem. Concr. Res.* **4**, 925.

G1 Guinier, A. and Regourd, M., in *5th ISCC*, Vol. 1, p. 1 (1969).
G2 Golovastikov, N. I., Matveeva, R. G. and Belov, N. V. (1975). *Sov. Phys. Cryst.* **20**, 441 (translated from *Kristallografiya* **20**, 721).
G3 Ghose, A. and Barnes, P. (1979). *Cem. Concr. Res.* **9**, 747.
G4 Ghose, A. (1980). Ph.D. Thesis, University of London.
G5 Gross, S. (1977). *Bull. Geol. Surv. Israel* (70), 1.
G6 Gazzozi, G. and Chiari, G. (1986). *Acta Cryst.* **B42**, 146.
G7 Groves, G. W. (1982). *Cem. Concr. Res.* **12**, 619.
G8 Grant, R. W., Geller, S., Wiedersich, H., Gonser, U. and Fullmer, L. D. (1968). *J. Appl. Phys.* **39**, 1122.
G9 Geller, S., Grant, R. W. and Gonser, U. (1971). *Prog. Solid State Chem.* **5**, 1.
G10 Gutt, W., in *Manufacture of Portland Cement from Phosphate Raw Materials* (Building Research Station Paper CP90/68), p. 5, HMSO, London (1968).
G11 Goodwin, D. W. and Lindop, A. J. (1970). *Acta Cryst.* **B26**, 1230.
G12 Gutt, W. (1961). *Nature* **190**, 339.
G13 Glasser, F. P. and Marr, J. (1975). *Trans J. Br. Ceram. Soc.* **74**, 113.
G14 Gutt, W. and Smith, M. A. (1967). *Trans Br. Ceram. Soc.* **66**, 557.
G15 Glasser, F. P. and Marr, J. (1980). *Cem. Concr. Res.* **10**, 753.
G16 Gutt, W. and Smith, M. A. (1968). *Trans Br. Ceram. Soc.* **67**, 487.
G17 Gutt, W. and Osborne, G. J. (1967). *Trans Br. Ceram. Soc.* **65**, 521.
G18 Gutt, W. and Osborne, G. J. (1968). *Trans Br. Ceram. Soc.* **67**, 125.
G19 Gilioli, C., Massazza, F. and Pezzuoli, H. (1979). *Cem. Concr. Res.* **9**, 295.
G20 Gutt, W. and Osborne, G. J. (1970). *Trans J. Br. Ceram. Soc.* **69**, 125.
G21 Gutt, W., in *5th ISCC*, Vol. 1, p. 93 (1969).
G22 Gutt, W., Chatterjee, A. K. and Zhmoidin, G. I. (1970). *J. Mater. Sci.* **5**, 960.
G23 Ghosh, S. N. (ed.) *Advances in Cement Technology*, 804 pp., Pergamon, Oxford (1983).
G24 Goldshtein, L. Y. A. *Cement Production from Industrial By-Products (Wastes) Utilization*, 80 pp., Stroyzdat, Leningrad (1985).
G25 Glasser, F. P., in *Structures and Properties of Cements* (ed. P. Barnes), p. 69, Applied Science Publishers, London (1983).
G26 Glasser, F. P., in *8th ICCC*, Vol. 6, p. 15 (1986).
G27 Gilioli, C., Massazza, F. and Pezzuoli, M. (1972). *Cemento* **69**, 19.
G28 Gilioli, C., Massazza, F. and Pezzuoli, M. (1973). *Cemento* **70**, 137.
G29 Gutt, W. (1963). *Nature* **197**, 142.
G30 Gutteridge, W. A. (1984). *Br. Ceram. Proc.* **35**, 11.

G31 Gille, F., Dreizler, I., Grade, K., Krämer, H. and Woermann, E. *Mikroskopie des Zementklinkers—Bilderatlas*, 75 pp., with English translation of text, 20 pp., Beton-Verlag Gmbh, Düsseldorf (1965).
G32 Goldstein, J. I., Newbury, D. E., Echlin, P., Joy, D. C., Fiori, C. and Lifshin, E. *Scanning Electron Microscopy and X-ray Microanalysis*, 673 pp., Plenum Press, New York (1981).
G33 Gutteridge, W. A. (1979). *Cem. Concr. Res.* **9**, 319.
G34 Gutteridge, W. A. (1986). Private communication.
G35 Groves, G. W., Le Sueur, P. J. and Sinclair, W. (1986). *J. Am. Ceram. Soc.* **69**, 353.
G36 Groves, G. W. (1987). *Mater. Res. Soc. Symp. Proc.* **85**, 3.
G37 Groves, G. W. (1981). *Cem. Concr. Res.* **11**, 713.
G38 Grudemo, Å. (1984). *Cem. Concr. Res.* **14**, 123.
G39 Greenberg, S. A. and Copeland, L. E. (1960). *J. Phys. Chem.* **64**, 1057.
G40 Gard, J. A., Mohan, K., Taylor, H. F. W. and Cliff, G. (1981). *J. Am. Ceram. Soc.* **63**, 336.
G41 Grudemo, A., in *4th ISCC*, Vol. 2, p. 615 (1962).
G42 Goto, S., Daimon, M., Hosaka, G. and Kondo, R. (1976). *J. Am. Ceram. Soc.* **59**, 281.
G43 Groves, G. W. (1981). *J. Mater. Sci.* **16**, 1063.
G44 Groves, G. W. (1983). *Phil. Trans. R. Soc. London* **A310**, 79.
G45 Groves, G. W. (1990). *Cem. Concr. Res.*, in press.
G46 Gard, J. A., Taylor, H. F. W., Cliff, G. and Lorimer, G. W. (1977). *Am. Mineral.* **62**, 365.
G47 Gard, J. A. and Taylor, H. F. W. (1976). *Cem. Concr. Res.* **6**, 667.
G48 Gartner, E. M. and Jennings, H. M. (1987). *J. Am. Ceram. Soc.* **70**, 743.
G49 Gartner, E. M. and Gaidis, J. M., in *Materials Science of Concrete I* (ed. J. P. Skalny), p. 95, American Ceramic Society, Westerville, OH, USA (1989).
G50 Glasser, F. P., Lachowski, E. E. and Macphee, D. E. (1987). *J. Am. Ceram. Soc.* **70**, 481.
G51 Grutzeck, M. W. and Ramachandran, A. R. (1987). *Cem. Concr. Res.* **17**, 164.
G52 Gerhard, W. and Nägele, E. (1983). *Cem. Concr. Res.* **13**, 849.
G53 Greening, N. R. (1972), quoted by J. F. Young in Ref. Y8.
G54 Gessner, W., Müller, D., Behrens, H.-J. and Scheler, G. (1982). *Zeit. Anorg. Allg. Chem.* **486**, 193.
G55 Granger, M.-M. and Protas, J. (1969). *Acta Cryst.* **B25**, 1943.
G56 Gard, J. A., in *Chemistry of Cements* (ed. H. F. W. Taylor), Vol. 2, p. 243, Academic Press, London (1964).
G57 Gaudefroy, C., Granger, M.-M., Permingeat, F. and Protas, J. (1968). *Bull. Soc. Franç. Mineral. Crist.* **91**, 43.
G58 Gartner, E. M., Tang, F. J. and Weiss, S. J. (1985). *J. Am. Ceram. Soc.* **68**, 667.
G59 Glasser, F. P. and Marinho, M. B. (1964). *Br. Ceram. Proc.* **35**, 221.
G60 Gupta, P., Chatterji, S. and Jeffery, J. W. (1970). *Cem. Technol.* **1**, 3.
G61 Gupta, P., Chatterji, S. and Jeffery, J. W. (1973). *Cem. Technol.* **4**, 63.
G62 Ghorab, H. Y. and El Fetouh, S. H. A. (1987). *Mater. Res. Soc. Symp. Proc.* **85**, 255.
G63 Glasser, F. P. and Marr, J. (1985). *Cemento* **82**, 85.

G64 Gregg, S. J. and Sing, K. S. W. (1982). *Adsorption, Surface Area and Porosity* (2nd Ed.), 303 pp., Academic Press, London (1985).

G65 Grudemo, Å., *Strength-structure relationships of cement paste materials. Part 1. Methods and basic data for studying phase composition and microstructure* (CBI Research, 6:77), 101 pp., Swedish Cement and Concrete Research Institute, Stockholm (1977); also private communication quoted by L.-O. Nilsson in Report TVBM-1003, Division of Building Materials, University of Lund, Sweden (1980).

G66 Good, R. J., in *Surface and Colloid Science* (eds E. Matijević and R. J. Good), Vol. 13, p. 283, Plenum Press, New York (1984).

G67 Goto, S. and Roy, D. M. (1981). *Cem. Concr. Res.* **11**, 575.

G68 Goto, S. and Roy, D. M. (1981). *Cem. Concr. Res.* **11**, 751.

G69 Goto, S., Yoshii, T. and Daimon, M. (1985). *Cem. Concr. Res.* **15**, 964.

G70 Ghose, A. and Pratt, P. L., in *Effects of Flyash Incorporation in Cement and Concrete* (ed. S. Diamond), p. 82, Mater. Res. Soc., University Park, PA, USA (1981).

G71 Grutzeck, M. W., Roy, D. M. and Scheetz, B. E., in *Effects of Flyash Incorporation in Cement and Concrete* (ed. S. Diamond), p. 92, Mater. Res. Soc., University Park, PA, USA (1981).

G72 Gegout, P., Hornain, H., Thuret, B., Mortureux, B., Volant, J. and Regourd, M., in *8th ICCC*, Vol. 4, p. 197 (1986).

G73 George, C. M., in *Structure and Performance of Cements* (ed. P. Barnes), p. 415, Applied Science Publishers, London (1983).

G74 Galtier, P. and Guilhot, B. (1984). *Cem. Concr. Res.* **14**, 679.

G75 Gill, S. M., Banfill, P. F. G. and El-Jazairi, B., in *8th ICCC*, Vol. 4, p. 322 (1986).

G76 George, C. M., in *7th ICCC*, Vol. 1, p. V-1/3 (1980).

G77 Ghosh, S. N., Mathur, V. K. and Chopra, S. K. (1984). *Cem. Concr. Res.* **14**, 437.

G78 Gard, J. A., Luke, K. and Taylor, H. F. W. (1981). *Cem. Concr. Res.* **11**, 659.

G79 Gard, J. A. and Taylor, H. F. W. (1957). *Mineral. Mag.* **31**, 361.

G80 Gouda, G. R. and Roy, D. M. (1975). *Cem. Concr. Res.* **5**, 551.

G81 Grandet, J. and Ollivier, J. P., in *7th ICCC*, Vol. 3, pp. VII-63, VII-85 (1980).

G82 Grandet, J. and Ollivier, J. P. (1980). *Cem. Concr. Res.* **10**, 759.

G83 Gard, J. A., Taylor, H. F. W. and Chalmers, R. A. (1957). *Mineral. Mag.* **31**, 611.

G84 Gillott, J. E. and Swenson, E. G. (1967). *J. Engnrg Geol.* **2**, 7.

G85 Greening, N. R. (1968). Private communication to D. W. Hadley, *J. PCA Res. Dev. Labs*, **10** (1), 17.

H1 Hudson, K. E. and Groves, G. W. (1982). *Cem. Concr. Res.* **12**, 61.

H2 Hahn, T., Eysel, W. and Woermann, E., in *5th ISCC*, Vol. 1, p. 61 (1969); summarized from E. Woermann, T. Hahn and W. Eysel, *Zem.-Kalk-Gips* **16**, 370 (1963); **20**, 385 (1967); **21**, 241 (1968); **22**, 235, 412 (1969).

H3 Harrisson, A. M., Taylor, H. F. W. and Winter, N. B. (1985). *Cem. Concr. Res.* **15**, 775.

H4 Harrisson, A. M., Winter, N. B. and Taylor, H. F. W., in *8th ICCC*, Vol. 4, p. 170 (1986).

H5 Hyde, B. G., Sellar, J. R. and Stenberg, L. (1986). *Acta Cryst.* **B42**, 423.

H6 Han, K. S., Gard, J. A. and Glasser, F. P. (1981). *Cem. Concr. Res.* **11**, 79.

H7	Hansen, W. C., Brownmiller, L. T. and Bogue, R. H. (1928). *J. Am. Chem. Soc.* **50**, 396.
H8	Hornain, H. (1973). *Rev. Matér. Constr.* (680) 4.
H9	Henmi, C., Kusachi, I., Kawahara, A. and Henmi, K. (1978). *Mineral. J. (Japan)* **9**, 169.
H10	Henmi, C., Kawahara, A., Henmi, K., Kusachi, I. and Takéuchi, Y. (1983). *Am. Mineral.* **68**, 156.
H11	Hörkner, W. and Müller-Buschbaum, H. (1976). *J. Inorg. Nucl. Chem.* **38**, 983.
H12	Hentschel, G. (1964). *Neues Jahrb. Mineral. Monatsh.* 22.
H13	Hansen, W. C. (1928). *J. Am. Chem. Soc.* **50**, 2155.
H14	Hanic, F., Handlovič, M. and Kaprálik, I. (1980). *Acta Cryst.* **B36**, 2863.
H15	Halstead, P. E. and Moore, A. E. (1962). *J. Appl. Chem.* **12**, 413.
H16	Hanic, F., Havlica, J., Kaprálik, I., Ambrúz, V., Gáliková, L. and Urbanova, O. (1986). *Trans J. Br. Ceram. Soc.* **85**, 52.
H17	Henmi, C. and Henmi, K. (1978). *Mineral. J. (Japan)* **9**, 106.
H18	Harker, R. I. and Tuttle, O. F. (1956). *Am. J. Sci.* **254**, 239.
H19	Hills, A. W. D. (1968). *Chem. Engnrg Sci.* **23**, 297.
H20	Heilmann, T., in *3rd ISCC*, p. 711 (1954).
H21	Hofmänner, F. *Microstructure of Portland Cement Clinker*, 48 pp., Rheintaler Druckerei und Verlag, Heerbrugg, Switzerland (1975).
H22	Hornain, H. and Regourd, M., in *7th ICCC*, Vol. 2, p. I-276 (1980).
H23	Hubbard, C. R., Evans, E. H. and Smith, D. K. (1976). *J. Appl. Cryst.* **9**, 169.
H24	Hjorth, L. and Laurén, K.-G. (1971). *Cem. Concr. Res.* **1**, 27.
H25	Halstead, P. E. and Moore, A. E. (1957). *J. Chem. Soc.* 3873.
H26	Harrisson, A. M., Winter, N. B. and Taylor, H. F. W. (1987). *J. Mater. Sci. Lett.* **6**, 1339.
H27	Hirljac, J., Wu, Z.-Q. and Young, J. F. (1983). *Cem. Concr. Res.* **13**, 877.
H28	Hansen, W. (1987), quoted by J. F. Young and W. Hansen in Ref. Y5.
H29	Hunt, C. M., in *4th ISCC*, Vol. 1, p. 297 (1962).
H30	Hara, N., Chan, C. F. and Mitsuda, T. (1978). *Cem. Concr. Res.* **8**, 113.
H31	Hamid, S. A. (1981). *Zeit. Krist.* **154**, 189.
H32	Hara, N. and Inoue, N. (1980). *Cem. Concr. Res.* **10**, 677.
H33	Hara, N., Inoue, N. and Noma, H., in *8th ICCC*, Vol. 3, p. 160 (1986).
H34	Henning, O., in *The Infrared Spectra of Minerals* (ed. V. C. Farmer), p. 445, Mineralogical Society, London (1974).
H35	Hernandez-Moreno, M. J., Ulibarri, M. A., Rendon, J. L. and Serna, C. J. (1985). *Phys. Chem. Minerals* **12**, 34.
H36	Houtepen, C. J. M. and Stein, H. N. (1976). *Cem. Concr. Res.* **6**, 651.
H37	Halse, Y., Goult, D. J. and Pratt, P. L. (1984). *Br. Ceram. Proc.* **35**, 403.
H38	Harchand, K. S., Vishwamittar and Chandra, K. (1980). *Cem. Concr. Res.* **10**, 243.
H39	Hobbs, D. W. (1978). *Cem. Concr. Res.* **8**, 211.
H40	Helmuth, R. A., in *7th ICCC*, Vol. 3, p. VI-0/16 (1980).
H41	Hansen, W. and Almudaiheem, J. (1987). *Mater. Res. Soc. Symp. Proc.* **85**, 105.
H42	Hagymassy, J., Odler, I., Yudenfreund, M., Skalny, J. and Brunauer, S. (1972). *J. Coll. Interface Sci.* **38**, 20.
H43	Hansen, W. (1987). *J. Am. Ceram. Soc.* **70**, 323.
H44	Hasselmann, D. P. H. (1962). *J. Am. Ceram. Soc.* **46**, 564.

H45 Helmuth, R. A. and Turk, D. A., in *Symposium on Structure of Portland Cement Paste and Concrete* (Sp. Rpt 90), p. 135, Highway Research Board, Washington (1966).
H46 Helmuth, R. A. and Turk, D. H. (1967). *J. PCA Res. Dev. Labs* **9** (2), 8.
H47 Hooton, R. D., in *Blended Cements* (ASTM Spec. Tech. Publ. 897), p. 128, Am. Soc. Testing Mater., Philadelphia (1986).
H48 Hooton, R. D., in *Advances in Cement Manufacture and Use*, p. 143, Engineering Foundation, New York (1989).
H49 Harrisson, A. M., Winter, N. B. and Taylor, H. F. W. (1987). *Mater. Res. Soc. Symp. Proc.* **85**, 213.
H50 Hubbard, F. H., Dhir, R. K. and Ellis, M. S. (1985). *Cem. Concr. Res.* **15**, 185.
H51 Halse, Y., Pratt, P. L., Dalziel, J. A. and Gutteridge, W. A. (1984). *Cem. Concr. Res.* **14**, 491.
H52 Halse, Y., Jensen, H.-U. and Pratt, P. L., in *8th ICCC*, Vol. 4, p. 176 (1986).
H53 Hjorth, L. *Microsilica in Concrete* (Nordic Concrete Res., No. 1, Paper 9), 18 pp., Aalborg Portland, Denmark (1982).
H54 Huang, C.-Y. and Feldman, R. F. (1985). *Cem. Concr. Res.* **15**, 285.
H55 Huang, C.-Y. and Feldman, R. F. (1985). *Cem. Concr. Res.* **15**, 585.
H56 Hemmings, R. T. and Berry, E. E. (1988). *Mater. Res. Soc. Symp. Proc.* **113**, 3.
H57 Halse, Y. and Pratt, P. L., in *8th ICCC*, Vol. 4, p. 317 (1986).
H58 Harchand, K. S., Vishwamittar and Chandra, K. (1984). *Cem. Concr. Res.* **14**, 19.
H59 Heller, L. and Ben-Yair, M. (1964). *J. Appl. Chem.* **14**, 20.
H60 Heller, L. (1952). *Acta Cryst.* **5**, 724.
H61 Hara, N. and Inoue, N. (1980). *Cem. Concr. Res.* **10**, 53.
H62 Hjorth, L. (1983). *Phil. Trans R. Soc. Lond.* **A310**, 167.
H63 Hansson, C. M. (1984). *Cem. Concr. Res.* **14**, 574.
H64 Harrisson, A. R. and Winter, N. B. (1989). Private communication.
H65 Hobbs, D. W. (1986). *Structural Engineer* **64A**, 381.
H66 Hobbs, D. W. (1981). *Mag. Concr. Res.* **33**, 208.
H67 Hobbs, D. W. and Gutteridge, W. A. (1980). *Mag. Concr. Res.* **32**, 235.
H68 Hansen, W. C. (1944). *J. Am. Concr. Inst. (Proc.)* **40**, 213.
H69 Hobbs, D. W. (1986). *Mag. Concr. Res.* **38**, 191.
H70 Hughes, D. C. (1985). *Cem. Concr. Res.* **15**, 1003.
H71 Hamada, M., in *5th ISCC*, Vol. 3, p. 343 (1962).

I1 Il'inets, A. M., Malinovskii, Yu. A. and Nevskii, N. N. (1985). *Dokl. Akad. Nauk SSSR* **281**, 332.
I2 Ilyinets, A. M., Simonov, V. I. and Bikbau, M. Ya., in *8th ICCC*, Vol. 6, p. 489 (1986).
I3 Il'inets, A. M., Nevskii, N. N., Ilyukhin, V. V., Bikbau, M. Ya. and Belov, N. V. (1982). *Dokl. Akad. Nauk SSSR* **267**, 641.
I4 Imlach, J. A., Dent Glasser, L. S. and Glasser, F. P. (1971). *Cem. Concr. Res.* **1**, 57.
I5 Insley, H. and McMurdie, H. F. (1938). *J. Res. Natl Bur. Stand.* **20**, 173.
I6 Imlach, J. A. and Hofmänner, F., in *6th ICCC*, Vol. 1, p. 281 (1976).
I7 Ings, J. B., Brown, P. W. and Frohnsdorff, G. (1983). *Cem. Concr. Res.* **13**, 843.

I8	Ingram, L. and Taylor, H. F. W. (1967). *Mineral. Mag.* **36**, 465.
I9	Ish-Shalom, M. and Bentur, A. (1974). *Cem. Concr. Res.* **4**, 519.
I10	Ish-Shalom, M. and Bentur, A. (1975). *Cem. Concr. Res.* **5**, 139.
I11	Ilyukhin, V. V., Nevsky, N. N., Bickbau, M. J. and Howie, R. A. (1977). *Nature* **269**, 397.
I12	Idorn, G., in *5th ISCC*, Vol. 3, p. 411 (1969).
I13	Imlach, B. V. and Taylor, H. F. W. (1972). *Trans J. Br. Ceram. Soc.* **71**, 71.
I14	Idorn, G. and Fördös, Z., in *6th ICCC*, Vol. 3, p. 287 (1976).
I15	Idorn, G. M. (1961). *Danish Inst. Bldg Res. Ser., Progr. Rpt (2)*; quoted by T. Knudsen and N. Thaulow in Ref. K67.
J1	Jeffery, J. W. (1952). *Acta Cryst.* **5**, 26.
J2	Jost, K. H., Ziemer, B. and Seydel, R. (1977). *Acta Cryst.* **B33**, 1696.
J3	Jeevaratnam, J., Glasser, F. P. and Dent Glasser, L. S. (1964). *J. Am. Ceram. Soc.* **47**, 105.
J4	Jøns, E. C. (1980). *Cem. Concr. Res.* **10**, 103.
J5	Johansen, V. (1973). *J. Am. Ceram. Soc.* **56**, 450.
J6	Johansen, V. (1978). *Cem. Concr. Res.* **8**, 245.
J7	Javellana, M. P. and Jawed, I. (1982). *Cem. Concr. Res.* **12**, 399.
J8	Jost, K. H. and Ziemer, B. (1984). *Cem. Concr. Res.* **14**, 177.
J9	Jones, F. E. (1940). *J. Soc. Chem. Ind.* **51**, 29.
J10	Jennings, H. M., Dalgleish, B. J. and Pratt, P. L. (1981). *J. Am. Ceram. Soc.* **64**, 567.
J11	Jennings, H. M. and Pratt, P. L. (1980). *J. Mater. Sci. Lett.* **15**, 250.
J12	Jennings, H. M. and Parrott, L. J. (1986). *J. Mater. Sci.* **21**, 4048.
J13	Jennings, H. M. and Parrott, L. J. (1986). *J. Mater. Sci.* **21**, 4053.
J14	Jawed, I. and Skalny, J. (1982). *J. Coll. Interface Sci.* **85**, 235.
J15	Jennings, H. M. (1986). *J. Am. Ceram. Soc.* **69**, 614.
J16	Jennings, H. M. (1988). *J. Am. Ceram. Soc.* **71**, C115.
J17	Jones, F. E. and Roberts, M. H., quoted by F. E. Jones in Ref. J19.
J18	Jones, F. E., in *2nd SCC*, p. 231 (1939).
J19	Jones, F. E., in *4th ISCC*, Vol. 1, p. 205 (1962).
J20	Jones, F. E. (1944). *J. Phys. Chem.* **48**, 311.
J21	Jones, F. E. (1944). *J. Phys. Chem.* **48**, 356, 379.
J22	Jawed, I. and Skalny, J., in *International Seminary on Calcium Aluminates* (ed. M. Murat *et al.*), Politecnico di Torino, Turin (1982).
J23	Jennings, H. M. and Pratt, P. L., in *7th ICCC*, Vol. 2, p. II-141 (1980).
J24	Jawed, I. and Skalny, J. (1978). *Cem. Concr. Res.* **8**, 37.
J25	Jelenić, I., Panović, A., Halle, R. and Gaćeša, T. (1977). *Cem. Concr. Res.* **7**, 239.
J26	Jelenić, I., Panović, A. and Bezjak, A. (1980). *Cem. Concr. Res.* **10**, 463.
J27	Jelenić-Bezjak, I., in *Advances in Cement Technology* (ed. S. N. Ghosh), p. 397, Pergamon Press, Oxford (1983).
J28	Jennings, H. M. and Johnson, S. K. (1986). *J. Am. Ceram. Soc.* **69**, 790.
J29	Jambor, J., in *6th ICCC*, Vol. 2, Part 1, p. 315 (1976).
J30	Jawed, I. and Skalny, J. in *Effects of Flyash Incorporation in Cement and Concrete* (ed. S. Diamond), p. 60, Mater. Res. Soc., University Park, PA, USA (1981).
J31	James, J. and Subba Rao, M. (1986). *Cem. Concr. Res.* **16**, 296.
J32	Jeanne, M. (1968). *Rev. Matér. Constr.* (629) 53.

J33 Jennings, H. M., Taleb, H., Frohnsdorff, G. and Clifton, J. R., in *8th ICCC*, Vol. 3, p. 239 (1986).
J34 Javelas, R., Maso, J. C., Ollivier, J. P. and Thenoz, B. (1975). *Cem. Concr. Res.* **5**, 285.
K1 Kristmann, M. (1978). *Cem. Concr. Res.* **8**, 93.
K2 Kondo, R.-I., Goto, S. and Fukuhara, M., in *Review of the 32nd General Meeting*, p. 38, Cement Association of Japan, Tokyo (1978).
K3 Kaprálik, I. and Hanic, F. (1980). *Trans J. Br. Ceram. Soc.* **79**, 128.
K4 Kirfel, A. and Will, G. (1980). *Acta Cryst.* **B36**, 2881.
K5 Kaprálik, I., Hanic, F., Havlica, J. and Ambrúz, V. (1986). *Trans J. Br. Ceram. Soc.* **85**, 107.
K6 Kaprálik, I. and Hanic, F. (1986). *Trans J. Br. Ceram. Soc.* **85**, 131.
K7 Kerton, C. P. and Murray, R. J., in *Structures and Properties of Cements* (ed. P. Barnes), p. 205, Applied Science Publishers, London (1983).
K8 Kirchner, G. (1986). *Zem.-Kalk-Gips* **39**, 555 (partial English translation, p. 368).
K9 Kuhlmann, K., Ellerbrock, H.-G. and Sprung, S. (1985). *Zem.-Kalk-Gips* **38**, 169 (partial English translation, p. 136).
K10 Klug, H. P. and Alexander, L. E. *X-ray Diffraction Procedures for Polycrystalline and Amorphous Materials* (2nd Ed.), 966 pp., Wiley, New York (1974).
K11 Knudsen, T. (1976). *Am. Ceram. Soc. Bull.* **55**, 1052.
K12 Kristmann, M. (1977). *Cem. Concr. Res.* **7**, 649.
K13 Kantro, D. L. and Weise, C. H. (1979). *J. Am. Ceram. Soc.* **62**, 621.
K14 Kantro, D. L., Brunauer, S. and Weise, C. H. (1959). *J. Colloid Sci.* **14**, 363.
K15 Kalousek, G. L. and Roy, R. (1957). *J. Am. Ceram. Soc.* **40**, 236.
K16 Komarneni, S., Roy, D. M., Fyfe, C. A. and Kennedy, G. J. (1987). *Cem. Concr. Res.* **17**, 891.
K17 Kalousek, G. L., in *3rd ISCC*, p. 296 (1954).
K18 Kusachi, I., Henmi, C. and Henmi, K. (1984). *J. Japan. Assoc. Mineral. Petr. Econ. Geol.* **79**, 267.
K19 Kalousek, G. L. (1944). *J. Res. Natl Bur. Stand.* **32**, 285.
K20 Kondo, R. and Ueda, S., in *5th ISCC*, Vol. 2, p. 203 (1969).
K21 Kantro, D. L., Brunauer, S. and Weise, C. H. (1962). *J. Phys. Chem.* **66**, 1804.
K22 Kondo, R., Ueda, S. and Kodama, M. (1967). *Semento Gijutsu Nenpo* **21**, 83.
K23 Kuzel, H.-J. (1966). *Neues Jahrb. Mineral. Monatsh.* 193.
K24 Kuzel, H.-J. (1970). *Neues Jahrb. Mineral. Monatsh.* 363.
K25 Kuzel, H.-J. (1976). *Neues Jahrb. Mineral. Monatsh.* 319.
K26 Kuzel, H.-J. (1968). *Zem.-Kalk-Gips* **21**, 493.
K27 Kalousek, G. L., Davis, C. W. and Schmertz, W. E. (1949). *J. Am. Concr. Inst. (Proc.)* **45**, 693.
K28 Kuzel, H.-J. (1969). *Neues Jahrb. Mineral. Monatsh.* 397.
K29 Kiriyama, R., Kiriyama, H. and Takagawa, M., in *5th ISCC*, Vol. 2, p. 98 (1969).
K30 Koritnig, S. and Süsse, P. (1975). *Tschermaks Mineral. Petr. Mitt.* **22**, 79.
K31 Kuzel, H.-J. (1987). *Neues Jahrb. Mineral. Abhandl.* **156**, 155.
K32 Kalousek, G. L. and Adams, M. (1951). *J. Am. Concr. Inst. (Proc.)* **48**, 77.
K33 Kalousek, G. M. (1965). *Mater. Res. Stand.* **5**, 292.
K34 Komarneni, S., Roy, R., Roy, D. M., Fyfe, C. A. and Kennedy, C. J. (1985). *Cem. Concr. Res.* **15**, 723.

K35 Kanare, H. M. and Gartner, E. M. (1985). *Cem. Res. Prog.* [for] 1984, 213.
K36 Knudsen, T. and Geiker, M. (1985). *Cem. Concr. Res.* **15**, 381.
K37 Knudsen, T., in *7th ICCC*, Vol. 2, p. I-170 (1980).
K38 Knudsen, T., in *Characterization and Performance Prediction of Cement and Concrete*, p. 125, Engineering Foundation, New York (1983).
K39 Knudsen, T. (1984). *Cem. Concr. Res.* **14**, 622.
K40 Knudsen, T., in *8th ICCC*, Vol. 3, p. 369 (1986).
K41 Kumar, A. and Roy, D. M. (1986). *Cem. Concr. Res.* **16**, 74.
K42 Kondo, R. and Ohsawa, S., in *5th ISCC*, Vol. 4, p. 255 (1969).
K43 Kühle, K. and Ludwig, U. (1972). *Sprechsaal Keram. Glas Email Silik.* **105**, 421.
K44 Kokubu, M., in *5th ISCC*, Vol. 4, p. 75 (1969).
K45 Kovács, R. (1975). *Cem. Concr. Res.* **5**, 73.
K46 Kawada, N. and Nemoto, A. (1968). *Semento Gijutsu Nenpo* **22**, 124.
K47 Kovács, R., in *6th ICCC*, Vol. 3, p. 99 (1976).
K48 Khan, M. H., Mohan, K. and Taylor, H. F. W. (1985). *Cem. Concr. Res.* **15**, 89.
K49 Klein, A. and Troxell, G. E. (1958). *Proc. Am. Soc. Testing Mater.* **58**, 986.
K50 Kawano, T., Hitotsuya, K. and Mori, T., in *6th ICCC*, Vol. 3, p. 179 (1976).
K51 Kurdowski, W. and Thiel, A. (1981). *Cem. Concr. Res.* **11**, 29.
K52 Klemm, W. A. and Skalny, J. (1977). *Cem. Res. Progr.* [for] 1976, 259.
K53 Kurdowski, W. (1987). *Cem. Concr. Res.* **17**, 361.
K54 Kurdowski, W. and Miskiewicz, K. (1985). *Cem. Concr. Res.* **15**, 785.
K55 Klieger, P. (1966), in *Significance of Tests and Properties of Concrete and Concrete-Making Materials* (Sp. Tech. Publ. 169A), p. 530, Am. Soc. Testing Materials, Philadelphia (1966).
K56 Kondo, R., Daimon, M. and Sakai, E. (1978). *Cemento* **75**, 225.
K57 Kantro, D. L. (1975). *J. Testing Evaln* **3**, 312.
K58 Kondo, R., Daimon, M., Sakai, E. and Ushiyama, H. (1977). *J. Appl. Chem. Biotechnol.* **27**, 191.
K59 Kalousek, G. L. (1954). *J. Am. Concr. Inst. (Proc.)* **50**, 365.
K60 Kalousek, G. L. (1966). *J. Am. Concr. Inst. (Proc.)* **63**, 817.
K61 Kalousek, G. L., in *5th ISCC*, Vol. 3, p. 523 (1969).
K62 Kalousek, G. L. (1957). *J. Am. Ceram. Soc.* **40**, 74.
K63 Komarneni, S., Roy, D. M. and Roy, R. (1982). *Cem. Concr. Res.* **12**, 773.
K64 Kendall, K., Howard, A. J. and Birchall, J. D. (1983). *Phil. Trans R. Soc. London* **A310**, 139.
K65 Knab, L. I., Clifton, J. R. and Ings, J. B. (1983). *Cem. Concr. Res.* **13**, 383.
K66 Kondo, R., Daimon, M. and Akiba, T., in *5th ISCC*, Vol. 3, p. 402 (1969).
K67 Knudsen, T. and Thaulow, N. (1975). *Cem. Concr. Res.* **5**, 443.
K68 Kollek, J. J., Varma, S. P. and Zaris, C., in *8th ICCC*, Vol. 4, p. 183 (1986).
K69 Kuenning, W. H. (1966). *Highway Res. Record* (113) 43.

L1 Lee, F. C., Banda, H. M. and Glasser, F. P. (1982). *Cem. Concr. Res.* **12**, 237.
L2 Lee, F. C. and Glasser, F. P. (1979). *J. Appl. Cryst.* **12**, 407.
L3 Louisnathan, S. J. (1971). *Canad. Mineral.* **10**, 822.
L4 Lea, F. M. and Parker, T. W. (1934). *Phil. Trans R. Soc. London* **A234**, 1.
L5 Lea, F. M. and Parker, T. W. *The Quaternary System $CaO–Al_2O_3–SiO_2–Fe_2O_3$ in Relation to Cement Technology* (Building Res. Techn. Paper 16), 52 pp., HMSO, London (1935).

L6	Lea, F. M. *The Chemistry of Cement and Concrete* (3rd Ed.), 727 pp., Arnold, London (1970).
L7	Leary, J. K. (1962). *Nature* **194**, 79.
L8	Lehmann, H., Locher, F. W. and Thormann, P. (1964). *TIZ* **88**, 489.
L9	Ludwig, U. and Ruckensteiner, G. (1974). *Cem. Concr. Res.* **4**, 239.
L10	Long, G. R., in *4th ICCem. Microsc.*, p. 92 (1982).
L11	Long, G. R. (1983). *Phil. Trans R. Soc. London* **A310**, 43.
L12	Locher, F. W. (1980). *World Cem. Technol.* **11**, 67.
L13	Lerch, W. and Brownmiller, L. T. (1937). *J. Res. Natl Bur. Stand.* **18**, 609.
L14	Long, G. R., in *4th ICCem. Microsc.*, p. 128 (1982).
L15	Lea, F. M. and Nurse, R. W. (1939). *J. Soc. Chem. Ind.* **58**, 227.
L16	Locher, F. W., Sprung, S. and Korf, P. (1973). *Zem.-Kalk-Gips* **26**, 349.
L17	Lebiedzik, J., Lebiedzik, J. and Gouda, G. R., in *3rd ICCem. Microsc.*, p. 154 (1981).
L18	Lerch, W. and Bogue, R. H. (1926). *Ind. Engnrg Chem.* **18**, 739.
L19	Le Sueur, P. J., Double, D. D. and Groves, G. W. (1984). *Br. Ceram. Proc.* **35**, 177.
L20	Lentz, C. W., in *Symposium on Structure of Portland Cement Paste and Concrete* (Sp. Rpt 90), p. 269, Highway Research Board, Washington (1966).
L21	Lipmaa, E., Mägi, M., Tarmak, M., Wieker, W. and Grimmer, A.-R. (1982). *Cem. Concr. Res.* **12**, 597.
L22	Lehmann, H. and Dutz, H., in *4th ISCC*, Vol. 1, p. 513 (1962).
L23	Long, J. V. P. and McConnell, J. D. C. (1959). *Mineral. Mag.* **32**, 117.
L24	Lerch, W. and Bogue, R. H. (1934). *J. Res. Natl Bur. Stand.* **12**, 645.
L25	Longuet, P., in *3rd ISCC*, p. 328 (1954).
L26	Lager, G. A., Armbruster, T., Rotella, F. J., Jorgensen, J. D. and Hinks, D. G. (1984). *Am. Mineral.* **69**, 910.
L27	Luke, K. and Glasser, F. P. (1987). *Cem. Concr. Res.* **17**, 273.
L28	Lachowski, E. E., Mohan, K., Taylor, H. F. W. and Moore, A. E. (1980). *J. Am. Ceram. Soc.* **63**, 447.
L29	Lachowski, E. E., Mohan, K., Taylor, H. F. W., Lawrence, C. D. and Moore, A. E. (1981). *J. Am. Ceram. Soc.* **64**, 319.
L30	Lachowski, E. E. and Diamond, S. (1983). *Cem. Concr. Res.* **13**, 177.
L31	Lachowski, E. E. (1979). *Cem. Concr. Res.* **9**, 337.
L32	Lawrence, C. D., in *Symposium on Structure of Portland Cement Paste and Concrete* (Sp. Rpt 90), p. 378, Highway Research Board, Washington (1966).
L33	Locher, F. W., Richartz, W. and Sprung, S. (1976). *Zem.-Kalk-Gips* **29**, 435 (partial English translation, p. 257).
L34	Longuet, P., Burglen, L. and Zelwer, A. (1973). *Rev. Matér. Constr.* (676) 35.
L35	Lashchenko, V. A. and Loganina, V. I. (1974). *Zh. Prikl. Khim.* **47**, 645 (p. 646 in English translation).
L36	Locher, F. W., in *7th ICCC*, Vol. 4, p. 49 (1981).
L37	Locher, F. W., Richartz, W. and Sprung, S. (1980). *Zem.-Kalk-Gips* **33**, 271 (partial English translation, p. 150).
L38	Lerch, W. (1946). *Proc. Am. Soc. Testing Mater.* **46**, 1252.
L39	Locher, F. W., in *Symposium on Structure of Portland Cement Paste and Concrete* (Sp. Rpt 90), p. 300, Highway Research Board, Washington (1966).
L40	Lapasin, R. (1982). *Cemento* **79**, 243.
L41	Lawrence, C. D. *An Examination of Possible Errors in the Determination of Nitrogen Isotherms on Hydrated Cements* (Techn. Rpt 520), 29 pp., Cement and Concrete Association, Slough, UK (1978).

L42 Litvan, G. G. (1976). *Cem. Concr. Res.* **6**, 139.
L43 Lawrence, C. D., Gimblett, F. G. R. and Sing, K. S. W., in *7th ICCC*, Vol. 3, p. VI-141 (1980).
L44 Lawrence, C. D. *The Interpretation of Nitrogen Sorption Isotherms on Hydrated Cement* (Techn. Rpt 530), 28 pp., Cement and Concrete Association, Slough, UK (1980).
L45 Lawrence, C. D., in *8th ICCC*, Vol. 5, p. 29 (1986).
L46 Luke, K. and Glasser, F. P. (1988). *Cem. Concr. Res.* **18**, 495.
L47 Lukas, W. (1976). *Matér. Constr. (Paris)* **9**, 331.
L48 Lossier, H. (1936). *Génie Civil.* (109) 285.
L49 Long, G. R., in *5th ICCem. Microsc.*, p. 86 (1983).
L50 Long, G. R., Longman, P. A. and Gartshore, G. C., in *9th ICCem. Microsc.*, p. 263 (1987).
L51 Locher, F., in *8th ICCC*, Vol. 1, p. 57 (1986).
L52 Libermann, G. V. and Kireev, V. A. (1964). *Zh. Prikl. Khim.* **37**, 194.
L53 Langton, C. A., White, E. L., Grutzeck, M. W. and Roy, D. M., in *7th ICCC*, Vol. 3, p. V-145 (1980).
L54 Lyubimova, T. Yu. and Pinus, E. R. (1962). *Kolloidn. Zh.* **24**, 491.
L55 Locher, F. W. (1966). *Zem.-Kalk-Gips* **19**, 395.
L56 Ludwig, U. and Mehr, S., in *8th ICCC*, Vol. 5, p. 181 (1986).
L57 Litvan, G. G. (1972). *J. Am. Ceram. Soc.* **55**, 38.

M1 Maki, I. and Chromý, S. (1978). *Cem. Concr. Res.* **8**, 407.
M2 Maki, I. and Chromý, S. (1978). *Cemento* **75**, 247.
M3 Maki, I. (1979). *Cemento* **76**, 167.
M4 Maki, I. and Kato, K. (1982). *Cem. Concr. Res.* **12**, 93.
M5 Maki, I., in *8th ICCC*, Vol. 1, p. 34 (1986).
M6 Maki, I. and Goto, K. (1982). *Cem. Concr. Res.* **12**, 301.
M7 Moir, G. K. (1983). *Phil. Trans. R. Soc. Lond.* **A310**, 127.
M8 Moore, P. B. (1973). *Am. Mineral.* **58**, 32.
M9 Midgley, C. M. (1952). *Acta Cryst.* **5**, 307.
M10 Moore, P. B. and Araki, T. (1976). *Am. Mineral.* **61**, 74.
M11 Mondal, P. and Jeffery, J. W. (1975). *Acta Cryst.* **B31**, 689.
M12 Maki, I. (1973). *Cem. Concr. Res.* **3**, 295.
M13 Maki, I. (1976). *Cem. Concr. Res.* **6**, 183.
M14 Maki, I. (1976). *Cem. Concr. Res.* **6**, 797.
M15 Maki, I. (1974). *Cem. Concr. Res.* **4**, 87.
M16 Morris, M. C., McMurdie, H. F., Evans, E. H., Paretzkin, B. and de Groot, J. H. *Standard X-ray Diffraction Powder Patterns* (NBS Monograph 25, Section 16), 186 pp., US Department of Commerce, Washington (1979).
M17 Marinho, M. B. and Glasser, F. P. (1984). *Cem. Concr. Res.* **14**, 360.
M18 Miyazawa, K. and Tomita, K., in *5th ISCC*, Vol. 1, p. 252 (1969).
M19 Muan, A. and Osborn, E. F. *Phase Equilibria among Oxides in Steelmaking*, 236 pp., Addison-Wesley, Reading, MA, USA (1965).
M20 Majumdar, A. J. (1965). *Trans Br. Ceram. Soc.* **64**, 105.
M21 Moseley, D. and Glasser, F. P. (1981). *Cem. Concr. Res.* **11**, 559.
M22 McMurdie, H. F. and Insley, H. (1936). *J. Res. Natl Bur. Stand.* **16**, 467.
M23 Majumdar, A. J. (1964). *Trans Br. Ceram. Soc.* **63**, 347.
M24 Midgley, H. G. (1979). *Cem. Concr. Res.* **9**, 623.
M25 McGinnety, J. A. (1972). *Acta Cryst.* **B28**, 2845.
M26 Mehrotra, B. N., Hahn, T., Eysel, W., Röpke, H. and Illguth, A. (1978). *Neues Jahrb. Mineral. Monatsh.* 408.

M27 Maki, I., Nakagawa, K., Hiraiwa, K. and Nonami, T. (1984). *Cemento* **81**, 3.
M28 Moore, A. E. (1976). *Cem. Technol.* **7**, 85, 134.
M29 Maki, I., Haba, H. and Takahashi, S. (1983). *Cem. Concr. Res.* **13**, 689.
M30 Maki, I. and Takahashi, S. (1984). *Cem. Concr. Res.* **14**, 413.
M31 Miller, F. M., in *Research on the Manufacture and Use of Cements*, p. 21, Engineering Foundation, New York (1986).
M32 Mohan, K. and Glasser, F. P. (1977). *Cem. Concr. Res.* **7**, 269, 379.
M33 Méric, J. P., in *7th ICCC*, Vol. 1, p. I-4/1 (1980).
M34 Maultzsch, M., Gierloff, M. and Schimmelwitz, P., in *7th ICCC*, Vol. 2, p. I-128 (1980).
M35 Midgley, H. G., in *3rd ISCC*, p. 140 (1954).
M36 Marchese, B. (1980). *Cem. Concr. Res.* **10**, 861.
M37 Midgley, H. G. (1979). *Cem. Concr. Res.* **9**, 77.
M38 Mohan, K. and Taylor, H. F. W. (1981). *J. Am. Ceram. Soc.* **64**, 717.
M39 Ménétrier, D., Jawed, I., Sun, T. S. and Skalny, J. (1979). *Cem. Concr. Res.* **9**, 473.
M40 Ménétrier, D., Jawed, I. and Skalny, J. (1980). *Silicates Ind.* **45**, 243.
M41 Ménétrier, D., Jawed, I. and Skalny, J. (1980). *Cem. Concr. Res.* **10**, 697.
M42 Ménétrier, D., McNamara, D. K., Jawed, I. and Skalny, J. (1980). *Cem. Concr. Res.* **10**, 107.
M43 Mohan, K. and Taylor, H. F. W. (1982). *Cem. Concr. Res.* **12**, 25.
M44 Massazza, F. and Testolin, M. (1983). *Cemento* **80**, 49.
M45 Michaux, M., Ménétrier, D. and Barret, P. (1983). *CR Acad. Sci. Paris Ser. 2*, **296**, 1043.
M46 Milestone, N. B., in *7th ICCC*, Vol. 3, p. VI-61 (1980).
M47 Macphee, D. E., Lachowski, E. E. and Glasser, F. P. (1988). *Adv. Cem. Res.* **1**, 131.
M48 Melzer, R. and Eberhard, E. (1989). *Cem. Concr. Res.* **19**, 411.
M49 McConnell, J. D. C. (1955). *Mineral. Mag.* **30**, 293, 672.
M50 Megaw, H. D. and Kelsey, C. H. (1956). *Nature* **177**, 390.
M51 Mitsuda, T. and Taylor, H. F. W. (1978). *Mineral. Mag.* **42**, 229.
M52 Macphee, D. E., Luke, K., Glasser, F. P. and Lachowski, E. E. (1989). *J. Am. Ceram. Soc.* **72**, 646.
M53 Maycock, J. N., Skalny, J. and Kalyoncu, R. (1974). *Cem. Concr. Res.* **4**, 835.
M54 Ménétrier, D., Jawed, I., Sun, T. S. and Skalny, J. (1980). *Cem. Concr. Res.* **10**, 425.
M55 Mehta, P. K. and Klein, A., in *Symposium on Structure of Portland Cement Paste and Concrete* (Sp. Rpt 90), p. 328, Highway Research Board, Washington (1966).
M56 Malquori, G. and Caruso, E. (1938). *Atti 10° Congr. Int. Chim. (Rome)* **2**, 713; *Chem. Abstr.* **33**, 8134 (1939).
M57 Midgley, H. G. and Rosaman, D., in *4th ISCC*, Vol. 1, p. 259 (1962).
M58 Moore, A. E. and Taylor, H. F. W. (1968). *Nature* **218**, 1048.
M59 Moore, A. E. and Taylor, H. F. W. (1970). *Acta Cryst.* **B26**, 386.
M60 Mylius, C. R. W. (1933). *Acta Acad. Aboensis, Math. Phys.* **7** (3), 147 pp.
M61 Moenke, H. (1964). *Naturwiss.* **51**, 239.
M62 Millet, J., Bernard, A., Hommey, R., Poindefert, A. and Voinovitch, I. A. (1980). *Bull. Liaison Lab. Ponts Chaussées* **109**, 91.
M63 Majumdar, A. J. and Roy, R. (1956). *J. Am. Ceram. Soc.* **39**, 434.
M64 Mascolo, G. and Marino, O. (1980). *Mineral. Mag.* **43**, 619.

M65 Murakami, K., in *5th ISCC*, Vol. 4, p. 457 (1969).
M66 Mehta, P. K. (1976). *Cem. Concr. Res.* **6**, 169.
M67 Mather, K., in *Evaluation of Methods of Identifying Phases of Cement Paste* (ed. W. L. Dolch), Transportation Res. Circ. 176, p. 9, Transportation Research Board, Washington (1976).
M68 Mehta, P. K. (1981). *Cem. Concr. Res.* **11**, 507.
M69 Mikhail, R. Sh., Copeland, L. E. and Brunauer, S. (1964). *Canad. J. Chem.* **42**, 426.
M70 Marsh, B. K., Day, R. L., Bonner, D. G. and Illston, J. M., in *Principles and Applications of Pore Structural Characterization* (eds J. M. Haynes and P. Rossi-Doria), Proc. RILEM/CNR Int. Symp., p. 365, Arrowsmith, Bristol, UK (1985).
M71 MacTavish, J. C., Miljkovic, L., Pintar, M. M., Blinc, R. and Lahajnar, G. (1985). *Cem. Concr. Res.* **15**, 367.
M72 MacTavish, J. C., Miljković, L., Schreiner, L. J., Pintar, M. M., Blinc, R. and Lahajnar, G. (1985). *Zeit. Naturforschung* **A40**, 32.
M73 Mindess, S. and Young, J. F. *Concrete*, 671 pp., Prentice-Hall, Englewood Cliffs, NJ, USA (1981).
M74 Mindess, S. (1970). *J. Am. Ceram. Soc.* **53**, 621.
M75 Mehta, P. K. and Manmohan, D., in *7th ICCC*, Vol. 3, p. VII-1 (1980).
M76 Marsh, B. K. (1986). Ph.D. thesis, quoted by R. F. Feldman in Ref. F39.
M77 Midgley, H. G. and Pettifer, K. (1971). *Cem. Concr. Res.* **1**, 101.
M78 Mather, B., in *5th ISCC*, Vol. 4, p. 113 (1969).
M79 Monk, M. (1983). *Mag. Concr. Res.* **35**, 131.
M80 Mohan, K. and Taylor, H. F. W., in *Effects of Flyash Incorporation in Cement and Concrete* (ed. S. Diamond), p. 54, Mater. Res. Soc., University Park, PA, USA (1981).
M81 Malquori, G., in *4th ISCC*, Vol. 2, p. 983 (1962).
M82 Massazza, F. (1976). *Cemento* **73**, 3.
M83 Massazza, F. and Costa, U. (1979). *Cemento* **76**, 3.
M84 McCarthy, G. J., Swanson, K. D., Keller, L. P. and Blatter, W. C. (1984). *Cem. Concr. Res.* **14**, 471.
M85 Mehta, P. K. and Groney, P. R. (1977). *J. Am. Concr. Inst. (Proc.)* **74**, 440.
M86 Manmohan, D. and Mehta, P. K. (1981). *Cem. Concr. Aggr.* **3**, 63.
M87 Marsh, B. K., Day, R. L. and Bonner, D. G. (1985). *Cem. Concr. Res.* **15**, 1027.
M88 Ménétrier-Sorrentino, D. (1989). Private communication.
M89 Ménétrier-Sorrentino, D., George, C. M. and Sorrentino, F. P., in *8th ICCC*, Vol. 4, p. 334 (1986).
M90 Müller, D., Rettel, A., Gessner, W. and Scheler, G. (1984). *J. Magn. Res.* **57**, 152.
M91 Midgley, H. G., in *International Seminary on Calcium Aluminates* (eds M. Murat *et al.*), Politecnico di Torino, Turin (1982).
M92 Ministère de l'Environnement, France (1979). Circular 79-34 of 27 March 1979.
M93 Marcdargent, S. and Mathieu, A. (1989). Private communication.
M94 Midgley, H. G. (1984). *Clay Minerals* **19**, 857.
M95 Majumdar, A. J., Singh, B. and Edmonds, R. N. (1990). *Cem. Concr. Res.*, in press.
M96 Midgley, H. G., in *7th ICCC*, Vol. 3, p. V-85 (1980).

M97 Mehta, P. and Polivka, M., in *6th ICCC*, Vol. 3, p. 158 (1976).
M98 Mikhailov, V. V., in *4th ISCC*, Vol. 2, p. 927 (1962).
M99 Mikhailov, V. V., in *Klein Symposium on Expansive Cement Concretes* (Sp. Publ. SP38), p. 415, American Concrete Institute, Detroit (1973).
M100 Mehta, P. K. (1973). *Cem. Concr. Res.* **3**, 1.
M101 Mehta, P. K. (1980). *World Cem. Technol.* **11**, 166.
M102 Massazza, F. and Gilioli, C. (1983). *Cemento* **80**, 101.
M103 Massazza, F. and Testolin, M. (1980). *Cemento* **77**, 73.
M104 Milestone, N. B. (1976). *Cem. Concr. Res.* **6**, 89.
M105 Massazza, F. and Costa, U., in *7th ICCC*, Vol. 4, p. 529 (1981).
M106 Murakami, K. and Tanaka, H., in *5th ISCC*, Vol. 2, p. 422 (1969).
M107 Murat, M., El Hajjouji, A. and Comel, C. (1987). *Cem. Concr. Res.* **17**, 633.
M108 Mironov, S. A., in *6th ICCC*, Vol. 2, Part 1, p. 182 (1976).
M109 Mironov, S. A., Kourbatova, I. I., Ivanova, O. S. and Vyssotosky, S. A., in *7th ICCC*, Vol. 2, p. II-52 (1980).
M110 Menzel, C. A. (1934). *J. Am. Concr. Inst. (Proc.)* **31**, 125.
M111 Mitsuda, T., Kobayakawa, S. and Toraya, H., in *8th ICCC*, Vol. 3, p. 173 (1986).
M112 Mitsuda, T. and Chan, C. F. (1977). *Cem. Concr. Res.* **7**, 191.
M113 Maso, J. C., in *7th ICCC*, Vol. 1, p. VII-1/3 (1980).
M114 Massazza, F. and Costa, U., in *8th ICCC*, Vol. 1, p. 158 (1986).
M115 Monteiro, P. J. M., Maso, J. C. and Ollivier, J. P. (1985). *Cem. Concr. Res.* **15**, 953.
M116 Monteiro, P. J. M. and Mehta, P. K. (1985). *Cem. Concr. Res.* **15**, 378.
M117 Monteiro, P. J. M. and Mehta, P. K. (1986). *Cem. Concr. Res.* **16**, 127.
M118 Manns, W. and Wesche, K., in *5th ISCC*, Vol. 3, p. 385 (1969).
M119 Midgley, H. G. and Illston, J. M. (1984). *Cem. Concr. Res.* **14**, 546.
M120 Monteiro, P. J. M., Gjorv, O. E. and Mehta, P. K. (1985). *Cem. Concr. Res.* **15**, 781.
M121 Merlino, S. (1983). *Am. Mineral.* **68**, 614.
M122 Mielenz, R. C., Greene, K. T. and Benton, E. J. (1947). *J. Am. Concr. Inst. (Proc.)* **44**, 193.
M123 Marchese, B. and Sersale, R., in *5th ISCC*, Vol. 2, p. 133 (1969).
M124 Massazza, F. (1985). *Cemento* **82**, 3.
M125 Majumdar, A. J. (1974). *Cem. Concr. Res.* **4**, 247.

N1 Nishi, F., Takéuchi, Y. and Maki, I. (1984). *Zeit. Krist.* **172**, 297.
N2 Nishi, F. and Takéuchi, Y. (1984). *Zeit. Krist.* **168**, 197.
N3 Niesel, K. and Thormann, P. (1967). *Tonind. Zeit.* **91**, 362.
N4 Nishi, F. and Takéuchi, Y. (1975). *Acta Cryst.* **B31**, 1169.
N5 Nurse, R. W. and Welch, J. H., quoted by R. W. Nurse in *4th ISCC*, Vol. 1, p. 9 (1962).
N6 Nurse, R. W., Welch, J. H. and Majumdar, A. J. (1965). *Trans Br. Ceram. Soc.* **64**, 409.
N7 Nurse, R. W., Welch, J. H. and Majumdar, A. J. (1965). *Trans Br. Ceram. Soc.* **64**, 323.
N8 Newkirk, T. F. and Thwaite, R. D. (1958). *J. Res. Natl Bur. Stand.* **61**, 233.
N9 Newkirk, T. F., in 3rd *ISCC*, p. 151 (1954).
N10 Nurse, R. W. (1949). *J. Sci. Instrum.* **26**, 102
N11 Newman, E. S. and Hoffman, R. (1956). *J. Res. Natl Bur. Stand.* **56**, 319.
N12 Nurse, R. W., Welch, J. H. and Gutt, W. (1959). *J. Chem. Soc.* 1077.

N13 Norris, A. C. *Computational Chemistry*, 454 pp., Wiley, Chichester, UK (1981).
N14 Newton, R. G. and Sharp, J. H. (1987). *Cem. Concr. Res.* **17**, 77.
N15 Neal, C. and Stanger, G. (1984). *Mineral. Mag.* **48**, 237.
N16 Nyame, B. K. and Illston, J. M., in *7th ICCC*, Vol. 3, p. VI-181 (1980).
N17 Nixon, P. J., Osborne, G. J. and Shepperd, C. N. (1983). *Silicates Ind.* **12**, 253.
N18 Nikushchenko, V. M., Khotimchenko, V. S., Rumyantsev, P. F. and Kalinin, A. I. (1973). *Cem. Concr. Res.* **3**, 625.
N19 Nakamura, T., Sudoh, G. and Akaiwa, S., in *5th ISCC*, Vol. 4, p. 351 (1969).
N20 Noudelman, B. I. and Gadaev, A. I., in *8th ICCC*, Vol. 2, p. 347 (1986).
N21 Nixon, P. J., Canham, I. and Bollinghaus, R., in *Concrete Alkali–Aggregate Reactions* (ed. P. E. Grattan-Bellew), p. 110, Noyes Publ., Park Ridge, NJ, USA (1987).

O1 Ono, Y., Kawamura, S. and Soda, Y., in *5th ISCC*, Vol. 1, p. 275 (1969).
O2 Ono, M. and Nagashima, N., in *6th ICCC*, Vol. 1, p. 170 (1976).
O3 Ol'shanskii, Ya. I. (1951). *Dokl. Akad. Nauk SSSR* **76**, 93.
O4 Ohashi, Y. (1984). *Phys. Chem. Minerals* **10**, 217.
O5 Okada, K. and Ossaka, J. (1980). *Acta Cryst.* **B36**, 919.
O6 Odler, I. and Dörr, H. (1977). *Am. Ceram. Soc. Bull.* **56**, 1086.
O7 Ono, Y., in *3rd ICCem Microsc.*, p. 198 (1981).
O8 Osbaeck, B. and Jøns, E. S., in *7th ICCC*, Vol. 2, p. II-135 (1980).
O9 Odler, I., Abdul-Maula, S., Nudling, P. and Richter, T. (1981). *Zem.-Kalk-Gips* **34**, 445 (partial English translation, p. 240).
O10 Odler, I. and Dörr, H. (1979). *Cem. Concr. Res.* **9**, 239.
O11 Odler, I. and Stassinopoulos, E. N. (1982). *TIZ* **106**, 394.
O12 Odler, I. and Schüppstuhl, J. (1981). *Cem. Concr. Res.* **11**, 765.
O13 Odler, I. and Dörr, H. (1979). *Cem. Concr. Res.* **9**, 277.
O14 Odler, I. and Schüppstuhl, J. (1982). *Cem. Concr. Res.* **12**, 13.
O15 Odler, I. and Abdul-Maula, S. (1984). *Cem. Concr. Res.* **14**, 133.
O16 Ogawa, K. and Roy, D. M. (1981). *Cem. Concr. Res.* **11**, 741.
O17 Odler, I., in *7th ICCC*, Vol. 4, p. 493 (1981).
O18 Odler, I. and Wonneman, R., in *7th ICCC*, Vol. 4, p. 510 (1981).
O19 Odler, I. and Köster, H. (1986). *Cem. Concr. Res.* **16**, 893.
O20 Odler, I. and Rössler, M. (1985). *Cem. Concr. Res.* **15**, 401.
O21 Ogawa, K., Uchikawa, H., Takemoto, K. and Yasui, I. (1980). *Cem. Concr. Res.* **10**, 683.
O22 Odler, I. and Zysk, K.-H. (1988). *Mater. Res. Soc. Symp. Proc.* **113**, 179.
O23 Okushima, M., Kondo, R., Muguruma, H. and Ono, Y., in *5th ISCC*, Vol. 4, p. 419 (1969).
O24 Ogawa, K. and Roy, D. M. (1982). *Cem. Concr. Res.* **12**, 101, 247.
O25 Odler, I., Abdul-Maula, S. and Lu Z.-Y. (1987). *Mater. Res. Soc. Symp. Proc.* **85**, 139.
O26 Odler, I. and Skalny, J. (1973). *J. Chem. Technol. Biotechnol.* **23**, 661.
O27 Oyefesobi, S. O. and Roy, D. M. (1976). *Cem. Concr. Res.* **6**, 803.
O28 Ozol, M. A., in *Alkali–Aggregate Reaction—Preventive Measures*, p. 113, Icelandic Building Research Institute, Reykjavik (1975).
O29 Odler, I. and Gasser, M. (1988). *J. Am. Ceram. Soc.* **71**, 1015.
O30 Oberholster, R. E., Du Toit, P. and Pretorius, J. L., in *6th ICCem. Microsc.*, p. 360 (1984).

P1 *Powder Diffraction File* (cards, magnetic tape and indices). JCPDS International Centre for Diffraction Data, Swarthmore, PA, USA (1989).
P2 Pollitt, H. W. W. and Brown, A. W., in *5th ISCC*, Vol. 1, p. 322 (1969).
P3 Ponomarev, V. I., Kheiker, D. M. and Belov, N. V. (1971). *Sov. Phys. Cryst.* **15**, 799 (translated from *Kristallografiya* **15**, 918).
P4 Phillips, B. and Muan, A. (1959). *J. Am. Ceram. Soc.* **42**, 413.
P5 Parker, T. W., in *3rd ISCC*, p. 485 (1954).
P6 Pryce, M. W. (1972). *Mineral. Mag.* **38**, 968.
P7 Pliego-Cuervo, Y. B. and Glasser, F. P. (1978). *Cem. Concr. Res.* **8**, 455.
P8 Pliego-Cuervo, Y. B. and Glasser, F. P. (1979). *Cem. Concr. Res.* **9**, 51.
P9 Pliego-Cuervo, Y. B. and Glasser, F. P. (1977). *Cem. Concr. Res.* **7**, 477.
P10 Pliego-Cuervo, Y. B. and Glasser, F. P. (1979). *Cem. Concr. Res.* **9**, 573.
P11 Peréz-Mendez, M., Howie, R. A. and Glasser, F. P. (1984). *Cem. Concr. Res.* **14**, 57.
P12 Peréz-Mendez, M., Fayos, J., Howie, R. A., Gard, J. A. and Glasser, F. P. (1985). *Cem. Concr. Res.* **15**, 600.
P13 Peray, K. *The Rotary Cement Kiln* (2nd Ed.), 389 pp., Arnold, London (1986).
P14 Peréz-Mendez, M. and Fayos, J., in *8th ICCC*, Vol. 2, p. 223 (1986).
P15 Petersen, I. F. and Johansen, V. (1979). *Cem. Concr. Res.* **9**, 631.
P16 Petersen, I. F. (1983). *World Cem. Technol.* **14**, 188, 220.
P17 Petersen, I. F., in *8th ICCC*, Vol. 2, p. 21 (1986).
P18 Petch, H. E. (1961). *Acta Cryst.* **14**, 950.
P19 Parrott, L. J., Patel, R. G., Killoh, D. C. and Jennings, H. M. (1984). *J. Am. Ceram. Soc.* **67**, 233.
P20 Powers, T. C. and Brownyard, T. L. *Studies of the Physical Properties of Hardened Portland Cement Paste* (Bull. 22), 992 pp., Portland Cement Association, Chicago (1948); reprinted from *J. Am. Concr. Inst. (Proc.)* **43**, 101, 249, 469, 549, 669, 845, 993 (1947).
P21 Parrott, L. J. and Young, J. F. (1981). *Cem. Concr. Res.* **11**, 11.
P22 Prosen, E. J., Brown, P. W., Frohnsdorff, G. and Davis, F. (1985). *Cem. Concr. Res.* **15**, 703.
P22a Pöllmann, H., Kuzel, H.-J. and Wenda, R. (1989). *Neues Jahrb. Mineral. Abhandl.* **160**, 133.
P23 Passaglia, E. and Rinaldi, R. (1984). *Bull. Minéral. (Paris)* **107**, 605.
P24 Pedersen, B. F. and Semmingsen, D. (1982). *Acta Cryst.* **B38**, 1074.
P25 Posnjak, E. (1938). *Am. J. Sci.* **A35**, 247.
P26 Percival, A. and Taylor, H. F. W. (1959). *J. Chem. Soc.* 2629.
P27 Plowman, C. and Cabrera, J. G. (1984). *Cem. Concr. Res.* **14**, 238.
P28 Patel, R. G., Killoh, D. C., Parrott, L. J. and Gutteridge, W. A. (1988). *Matér. Constr. (Paris)* **21**, 192.
P29 Pressler, E. E., Brunauer, S., Kantro, D. L. and Weise, C. H. (1961). *Analyt. Chem.* **33**, 877.
P30 Parrott, L. J. and Killoh, D. C. (1984). *Br. Ceram. Proc.* **35**, 41.
P31 Pratt, P. L. and Ghose, A. (1983). *Phil. Trans R. Soc. Lond.* **A310**, 93.
P32 Parrott, L. J., in *Research on the Manufacture and Use of Cements*, p. 43, Engineering Foundation, New York (1986).
P33 Parrott, L. J., in *Materials Science of Concrete I* (ed. J. P. Skalny), p. 181, American Ceramics Society, Westerville, OH, USA (1989).
P34 Powers, T. C., in *4th ISCC*, Vol. 2, p. 577 (1962).
P35 Parrott, L. (1987). *Mater. Res. Soc. Symp. Proc.* **85**, 91.

P36 Parrott, L. J. (1981). *Cem. Concr. Res.* **11**, 651.
P37 Parrott, L. J., Hansen, W. and Berger, R. L. (1980). *Cem. Concr. Res.* **10**, 647.
P38 Parrott, L. J. and Young, J. F., in *Fundamental Research on Creep and Shrinkage of Concrete* (ed. F. H. Wittmann), p. 35, Martinus Nijhoff, The Hague (1982).
P39 Pearson, D., Allen, A., Windsor, C. G., Alford, N. McN. and Double, D. D. (1983). *J. Mater. Sci.* **18**, 430.
P40 Pratt, P. L. (1987). *Mater. Res. Soc. Symp. Proc.* **85**, 145.
P41 Parrott, L. J. (1977). *Cem. Concr. Res.* **7**, 597.
P42 Parrott, L. J. (1977). *Mag. Concr. Res.* **29**, 26.
P43 Powers, T. C., Copeland, L. E. and Mann, H. M. (1959). *J. PCA Res. Dev. Labs* **1** (2), 38.
P44 Page, C. L., Short, N. R. and El Tarras, A. (1981). *Cem. Concr. Res.* **11**, 395.
P45 Pérez, M., Vázquez, T. and Triviño, F. (1983). *Cem. Concr. Res.* **13**, 759.
P46 Percival, A., Buttler, F. P. and Taylor, H. F. W., in *4th ISCC*, Vol. 1, p. 277 (1962).
P47 Pérez Mendez, M. and Triviño Vázquez, F. (1984). *Cem. Concr. Res.* **14**, 161.
P48 Page, C. L. and Vennesland, Ø. (1983). *Matér. Constr. (Paris)* **16**, 19.
P49 Purton, M. J. (1973). *J. Appl. Chem. Biotechnol.* **23**, 871.
P50 Peacor, D. R., Sarp, H., Dunn, P. J., Innes, J. and Nelen, J. A. (1988). *Am. Mineral.* **73**, 888.
P51 Patel, R. G., Parrott, L. J., Martin, J. A. and Killoh, D. C. (1985). *Cem. Concr. Res.* **15**, 343.
P52 Page, C. L., Short, N. R. and Holden, W. R. (1986). *Cem. Concr. Res.* **16**, 79.
P53 Pettifer, K. and Nixon, P. J. (1980). *Cem. Concr. Res.* **10**, 173.
P54 Powers, T. C. and Steinour, H. H. (1955). *J. Am. Concr. Inst (Proc.)* **51**, 497, 785.
P55 Powers, T. C. (1945). *J. Am. Concr. Inst. (Proc.)* **41**, 245.
P56 Powers, T. C. and Helmuth, R. A. (1953). *Proc. US Highway Res. Board* **32**, 285.
P57 Pagenkopf, G. K. *Introduction to Natural Water Chemistry*, 272 pp., Marcel Dekker, New York (1978).
P58 Penner, E., Gillott, J. E. and Eden, W. J. (1970). *Canad. Geotech. J.* **7**, 333.
P59 Philleo, R. (1958). *J. Am. Concr. Inst. (Proc.)* **54**, 857.
P60 Piasta, J., Sawicz, Z. and Rudzinski, L. (1984). *Matér. Constr. (Paris)* **17**, 291.

Q1 Qian, J. C., Lachowski, E. E. and Glasser, F. P. (1988). *Mater. Res. Soc. Symp. Proc.* **113**, 45.

R1 Regourd, M. and Guinier, A., in *6th ICCC*, Vol. 1, p. 25 (1976).
R2 Regourd, M. (1967). *Rev. Matér. Constr.* (620) 167.
R3 Regourd, M., Bigaré, M., Forest, J. and Guinier, A., in *5th ISCC*, Vol. 1, p. 44 (1969).
R4 Rankin, G. A. and Wright, F. E. (1915). *Am. J. Sci. (4th series)* **39**, 31.
R5 Regourd, M., Chromý, S., Hjorth, L., Mortureux, B. and Guinier, A. (1973). *J. Appl. Cryst.* **6**, 355.
R6 Roy, D. M. and Roy, R., in *4th ISCC*, Vol. 1, p. 307 (1962).
R7 Rowe, J. J., Morey, G. W. and Hansen, I. D. (1965). *J. Inorg. Nucl. Chem.* **27**, 53.
R8 Ragozina, T. A. (1957). *Zh. Prikl. Khim.* **30**, 1682.

R9 Ritzmann, H. (1971). *Zem.-Kalk-Gips* **24**, 338.
R10 Rivera, M., Odler, I. and Abdul-Maula, S. (1987). *Adv. Cem. Res.* **1**, 52.
R11 Rouanet, A., in *8th ICCC*, Vol. 2, p. 25 (1986).
R12 Ritzmann, H. (1968). *Zem.-Kalk-Gips* **21**, 390.
R13 Regourd, M., Hornain, H. and Mortureux, B., in *7th ICCC*, Vol. 4, p. 477 (1981).
R14 Ramachandran, V. S. (1979). *Cem. Concr. Res.* **9**, 677.
R15 Rayment, D. L. (1988). Private communication.
R16 Rodger, S. A., Groves, G. W., Clayden, N. J. and Dobson, C. M. (1988). *J. Am. Ceram. Soc.* **71**, 91.
R17 Relis, M. and Soroka, J. (1980). *Cem. Concr. Res.* **10**, 499.
R18 Regourd, M., Thomassin, J. H., Baillif, P. and Touray, J. C. (1980). *Cem. Concr. Res.* **10**, 223.
R19 Roberts, M. H., in *5th ISCC*, Vol. 2, p. 104 (1969).
R20 Regourd, M., Hornain, H. and Mortureux, B. (1976). *Cem. Concr. Res.* **6**, 733.
R21 Rogers, D. E. and Aldridge, L. P. (1977). *Cem. Concr. Res.* **7**, 399.
R22 Roberts, M. H. (1970), quoted by F. M. Lea in Ref. L6.
R23 Reisdorf, K. and Abriel, W. (1988). *Zem.-Kalk-Gips* **41**, 356 (partial English translation, p. 218).
R24 Rodger, S. A. and Groves, G. W. (1989). *J. Am. Ceram. Soc.* **72**, 1037.
R25 Rayment, P. L. (1982). *Cem. Concr. Res.* **12**, 133.
R26 Regourd, M., Mortureux, B. and Hornain, H., in *Fly Ash, Silica Fume, Slag and Other Mineral By-Products in Concrete* (ed. V. M. Malhotra), Sp. Publ. SP79, Vol. 2, p. 847, American Concrete Institute, Detroit (1983).
R27 Rayment, D. L. (1986). *Cem. Concr. Res.* **16**, 341.
R28 Rayment, D. L. and Majumdar, A. J. (1982). *Cem. Concr. Res.* **12**, 753.
R29 Rayment, D. L. and Lachowski, E. E. (1984). *Cem. Concr. Res.* **14**, 43.
R30 Roy, D. M. and Asaga, K. (1979). *Cem. Concr. Res.* **9**, 731.
R31 Rössler, M. and Odler, I. (1985). *Cem. Concr. Res.* **15**, 320.
R32 Ramachandran, V. S., Feldman, R. F. and Beaudoin, J. J. *Concrete Science*, 427 pp., Heyden, London (1981).
R33 Ryshkewitch, E. (1953). *J. Am. Ceram. Soc.* **36**, 65.
R34 Regourd, M., in *8th ICCC*, Vol. 1, p. 199 (1986).
R35 Regourd, M. (1985). *Chem. Scripta* **26A**, 37.
R36 Royak, S. M. and Chkolnik, J. Ch., in *7th ICCC*, Vol. 2, p. III-74 (1980).
R37 Regourd, M., in *7th ICCC*, Vol. 1, p. III-2/10 (1980).
R38 Regourd, M., Mortureux, B., Gautier, E., Hornain, H. and Volant, J., in *7th ICCC*, Vol. 2, p. III-105 (1980).
R39 Regourd, M., Thomassin, J. H., Baillif, P. and Touray, J. C. (1983). *Cem. Concr. Res.* **13**, 549.
R40 Ravina, D., in *Effects of Flyash Incorporation in Cement and Concrete* (ed. S. Diamond), p. 2, Mater. Res. Soc., University Park, PA, USA (1981).
R41 Richartz, E. (1984). *Zem.-Kalk-Gips* **37**, 62 (partial English translation, p. 65).
R42 Rodger, S. A. and Groves, G. W. (1988). *Adv. Cem. Res.* **1**, 84.
R43 Regourd, M., Mortureux, B., Aitcin, P. C. and Pinsonneault, P., in *4th ICCem. Microsc.*, p. 249 (1982).
R44 Roy, D. M. and Parker, K. M., in *Fly Ash, Silica Fume, Slag and Other Mineral By-Products in Concrete* (ed. V. M. Malhotra), Sp. Publ. SP79, Vol. 1, p. 397, American Concrete Institute, Detroit (1983).

R45 Rodger, S. A. and Double, D. D. (1984). *Cem. Concr. Res.* **14**, 73.
R46 Rettel, A., Gessner, W., Müller, D. and Scheler, G. (1985). *Trans J. Br. Ceram. Soc.* **84**, 25.
R47 Robson, T. D. *High-Alumina Cements and Concretes*, 263 pp., Contractors Record Ltd, London and John Wiley and Sons Inc., New York (1962).
R48 Rengade, E., L'Hopitallier, P. and Durand de Fontmagne, P. (1936). *Rev. Matér. Constr.* (318), 52; (319), 78.
R49 Raask, E., in *International Symposium on the Carbonation of Concrete* (Proc. RILEM), paper 5.6, Cement and Concrete Association, Slough, UK (1976).
R50 Roy, D. M. and Oyefosobi, S. O. (1977). *J. Am. Ceram. Soc.* **60**, 178.
R51 Ramachandran, V. S. (1978). *Zem.-Kalk-Gips* **31**, 206 (partial English translation, p. 144).
R52 Ramachandran, V. S. (1976). *Cem. Concr. Res.* **6**, 623.
R53 Roy, D. M. and Asaga, K. (1980). *Cem. Concr. Res.* **10**, 387.
R54 Ramachandran, V. S. and Feldman, R. F. (1971). *Cem. Technol.* **2**, 121.
R55 Ramachandran, V. S. (1972). *Cem. Concr. Res.* **2**, 179.
R56 Ramachandran, V. S. and Feldman, R. F. (1972). *Matér. Constr. (Paris)* **5**, 67.
R57 Ramachandran, V. S. and Feldman, R. F. (1978). *Cemento* **75**, 311.
R58 Ramachandran, V. S. (1971). *Matér. Constr. (Paris)* **4**, 3.
R59 Regourd, M., Hornain, H. and Aitcin, P.-C. (1987). *Mater. Res. Soc. Symp. Proc.* **85**, 77.
R60 Roy, D. M. and Harker, R. I., in *4th ISCC*, Vol. 1, p. 196 (1962).
R61 Royak, S. M., in *6th ICCC*, Vol. 3, p. 231 (1976).
R62 Roy, D. M., Gouda, G. R. and Bobrowsky, A. (1972). *Cem. Concr. Res.* **2**, 349.
R63 Roy, D. M. and Gouda, G. R. (1973). *Cem. Concr. Res.* **3**, 807.
R64 Rodger, S. A., Brooks, S. A., Sinclair, W., Groves, G. W. and Double, D. D. (1985). *J. Mater. Sci.* **20**, 2853.
R65 Regourd, M. (1985). *Mater. Res. Soc. Symp. Proc.* **42**, 3.
R66 Rosenberg, A. M. and Gaidis, J. M. (1989). *Concrete International: Design and Construction*, **2** (4), 31.
R67 Roy, D. M., in *8th ICCC*, Vol. 1, p. 362 (1986).
R68 Roy, D. M., Malek, R. I. A., Rattanussorn, M. and Grutzeck, M. W. (1986). *Mater. Res. Soc. Symp. Proc.* **65**, 219.
R69 Regourd, M., Hornain, H. and Poitevan, P., in *Alkali–Aggregate Reaction in Concrete*, Paper S252/35, CSIR, Pretoria (1981).
R70 Roy, D. M., Sonnenthal, E. and Prave, R. (1985). *Cem. Concr. Res.* **15**, 914.
R71 Regourd, M., Hornain, H. and Mortureux, B., in *Durability of Building Materials and Components* (eds P. J. Sereda and G. G. Litvan), p. 253, Am. Soc. Testing Mater., Philadelphia (1980).
R72 Rogers, D. E. (1973). *New Zealand J. Sci.* **16**, 875.

S1 Sarkar, S. L. and Roy, D. M., in *6th ICCem. Microsc.*, p. 37 (1984).
S2 Sinclair, W. and Groves, G. W. (1984). *J. Am. Ceram. Soc.* **67**, 325.
S3 Saalfeld, H. (1975). *Am. Mineral.* **60**, 824.
S4 Smith, D. K. (1962). *Acta Cryst.* **15**, 1146.
S5 Swanson, H. E. and Tatge, E. *Standard X-ray Powder Diffraction Patterns* (NBS Circular 539, Vol. 1), 95 pp., US Department of Commerce, Washington (1953).
S6 Saburi, S., Kusachi, I., Henmi, C., Kawahara, A., Henmi, K. and Kawada, I. (1976). *Mineral. J. (Japan)* **8**, 240.

S7	Smirnov, G. S., Chatterjee, A. K. and Zhmoidin, G. I. (1973). *J. Mater. Sci.* **8**, 1278.
S8	Swayze, M. A. (1946). *Am. J. Sci.* **244**, 1.
S9	Sorrentino, F. and Glasser, F. P. (1976). *Trans J. Br. Ceram. Soc.* **75**, 95.
S10	Swayze, M. A. (1946). *Am. J. Sci.* **244**, 65.
S11	Sharp, J. D., Johnson, W. and Andrews, W. (1960). *J. Iron Steel Inst.* **195**, 83.
S12	Schlaudt, C. M. and Roy, D. M. (1966). *J. Am. Ceram. Soc.* **49**, 430.
S13	Sarkar, S. L. and Jeffery, J. W. (1979). *J. Am. Ceram. Soc.* **62**, 630.
S14	Swanson, H. E., McMurdie, H. F., Morris, M. C. and Evans, E. H. *Standard X-Ray Diffraction Patterns* (NBS Monograph No. 25, Section 7), 186 pp., US Department of Commerce, Washington (1969).
S15	Shame, E. G. and Glasser, F. P. (1987). *Trans J. Br. Ceram. Soc.* **86**, 13.
S16	Smart, R. M. and Roy, D. M. (1979). *Cem. Concr. Res.* **9**, 269.
S17	Smith, J. V., Karle, I. L., Hauptman, H. and Karle, J. (1960). *Acta Cryst.* **13**, 454.
S18	Shibata, S., Kishi, K., Asaga, K., Daimon, M. and Shrestha, P. R. (1984). *Cem. Concr. Res.* **14**, 323.
S19	Spohn, E., Woermann, E. and Knoefel, D., in *5th ISCC*, Vol. 1, p. 172 (1969).
S20	Southard, J. C. and Royster, P. H. (1936). *J. Phys. Chem.* **40**, 435.
S21	Sprung, S., in *7th ICCC*, Vol. 1, p. I-2/1 (1980).
S22	Sylla, H.-M. (1974). *Zem.-Kalk-Gips* **27**, 499.
S23	Sylla, H.-M. (1977). *Zem.-Kalk-Gips* **30**, 487 (partial English translation, p. 264).
S24	Sprung, S. (1985). *Zem.-Kalk-Gips* **38**, 577 (partial English translation, p. 309).
S25	Sumner, M. S., Hepher, N. M. and Moir, G. K., in *8th ICCC*, Vol. 2, p. 310 (1986), and private communication.
S26	Sprung, S., Kuhlmann, K. and Ellerbrock, H.-G. (1985). *Zem.-Kalk-Gips* **38**, 528 (partial English translation, p. 275).
S27	St. John, D. A. and McGivin, P. N., in *3rd ICCem. Microsc.*, p. 193 (1981).
S28	Scrivener, K. L. and Pratt, P. L., in *6th ICCem. Microsc.*, p. 145 (1984).
S29	Scrivener, K. L. (1987). *Mater. Res. Soc. Symp. Proc.* **85**, 39.
S30	Struble, L. J. (1983). *Cem. Concr. Aggr.* **5**, 62.
S31	Struble, L. J. (1985). *Cem. Concr. Res.* **15**, 631.
S32	Smallwood, T. B. and Wall, C. D. (1981). *Talanta* **28**, 265.
S33	Sakurai, T., Sato, T. and Yoshinaga, A., in *5th ISCC*, Vol. 1, p. 300 (1969).
S34	Skalny, J. P. and Young, J. F., in *7th ICCC*, Vol. 1, p. II-1/3 (1980).
S35	Spierings, G. A. C. M. and Stein, H. N. (1976). *Cem. Concr. Res.* **6**, 265, 487.
S36	Spinolo, G. and Tamburini, U. A. (1985). *Zeit. Naturforschung* **A40**, 73.
S37	Shebl, F. A., Helmy, F. M. and Ludwig, U. (1985). *Cem. Concr. Res.* **15**, 747.
S38	Scrivener, K. L., Patel, H. H., Pratt, P. L. and Parrott, L. J. (1987). *Mater. Res. Soc. Symp. Proc.* **85**, 67.
S39	Scrivener, K. L. (1984). Ph.D. Thesis, University of London.
S40	Scrivener, K. L. and Pratt, P. L. (1984). *Mater. Res. Soc. Symp. Proc.* **31**, 351.
S41	Scrivener, K. L. and Pratt, P. L. (1984). *Br. Ceram. Proc.* **35**, 207.
S42	Stewart, H. R. and Bailey, J. E. (1983). *J. Mater. Sci.* **18**, 3686.
S43	Stade, H., Wieker, W. and Garzo, G. (1983). *Zeit. Anorg. Allg. Chem.* **500**, 123.

S44	Slegers, P. A., Genet, M., Léonard, A. J. and Fripiat, J. J. (1977). *J. Appl. Cryst.* **10**, 270.
S45	Stade, H. and Wieker, W. (1980). *Zeit. Anorg. Allg. Chem.* **466**, 55.
S46	Stade, H. (1980). *Zeit. Anorg. Allg. Chem.* **470**, 69.
S47	Stade, H., Grimmer, A.-R., Engelhardt, G., Mägi, M. and Lipmaa, E. (1985). *Zeit Anorg. Allg. Chem.* **528**, 147.
S48	Steinour, H. H. (1947). *Chem. Rev.* **40**, 391.
S49	Steinour, H. H., in *3rd ISCC*, p. 261 (1954).
S50	Stein, H. N. (1972). *Cem. Concr. Res.* **2**, 167.
S51	Smith, R. M. and Martell, A. E. *Critical Stability Constants*, Vol.4: *Inorganic Complexes*, 257 pp., Plenum Press, New York (1976).
S52	Suzuki, K., Nishikawa, T., Ikenaga, H. and Ito, S. (1986). *Cem. Concr. Res.* **16**, 333.
S53	Stein, H. N. and Stevels, J. M. (1964). *J. Appl. Chem.* **14**, 338.
S54	Sierra, R., in *6th ICCC*, Vol. 2, Part 1, p. 138 (1976).
S55	Scheller, T. and Kuzel, H.-J., in *6th ICCC*, Vol. 2, Part 1, p. 217 (1976).
S56	Seligmann, P. and Greening, N. R., in *5th ISCC*, Vol. 2, p. 179 (1969).
S57	Schwiete, H. E. and Iwai, T. (1964). *Zem.-Kalk-Gips* **17**, 379.
S58	Schwiete, H. E., Ludwig, U. and Jäger, P., quoted by H. E. Schwiete and U. Ludwig in *5th ISCC*, Vol. 2, p. 37 (1969).
S59	Struble, L. J., in *8th ICCC*, Vol. 6, p. 582 (1986).
S60	Serb-Serbina, N. N., Savvina, Yu. A. and Zhurina, V. S. (1956). *Dokl. Akad. Nauk SSSR* **111**, 659.
S61	Schwiete, H. E., Ludwig, U. and Albeck, J. (1968). *Naturwiss.* **55**, 179.
S62	Sacerdoti, M. and Passaglia, E. (1985). *Bull. Minéral. (Paris)* **108**, 1.
S63	Swanson, H. E., Gilfrich, N. T. and Cook, M. I. *Standard X-ray Powder Diffraction Patterns* (NBS Circular 539, Vol. 6), 62 pp., US Department of Commerce, Washington (1956).
S64	Stein, H. N. (1963). *Silicates Ind.* **28**, 141.
S65	Schwiete, H. E., Ludwig, U. and Jäger, P., in *Symposium on Structure of Portland Cement Paste and Concrete* (Sp. Rpt 90), p. 353, Highway Research Board, Washington (1966).
S66	Skalny, J. and Tadros, M. E. (1977). *J. Am. Ceram. Soc.* **60**, 174.
S67	Šatava, V. (1988). *Silikaty* **32**, 203.
S68	Scrivener, K. L., in *Materials Science of Concrete I* (ed. J. P. Skalny), p. 127, American Ceramic Society, Westerville, OH, USA (1989).
S69	Sarkar, A. K. and Roy, D. M. (1979). *Cem. Concr. Res.* **9**, 343.
S70	Stade, H., Müller, D. and Scheler, G. (1984). *Zeit. Anorg. Allg. Chem.* **510**, 16.
S71	Stade, H. and Müller, D. (1987). *Cem. Concr. Res.* **17**, 553.
S72	Stade, H., Ulbricht, K. and Mehner, H. (1984). *Zeit. Anorg. Allg. Chem.* **514**, 149.
S73	Stade, H. and Wieker, W. (1982). *Zeit. Anorg. Allg. Chem.* **494**, 179.
S74	Scrivener, K. L., in *8th ICCC*, Vol. 3, p. 389 (1986).
S75	Stein, H. H. (1961). *J. Appl. Chem.* **11**, 474.
S76	Shaughnessy III, R. and Clark, P. E. (1988). *Cem. Concr. Res.* **18**, 327.
S77	Sellevold, E. J. (1974). *Cem. Concr. Res.* **4**, 399.
S78	Schwiete, H.-E. and Ludwig, U. (1966). *Tonind. Zeit.* **90**, 562.
S79	Sereda, P. J., Feldman, R. F. and Swenson, E. G., in *Symposium on Structure of Portland Cement Paste and Concrete* (Sp. Rpt 90), p. 58, Highway Research Board, Washington (1966).

S80 Scrivener, K. L. (1989). *Mater. Res. Soc. Symp. Proc.* **137**, 129
S81 Seligmann, P. (1968). *J. PCA Res. Dev. Labs* **10** (1), 52.
S82 Schreiner, L. J., MacTavish, J. C., Miljković, L., Pintar, M. M., Blinc, R., Lahajnar, G., Lasic, D. and Reeves, L. W. (1985). *J. Am. Ceram. Soc.* **68**, 10.
S83 Sellevold, E. J. and Bager, D. H., in *7th ICCC*, Vol. 4, p. 394 (1981).
S84 Schiller, K. K. (1971). *Cem. Concr. Res.* **1**, 419.
S85 Sereda, P. J., Feldman, R. F. and Ramachandran, V. S., in *7th ICCC*, Vol. 1, p. VI-1/3 (1980).
S86 Smolczyk, H.-G., in *7th ICCC*, Vol. 1, p. III-1/3 (1980).
S87 Stutterheim, N., in *5th ISCC*, Vol. 4, p. 270 (1969).
S88 Smolczyk, H.-G. (1978). *Zem.-Kalk-Gips* **31**, 294 (partial English translation, p. 180).
S89 Stutterheim, N., in *4th ISCC*, Vol. 2, p. 1035 (1962).
S90 Schröder, F., in *5th ISCC*, Vol. 4, p. 149 (1969).
S91 Smolczyk, H. G. (1965). *Zem.-Kalk-Gips* **18**, 238.
S92 Sato, T. and Furuhashi, I., in *Review of the 36th General Meeting*, p. 42, Cement Association of Japan, Tokyo (1982).
S93 Sersale, R. and Orsini, P. G., in *5th ISCC*, Vol. 4, p. 114 (1969).
S94 Stevenson, R. J., Collier, J. C., Crashell, J. J. and Quandt, L. R. (1988). *Mater. Res. Soc. Symp. Proc.* **113**, 87.
S95 Scian, A. N., Porto López, J. M. and Pereira, E. (1987). *Cem. Concr. Res.* **17**, 198.
S96 Sorrentino F. P. and Glasser, F. P., in *International Seminary on Calcium Aluminates* (eds M. Murat et al.), Politecnico di Torino, Turin (1982).
S97 Sorrentino, F. P. (1989). Private communication.
S98 Sorrentino, F. P. and Glasser, F. P. (1975). *Trans J. Br. Ceram. Soc.* **74**, 253.
S99 Sorrentino, F. P. (1973). Ph.D. thesis, University of Aberdeen, UK.
S100 Soustelle, M., Guilhot, B., Fournier, A. A., Murat, M. and Negro, A. (1985). *Cem. Concr. Res.* **15**, 421.
S101 Sudoh, G., Ohta, T. and Harada, H., in *7th ICCC*, Vol. 3, p. V-152 (1980).
S102 Stark, J., Müller, A., Seydel, R. and Jost, K., in *8th ICCC*, Vol. 2, p. 306 (1986).
S103 Stadelmann, C. and Wieker, W. (1985). *Cemento* **82**, 203.
S104 Skalny, J. and Odler, I. (1972). *J. Coll. Interface Sci.* **40**, 199.
S105 Sarp, H. and Burri, G. (1986). *Schweiz. Mineral. Petrogr. Mitt.* **66**, 453.
S106 Smith, D. K. *Cementing* (2nd Ed.), 254 pp., American Institute of Mining, Metallurgical and Petroleum Engineers, New York (1987).
S107 Sinclair, W. and Groves, G. W. (1985). *J. Mater. Sci.* **20**, 2846.
S108 Sugama, T. and Kukacka, L. E. (1980). *Cem. Concr. Res.* **10**, 303.
S109 Struble, L., Skalny, J. and Mindess, S. (1980). *Cem. Concr. Res.* **10**, 277.
S110 Struble, L. and Mindess, S. (1983). *J. Cem. Comp. Lightwt Concr.* **5**, 79.
S111 Scrivener, K. L. and Gartner, E. M. (1988). *Mater. Res. Soc. Symp. Proc.* **114**, 77.
S112 Scrivener, K. L. and Pratt, P. L., in *8th ICCC*, Vol. 3, p. 466 (1986).
S113 Scrivener, K. L. and Pratt, P. L., in *From Materials Science to Construction Materials Engineering* (ed. J.-C. Maso), Vol. 1, p. 61, Chapman and Hall, London (1987).
S114 Scrivener, K. L., Crumbie, A. K. and Pratt, P. L. (1988). *Mater. Res. Soc. Symp. Proc.* **114**, 87.
S115 Scrivener, K. L., Bentur, A. and Pratt, P. L. (1988). *Adv. Cem. Res.* **1**, 230.

S116 Schimmel, G. (1970). *Naturwiss.* **57**, 38.
S117 Somerville, G. (1986). *Structural Engineer* **64A**, 60, 233.
S118 Suzuki, K., Nishikawa, T. and Ito, S. (1985). *Cem. Concr. Res.* **15**, 213.
S119 Stanton, T. E. (1940). *Proc. Am. Soc. Civil Eng.* **66**, 1781.
S120 Struble, L. J. and Diamond, S. (1982). *J. Am. Ceram. Soc.* **64**, 652.
S121 Struble, L. and Diamond, S. (1981). *Cem. Concr. Res.* **11**, 611.
S122 Strohbauch, G. and Kuzel, H.-J. (1988). *Zem.-Kalk-Gips* **41**, 358 (partial English translation, p. 219).
S123 St John, D. A. (1982). *Cem. Concr. Res.* **12**, 633.
S124 Smolczyk, H. G., in *5th ISCC*, Vol. 3, p. 274 (1969).
S125 Stucke, M. S. and Majumdar, A. J. (1976). *J. Mater. Sci.* **11**, 1019.

T1 Timashev, V. V., in *7th ICCC*, Vol. 1, p. I-3/1 (1980).
T2 Terrier, P., Hornain, H. and Socroun, G. (1968). *Rev. Matér. Constr.* **630**, 109.
T3 Terrier, P., in *5th ISCC*, Vol. 2, p. 278 (1969).
T4 Takéuchi, Y., Nishi, F. and Maki, I. (1980). *Zeit. Krist.* **152**, 259.
T5 Taylor, H. F. W. (1989). New data.
T6 Taylor, W. C. (1942). *J. Res. Natl Bur. Stand.* **29**, 437.
T7 Tanaka, M., Sudoh, G. and Akaiwa, S., in *5th ISCC*, Vol. 1, p. 122 (1969).
T8 Thorvaldson, T., quoted by F. M. Lee in Ref. L6.
T9 Triviño Vázquez, F. (1985). *Cem. Concr. Res.* **15**, 581.
T10 Takashima, S., in *Review of the 12th General Meeting*, p. 12, Cement Association of Japan, Tokyo (1958).
T11 Tabikh, A. and Weht, R. J. (1971). *Cem. Concr. Res.* **1**, 317.
T12 Tamás, F. D., Sarkar, A. K. and Roy, D. M., in *Hydraulic Cement Pastes: Their Structure and Properties*, p. 55, Cement and Concrete Association, Slough, UK (1976).
T13 Taylor, H. F. W. (1989). *Adv. Cem. Res.* **2**, 73.
T14 Taylor, H. F. W. and Turner, A. B. (1987). *Cem. Concr. Res.* **17**, 613.
T15 Taylor, H. F. W. (1981). *Chem. Ind.* 620.
T16 Taylor, H. F. W. (1984). *Br. Ceram. Proc.* **35**, 65.
T17 Taylor, H. F. W., Mohan, K. and Moir, G. K. (1985). *J. Am. Ceram. Soc.* **68**, 680.
T18 Taylor, H. F. W. and Newbury, D. E. (1984). *Cem. Concr. Res.* **14**, 93.
T19 Taylor, H. F. W. and Newbury, D. E. (1984). *Cem. Concr. Res.* **14**, 565.
T20 Taylor, H. F. W. (1950). *J. Chem. Soc.* 3682.
T21 Taylor, H. F. W., in *5th ISCC*, Vol. 2, p. 1 (1969).
T22 Taylor, H. F. W. (1960). *Prog. Ceram. Sci.* **1**, 89.
T23 Taylor, H. F. W. and Howison, J. W. (1956). *Clay Minerals Bull.* **3**, 98.
T24 Taylor, H. F. W. (1986). *J. Am. Ceram. Soc.* **69**, 464.
T25 Taylor, H. F. W., Barret, P., Brown, P. W., Double, D. D., Frohnsdorff, G., Johansen, V., Ménétrier-Sorrentino, D., Odler, I., Parrott, L. J., Pommersheim, J. M., Regourd, M. and Young, J. F. (1984). *Matér. Constr. (Paris)* **17**, 457.
T26 Taylor, H. F. W. (1985). *Mater. Sci. Monogr.* **28A** (Reactivity of Solids, Part A), 39.
T27 Thomassin, J.-H., Regourd, M., Baillif, P. and Touray, J.-C. (1979). *CR Acad. Sci. Paris* **C288**, 93.

T28 Tadros, M. E., Skalny, J. and Kalyoncu, R. S. (1976). *J. Am. Ceram. Soc.* **59**, 344.
T29 Tong, H. S. and Young, J. F. (1977). *J. Am. Ceram. Soc.* **60**, 321.
T30 Turriziani, R. and Schippa, G. (1954). *Ricerca Sci.* **24**, 2356.
T31 Tilley, C. E., Megaw, H. D. and Hey, M. H. (1934). *Mineral. Mag.* **23**, 607.
T32 Turriziani, R. and Schippa, G. (1956). *Ricerca Sci.* **26**, 2792.
T33 Teoreanu, I., Filoti, G., Hritcu, C., Bucea, L., Spânu, V., Ciocanel, S. and Ivascu, M. (1979). *Cemento* **76**, 19.
T34 Tang, F. J. and Gartner, E. M. (1988). *Adv. Cem. Res.* **1**, 67.
T35 Taylor, H. F. W. (1987). *Mater. Res. Soc. Symp. Proc.* **85**, 47.
T36 Tamás, F. and Amrich, L. (1978). *Cemento* **75**, 357.
T37 Taylor, H. F. W. (1987). *Adv. Cem. Res.* **1**, 5.
T38 Thomas, N. L. and Double, D. D. (1981). *Cem. Concr. Res.* **11**, 675.
T39 Tamás, F. D., Farkas, E., Vörös, M. and Roy, D. M., in *8th ICCC*, Vol. 6, p. 374 (1986).
T40 Taplin, J. H. (1959). *Austral. J. Appl. Sci.* **10**, 329.
T41 Taplin, J. H., in *5th ISCC*, Vol. 2, pp. 249, 337 (1969).
T42 Tattersall, G. H. and Banfill, P. F. G. *The Rheology of Fresh Concrete*, 356 pp., Pitman Books, London (1983).
T43 Tattersall, G. H. (1955). *Br. J. Appl. Phys.* **6**, 165.
T44 Taylor, H. F. W., Mohan, K. and Moir, G. K. (1985). *J. Am. Ceram. Soc.* **68**, 685.
T45 Tanaka, H., Totani, Y. and Saito, Y., in *Fly Ash, Silica Fume, Slag and Other Mineral By-Products in Concrete* (ed. V. M. Malhotra), Sp. Publ. SP79, Vol. 2, p. 963, American Concrete Institute, Detroit (1983).
T46 Thorne, D. J. and Watt, J. D. (1965). *J. Appl. Chem.* **15**, 595.
T47 Traetteberg, A. (1978). *Cemento* **75**, 259.
T48 Taplin, J. H., in *4th ISCC*, Vol. 2, p. 924 (1962).
T49 Thomas, N. L. and Birchall, J. D. (1983). *Cem. Concr. Res.* **13**, 830.
T50 Thomas, N. L. and Birchall, J. D. (1984). *Cem. Concr. Res.* **14**, 761.
T51 Thomas, N. L. and Double, D. D. (1983). *Cem. Concr. Res.* **13**, 391.
T52 Taylor, H. F. W., in *Chemistry of Cements* (ed. H. F. W. Taylor), Vol. 1, p. 167, Academic Press, London (1964).
T53 Tenoutasse, N., in *5th ISCC*, Vol. 2, p. 372 (1969).
T54 Thomas, N. L., Jameson, D. A. and Double, D. D. (1981). *Cem. Concr. Res.* **11**, 143.
T55 Taylor, H. F. W. and Roy, D. M., in *7th ICCC*, Vol. 1, p. II-2/1 (1980).
T56 Taylor, H. F. W., in *6th ICCC*, Vol. 2, Part 1, p. 192 (1976).
T57 Taylor, W. H., Moorehead, D. R. and Cole, W. F., in *Autoclaved Calcium Silicate Building Products*, p. 130, Society of Chemical Industry, London (1967).
T58 Tashiro, C. and Tatibana, S. (1983). *Cem. Concr. Res.* **13**, 377.
T59 Turriziani, R., in *8th ICCC*, Vol. 1, p. 388 (1986).
T60 Thorvaldson, T., in *3rd ISCC*, p. 436 (1954).
T61 Tang, M.-S., Liu, Z. and Han, S.-F., in *Concrete Alkali–Aggregate Reactions* (ed. P. E. Grattan-Bellew), p. 275, Noyes Publ., Park Ridge, NJ, USA (1987).
U1 Uchikawa, H., Ogawa, K. and Uchida, S. (1985). *Cem. Concr. Res.* **15**, 561.
U2 Udagawa, S. and Urabe, K., in *Review of the 32nd General Meeting*, p. 31, Cement Association of Japan, Tokyo (1978).
U3 Udagawa, S., Urabe, K. and Yano, T., in *Review of the 34th General Meeting*, p. 37, Cement Association of Japan, Tokyo (1980).

U4	Udagawa, S., Urabe, K., Natsume, M. and Yano, T. (1980). *Cem. Concr. Res.* **10**, 139.
U5	Udagawa, S., Urabe, K., Yano, T. and Natsume, M. (1980). *Yogyo-Kyokai-Shi* **88**, 285.
U6	Udagawa, S., Urabe, K. and Yano, T., in *Review of the 31st General Meeting*, p. 23, Cement Association of Japan, Tokyo (1977).
U7	Urabe, K., Yano, T., Iwai, A., Udagawa, S. and Ikawa, H., in *Review of the 36th General Meeting*, p. 27, Cement Association of Japan, Tokyo (1982).
U8	Udagawa, S., Urabe, K., Yano, T., Takada, K. and Natsume, M., in *Review of the 33rd General Meeting*, p. 35, Cement Association of Japan, Tokyo (1979).
U9	Uchikawa, H. and Furuta, R. (1981). *Cem. Concr. Res.* **11**, 65.
U10	Ueda, S., Hashimoto, M. and Kondo, R. (1968). *Semento Gijutsu Nenpo* **22**, 67.
U11	Uchikawa, H. (1984). *Am. Ceram. Soc. Bull.* **63**, 1143.
U12	Uchikawa, H. (1986). *J. Res. Onoda Cement Company* **38**, 1.
U13	Uchikawa, H., Uchida, S., Ogawa, K. and Hanehara, S. (1984). *Cem. Concr. Res.* **14**, 645.
U14	Uchikawa, H., Uchida, S. and Hanehara, S. (1987). *Cemento* **84**, 3.
U15	Ushiyama, H. and Goto, S., in *6th ICCC*, Vol. 2, Part 1, p. 331 (1976).
U16	Uchikawa, H., Uchida, S. and Hanehara, S., in *8th ICCC*, Vol. 4, p. 245 (1986).
U17	Uchikawa, H., in *8th ICCC*, Vol. 1, p. 249 (1986).
U18	Uchikawa, H., Uchida, S. and Ogawa, K., in *8th ICCC*, Vol. 4, p. 251 (1986).
U19	Uchikawa, H., Uchida, S. and Ogawa, K., in *The Use of PFA in Concrete* (eds J. G. Cabrera and A. R. Cusens), p. 83, Leeds University, Concrete Society and CEGB, UK (1982).
U20	Uchikawa, H. (1985). *Cem. Concr. (Japan)* (460), 20.
U21	Uchikawa, H. and Tsukiyama, K. (1973). *Cem. Concr. Res.* **3**, 263.
U22	Uchikawa, H. and Uchida, S. (1972). *Cem. Concr. Res.* **2**, 681.
U23	Uchikawa, H. and Uchida, S. (1973). *Cem. Concr. Res.* **3**, 607.
U24	Ushiyama, H., Kawano, Y. and Kamegai, N., in *8th ICCC*, Vol. 3, p. 154 (1986).
U25	Udagawa, S., Urabe, K. and Hiyama, K. (1979). Abstract of paper presented at Annual Meeting, Ceramic Society of Japan.
V1	Vincent, M. G. and Jeffery, J. W. (1978). *Acta Cryst.* **B34**, 1422.
V2	Vosteen, B. (1974). *Zem.-Kalk-Gips* **27**, 443.
V3	Vásquez, T., Blanco-Varela, M. T. and Palomo, A. (1984). *Br. Ceram. Proc.* **35**, 115.
V4	Van Aardt, J. H. P. and Visser, S., in *7th ICCC*, Vol. 4, p. 483 (1981).
V5	Verbeck, G. J. and Helmuth, R. H., in *5th ISCC*, Vol. 3, p. 1 (1969).
V6	Volkov, V. V., Kolyovski, V. P. and Yanev, Ya. D., in *6th ICCC*, Vol. 3, p. 182 (1976).
V7	Vernet, C., in *8th ICCC*, Vol. 2, p. 316 (1986).
V8	Viswanathan, V. N., Raina, S. J. and Chatterjee, A. K. (1978). *World Cem. Technol.* **9**, 109.
V9	Von Lampe, F., Hilmer, W., Jost, K.-H., Reck, G. and Boikova, A. I. (1986). *Cem. Concr. Res.* **16**, 505.
V10	Vivian, H. E. (1950). *CSIRO Bull.* (256), 60.
V11	Van Aardt, J. H. P. and Visser, S. (1977). *Cem. Concr. Res.* **7**, 643.
V12	Van Aardt, J. H. P. and Visser, S. (1978). *Cem. Concr. Res.* **8**, 677.

W1   Woermann, E., Hahn, T. and Eysel, W. (1979). *Cem. Concr. Res.* **9**, 701.
W2   Welch, J. H. and Gutt, W. (1959). *J. Am. Ceram. Soc.* **42**, 11.
W3   Winchell, A. N. and Winchell, H. *The Microscopical Characters of Artificial Inorganic Solid Substances*, 439 pp., Academic Press, London and New York (1964).
W4   Williams, P. P. (1968). *J. Am. Ceram. Soc.* **51**, 531.
W5   Williams, P. P. (1973). *Acta Cryst.* **B29**, 1550.
W6   Welch, J. H., in *Chemistry of Cements* (ed. H. F. W. Taylor), Vol. 1, p. 49, Academic Press, London (1964).
W7   Welch, J. H. (1961). *Nature* **191**, 559.
W8   Welch, J. H. and Gutt, W., in *4th ISCC*, Vol. 1, p. 59 (1962).
W9   Wolter, A. (1985). *Zem.-Kalk-Gips* **38**, 612 (partial English translation, p. 327).
W10 Wagman, D. D., Evans, W. H., Parker, V. B., Schumm, R. H., Halow, I., Bailey, S. M., Churney, K. L. and Nuttall, R. L. (1982) The NBS Tables of Chemical Thermodynamic Properties, *J. Phys. Chem. Ref. Data* **11**, Suppl. 2, 392 pp., American Chemical Society, Washington and American Institute of Physics, New York.
W11 Williamson, J. and Glasser, F. P. (1962). *J. Appl. Chem.* **12**, 535.
W12 Weyer, I. (1931). *Zement* **20**, 560, 608, 692.
W13 Weber, P. *Heat Transfer in Rotary Kilns*, 98 pp., Bauverlag, Wiesbaden and Berlin (1963).
W14 Wolter, A., in *8th ICCC*, Vol. 2, p. 89 (1986).
W15 Weisweiler, W., Osen, E., Eck, J. and Höfer, H. (1986). *Cem. Concr. Res.* **16**, 283.
W16 Wagner, L. A. (1933). *ASTM Bull.* **33**, 553.
W17 Webb, T. L. and Heystek, H., in *The Differential Thermal Investigation of Clays* (ed. R. C. Mackenzie), p. 329, Mineralogical Society, London (1957).
W18 Williamson, R. B. (1972). *Prog. Mater. Sci.* **15**, 189.
W19 Wu, Z.-Q., Hriljac, J., Hwang, C.-L. and Young, J. F. (1983). *J. Am. Ceram. Soc.* **66**, C86.
W20 Wieker, W. (1968). *Zeit. Anorg. Allg. Chem.* **360**, 307.
W21 Wieker, W., Grimmer, A.-R., Winkler, A., Mägi, M., Tarmak, M. and Lipmaa, E. (1982). *Cem. Concr. Res.* **12**, 333.
W22 Wu, Z.-Q. and Young, J. F. (1984). *J. Am. Ceram. Soc.* **67**, 48.
W23 Wittmann, F. H., in *Hydraulic Cement Pastes: Their Structure and Properties*, p. 96, Cement and Concrete Association, Slough, UK (1976).
W24 Wittmann, F. H., in *7th ICCC*, Vol. 1, p. VI-2/1 (1980).
W25 Winslow, D. N. and Diamond, S. (1973). *J. Coll. Interface Sci.* **45**, 425.
W26 Winslow, D. N. and Diamond, S. (1974). *J. Am. Ceram. Soc.* **57**, 193.
W27 Watt, J. D. and Thorne, D. J. (1965). *J. Appl. Chem.* **15**, 585.
W28 Watt, J. D. and Thorne, D. J. (1966). *J. Appl. Chem.* **16**, 33.
W29 Wei, F.-J., Grutzeck, M. W. and Roy, D. M. (1985). *Cem. Concr. Res.* **15**, 174.
W30 Wu, C.-W. and Wang, Y.-S., in *7th ICCC*, Vol. 3, p. V-27 (1980).
W31 Wilding, C. R., Walter, A. and Double, D. D. (1984). *Cem. Concr. Res.* **14**, 185.
W32 Winkler, A. and Wieker, W. (1979). *Zeit. Anorg. Allg. Chem.* **451**, 45.
W33 Wu, Z.-W., Liu, B.-Y. and Xie, S.-S., in *Liaisons Pâtes de Ciment Matériaux Associés* (Proc. RILEM Colloq.), p. A28, Laboratoire de Génie Civil, Toulouse (1982).

W34 Way, S. J. and Cole, W. F. (1982). *Cem. Concr. Res.* **12**, 611.
X1 Xue, J.-G., Xu, W.-X. and Ye, M.-X. (1983). *Guisuanyan Xuebao* **11**, 276; *Chem. Abstr.* **100**, 73009c (1984).
X2 Xue, J.-G., Tong, X.-L., Chen, W.-H., Xu, J.-Zh., Xi, Y.-Zh. and Zhang, Y.-Y., in *8th ICCC*, Vol. 4, p. 306 (1986).
Y1 Yamaguchi, G. and Takagi, S., in *5th ISCC*, Vol. 1, p. 181 (1969).
Y2 Yannaquis, N. and Guinier, A. (1959). *Bull. Soc. Franç. Minéral. Cryst.* **82**, 126.
Y3 Yamanaka, T. and Mori, H. (1981). *Acta Cryst.* **B37**, 1010.
Y4 Yoshioka, T. (1970). *Bull. Chem. Soc. Japan* **43**, 1981, 2317.
Y5 Young, J. F. and Hansen, W. (1987). *Mater. Res. Soc. Symp. Proc.* **85**, 313.
Y6 Young, J. F. (1988). *J. Am. Ceram. Soc.* **71**, C118.
Y7 Yan, R.-Zh., Ouyang, Sh.-X. and Gao, Q.-Y. (1983). *Silicates Ind.* **12**, 3.
Y8 Young, J. F. (1972). *Cem. Concr. Res.* **2**, 415.

Z1 Zheng, Y. and Yang, N., in *8th ICCC*, Vol. 2, p. 293 (1986).
Z2 Zhmoidin, G. I. and Chatterjee, A. K. (1984). *Cem. Concr. Res.* **14**, 386.
Z3 Zhmoidin, G. I. and Chatterjee, A. K. (1985). *Cem. Concr. Res.* **15**, 442.
Z4 Zhmoidin, G. I. and Chatterjee, A. K. (1974). *Inorg. Mater.* **10**, 1584 (translated from *Neorg. Mater.* **10**, 1846).
Z5 Zur Strassen, H. (1957). *Zem.-Kalk-Gips* **10**, 1.
Z6 Ziegler, E. (1971). *Zem.-Kalk-Gips* **24**, 543.
Z7 Zur Strassen, H. (1959). *Zement Beton* (16) 32.
Z8 Zürz, A. and Odler, I. (1987). *Adv. Cem. Res.* **1**, 27.
Z9 Zur Strassen, H. and Strätling, W. (1940). *Zeit. Anorg. Allg. Chem.* **245**, 267.
Z10 Zur Strassen, H. (1958). *Zem.-Kalk-Gips* **11**, 137.
Z11 Zelwer, A., in *7th ICCC*, Vol. 2, p. II-147 (1980).
Z12 Zakharov, L. A., in *6th ICCC*, Vol. 3, p. 153 (1976).

Appendix

Calculated X-ray powder diffraction patterns for tricalcium silicate and clinker phases

The patterns (T5) were calculated as described by Morris *et al.* (M16). The columns headed d(nm), I_{int}, h, k and l give the integrated intensity pattern as defined by these authors, i.e. they refer to individual reflections, which are liable to be components of overlaps. I_{int} is the integrated intensity relative to a value of 100 for the strongest. Reflections with $I_{int} < 1.5$ are not listed. The column headed I_{int} gives an indication of relative peak heights assuming typical experimental conditions using a diffractometer, but these depend on resolution and on any variation in cell parameters, because these affect separations between overlapping peaks. They take into account contributions from reflections too weak to appear in the list of integrated intensities, and their spacings sometimes differ significantly from those of the individual components. Peaks of significant intensity can arise from groups of reflections all of which are too weak to be listed individually; the more prominent are noted in footnotes to the tables. Except with T_1 C_3S, the cell parameters and compositions assumed are typical ones for phases in Portland cement clinkers. No data are available from which it would be possible to calculate a pattern for an M_1 alite. Reference intensity ratios, I_{hkl}/I_c, (H23,M16) are also given; their use is explained in Chapter 4.

Table A.1 Calculated XRD powder pattern for T_1 tricalcium silicate*

d(nm)	I_{int}	I_{pk}	h	k	l	d(nm)	I_{int}	I_{pk}	h	k	l
1.1108	2	3	0	−1	1	0.2207	3		3	−4	4
0.5979	2		1	2	−1	0.2204	2		1	−2	6
0.5964	3	8	−1	2	0	0.2201	13		4	−4	2
0.5937	4		−1	0	2	0.2197	19	44	0	0	6
0.3916	5	6	1	−2	3	0.2196	12		4	4	0
0.3890	4		1	2	2	0.2195	3		3	4	2
		7				0.2194	2		1	2	5
0.3878	2		3	0	0	0.2191	3		5	−2	1
0.3560	3	3	0	−4	1	0.2190	3		−1	0	6
0.3523	2	3	−2	2	2	0.2190	2		5	2	0
0.3459	2	3	3	1	−2	0.2186	4	11	2	−6	2
0.3063	34	38	1	4	−3	0.2183	4		2	6	−1
0.3044	33	43	−1	4	1	0.2176	3		2	2	−6
0.3027	35	44	−3	0	3	0.2175	2		4	4	−3
0.2990	8		2	4	−2	0.2169	2		−4	4	1
		19						8			
0.2982	9		−2	4	0	0.2166	3		−2	2	5
0.2969	10	15	−2	0	4	0.2137	2	3	5	1	−3
0.2805	2		−4	1	1	0.2076	3	3	−1	5	3
		100				0.2000	4	6	1	6	−5
0.2801	100		3	0	3	0.1988	2	6	−3	6	0
0.2777	42	51	0	−4	4	0.1986	4	7	−1	6	2
0.2755	45	53	0	4	2	0.1979	2	6	−3	0	6
0.2741	44	54	−4	0	2	0.1958	10	10	2	−4	6
0.2623	40		3	−4	1	0.1945	8	9	2	4	4
		85				0.1939	9	12	6	0	0
0.2622	40		3	4	−1	0.1916	2	3	3	4	−6
0.2615	31	84	−1	0	5	0.1851	3	4	0	−6	6
0.2588	2	2	1	−5	3	0.1838	3		5	−4	3
0.2473	3	5	3	−1	4			7			
0.2462	2	4	2	5	0	0.1837	2		0	6	3
0.2454	2		−1	1	5	0.1833	2		5	4	1
		6				0.1830	2	7	4	3	−6
0.2453	3		4	3	−2	0.1828	2		−6	0	3
0.2382	2	3	4	−3	2	0.1780	34	32	0	−8	2
0.2350	7	8	1	−4	5	0.1771	26	27	4	4	−6
0.2342	3	8	4	0	3	0.1762	28	28	−4	4	4
0.2333	2		2	0	5	0.1653	3		5	−6	2
		10				0.1651	2		5	6	−1
0.2332	6		1	4	3			18			
0.2323	9	12	−5	0	1	0.1649	3		−1	0	8
0.2307	2	6	4	3	1	0.1647	18		3	−4	7
0.2291	2	6	3	5	−2	0.1637	18	18	3	4	5
0.2252	2	5	4	4	−1	0.1633	20	20	7	0	1
						0.1554	10		6	−4	4
								15			
						0.1554	8		2	0	8
						0.1550	25	15	6	4	2

*Pattern calculated from the data of Golovastikov et al. (G2); $a = 1.167$ nm, $b = 1.424$ nm, $c = 1.372$ nm, $\alpha = 105.50°$, $\beta = 94.33°$, $\gamma = 90.00°$. Other peaks of significant height; 0.3717 nm (3), 3 1 −1 +; 0.2099 nm (4), −1 4 4 +. $I_{303}/I_c = 0.798$.

Table A.2 Calculated XRD powder pattern for M_3 alite*

d(nm)	I_{int}	I_{pk}	h	k	l	d(nm)	I_{int}	I_{pk}	h	k	l
0.5949	4 } 11	{	4	0	2	0.2186	15 }	{	4	0	8
						0.2186	32		−10	2	4
0.5940	7		3	1	0	0.2186	5 } 48	{	−13	1	4
0.5490	2	3	1	1	2	0.2183	6		1	1	8
0.3877	4 } 8	{	−8	0	2	0.2182	6		−6	2	6
						0.2170	5 }		9	1	6
0.3869	6		−1	1	4	0.2170	6 } 17	{	12	2	0
0.3535	5 } 6	{	7	1	2	0.2166	6		−5	3	2
						0.2071	2	2	−10	2	5
0.3527	3		0	2	0	0.1984	4 }		16	0	2
0.3180	2	3	−6	0	5	0.1980	4 } 11	{	9	3	0
0.3042	34 } 90	{	10	0	2	0.1979	7		5	3	4
						0.1938	7 } 18	{	−16	0	4
0.3034	78		4	2	2						
0.2975	10 } 27	{	8	0	4	0.1934	17		−2	2	8
						0.1837	3 } 5	{	14	2	3
0.2970	20		6	2	0						
0.2778	100	90	−6	0	6	0.1835	2		18	0	0
0.2753	43 } 99	{	12	0	0	0.1830	5	7	3	3	6
						0.1825	4 } 7	{	−12	2	6
0.2745	84		2	2	4						
0.2697	2 }		−11	1	2	0.1824	2		2	0	10
0.2694	2 } 8	{	3	1	6	0.1768	58	50	14	2	4
0.2692	3		−4	2	4	0.1763	31	52	0	4	0
0.2611	41 } 100	{	6	0	6	0.1644	2 } 5	{	8	0	10
0.2609	80		−8	2	2	0.1644	3		−13	3	4
0.2453	5	5	−6	0	7	0.1630	14 } 26	{	−18	0	6
0.2327	6 }		−14	0	2						
0.2321	16 } 18	{	0	2	6	0.1627	27		−4	2	10
0.2318	2		6	0	7	0.1542	14 } 16	{	−14	2	8
0.2299	2	1	3	3	0						
0.2287	2	3	−10	2	3	0.1542	8		0	0	12

* Pattern calculated assuming $a = 3.312$ nm, $b = 0.7054$ nm, $c = 1.855$ nm, $\beta = 94.17°$; atomic parameters given by Nishi et al. (N1); site occupancies $(Ca_{0.98}Mg_{0.01}Al_{0.0067}Fe_{0.003})_3 (Si_{0.97}Al_{0.03})O_5$. Other peaks of significant height: 0.3248 nm (2), −2 2 2 +; 0.3211 nm (1), 10 0 1+; 0.2673 nm (3), −8 2 1; 0.2562 nm (2), −8 0 6+; 0.2504 nm (1), 2 2 5+; 0.2276 nm (2), −1 3 4 +; 0.2097 nm (2), −1 3 4+; 0.2036 nm (1), 16 0 1+; 0.1861 nm (1), −16 0 5+; 0.1794 nm (2), −8 2 8+; 0.1714 nm (1), −7 1 10+; 0.1693 nm (2), −18 0 5; 0.1691 nm (2), 14 2 5+; 0.1594 nm (1), −13 3 5+; 0.1561 nm (1), −18 0 7. $I_{-606}/I_c = 0.742$.

450   Cement Chemistry

Table A.3 Calculated XRD powder pattern for a belite*

d(nm)	I_{int}	I_{pk}	h	k	l	d(nm)	I_{int}	I_{pk}	h	k	l
0.4891	3	3	−1	0	1	0.1915	9	8	0	2	4
0.4651	10	10	0	0	2	0.1900	8	8	2	1	3
0.3830	7	7	0	1	2	0.1893	16	15	2	2	2
0.3786	6	6	1	1	1	0.1845	6	5	−2	0	4
0.3375	8	7	0	2	0	0.1821	4	4	0	3	3
0.3237	9	9	−1	1	2	0.1805	12 ⎫	⎧	−2	2	3
0.3173	6	6	0	2	1		⎬ 16 ⎨				
0.3051	12	11	1	1	2	0.1804	8 ⎭	⎩	−1	0	5
0.2874	26	24	1	2	0	0.1794	13	12	0	1	5
0.2818	14	19	0	1	3	0.1764	5	4	3	1	0
0.2793	100	100	−1	0	3	0.1743	2	2	−1	1	5
0.2778	81	94	−1	2	1	0.1725	7	7	−2	3	1
0.2742	88 ⎫	⎧	2	0	0	0.1722	4	7	1	0	5
	⎬ 88 ⎨					0.1710	3 ⎫	⎧	2	0	4
0.2732	28 ⎭	⎩	0	2	2		⎬ 14 ⎨				
0.2716	32	38	1	2	1	0.1708	15 ⎭	⎩	2	2	3
0.2615	69	60	1	0	3	0.1695	5	5	2	3	1
0.2540	16	14	2	1	0	0.1687	5	5	0	4	0
0.2446	17	16	−2	0	2	0.1660	2	2	0	4	1
0.2438	8	16	1	1	3	0.1630	21 ⎫	⎧	−3	0	3
0.2407	13 ⎫	⎧	2	1	1		⎬ 23 ⎨				
	⎬ 24 ⎨					0.1629	10 ⎭	⎩	0	2	5
0.2402	18 ⎭	⎩	1	2	2	0.1619	4	6	−2	2	4
0.2326	3	3	0	0	4	0.1613	5 ⎫	⎧	1	4	0
0.2299	3	4	−2	1	2		⎬ 8 ⎨				
0.2287	3 ⎫	⎧	2	0	2	0.1611	5 ⎭	⎩	3	1	2
	⎬ 28 ⎨					0.1607	8	11	3	2	0
0.2283	31 ⎭	⎩	0	2	3	0.1604	2 ⎫	⎧	2	3	2
0.2199	18	18	0	1	4		⎬ 15 ⎨				
0.2187	51	45	0	3	1	0.1602	14 ⎭	⎩	−3	2	1
0.2166	19 ⎫	⎧	2	1	2	0.1595	2 ⎫	⎧	−1	4	1
	⎬ 18 ⎨						⎬ 5 ⎨				
0.2152	2 ⎭	⎩	−1	2	3	0.1591	3 ⎭	⎩	−1	2	5
0.2128	8	7	2	2	0	0.1586	6 ⎫	⎧	0	4	2
0.2102	2	<1	−2	2	1		⎬ 8 ⎨				
0.2094	8	8	−1	1	4	0.1585	2 ⎭	⎩	−3	1	3
0.2082	5	5	1	3	0	0.1583	5	8	1	4	1
0.2048	17	16	2	2	1	0.1574	16	13	−1	3	4
0.2044	3	13	−1	3	1	0.1555	4 ⎫	⎧	−2	1	5
0.2037	10	12	−2	1	3	0.1552	4 ⎪	⎪	−3	2	2
0.2025	11	13	0	3	2		⎬ 10 ⎨				
0.2019	15	18	1	3	1	0.1550	6 ⎪	⎪	0	0	6
0.1992	18	16	1	1	4	0.1549	5 ⎭	⎩	−2	3	3
0.1980	31	28	−2	2	2						

* Pattern calculated assuming $a = 0.550$ nm, $b = 0.675$ nm, $c = 0.933$ nm, $\beta = 94.40°$; atomic parameters given by Jost *et al.* (J2); site occupancies $(K_{0.01}Na_{0.005}Ca_{0.975}Mg_{0.01})_2(Fe_{0.02}Al_{0.06}Si_{0.90}P_{0.01}S_{0.01})O_{3.96}$. Other peaks of significant height: 0.1760 nm (3), −3 1 1; 0.1656 nm (2), −2 3 2 +. $I_{-103}/I_c = 0.724$.

Table A.4 Calculated XRD powder pattern for a cubic aluminate phase*

d(nm)	I_{int}	I_{pk}	h	k	l	d(nm)	I_{int}	I_{pk}	h	k	l
0.6815	2	2	1	0	2	0.2694	100	100	0	4	4
0.5080	3	4	1	2	2	0.2410	2	3	0	2	6
0.4595	2	2	1	1	3	0.2380	2	3	0	4	5
0.4227	5	6	0	2	3	0.2200	7	7	4	4	4
0.4073	5	} 14 {	1	2	3	0.2037	2	2	2	4	6
						0.1951	3	3	3	4	6
0.4073	7		2	1	3	0.1905	31	29	0	0	8
0.3810	3	3	0	0	4	0.1890	2	4	1	0	8
0.3326	3	3	2	1	4	0.1822	2	3	3	5	6
0.3048	2	2	3	0	4	0.1555	34	30	4	4	8
0.2830	4	5	0	2	5						
0.2782	5	} 8 {	1	2	5						
0.2782	2		2	1	5						

* Pattern calculated assuming $a = 1.5240$ nm, space group Pa3, atomic parameters given by Mondal and Jeffery (M11) and composition $(K_{0.03}Na_{0.06}Ca_{2.76}Mg_{0.08}Ti_{0.01})(Fe_{0.22}Al_{1.60}Si_{0.18})O_6$. Other peaks of significant height: 0.6215 nm (2), 1 1 2; 0.2272 nm (2), 4 2 5+; 0.2177 nm (2), 2 3 6+; 0.1984 nm (2), 3 1 7+; 0.1643 nm (2), 5 5 6+. $I_{044}/I_c = 3.06$.

Table A.5 Calculated XRD powder pattern for an orthorhombic aluminate phase*

d(nm)	I_{int}	I_{pk}	h	k	l	d(nm)	I_{int}	I_{pk}	h	k	l
0.7553	2	2	0	0	2	0.2282	2	2	3	3	3
0.6846	3	4	1	1	1	0.2270	2	3	3	1	5
0.5104	5	6	0	2	1	0.2207	5 ⎫ ⎬ 9 ⎧		4	0	4
0.4414	5	5	2	0	2						
0.4211	8	8	1	1	3	0.2202	6 ⎭ ⎩		0	4	4
0.3840	2	2	2	2	0	0.2186	2	3	2	4	3
0.3690	2	2	0	2	3	0.2030	2	2	1	3	6
0.3423	9	9	2	2	2	0.1964	3	3	5	1	3
0.3389	2	3	1	1	4	0.1920	32	29	4	4	0
0.3353	4 ⎫ ⎬ 7 ⎧		3	1	1	0.1905	4	5	4	4	1
						0.1892	2 ⎫ ⎬ 14 ⎧		2	4	5
0.3345	3 ⎭ ⎩		1	3	1						
0.3054	6	6	2	2	3	0.1888	15 ⎭ ⎩		0	0	8
0.2812	5	5	1	1	5	0.1848	2	2	4	0	6
0.2720	31 ⎫ ⎬ 49 ⎧		4	0	0	0.1749	2	2	4	2	6
						0.1688	2	1	0	4	7
0.2711	25 ⎭ ⎩		0	4	0	0.1624	3	3	2	6	3
0.2693	100	100	2	2	4	0.1565	17	17	6	2	4
0.2431	3	3	4	2	0	0.1562	15	21	2	6	4
0.2392	2	3	1	1	6	0.1551	9 ⎫ ⎬ 11 ⎧		4	0	8
0.2374	4	4	2	2	5						
0.2354	2	2	2	3	4	0.1549	6 ⎭ ⎩		0	4	8
0.2310	2	3	2	4	2						

*Pattern calculated assuming $a = 1.0879$ nm, $b = 1.0845$ nm, $c = 1.5106$ nm, space group Pbca, composition $(Na_{0.292}Ca_{2.792})(Fe_{0.150}Al_{1.725}Si_{0.125})O_6$, atomic parameters and site occupancies given by Nishi and Takéuchi (N4). Other peaks of significant height: 0.5414 nm (2), 0 2 0; 0.5388 nm (2), 1 1 2; 0.2639 nm (2), 0 2 5; 0.2586 nm (2), 3 2 3+; 0.2425 nm (3), 3 3 2; 0.2047 nm (2), 1 5 2; 0.2042 nm (2), 2 4 4; 0.1827 nm (2), 1 3 7+; 0.1616 nm (2), 6 1 4. $I_{224}/I_c = 1.94$.

Table A.6 Calculated XRD powder pattern for a clinker ferrite phase*

d(nm)	I_{int}	I_{pk}	h	k	l	d(nm)	I_{int}	I_{pk}	h	k	l
0.7321	19	21	0	2	0	0.1919	48	41	2	0	2
0.5177	4	4	1	1	0	0.1882	3	3	2	5	1
0.3839	2	2	1	0	1	0.1857	9	8	2	2	2
0.3661	16	17	1	3	0	0.1830	2	⎫	2	6	0
0.3400	7	7	1	2	1			⎬ 20			
0.2767	35	33	2	0	0	0.1830	21	⎭	0	8	0
0.2664	47	53	0	0	2	0.1799	3	3	0	6	2
0.2649	100	100	1	4	1	0.1726	10	8	3	3	0
0.2588	14	14	1	5	0	0.1648	2	2	1	2	3
0.2422	5	4	2	1	1	0.1592	4	3	2	7	1
0.2208	9	9	2	4	0	0.1574	17	15	3	4	1
0.2154	7	⎫	1	3	2	0.1561	3	4	3	5	0
		⎬ 14				0.1535	21	17	1	4	3
0.2154	9	⎭	0	4	2	0.1527	6	6	2	8	0
0.2059	23	20	1	6	1						

* Pattern calculated assuming $a = 0.5535$ nm, $b = 1.4642$ nm, $c = 0.5328$ nm, space group Ibm2, atomic parameters given for C_4AF by Colville and Geller (C3) and composition $Ca_2[Al_{0.7}Fe_{0.3}](Al_{0.7}Fe_{0.3})O_5$, where square and round brackets denote octahedral and tetrahedral sites, respectively. $I_{141}/I_c = 1.79$.

Index

Substances are indexed under the most commonly used designations; thus, the major phases of Portland cement clinker, natural minerals and common chemical compounds are usually indexed by name, and other compounds of specific importance to cement chemistry by cement chemical formulae. Broad categories of compounds (e.g. calcium magnesium silicates) are indexed by name.

Abbreviations, 4–5
Abrasion of concrete, 328, 403
Accelerators
  calcium aluminate cements, 330–331
  Portland cement, 357–362
Acceleratory period
  calcium silicate pastes, 160, 164–165
  Portland cement pastes, 223–224, 226–227
Acetone, reaction with C_3S pastes, 125
Acid attack
  calcium aluminate cements, 333
  Portland or composite cements, 405
Additions, mineral, definition, 276
Additives, definition, 345
Admixtures (see also under specific types), 345–362
  definition, 345
  for calcium aluminate cements, 330–331
  for oilwell cements, 372–373
Adsorption
  loop and train modes, 355
  of retarders, 345–349
  of water reducers and superplasticizers, 352, 354–357
Aerated concretes, autoclaved, 366
AFm phases (see also formulae of individual compounds)
  alkali substitution, 173
  as a hydration product of pure compounds, 193–197

AFm phases (*cont.*)
  as a normal hydration product of Portland cement, 199–202, 204, 206, 208–209, 211–212, 214–222, 225, 226, 232, 233
  content of Fe, 211, 212, 214, 215
  content of SO_3, 200–202, 209, 211, 216
  crystallinity, closeness of mixing with C-S-H, 199–200, 202, 204, 209, 211, 214, 215, 221
  morphological or microstructural aspects, 222, 225
  quantity formed, 209, 217–221
  relation to heat evolution, 226, 232
  relation to thermogravimetric curve, 208, 220, 221
  X-ray, DTA and infrared evidence, 199–202
  as a product resulting from flash set, 233
  crystal, optical, thermal and other data, 169–177
  crystal structures, compositions, 167–169
AFt phases (see also ettringite, thaumasite and formulae of other AFt phases)
  as a hydration product of Portland cement, 199–202, 208, 216–226, 228, 232, 233
  content of SO_3, 180

AFt phases (*cont.*)
  morphological or microstructural aspects, 216, 221–225, 400
  quantity formed, 208, 217–221, 228, 400
  relation to heat evolution, 226, 232
  relation to thermogravimetric curve, 220
  X-ray, DTA and infrared evidence, 199–202
  as a hydration product of pure compounds, 194–197
  crystal, optical, thermal and other data, 178–181
  crystal structures, compositions, 177–180
Afwillite, 142
Aggregate-cement paste interface, 377–381
  calcium aluminate cements, 328
  composite cements, 381
Air entraining agents, 351, 403
Åkermanite, 38, 49
Alcohols, polyhydric, as grinding aids, 352
Alinite and alinite cements, 343–344
Alite (see also tricalcium silicate), 5–15
  chromium substitution, defects, 121
  crystal structure and data for polymorphs, 5–8
  density, 15
  factors affecting crystal size, 82–83, 105
  factors affecting reactivity, 119–122
  fluoride substitution, 57–58
  formation in cement kiln, 60–61, 81–83
  hydration, see calcium silicate pastes, Portland cement pastes
  in clinkers
    compositions, 8–11
    effect of cooling rate, 88–89
    polymorphic types, 9, 12–14
    zoning, 12
  natural occurrence, 15
  optical and thermal data, 15
  polymorphic transitions in cement kiln, 12–13, 83
  polymorphism, 5–9, 12–13
  recrystallization and crystal growth, 13, 82–83

Alite (*cont.*)
  solid solutions or ionic substitutions, 8–13
  X-ray powder patterns, 13–14, 449
  zinc substitution, defects, 121–122
Alkali and alkali calcium sulphates
  data, equilibria, 52–56
  effects on cement hydration, 236–237
  in Portland cement clinker, 89–91, 103, 114–115, 117
Alkali attack
  calcium aluminate cements, 333–334
  Portland or composite cements, 405
Alkali carbonates
  effect on cement hydration, 361–362
  in cement clinkers, 54
Alkali silica reaction, 388–396
  similarity to pozzolanic reaction, 392–393
  sources of hydroxyl ions, 389, 396, 397
Alkaline hydrolysis, of calcium aluminate cements, 334
Alkalis, effects of
  on $CaO-Al_2O_3-H_2O$ system, 331
  on $CaO-SiO_2-H_2O$ system, 158
  on Portland cement hydration, 236–237
Alkalis in pore solutions
  composite cement pastes, 395–396
  Portland cement pastes, 227–230
Alkalis in Portland cement making, 74, 76, 78, 89, 92, 93
Alkyl aryl sulphonic acid salts, 351
Alumina, hydrous, from calcium aluminate cements, 320–325
Alumina ratio (alumina modulus), 61, 80, 85–87, 103
Aluminate (phase), 23–28
  alkali substitution, 24–25
  crystal structures and data for variants, 23–26
  density, 28
  enthalpy change on formation, 64
  enthalpy change on hydration, 197–198
  equilibria, 35, 39, 42, 45, 48
  factors affecting reactivity, 119–122
  formation in cement making, 60–61, 73–74, 84–88

Aluminate (phase) (*cont.*)
  hydration
    in Portland cement pastes, 205–206, 221–228, 231–237
    of pure compound, 193–194
    of pure compound in presence of sulphate, 194–196
  in clinkers, compositions and modifications, 10–11, 26–28
  ionic substitutions, 24–25, 42
  optical properties, 28
  structural variants, 24–26
  X-ray powder patterns, 26–27, 451–452
Aluminium hydroxide
  formation from calcium aluminate cements, 320–325
  solubility, 189–190, 324
  structure, polymorphism, 184–185
Aluminium salts, attack on concrete, 405
"Aluminosulphate", see $C_4A_3\bar{S}$
Aluminous cements, see calcium aluminate cements
Amines, as grinding aids, 352
Ammonium salts, attack on concrete, 405
Analytical electron microscopy
  general principles, 132–133
  of calcium silicate pastes, 132–133
  of pfa cement pastes, 296–297
  of Portland cement pastes, 211–212
  of rice husk ash–C_3S pastes, 311
  of slag cement pastes, 284
Andreason pipette, 97
Anhydrite
  data, equilibria, 52–56
  effect of admixtures of hydration, 359
  effect on cement hydration, 234
  in cement making or clinker, 74, 83, 89–91
  solubility, 188
  'soluble', see calcium sulphate phases
Aphthitalite
  data, equilibria, 52–56
  effect on cement hydration, 237
Aqueous solution equilibria, calculation, 323–324, 404–405
Aragonite, from sea water attack, 406

Arcanite, data, equilibria, 52–56
Ash deposition, and light microscopy, 104
Autoclaved materials, 365–371
  resistance to sulphate attack, 400
  very high strength, 375
Autogenous shrinkage, 270

Backscattered electron imaging (see also scanning electron microscopy)
  general principles, 105–106
Bacterial attack on concrete, 406–407
Belite (see also dicalcium silicate polymorphs), 15–23
  effect of pfa or pozzolanas on hydration, 293, 305
  formation in cement making, 60–61, 74, 75, 82, 86, 88
  hydration, see calcium silicate pastes, Portland cement pastes
  in clinkers
    cell parameters, 22
    compositions, 10–11, 21–22
    lamellar textures, 19–20
    optical properties, 22
    polymorphic types, 20
    X-ray powder patterns, 20, 450
  in low energy cements, 341–343
  modulated structures, 21
  natural occurrence, 23
  reactivities of different polymorphs, 1, 19, 122
Bentonite, in oilwell cementing, 373
BET method
  cement pastes, 259–260
  unhydrated cements, 98–99
Bingham model, 244
Blaine method, 98
Blastfurnace slag (see also slag cements; supersulphated cements)
  and alkali silica reaction, 395–396
  and sulphate attack, 401
  formation, treatment, composition, properties, 277–282
  mixtures with calcium aluminate cements, 332
  use in composite cements, 278–290
Bleeding, 246
Blended cement, definition, 276

Bogue calculation, 62–63, 114
  sources of error, 113, 118
Bond, cement-aggregate, see aggregate-cement paste interface
Borates, and Portland cement hydration, 361
Bottle hydration, 147
Bound water (chemically bound water)
  calcium silicate pastes, 130–131
  composite cement pastes, 289–301
  definition, 130
  Portland cement pastes, 206, 219
  effect of curing temperature, 364
Bredigite, 23, 49
Brownmillerite (see also ferrite (phase)), 32
Brucite
  crystal structure, data, 184
  formation by reactions of dolomitic aggregates, 407
  formation on hydration of Portland cement, 216
  in concrete attacked by ground or sea waters, 399, 406
Brunauer model (paste structure), 252
Brunauer–Emmett–Teller method, see BET method
Burnability, of cement raw mixes, 79–81
Burning conditions of clinker, and light microscopy, 104
Butyl phosphate, and air entrainment, 351

CA
  data, equilibria, structure, polymorphism, 35
  in calcium aluminate cements, hydration, 317–334
  superficially hydroxylated, 323
CAH_{10}
  crystal data, thermal behaviour, 183–184
  formation from calcium aluminate cements; equilibria, 320–326
CA_2
  data, equilibria, structure, 35
  in calcium aluminate cements, hydration, 318–319

CA_6, 36, 318
C_2A, 28
C_2AH_x phases (x = 5, 7.5 or 8; see also AFm phases)
  data, structures, thermal behaviour, 171, 172, 174–177
  equilibria, 189, 320–321, 323–324
  formation from aluminate phase, 194
  formation from calcium aluminate cements, 320 et seq.
  preparation, 192
C_2ASH_8 (strätlingite; see also AFm phases)
  data, structure, thermal behaviour, 171–177
  equilibria, 190–191
  formation from calcium aluminate cements, 328, 332
  formation from composite cements, 304, 311
  preparation, 304
C_3A and substituted modifications, see aluminate (phase)
C_3AH_6 (see also hydrogarnet phases)
  data, structure, solid solutions, 181
  equilibria, 189–190, 320–321, 323–324
  formation from aluminate phase, 194
  formation from calcium aluminate cements, 320, 326–328
  preparation, 193
$C_3A \cdot CaCl_2 \cdot H_6$ (see also AFm phases), 171–172
$C_3A \cdot CaCl_2 \cdot H_{10}$ (Friedel's salt; see also AFm phases)
  data, structure, solid solutions, thermal behaviour, 171–177
  in cement pastes, 386, 405
$C_3A \cdot 3CaCl_2 \cdot H_{30}$ (see also AFt phases), 180
$C_3A \cdot 3CaSO_4 \cdot H_{32}$, see AFt phases; ettringite
C_3A_2M, 49–51
$C_3A_3 \cdot CaF_2$, 58
$C_4A\bar{C}_{0.5}H_x$ phases (x = 6.5–12; see also AFm phases)
  data, structures, thermal behaviour, 169–176
  formation from Portland cement, 200

$C_4A\bar{C}H_x$ phases (x = 6–11; see also AFm phases)
　data, structures, thermal behaviour, 169–177
　equilibria, 324, 332
　formation from calcium aluminate cements, 328, 332
　formation from Portland cement, 200, 312, 381
C_4AF, see ferrite (phase)
$C_4(A,F)H_x$ phases (see also AFm phases), 175
C_4AH_x phases (x = 7–19; see also AFm phases)
　data, structures, thermal behaviour, 169–177
　equilibria, 189–190
　formation from aluminate or ferrite phase, 194–196
　formation from Portland cement, see under AFm phases
　preparation, 192
$C_4A\bar{S}H_x$ phases (see also AFm phases), 174
$C_4A\bar{S}H_x$ phases (x = 7–16; mono-sulphate; see also AFm phases)
　data, structures, 169–177
　equilibria, 190
　formation from aluminate or ferrite phase, 195–196
　formation from Portland cement, see under AFm phases
　preparation, 193
　solid solutions, 173, 175
C_4A_3, 38
$C_4A_3H_3$, 38, 177
$C_4A_3\bar{S}$ ("aluminosulphate")
　data, structure, preparation, equilibria, 52–56
　in cement making, 74–76
　in expansive cements, 336–339
　in low energy cements, 341–342
　in rapidly setting and hardening cements, 339–341
C_5A_3
　conditions of formation, structure, data, 37, 38
$C_6A\bar{C}_3H_{32}$ (see also AFt phases), 180
C_6AH_{36} (see also AFt phases), 180
$C_6AS_3H_{31}$ (see also AFt phases), 180

$C_6A\bar{S}_3H_{32}$, see AFt phases; ettringite
　formation from Portland cement, see under AFt phases
$C_6A\bar{S}_3H_{36}$ (see also AFt phases), 180
C_7A_5M, 49–51
$C_{11}A_7\cdot CaCl_2$, 36, 343
$C_{11}A_7\cdot CaF_2$
　in rapidly setting and hardening cements, 339–340
　structure, data, 36–37
$C_{11}A_7\cdot CaS$, 37
$C_{12}A_7$
　absence in Portland cement clinkers, 86
　data, structure, thermal behaviour, 36–37
　derived structures containing additional anions, 36–37
　in calcium aluminate cements, hydration, 318, 319, 322–325
　in cement making, 61, 74–75
$C_{12}A_7H$, 36, 176
CF, 41–42
CF_2, 41–42
C_2F (see also ferrite (phase)), 28, 41, 42, 49
C_3FH_6 (see also hydrogarnet phases), 181–183
$C_3F\cdot CaCl_2\cdot H_{10}$ (see also AFm phases), 175
$C_4F\bar{S}H_x$ phases (see also AFm phases), 175
CS, structures, polymorphism, data, 33
C_2S, see belite; dicalcium silicate polymorphs
C_2S pastes, see calcium silicate pastes
α-C_2SH (α-dicalcium silicate hydrate), 364, 367–369
C_3S, see alite; tricalcium silicate
C_3S pastes, see calcium silicate pastes
C_3S_2, structures, polymorphism, data, 33
$C_3S\bar{S}CH_{15}$, see AFt phases; thaumasite
$C_4S_2\cdot CaF_2$, structure, stability, 56–57
$C_5S_2\bar{S}$ ("silicosulphate")
　data, structure, preparation, equilibria, 52–56
　in cement making, 74–76
$C_{10}S_3\bar{S}_3H$ (hydroxyl-ellestadite), 74
$C_{10}S_4\cdot CaF_2$, structure, stability, compositional variation, 56–57

$C_{20}A_{13}M_3S_3$ ("Phase Q"), 49, 50
C–S–H (C–S–H gel)
  at paste–aggregate interface, 377–379
  carbonation, 124, 384–386
  decomposition on sulphate attack, 397–399
  definition, 123
  formation in autoclave, 367
  mechanism of growth, 348–349
  preparations with additional ions, 158, 213–215
  solubility curves, 153–156
  structural models, 142–152
  thermochemistry, thermodynamics, 157–158
  tobermorite and jennite type structures in, 150–152
C–S–H in calcium silicate pastes
  Ca/Si ratio, 128–130, 132–133
  density, 140–141
  infra red spectrum, 142
  kinetics and mechanism of formation, 159–166
  morphology, microstructure, 133–137
  silicate anion structure, 137–140
  thermogravimetry and DTA, 129
  water content, 130–131
  X-ray powder pattern, 140–141
C–S–H in composite cement pastes
  microsilica cements, 306, 308
  pfa cements, 296–301
  pozzolanic cements, 305
  rice husk ash cements, 311
  slag cements, 284–289
  supersulphated cements, 290
C–S–H in Portland cement pastes
  calorimetry, energetics of formation, 226, 230–232
  composition, 209–218, 220
  kinetics of formation, 228, 233, 241
  microstructure, 202–204, 221–225
  quantity formed, 218–219
  silicate anion structure, 212–213
  thermogravimetry, 207–208, 220–221
  X-ray pattern, DTA and infrared evidence, 199–202
C–S–H(di, poly), 146
C–S–H(I)
  equilibria, 153–156

C–S–H(I) (*cont.*)
  preparation, thermal properties, X-ray pattern, 142–143, 145–148
  relation to 1.4 nm tobermorite, silicate anion structure, 146–152
C–S–H(II), 142, 148, 155
Calcite, see calcium carbonate
Calcium aluminate cements, 316–334
  behaviour with aggressive agents, 316, 333–334
  hydration chemistry, 319–325
  manufacture, phase composition, 317–319
  setting, hardening, conversion, 325–330
  use in MDF cements, 375–376
  use in mixtures with Portland cement, etc., 330–333
  use in refractory concrete, 334
Calcium aluminate hydrates (see also formulae of individual compounds)
  preparative methods, 192–193
Calcium carbonate
  as addition to Portland cements, 312
  reaction with calcium aluminate cement paste, 328, 332
  reaction with Portland cement paste, 312, 381
  thermal decomposition, 64–65, 71–72
Calcium chloride
  and corrosion of reinforcement, 383, 387–388
  as accelerator for Portland cement, 357–361
  attack of concentrated solutions on concrete, 405
  in manufacture of alinite cements, 343–344
Calcium compounds (see also formulae or names of individual compounds)
  attack by atmospheric carbon dioxide, 124, 168, 200, 383–386
  attack by carbon dioxide solutions, 403–405
Calcium hydroxide (portlandite)
  at paste–aggregate interface, 377–379
  data, structure, solubility, thermal behaviour, 125–128

Calcium hydroxide (*cont.*)
  determination
    in calcium silicate pastes, 128–130
    in Portland cement pastes, 207–208
  formation
    from calcium silicates, 128, 132–133, 136–137, 140–141, 162–166
    from microsilica cements, 306–308
    from pfa cements, 293–294
    from Portland cements
      DTA, infrared and X-ray evidence, 199–202
      microstructure, 202–203, 209, 211, 214, 223, 242
      quantity formed, thermogravimetry, 207–208, 217–221
      kinetics of formation, 226, 233
    from pozzolanic cements, 304
    from rice husk ash cements, 311
    from slag cements, 282–289
Calcium langbeinite
  effect on cement hydration, 235
  structure, data, equilibria, 52–56
Calcium magnesium aluminates, 48–49
Calcium magnesium silicates, 48–49
Calcium oxide (see also free lime), structure, data, 33
Calcium potassium sulphate, see calcium langbeinite
  hydrated, see syngenite
Calcium silicate bricks, 366
Calcium silicate carbonates (see also spurrite), 58
Calcium silicate chlorides, 344
Calcium silicate fluorides, 56–58
Calcium silicate hydrates, formed hydrothermally, 367–371, 374
Calcium silicate pastes (see also C–S–H), 123–166
  action of accelerators, 358–362
  calorimetric curves, 160, 166
  early hydration reactions and products, 162–166
  hydration kinetics and mechanisms, 159–166
  microstructures, 133–137
  phase compositions, 123, 128–130
  thermogravimetry and DTA, 129
  X-ray diffraction patterns, 140

Calcium sulphate phases (see also anhydrite; gypsum; hemihydrate)
  attack on concrete, 399
  effects on cement hydration, 231–237
  effects on slag hydration, 290
  γ-$CaSO_4$ ('soluble anhydrite')
    data, 186–187
    effect of heating, 54
    effect on cement hydration, 234
    formation during clinker grinding, 92
Calorimetry
  aluminate phase, 194
  at low temperatures, 265, 403
  calcium aluminate cements, 319
  calcium silicates, 160
  cement or calcium silicates with accelerators, 358
  Portland cement, 226–227
Capillaries, and movement of water during freezing, 403
Capillary pores, 247–251, 254–256
  and modulus of elasticity, 270
  and permeability, 273–274
  and strength, 266, 268
Capillary stress, 271–272
Carbon dioxide, aggressive, 404–405
Carbonates, and Portland cement hydration, 312, 361–362
Carbonation
  avoidance in small samples, 124
  of cement paste in concrete, 383–386
  of hydrated aluminate phases, 168, 180, 185, 200
Carbonation shrinkage, 386
Cement based materials, very high strength, 374–376
Cement(s)
  alinite, 343–344
  autoclaved, 365–371
  calcium aluminate, 316–334
  composite, 276–315
  expansive, 335–339
  high in belite and $C_4A_3\bar{S}$, 341–343
  hydraulic, xv
  jet, 339–341
  latent hydraulic, 277
  low energy, 341–344
  microsilica, 305–308

462 *Index*

Cement(s) (*cont.*)
  oilwell, 371–374
  pfa (Class F fly ash), 290–301
  pozzolanic, 302–305
  rapid setting, 316, 333, 339–341, 361–362
  regulated set, 339–341
  self stressing, 335–336
  shrinkage compensated, 335–336
  slag, 277–290
  supersulphated, 290
  very high strength, 374–376
  with Class C fly ash, 308–311
  with limestone or other fillers, 312
  with rice husk ash or other wastes, 311
Cenospheres, 291, 293
Chemical attack on concrete, 405
Chemical shrinkage, 238, 250
Chemically bound water, see bound water
Chloride ion
  and corrosion, 383–384, 386–388
  as accelerator or retarder, 357–361, 372–373
  in cement making, 74, 76, 94, 343–344
  in cement paste or concrete, 360–361, 386–387
Chromium substitution in clinker phases, 121
Ciment Fondu (see also calcium aluminate cements), 316–334
Clay minerals, reactions in cement making, 60–61, 64, 73–74
Clays
  as raw materials for cement manufacture, 65–66
  heated, as pozzolanic materials, 302
Clinker formation in pure systems
  $CaO-Al_2O_3-Fe_2O_3-SiO_2$, 45–48
  $CaO-Al_2O_3-SiO_2$, 39–41
Clinker, Portland cement, see Portland cement clinker
Cohesion point, 223, 233
Composite cements, 276–315
  and alkali silica reaction, 394–396
  definition, 276
  hydration at temperatures up to 100°C, 363

Composite cements (*cont.*)
  in concrete, 381–382
  pore structure and physical properties of pastes, 312–315
Concrete
  cement paste in, 377–383
  durability, 383–408
  refractory, 316, 334
  structural variations near exposed surfaces, 382–383
Condensed silica fume, see microsilica
Conductivity, electrical
  calcium aluminate cement pastes, 319, 321
  Portland cement pastes, 228
Connectivity, of silicate tetrahedra, 139
Consistometer, high temperature, high pressure, 372
Conversion, in calcium aluminate cement pastes, 320, 328–330
Coolers, in Portland cement clinker making, 70
Cooling of Portland cement clinker
  reactions during, 84–91
  studied by light microscopy, 104
Corrosion of concrete reinforcement, 383–384, 387–388
Creep, 272–273
Cristobalite, 61, 73, 75
Curing, definition, 123
Cuspidine, 58

d'Arcy's law, 273
D-drying, definition, 131
Deceleratory period, 159–160, 226
Defects, and reactivities of clinker phases, 121–122
Deformation of cement pastes, 269–273, 314
Degree of hydration, ultimate, at elevated temperatures, 363–364
De-icing salts, and damage to concrete, 403, 405
Densities (see also under individual phases)
  Portland cement phases and hydration products, 220
Deposits, in cement kilns or preheaters, 76–77, 81

Design life, of concrete, 383
Despujolsite, 180
Diatomaceous earth, 302
Dicalcium ferrite (see also ferrite
 (phase)), 28–29, 41–42, 48–50
α-Dicalcium silicate hydrate, 364–365,
 367–369
Dicalcium silicate pastes, see calcium
 silicate pastes
Dicalcium silicate polymorphs (see also
 belite)
 data, structures, thermal behaviour,
  15–23
 dusting, 16
 enthalpy changes on formation, 64–65
 enthalpy changes on hydration, 157,
  232
 equilibria, 15, 33–34, 39, 43–50, 57,
  95
 in calcium aluminate cements, 318
 modulated structures, 21
 preparation, 59
 stabilization of β-polymorph, 16, 21
 thermal behaviour, 15, 22
Dielectric constants of cement pastes,
 228, 319
Differential thermal analysis
 AFm phases, 177
 AFt phases, 181
 CAH_{10}, 184
 C–S–H(I) and similar materials, 147
 calcium hydroxide, 129
 hydrogarnet phases, 183
 Portland cement clinker phases, 15,
  22–23
 Portland cement pastes, 200–201, 364
Diffusion
 in Portland cement pastes, 274–275
 of chloride ion in concretes, 386–387
Diopside, 49
Disjoining pressure, 253, 271
Dolomite
 aggregates, and expansion in
  concrete, 407–408
 in cement making, 72, 103
Dreierkette, 144–145, 149
Dry process, 66–70
Drying of C–S–H, effect on silicate
 anion structure, 140, 147

DSP materials, 374–375
Duplex films, 377–379, 382
Durability of concrete, general aspects,
 383
Dust, formation in cement kiln, 70, 81

Early product
 calcium silicate pastes or suspensions,
  136–137, 154–157, 162–166
 Portland cement pastes, 221–223, 228
Elasticity, modulus of, 269–270, 314
Electron diffraction
 alite, 13
 C_3S surface structures, 162
 C–S–H(I), hydrothermally prepared,
  370
 C–S–H(II), 148
 calcium silicate pastes, 140, 151
 pfa cement pastes, 296–297
Electron probe microanalysis, see X-ray
 microanalysis
Electron spectroscopy for chemical
 analysis (ESCA)
 aluminate phase, 194
 calcium silicate pastes, 162, 166
 ferrite phase, 196
Ellestadite, hydroxyl, 74
Enstatite, 49
Enthalpy changes, see thermochemistry
Equilibria (see also under system and
 phases concerned)
 in aqueous solutions, method of
  calculation, 323–324, 405
Etching, of clinker for microscopy, 101
Ethylene diamine tetraacetic acid, effect
 on cement hydration, 347–348
Ettringite (see also AFt phases)
 decomposition by magnesium ion,
  399
 effect of high temperatures, 181, 191,
  364
 from monosulphate–calcite reaction,
  407
 in concrete attacked by sulphates,
  397–398, 400
 in expansive cements, 335–339
 in Portland cement pastes, see under
  AFt phases
 in rapidly hardening cements, 339–341

464 Index

Ettringite (*cont.*)
  in steam cured products, 402
  in supersulphated cements, 290
  morphology, 178, 195, 221–224, 338, 400
  preparation, 193
  quantitative determination, 181, 208
Evaporable water, 247
Expansion (see also alkali silica reaction, sulphate attack, expansive cements)
  from excessive gypsum content in cement, 235
  in concrete containing dolomite aggregates, 407
Expansive cements, 335–339
Extraction methods
  for calcium silicate or cement pastes, 128
  for unreacted Portland cement or clinker, 111–112

Failure, mechanisms of, 268–269
False set, 233–234
Fatty acid salts, as air entraining agents, 351
Fe_2O_3 (oxide component), in Portland cement hydration products, 215
Feldman–Sereda model, 252–253
Felspars, reaction with calcium hydroxide and water, 408
Feret's law, 265
Ferrite (phase)
  colour, 32
  compositions in clinkers, 10–11, 30–31
  data, structure, 28–32
  effect of gypsum on hydration kinetics, 233
  electrical and magnetic properties, 32
  enthalpy changes on formation, 64
  enthalpy changes on hydration, 197–198
  equilibria, 41–50
  factors affecting reactivity, 121–122
  formation in cement making, 60–61, 74, 84–89
  hydration
    in Portland cement pastes, 215–220, 225–226, 233

Ferrite (phase) (*cont.*)
  in pure systems, 196–197
  ionic substitutions, 28–31
  present in calcium aluminate cements, 318–319
  X-ray powder patterns, 27, 32, 453
  zoning, 43
Fine powders, effect on cement hydration kinetics, 293, 312
Fire damage, 408
Flash set, 233
Floc structure, 245–246
  effect of water reducers and superplasticizers, 354–357
Fluoride phases in Portland cement clinker formation, 56–58, 76–77, 94–95
Fluorosilicates, as fluxes in cement making, 58
Fluxes, in Portland cement making, 92–93, 341
Fly ash
  Class C, 308–311
  Class F, see pulverized fuel ash
Foshagite, 370
Fracture mechanics, 269
Free lime (calcium oxide)
  data, structure, 33
  determination, 110, 112
  in cement making, 60–61, 74–75, 79–83
  in expansive cements, 337
  microscopic observation in clinker, 103
Free water porosity, 219, 255
  comparison with mercury porosities, 262–263
  in pastes of composite cements, 289, 301, 314
Freeze-thaw damage, 351, 402–403
Fresh pastes
  calcium aluminate cements, 325
  Portland cement, 243–246
    effects of water reducers and superplasticizers, 352–357
Friedel's salt, see $C_3A.CaCl_2.H_{10}$
Fuels, for cement making, 66

Gaize, 302

Garnet, 39, 181–183
Gases, diffusion in Portland cement pastes, 274–275
Gehlenite, 38–39, 50–51, 318
Gehlenite hydrate, see C_2ASH_8
Gel
  cement, 213–216, 247
  formed in early period of hydration, 221–223, 228, 245–246
  formed in alkali silica reaction, 388–394
Gel pores, 247–255, 260, 262
Gel/space ratio, 266
Gel water, 131, 247–252
Geothermal well cementing, 372, 374
Gibbsite (see also aluminium hydroxide), 184, 324, 326
Glaserite, 16, 54
Glass
  in Class C fly ash, 308–311
  in ground granulated blastfurnace slag, 277–282
  in microsilica, 305–306
  in natural pozzolanas, 299, 302
  in Portland cement clinkers, 85–86
  in pulverized fuel ash (Class F fly ash), 291–292
  in white cements, 86
Glass fibre reinforced cements, deterioration in moist environments, 408
Granulated blastfurnace slag (see blastfurnace slag, slag cements; supersulphated cements)
Grinding aids, 352
Grossular, 39, 181–183
Ground granulated blastfurnace slag, see blastfurnace slag; slag cements; supersulphated cements
Guniting, 362
Gypsum
  attack on concrete, 399
  behaviour on grinding and storage of cement, 92
  crystal and other data, 186–187
  effects on cement hydration, 231
  equilibria, 188–192, 228
  formation on sulphate attack, 397–398
  optimum, 234–236
  reactions with calcium aluminate cements, 333

Gypsum (cont.)
  secondary, 233
Gyrolite, 369, 374

Hadley grains, 225
Hardening
  calcium aluminate cements, 326–330
  definition, relation to hydration reactions 123
Hatrurite, 15
Heat evolution (see also calorimetry)
  calcium aluminate cements, 316
  modelling, 241–242
  Portland cements, 3, 230–231
Heavy metals, and Portland cement making, 77, 95
Hemihydrate
  effect of admixtures on hydration, 359
  effects on cement hydration, 234
  formation during clinker grinding, 92, 233, 236
  solubility, structure, data, 186–188
Hibschite, 181
High alumina cements, see calcium aluminate cements
High temperature phases, laboratory preparation, 58–59
High temperatures
  effect on Portland cement hydration, 362–365
  in autoclave, 365–371
  in oilwell cementing, 373–374
Hillebrandite, 369
Hot pressed cements, 375
Humidity, effect on cement hydration kinetics, 238
Hydration
  definition, 123
  reactions and products, see under relevant type of cement
  termination of, 124–125
Hydraulic cements, xv
  latent, definition, 276
Hydrocalumite, 175
Hydrogarnet phases (see also C_3AH_6)
  compositions, data, structures, 181–183

Hydrogarnet phases (*cont.*)
    formation
        from aluminate or ferrite phase, 194, 197–198
        from pfa cements, 296–297
        from Portland cements, 200–201, 204, 215–216
        from pozzolanic cements, 304
        in autoclave processes, 369
Hydrogrossular, 181
Hydrotalcite-type phases
    compositions, data, structures, 185
    formation
        from Portland cement, 201, 204, 216
        from slag cements, 282, 284–285
        through sulphate attack, 400
Hydroxyl-ellestadite, 74

Ice crystal formation, and freeze thaw damage, 402–403
Image analysis
    aggregate-cement paste interface, 379–380
    calcium hydroxide in pastes, 128
    Portland cement or clinker, 108
Imbibition
    in alkali silica reaction, 393–394
    in expansive cements, 338–339
Induction period
    calcium aluminate cements, 321–322, 325
    calcium silicate pastes, 159–164, 166
    Portland cement pastes, 226
Infra red absorption spectra
    AFm phases, 177
    AFt phases, 181
    calcium aluminate cements, 319
    calcium silicate pastes, 142
    hydrogarnet phases, 183
    Portland cement pastes, 201–202
    unhydrated Portland cement, 113, 201
Inner product
    definition, calcium silicate pastes, 133
    Portland cement pastes, 203, 222, 224–225
Inorganic salts, effect of
    on calcium aluminate cement hydration, 330–331

Inorganic salts (*cont.*)
    on calcium silicate or Portland cement hydration, 357–362
    on hemihydrate or anhydrite hydration, 359
Interfacial zone, cement paste and aggregate, 377–383
Interground cements, definition, 276
Interstitial material in Portland cement clinkers
    composition, 86–87
    effects of cooling rate, 87–88
Ionic substitutions, and reactivities of clinker phases, 121–122
Ions, diffusion in Portland cement pastes, 274–275
Iron(II) oxide, see wüstite
Iron(III) oxide or hydroxide, as product from ferrite phase, 196–197, 225

Jaffeite, 369
Jasmundite, 343
Jennite, data, structure, 142–145

$KC_{23}S_{12}$, non-existence of, 21
Katoite, 181
Kilchoanite, 34
Kiln rings and deposits, 76–77
Kinetics of hydration
    of calcium silicates, 159–166
    of Portland cement, 237–241
KOSH reagent (KOH and sucrose), 111–112

Larnite, 23
Laser granulometry, 97–98
Late product
    in calcium silicate pastes, 136–137
    in Portland cement pastes, 203, 224–225
Latent hydraulic cements, definition, 276
Lea and Nurse method, 98
Leaching, of concrete by water, 403–405
Lead salts, and Portland cement hydration, 361

Lepol process, 71
Light microscopy
  calcium silicate pastes, 133
  Portland cement clinker, 101–105
  Portland cement pastes, 203
Lignosulphonates
  as retarders, 349–350
  as water reducers, 352–353
Lime, free, see free lime
Lime saturation factor, 47, 61–62, 80–81, 103
Limestone
  as addition to Portland cement, 312
  as raw material for making cement, 65
Linear kinetics, 240
Liquid in formation of Portland cement clinker
  composition, 87
  physical properties, 81, 92–93
  quantity, 77–79
  solidification, 84–91
Lithium salts, and calcium aluminate cements, 331
Litre weight, 96
Low temperatures, effects on Portland cement hydration, 365
Lurgi process, residues from, 311

Macropores, definition, 253
Magnesium aluminium hydroxide hydrate and related phases, 185
Magnesium chloride solutions, concentrated, attack on concrete, 405
Magnesium compounds, anhydrous, of cement chemical interest, 49
Magnesium hydroxide, see brucite
Magnesium oxide, see periclase
Magnesium salts, and Portland cement hydration, 361
Magnesium silicate, hydrated, from sulphate attack, 400
Magnesium sulphate solutions, attack on concrete, 397, 399–400
Map cracking, 388
Mass balance
  clinkers, 116–117
  pfa cement pastes, 300

Mass balance (cont.)
  Portland cement pastes, 216–221
  during early hydration, 228
  slag cement pastes, 288
Maturity functions, 362
Mayenite, 38
MDF (Macro Defect Free) cements, 375–376
Meixnerite, data, structure, 185
Melamine formaldehyde sulphonate polymers, 353–354
Melilite structure, 38, 49
Mercury intrusion porosimetry
  composite cement pastes, 313–314
  general, Portland cement pastes, 261–263
Mercury porosity, 262–263, 313–314
Merwinite, 49
Mesopores, definition, 253
Metajennite, 145, 152
Metal cations, attack on concrete, 405
Methanol, reactions with C_3S or cement pastes, 124–125
MgO (as oxide component; see also periclase)
  in compounds, see under names
  in Portland cement, behaviour on hydration, 215–216
  in Portland cement making, 78–79, 87–89, 93–94
Micropores, definition, 253
Microscopy, see light microscopy; scanning electron microscopy; transmission electron microscopy
Microsilica (condensed silica fume)
  formation, composition, physical properties, 305–306
  in very high strength materials, 374–375
Microsilica cements
  cement-aggregate bond, 381
  concrete properties, 305–306
  hydration chemistry, 306–308
  pore structures, 313–314
Microstructure
  calcium aluminate cement pastes, 326–327
  calcium silicate pastes, 133–137
  concrete near exposed surfaces, 382–383

Microstructure (*cont.*)
  development in cement pastes, modelling, 241–242
  paste–aggregate interfaces, 377–379
  paste–glass interfaces, 377–378, 408
  paste–metal interfaces, 382
  pfa cement pastes, 296
  Portland cement clinker, 101–108
  Portland cement (ground), 107–108
  Portland cement pastes, 202–204, 221–225
    after carbonation, 385
    after sulphate attack, 397
  slag cement pastes, 282
Middle product
  calcium silicate pastes, 136–137
  Portland cement pastes, 223–224
Mineral additions, definition, 276
Mineralizers, in Portland cement making, 93
Modelling, mathematical, of Portland cement hydration, 241–242
Modulus of elasticity
  composite cement pastes, 314
  Portland cement pastes, 269–270
Moler, 302
Molybdate method (see also silicate anion structure), 137
Monosulphate (monosulphoaluminate), see $C_4A\bar{S}H_x$ phases
  as hydration product of Portland cement, see under AFm phases
Monticellite, 49
Mortars, failure through thaumasite formation, 401–402
Mössbauer spectra
  C–S–H preparations containing Fe^{3+}, 214
  calcium aluminate cements, 319
  ferrite phase hydration, 196–197
Munich model, 253

NC_8A_3, non-existence of, 25
Nagelschmidtite, 23
Naphthalene formaldehyde sulphonate polymers, 353–354
Neutron diffraction
  calcium aluminate hydration, 319
  porosity determination, 264

Newton's law (rheology), 243–244
Nitrogen sorption isotherms, 258–261
Nodulization, in clinker formation, 81
Nomenclature, cement chemical, 4
Non evaporable water
  and Powers–Brownyard model, 247–251
  definition, 131
  in calcium silicate pastes, 131
  in cement pastes with Class C fly ash, 310
  in microsilica cement pastes, 306–307
  in pfa cement pastes, 299, 301
  in Portland cement pastes, 206, 219, 247–251
  in slag cement pastes, 287, 289
  in supersulphated cement pastes, 290
  specific volume, 250
Nuclear magnetic resonance, ^{27}Al
  alite hydration products, aluminous C–S–H, 214
  aluminous tobermorite, 369
  CAH_{10}, 184
  C_2AH_8, 174
  calcium aluminate hydration, 319
Nuclear magnetic resonance, proton, 264
Nuclear magnetic resonance, ^{29}Si (see also silicate anion structure)
  calcium silicate pastes, etc., 137, 146–147, 162, 360
  Portland cement pastes, 213
  tobermorite, 143, 370

Oilwell cementing, 371–374
Okenite, and alkali silica reaction, 393
Ono's method (light microscopy), 105
Opal, as aggregate, 390
Organic compounds, attack on concrete, 405
Outer product
  calcium silicate pastes, 133
  definition, Portland cement pastes, 204, 222
Oxalic acid, effect on C_3S hydration, 350
Oyelite, 152

P_2O_5 (oxide component) in Portland cement making, 94–95
Parabolic kinetics, 240
Parawollastonite, 33
Particle size distribution of cement, 96–99
  effect on hydration kinetics, 99–100, 239–241
Passivation, of steel, 387
Paste (see also under starting material, e.g. calcium silicate pastes, Portland cement pastes)
  definition, 123
Pelletized blastfurnace slag (see also blastfurnace slag; slag cements; supersulphated cements), 277
Periclase (magnesium oxide)
  data, structure, equilibria, 48–51
  in expansive cements, 337
  in Portland cement clinkers, 2, 88, 94, 103
Permeability
  composite cement pastes, 313–315
  modelling for cement pastes, 241–242
  Portland cement pastes, 273–274
Pessimum composition, 390, 394
Pfa, see pulverized fuel ash
Phase composition, potential, 63
Phase equilibria, high temperature, 33–58, 95, 318
Phase Q ($C_{20}A_{13}M_3S_3$ approx.; see also pleochroite), 49–51
Phase T ($C_{1.7}M_{0.3}S$ approx.), 49
Phase X (hydration product of Portland cement), 201, 364
Phenolphthalein test, for carbonation, 384–386
Phosphates
  and Portland cement hydration, 361
  and Portland cement making, 94–95
Physical attack on concrete, 402–403
Pipette, Andreason, 97
Pleochroite, 51, 318
Plerospheres, 291
Plombierite, 142, 152
Point counting, 103
Polymer concrete, 375
Pore size distributions
  clinkers, 96
  composite cement pastes, 312–314

Pore size distributions (cont.)
  Portland cement pastes, 260–265
Pore solutions
  and alkali silica reaction, 389, 393, 395–396
  C_3S pastes, 156
  hydrating calcium aluminate cements, 320–325
  Portland cement pastes, 227–230
Pore structures
  composite cement pastes, 312–315
  and sulphate attack, 401
  Portland cement pastes, 254–265
  effect of curing temperature, 363, 365
Pores and porosities, definitions
  capillary, gel, 247
  free water, 256
  mercury, 262
  micro, meso, macro, 253
  total, total water, 255
Porosity
  calculated, 219, 221, 254–256, 289, 301, 314
  experimentally determined, 255–258, 261–265, 312–314
  of 'cement gel' (Powers–Brownyard), 247, 251
Porsal cement, 342
Portland cement (see also Portland cement clinker)
  chemical analysis, 100–101
  constituent phases, 1–2
  depth of reaction as function of time, 99–100
  high early strength, 3
  hydration, 199–242
    at high temperatures, 362–365, 373–374
    at low temperatures, 365, 374
    in autoclave processes, 365–371
  infra red and Raman spectroscopy, 113
  low heat, 3
  mixtures with calcium aluminate cements, 332–333
  moderate heat of hardening, 3
  oilwell, 372
  particle size distribution, 96–99
  phase compositions of particle size fractions, 99

Portland cement (*cont.*)
  phase distributions within grains, 106–108
  rapid hardening, 3
  reactions during storage, 71, 91–92
  specific surface area, 98–99
  specifications, 2
  sulphate resisting, 3, 10–11, 31, 400
  thermogravimetry, 112–113
  types of, 2
  white, 3, 10–11, 27, 83, 84, 86, 87
  X-ray powder diffraction patterns, 108–111
Portland cement clinker
  calculation of phase composition, 62–63, 113–119
  colour, 32, 96
  density, 96
  effects of minor components, 92–95
  general description, 96
  light microscopy, 101–105
  manufacture, 65–71
    effect of reducing conditions, 84, 104
    enthalpy changes in formation, 63–65
    grinding, 70–71, 91–92
    raw materials, 65–66
    reactions during formation, 60–61, 71–92
  physical properties, pore structure, 96
  quantitative phase determination, 113–119
  reactivities of individual phases, 119–122
  scanning electron microscopy, 105–108
  typical compositions of phases, 10–11, 114
Portland cement pastes
  calorimetry, 226–227
  compositions of constituent phases, 209–219
  determination of unreacted clinker phases, 204–206
  effects of admixtures, 345–362
  enthalpy changes during hydration, 230–232
  fresh, 243–246
  kinetics of hydration, 231, 235–241

Portland cement pastes (*cont.*)
  microstructure, 202–204, 221–225
  models of paste structure, 246–254
  phase compositions, 199–204, 207–209, 216–221
  pore structure, 254–265
    relations to physical properties, 265–275
  setting, 231–234
  volume percentages of phases, 219, 221
Portlandite, see calcium hydroxide
Potassium aluminate, in cement clinkers, 54
Potassium calcium sulphate, see calcium langbeinite
  hydrated, see syngenite
Potassium iron sulphide, in reduced clinkers, 84
Potassium sodium sulphate, see aphthitalite
Potassium sulphate (arcanite), data, equilibria, 52–56
Potassium sulphide, in reduced clinkers, 84
Powers–Brownyard model, 246–251
Pozzolanas and pozzolanic cements, 299–305
  resistance to sea water, 406
Pozzolanic materials, definition, 276
Pozzolanic reaction, 298
  in autoclave processes, 367–371
  similarity to alkali silica reaction, 392–393
Precalciners, for cement making, 67–69
Precast concrete products, thaumasite formation in, 402
Preheaters, for cement making, 67–68
Preparation in laboratory
  calcium aluminate hydrates, etc., 192–193
  high temperature phases, 58–59
  hydrated phases in general, 124–125
Primary air, 69
Primary flue gas desulphurization, residues from, 311
Protected phases, 43
Proto C_3A, 26, 86
Pseudowollastonite, 33

Pulverized fuel ash (pfa; fly ash)
 and alkali silica reaction, 395–396
 Class C, 291, 308–310
 Class F, 290–292
 reactivity, 294–295
Pulverized fuel ash cements, 292–301
 pore structure and physical properties of pastes, 312–315
Pyknometry, 257–258
Pyroaurite, 185

Quantitative X-ray diffraction analysis
 AFt phases, 181
 calcium silicate pastes, 128–129
 cement pastes, 204–206
 Portland cement clinker, 108–111
Quartz
 reactions in cement making, 60–61, 74–75, 80–81
 thermal changes, 73
Quick set, 233

Raman spectroscopy, of unhydrated cements, 113
Rankinite, 33
Rate equations, in cement hydration kinetics, 164, 239–241
Raw feed preparation, in cement making, 67, 103
Reactivities of phases
 calcium aluminate cements, 319
 Portland cements, 119–122
Reducing conditions, in making Portland cement clinker, 84, 96, 104
Reference intensity ratio, 110
Renderings, and thaumasite formation, 402
Retarders, 345–350
 calcium aluminate cements, 330–331
 oilwell cementing, 372–373
Rheology, 243–246
 calcium aluminate cements, 325
 effects of water reducers and superplasticizers, 352–357
Rice husk ash, 311
Rosin–Rammler function, 97–99
Rotary kiln, 67, 69–71

Salt scaling, 403
SAM reagent (salicylic acid in methanol), 111
Sampling of clinker, 101
Santorin Earth, 299, 303
Saturation factor (see also lime saturation factor), 228
Scanning electron microscopy
 calcium aluminate cement pastes, 326–327
 calcium silicate pastes, 133–137
 concrete
  at aggregate–cement paste interface, 377–383
  attacked by alkali silica reaction, 389
  attacked by carbon dioxide, 385
  attacked by sulphates, 397
  near exposed surfaces, 382–383
 fresh pastes, 246
 general considerations, 105–106, 136
 porosity determination, 263–264
 Portland cement or clinker, 105–108
 Portland cement pastes, 202–204, 221–225
Sea water attack
 calcium aluminate cements, 333
 concrete, 406
Secondary air, 69
Secondary ion mass spectroscopy (SIMS), 162
Sedigraph, X-ray, 97–99
Self desiccation, 250
Separation of phases in Portland cement, 111–112
Setting
 calcium aluminate cements, 320, 325
 definition, calcium silicate pastes, 123
 Portland cement pastes, 231–234
Sewage, attack on concrete, 406–407
Shales, as raw materials for cement making, 65–66
Shotcreting, 362
Shrinkage
 carbonation, 386
 drying, in Portland cement pastes, 270–272
Sieve analysis, 97
Silica fume, see microsilica

Silica ratio (silica modulus), 61, 80–81, 103
Silicate anion structure
  calcium silicate pastes, 137–140, 214
  with admixtures, 360
  C–S–H(I), etc., 146–148, 214
  pfa-C_3S pastes, 297–298
  Portland cement pastes, 212–213
  pozzolanic cement pastes, 305
  1.1 nm tobermorite, 370
  1.4 nm tobermorite and jennite, 143–145
Silicate anions, species in solution, 156–157
'Silicate garden' mechanism, 223
Siliceous aggregates, potentially reactive, 390
Silicocarnotite, 54
"Silicosulphate" ($C_5S_2\bar{S}$)
  data, equilibria, 52–56
  in cement making, 74–76
Sjögrenite, 185
Slag cments (see also blastfurnace slag)
  hydration chemistry, microstructure, 282–290
  pore structures and physical properties, 312–315
  sea water attack, 406
  sulphate attack, 401
Slump, 243
Slump loss, 353, 354
SO_3 (as oxide component; see also individual phases)
  in clinker formation, 55–56, 94
  in Portland cement hydration products, 216
Sodium chloride, in oilwell cementing, 372–373
Sodium sulphate, data, equilibria, 52–55
Sodium sulphate solutions, attack on concrete, 396–401
Soft water, attack on concrete, 403–405
Solubility curves
  CaO–Al_2O_3–H_2O and related systems, 189–192, 321–324
  CaO–SiO_2–H_2O system, 153–156
  $CaSO_4$–H_2O system, 188
Soluble salts, and damage to concrete, 403

Sorption isotherms, 247, 258–261
Specific surface areas,
  calcium silicate and cement pastes, 257, 259–260, 264
  cements, 98–99
  effect on hydration kinetics, 238
Spinel, 49
Spurrite
  data, structure, equilibria, 58
  formation in cement kiln, 74, 76
Steam curing
  high pressure, 365–371
  low pressure, 362–365
Steel
  corrosion, 383–384, 387–388
  interface with paste, 382
Stoichiometry of hydration
  pfa cement pastes, 298–301
  Portland cement pastes, 216–221
  slag cement pastes, 286–289
Strätlingite, see C_2ASH_8
Strength development
  and pore structure, 265–268, 314
  calcium aluminate cements, 326–330
  microsilica cements, 306
  modelling, 241–242
  pfa cements, 292–293
  Portland cements, 234–237, 265–269
  slag cements, 278–279
Strength retrogression, 373
Strontium compounds, and cement making, 94
Sucrose and other organic compounds, as retarders, 345–350
Sulphate attack
  autoclaved materials, 400
  calcium aluminate cements, 316, 333
  composite cements, 401
  expansive cements, 336
  internal, 399
  Portland cements, 396–400
  supersulphated cements, 400
Sulphate phases in Portland cement clinker
  effects on hydration, 237
  estimation from bulk analysis, 114–115
  formation, 74, 76, 89–91, 94
  light microscopy, 103
Sulphate resisting Portland cements, 3

Index 473

Sulphate resisting Portland cements (*cont.*)
  ferrite phase compositions, 10, 11, 31
  mechanism of resistance, 400
Sulphide ion
  attack on concrete, 407
  in reduced clinkers, 84
  in slag and slag cements, 278, 281, 284
"Sulphospurrite" ($C_5S_2\bar{S}$), see silico-sulphate
Sulphur compounds, organic, attack on concrete, 406
Sulphur impregnated cement materials, 375, 376
Sulphuric acid, attack on concrete, 405–407
Superplasticizers, 353–357
Supersulphated cements, 290
  resistance to sulphate attack, 400
Surkhi, 302
Syngenite
  data, equilibria, 186–188, 228
  detection by thermal analysis, 112
  formation and effects on cement hydration, 228, 233, 237
  formation during cement storage, 92, 93
System(s)
  $CaO–Al_2O_3$, 34–36
  $CaO–Al_2O_3–Fe_2O_3$, 41–43
  $CaO–Al_2O_3–Fe_2O_3–SiO_2$, 43–45, 318
    under mildly reducing conditions, 318
  $CaO–Al_2O_3–SiO_2$, 38–41
  $CaO–C_2S–C_4AF$, 43–44
  $CaO–C_2S–C_{12}A_7–C_2F–MgO$, 48–50
  $CaO–C_2S–C_{12}A_7–C_4AF$, 43–48
  $CaO–C_2S–C_3P$, 95
  $CaO–Fe_2O_3$, 41–42
  $CaO–MgO–Al_2O_3–SiO_2$, 48, 51
  $CaO–SiO_2$, 33–34
  containing alkalis, 55
  containing CaF_2, 57
  containing CO_2 or $CaCO_3$, 58
  containing MgO, 48
  containing P_2O_5, 95
  containing SO_3 or sulphates, 55–56
System(s), hydrated
  $CaO–Al_2O_3–H_2O$, 189–190, 320–324
    plus Na_2O, 331

System(s), hydrated (*cont.*)
  $CaO–Al_2O_3–SiO_2–H_2O$, 190–191
  $CaO–Al_2O_3–SO_3–H_2O$, 190–192
  $CaO–SiO_2–H_2O$, 153–156
    at high temperatures and pressures, 369
    plus Na_2O or K_2O, 158
  $CaSO_4–H_2O$ and related, 188–189

Tacharanite, 152
Temperature, effect on cement hydration kinetics, 239
Tertiary air, 70
Tetracalcium aluminoferrite, see ferrite (phase)
Tetrahedra, paired and bridging, 149
Thaumasite
  data, structure, thermal behaviour, 180, 181
  from sulphate attack with carbonation, 401–402
  preparation, 193
Thenardite, data, equilibria, 52–55
Thermal insulation materials, autoclaved, 366
Thermochemistry
  aluminate and ferrite phase hydration, 197–198
  calcium hydroxide dissolution and dehydroxylation, 127
  calcium silicate hydration, 157
  clinker formation, 63–65
  Portland cement hydration, 230–232
Thermodynamics
  $CaO–Al_2O_3–H_2O$ system, 323–324
  $CaO–SiO_2–H_2O$ system, 157–158
  $CaSO_4–CH–K_2SO_4–H_2O$ system, 188–189
Thermogravimetry
  AFm phases, 176
  AFt phases, 181
  CAH_{10}, 184
  calcium hydroxide, 127–128
  calcium silicate pastes, 129
  data for cement hydration products, 220
  hydrogarnet phases, 183
  Portland cement pastes, 207–208
  unhydrated cements, 112

Thickening time, 372
Tilleyite, 58
Tobermorite, 1.1 nm, 143, 367–371, 374
Tobermorite, 1.4 nm, 142–145
"Tobermorite gel", 149
Total (water) porosity, 255
Transition elements, and Portland cement making, 95
Transmission electron microscopy
  alite, 13
  C–S–H preparations, 146, 148
  calcium silicate pastes, 136
  pfa cement pastes, 296–297
  Portland cement pastes, 204, 221, 223
Tricalcium aluminate and substituted modifications, see aluminate (phase)
Tricalcium silicate (see also alite)
  enthalpy change on formation, 64–65
  enthalpy change on hydration, 157
  equilibria, 33–34, 38–41, 43–50, 56–57, 95
  hydration, 123–165
  preparation, 59
  surface composition, 162
  thermal stability, 33–34, 88–89
  X-ray powder patterns, 13–14, 448
Tricalcium silicate hydrate, 369
Tricalcium silicate pastes, see calcium silicate pastes
Tridymite, 73, 75
Triethanolamine, and cement hydration, 350
Trimethylsilylation (see also silicate anion structure), 137–138
Truscottite, 369
Tuffs, 299

Undesigned product, 204, 209
Units of pressure or stress, 5

Viscosity, definitions, 244
Void spacing factor, 351
Volatiles, circulation in cement manufacture, 70, 74–77, 83
Volume changes, in concrete on freezing, 403

Volumetric quantities, calculation on Powers–Brownyard model, 250–251

Wagner turbidimeter, 98
Washburn equation, 261
Waste material utilization
  composite cements, 276, 311
  low energy cements, 341–343
  Portland cement, 66
Water, categories of
  in calcium silicate pastes, 130–131
  in Portland cement pastes, 206–207
Water/cement ratio, minimum for complete hydration, 248–251
Water contents, see bound water; non evaporable water
Water reducers, 352–357
  calcium aluminate cements, 331
  oilwell cementing, 372
Water sorption isotherms, 247, 258–259
Wet process, 66–67, 70–71
White Portland cements, 3, 10–11, 27, 83, 86, 87
White's reagent, 103
Wollastonite, 33, 75
Workability, 243
  effect of pfa, 292–293
  effect of water reducers or superplasticizers, 352–353
Wüstite, 48, 318–319

X-ray microanalysis
  C_3S-pozzolana pastes, 305
  calcium silicate pastes, 132
  clinker phases, 8–9, 21–22, 27–28, 30–31
  general considerations, 132
  microsilica cement pastes, 308
  pfa cement pastes, 296–297
  Portland cement pastes, 209–212
  slag cement pastes, 284–286
X-ray powder diffraction
  C–S–H gel, C–S–H(I), C–S–H(II), 141, 146, 148–149
  calculated patterns, 110, 447–453

X-ray powder diffraction (*cont.*)
  clinker phases, 13–14, 20, 22, 26–27, 31–32, 447–453
  Portland cement pastes, 199–201, 204–206
  unhydrated Portland cement or clinker, 108–111
X-ray scattering, small angle, 264
Xerogel, definition, 246
Xonotlite, 369, 374

Young's modulus, see elasticity

Zeolites, as pozzolanic materials, 302–305
Zeta potential
  calcium aluminate cements, 325
  Portland cements; effect of superplasticizers, 355–357
Zinc salts
  in cement hydration, 361
  in cement making, 77, 95
Zinc substitution in alites, 121
Zoning
  in alite, 11–12
  in ferrite phase, 32, 43

X-ray powder diffraction (*cont.*)
 clinker phases, 13–14, 20, 22, 26–27, 31–32, 447–453
 Portland cement pastes, 199–201, 204–206
 unhydrated Portland cement or clinker, 108–111
X-ray scattering, small angle, 264
Xerogel, definition, 246
Xonotlite, 369, 374

Young's modulus, see elasticity

Zeolites, as pozzolanic materials, 302–305
Zeta potential
 calcium aluminate cements, 325
 Portland cements; effect of superplasticizers, 355–357
Zinc salts
 in cement hydration, 361
 in cement making, 77, 95
Zinc substitution in alites, 121
Zoning
 in alite, 11–12
 in ferrite phase, 32, 43